T3-CAO-885

Science Horizons Year Book

1992

P. F. COLLIER, INC.

NEW YORK TORONTO SYDNEY

Science Horizons Year Book

1992

Published 1992 by P.F. Collier, Inc.

This book is also published under the title Science Annual 1993

Library of Congress Catalog Card Number 64–7603

ISBN 0-02-527175-X

Printed in the United States of America

Contributors

RUSS ALLEN, Free-lance Writer
PUBLIC HEALTH REVIEW

NATALIE ANGIER, Science correspondent, *The New York Times*
THE WORLD'S BIGGEST FUNGUS

DAVIS BERREBY, Contributor, *Discover* magazine
THE GREAT BRIDGE CONTROVERSY

DANA BLANKENHORN, Free-lance writer based in Atlanta, GA
WHAT'S NEW IN THE NEWSROOM?

KAREN BOEHLER, Free-lance writer specializing in space science
ANIMALS IN SPACE

BRUCE BOWER, Behavioral sciences editor, *Science News*
BEHAVIORAL SCIENCES REVIEW

PETER BRITTON, Contributor, *Popular Science* magazine
OFFSHORE OIL: HOW DEEP CAN WE GO?

LINDA J. BROWN, Free-lance writer
MEDICAL BLOODSUCKERS
ENDANGERED SPECIES UPDATE

SHANNON BROWNLEE, Senior editor, *U.S. News & World Report*
Coauthor, THE AGE OF GENES

FRED BRUEMMER, Contributor, *International Wildlife*
ANCIENT SPELL OF THE SEA UNICORN

JOHN CAREY, Science correspondent, *Business Week*
SECRETS OF SNOW

ANTHONY J. CASTAGNO, Energy consultant; manager, nuclear information, Northeast Utilities, Hartford, CT
ENERGY REVIEW
IN MEMORIAM

CLARK R. CHAPMAN, Planetary Science Institute, Tucson, AZ
Coauthor, ASTEROID THREAT!

GLENN ALAN CHENEY, Adjunct professor, Fairfield University, Fairfield, CT
CHERNOBYL'S WORSENING AFTERMATH
THE 1991 NOBEL PRIZE FOR PHYSIOLOGY OR MEDICINE
THE 1991 NOBEL PRIZE FOR PHYSICS AND CHEMISTRY

THEODORE A. REES CHENEY, Associate professor, Fairfield University, Fairfield, CT
CHEAP POWER DOESN'T COME CHEAP

JAMES R. CHILES, Contributor, *Smithsonian*
THE NEW COMMUNICATIONS AGE

ROGER COHN, Senior editor, *Audubon* magazine
NATURE BLOOMS IN THE INNER CITY

PATRICK COOKE, Contributing editor, *Health* magazine
TV OR NOT TV

DONALD CUNNINGHAM, Washington, D.C.-based free-lance writer
TECHNOLOGY REVIEW

OWEN DAVIES, Contributor, *Omni* magazine; coauthor, *The Haves and the Have Nots in the New World Order*
VOLATILE VACUUMS

GODE DAVIS, Free-lance writer
OCEANOGRAPHY REVIEW
SNOWFLAKE

JAMES A. DAVIS, Department of mathematics, University of Richmond, Richmond, VA
MATHEMATICS REVIEW

JERRY DENNIS, Free-lance writer; author, *It's Raining Frogs and Fishes: Four Seasons of Natural Phenomena and Oddities of the Sky*
WHAT HAPPENED TO THE PASSENGER PIGEON?

JARED DIAMOND, Professor of physiology, University of California — Los Angeles School of Medicine
THE ATHLETE'S DILEMMA

EDWARD EDELSON, Free-lance science writer based in New York City
BUCKYBALLS: THE MAGIC MOLECULES

DAVID S. EPSTEIN, Assistant director, New England Weather Service, Hartford, CT
WEATHER REVIEW

JACK FINCHER, Contributor, *Smithsonian*
COMPANIONABLE CANINES

THOMAS G. FLYNN, Senior vice president and manager of public relations, Bechtel Group, Inc.
COOLNESS UNDER FIRE

TIM FOLGER, Associate news editor, *Discover* magazine
THE MAGIC OF CARD SHUFFLING
THE PHYSICS OF CAR ACCIDENTS

DEBORAH FRANKLIN, Staff writer, *Health* magazine
FEARS YOU CAN TURN OFF

LEE GALWAY, Free-lance writer
THE LOST CITY OF UBAR

BIL GILBERT, Contributor, *Sports Illustrated*
WILD HORSE REFUGE

DANIEL GOLEMAN, Behavorial science correspondent, *The New York Times*
THE QUIET COMEBACK OF ELECTROSHOCK THERAPY

TIMOTHY S. GREEN, Contributor, *Smithsonian;* author, *The Gold Companion*
THE MYSTERY OF MAKING SCENTS

ABIGAIL W. GRISSOM, Free-lance writer
COMPUTER REVIEW

JOHN GROSSMANN, Contributor, *Air & Space/Smithsonian*
THE BLIMP BOWL

KATHERINE HARAMUNDANIS, Free-lance writer on science and technology; member, American Astronomical Society
ASTRONOMY REVIEW

T. A. HEPPENHEIMER, Contributor, *American Heritage of Invention & Technology*
THE RISE OF THE INTERSTATES

GLADWIN HILL, National environment correspondent — retired, *The New York Times*
ENVIRONMENT REVIEW

ERIN HYNES, Free-lance writer
BOTANY REVIEW

ALEX KERSTITCH, Free-lance photographer and writer
PRIMATES OF THE SEA

CHRISTOPHER KING, Corporate Research, Institute for Scientific Information, Philadelphia, PA
PLEASE STAND BY: HDTV

MARC KUSINITZ, Assistant director, Office of public affairs, The Johns Hopkins Medical Institutions, Baltimore, MD
CHEMISTRY REVIEW
PHYSICS REVIEW

LOUIS LEVINE, Department of Biology, City College of New York, New York, NY; author, *Biology of the Gene*
GENETICS REVIEW

GRACE LICHTENSTEIN, Former Rocky Mountains bureau chief, *The New York Times*
TAKING BACK THE PAST

DENNIS L. MAMMANA, Resident astronomer, Reuben H. Fleet Space Theater and Science Center, San Diego, CA
MAGELLAN AT VENUS
ECLIPSE!
THE NEW PLANETARIUM
SPACE SCIENCE REVIEW

THOMAS H. MAUGH II, Science writer, *Los Angeles Times*
GENETIC FINGERPRINTING

JAMES McCOMMONS, Contributor, *Country Journal*
A MULTITUDE OF MOOSE

ELIZABETH McGOWAN, Free-lance writer
THE SEARCH FOR AMELIA EARHART

MARTIN M. McLAUGHLIN, Free-lance consultant; former vice president for education, Overseas Development Council
FOOD AND POPULATION REVIEW

WENDY J. MEYEROFF, Free-lance writer
THE GOLDEN HOUR: ADVANCES IN EMERGENCY MEDICINE

RICHARD MONASTERSKY, Earth sciences editor, *Science News*
GEOLOGY REVIEW
PALEONTOLOGY REVIEW
SEISMOLOGY REVIEW
VOLCANOLOGY REVIEW

DAVID MORRISON, Space-science division, NASA Ames Research Center
Coauthor, ASTEROID THREAT!

ELIZABETH PENNISI, Chemistry/materials science editor, *Science News*
BOLDER BAR CODES
ROBOTS GO BUGGY

IVARS PETERSON, Mathematics/physics editor, *Science News*
FROM GRAPHICS TO PLASTICS

DEVERA PINE, New York City-based free-lance writer and editor
HAIR: PERSONAL STATEMENT TO PERSONAL PROBLEM
COMEBACK DISEASES

NICK RAVO, Correspondent, *The New York Times*
STORM CHASERS

SYLVIA DURAN SHARNOFF, Free-lance writer/photographer specializing in natural history subjects
THE ENIGMATIC SLIME MOLD

JOANNE SILBERNER, Senior editor, *U.S. News & World Report*
Coauthor, THE AGE OF GENES

NANCY SHUTE, Contributor, *Smithsonian*
WILDLIFE CRIMEBUSTERS

DOUG STEWART, Contributor, *Air and Space/Smithsonian*
THE WORLD AT ZERO G

JENNY TESAR, Free-lance science and medical writer; author, *Scientific Crime Investigation; Global Warming; The Waste Crises*
HEALTH AND DISEASE REVIEW
NUTRITION REVIEW
SAVING HAWAII'S BOTANICAL HERITAGE

BRUCE WATSON, Contributor, *Smithsonian*
SALEM'S DARKEST HOUR

PETER S. WELLS, Director, Center for Ancient Studies, University of Minnesota, Minneapolis, MN
ANTHROPOLOGY REVIEW
ARCHAEOLOGY REVIEW

CURT WOHLEBER, Communications specialist, University of Pittsburgh, Pittsburgh, PA
NIKOLA TESLA: THE FORGOTTEN PHYSICIST

JUDITH YEAPLE, Assistant editor, *Popular Science* magazine
THE BOMB CATCHERS

Contents

Features

PAST, PRESENT, AND FUTURE 182

PHYSICAL SCIENCES 224

PLANTS AND ANIMALS 276

TECHNOLOGY 326

Reviews

ASTRONOMY and SPACE SCIENCE

CONTENTS

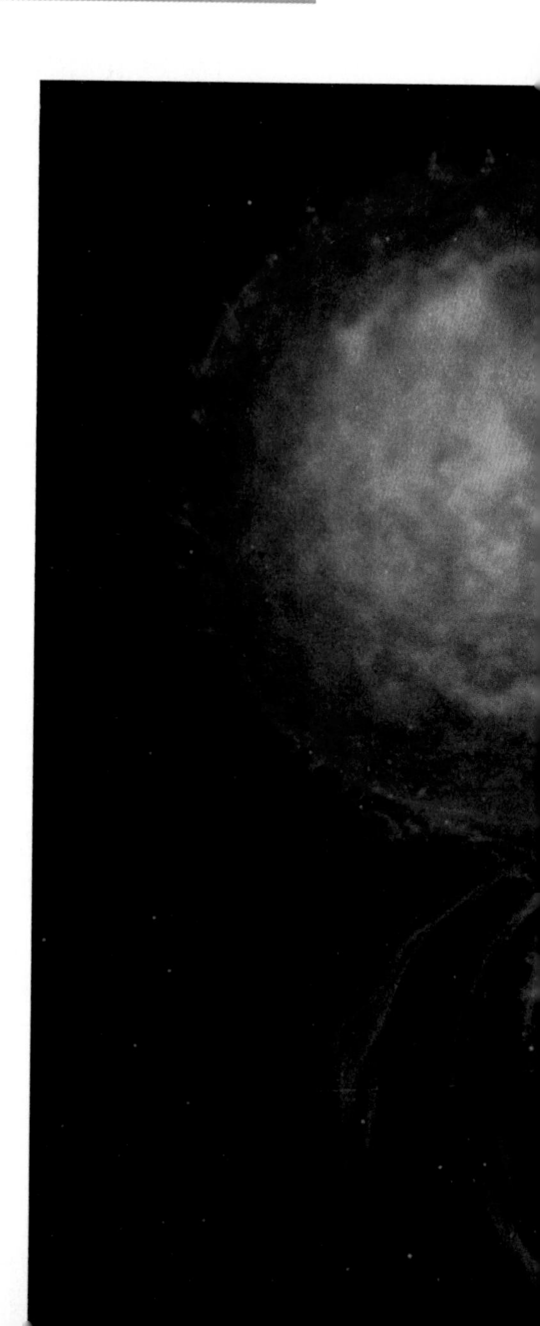

MAGELLAN AT VENUS

by Dennis Mammana

Of all the planets in our solar system, the most similar to Earth in size, mass, and distance from the Sun is Venus. In fact, for centuries, astronomers have described it as Earth's twin in space.

But no one could be sure what Venus was really like. That's because, unlike the other two terrestrial planets—Mercury and Mars—the surface of Venus is covered with thick clouds. These prevented scientists from viewing and photographing its surface directly. That didn't stop them from speculating, however. Some imagined Venus to be a world of tropical forests and steaming swamps, inhabited by a race of strange and exotic creatures. Perhaps even dinosaurs.

An Alien World

The mysteries beneath its clouds have made Venus one of the most frequently visited planets of our solar system. In the past three decades, some 20 spacecraft from the United States and the former Soviet Union have journeyed there. They have probed the planet's sulfur-yellow clouds and measured its atmospheric structure and composition. Robot landers photographed portions of its landscape and chemically analyzed its rocks. And they have revealed a world far more alien than anyone could have ever imagined.

As they descended into this veil of clouds, spacecraft encountered gases of carbon dioxide, rains of sulfuric acid, and temperatures of 900° F (480° C)—hotter than molten lead. Deep within this searing, suffocating atmosphere, the probes found a world of strange, distorted vision. Barely 2 percent of the Sun's light makes it through the Venusian clouds to illuminate the surface.

On the surface the atmospheric pressure is 90 times that on the Earth's surface—equivalent to the pressure experienced nearly 1 mile (1.6 kilometers) beneath terrestrial oceans. And the rocks the probes found on Venus were pitted and scarred from the constant pelting of acid rain.

Our First Look

Despite these few tantalizing glimpses we had received from the Venusian landscape, most of the planet's surface remained hidden. To learn more about the global nature of its surface, scientists from the United States and the former Soviet Union sent spacecraft there in the late 1970s and early 1980s.

In 1978 the National Aeronautics and Space Administration (NASA) launched to Venus the Pioneer Probe and Orbiter mission. While most of its attention was focused on studying the planet's atmosphere, the mission's orbiting spacecraft carried a radar system to pierce the thick clouds and provide us clues as to the nature of the Venusian surface. During its mission, Pioneer mapped 92 percent of the planet's surface with such great detail that objects as small as 30 miles (50 kilometers) across could be seen.

In 1983 the Soviet Union sent two Venera spacecraft to map Venus. Because of the nature of their orbits around the planet, the craft were able to map only one-quarter of the Venusian surface—that part near the planet's north pole. Nevertheless, they revealed objects as small as 1.2 miles (2 kilometers) across.

From these missions, scientists were able to piece together the first global map of the Venusian surface. They saw for the first time amazing features. Most of the planet appeared to consist of either rolling upland plains (apparently composed of older crustal rock) or smooth lowland areas.

Two major "continents" or elevated plateaus appeared—Aphrodite, named for the Greek equivalent of the goddess Venus, and Ishtar, named for the Babylonian equivalent. These regions appear to be younger than the surrounding areas. Ishtar is about the size of Australia; Aphrodite, about as large as South America.

Jutting up from the Ishtar highlands appears one of the highest mountains in the solar system. Mount Maxwell towers some 35,400 feet (10,800 meters) above the surrounding landscape—1 mile (1.6 kilometers) higher than Earth's Mount Everest. Two other highland areas of possible volcanic and tectonic origin, Alpha Regio and Beta Regio, also command attention.

As impressive as this first global map was, it showed only large-scale features. The fine details—the hills and valleys, the craters and lava flows—all remained unseen. And while these missions answered many questions about our "twin" planet, they raised even more.

How did Venus come to be such a hothouse, and were conditions there ever similar to those on Earth? Does Venus, like the Earth, have tectonic plates on which its continents ride? Did Venus ever have oceans similar to those on Earth? And how is its surface shaped by volcanoes, impact craters, and erosion?

Magellan

To answer these questions, NASA built and launched a new spacecraft to orbit and map the surface of Venus like never before.

Named for the 16th-century Portuguese explorer Ferdinand Magellan, this craft was designed to swing within 150 miles (240 kilometers) of the Venusian surface, and to provide waiting scientists on Earth with images 10 times better than that from any previous Venus-spacecraft mission.

To save costs, Magellan was constructed from hardware left over from other planetary projects—particularly from the Mariner 9, Voyagers 1 and 2, and Galileo missions.

With a weight of some 7,604 pounds (3,450 kilograms), Magellan carries only one scientific instrument—a radar sensor. But this instrument is the key to the entire mis-

Venus is often referred to as the Earth's sister planet for its similar size and mass. Scientists believe that Venus, like the Earth, has a solid inner core and a molten outer core.

sion, for it is responsible for three main jobs. First, it must collect data from which images may be constructed. Second, it must measure the elevation of the planet's features. And third, it must measure the natural heat emanating from the Venusian surface.

The craft must use its 12-foot (3.7-meter)-diameter antenna to beam radar waves toward the Venusian surface. Unlike visible light, radar waves penetrate the clouds and reflect off the solid surface back into space. The spacecraft then retrieves these "echoes" and stores them on tape. After the data is sent back to Earth, scientists can use computer-processing techniques to turn these radar reflections into remarkably precise photos of the hidden surface of this mysterious planet.

The Launch

Magellan was carried into space aboard the space shuttle *Atlantis* on May 4, 1989. After several revolutions around the Earth, the craft was deployed from the shuttle's cargo bay into its own orbit. Then, after two-thirds of another orbit around Earth, the spacecraft's Inertial Upper Stage (IUS) was fired, and Magellan headed outward toward its rendezvous with Venus.

Fifteen months—and one and a half circuits around the Sun—later, Magellan arrived at Venus. On August 10, 1990, a solid-rocket motor aboard the craft fired, and placed Magellan in its orbit. Completing one revolution every three hours and 15 minutes, Magellan swings around the planet in a highly elliptical orbit that ranges from as close as 182 miles (293 kilometers) to as far as 5,296 miles (8,523 kilometers).

Making a Map

Each time Magellan swings near the Venusian surface, its antenna looks downward along the left side of the spacecraft's path. For 37.2 minutes, Magellan emits several thousand radar pulses each second. Traveling at the speed of light, the pulses strike the planet and "illuminate" a 12-mile (20-kilometer)-wide strip of the planet's surface. The signals immediately bounce back and are received at the instrument.

The radar system—known as Synthetic Aperture Radar (SAR)—takes advantage of the spacecraft's motion to create an artificial aperture many times larger than the antenna's actual size. It collects many echoes as it moves along, looking at the target at an angle toward the side. At the same time, its altimeter looks straight down with a separate antenna to measure the elevation of ground features on the planet's surface.

By recording the returned pulses, scientists can use two measurements on each pulse to locate each point on the planet's surface. The first measures the time required for the radar signal to return to Magellan. This gives the spacecraft's accurate distance to that point.

The second measures the returned signals for their Doppler shift, a change in radar frequency caused by the spacecraft's motion over the surface. This gives the location of the point relative to the spacecraft's line of flight. The brightness of the image at that point also becomes an element of the map image.

Since Magellan is in elliptical orbit around Venus, it passes close enough to the surface to conduct mapping operations for only 37 minutes out of each orbital period. During the process the data are recorded onto two multitrack digital recorders.

Later, while the craft is farther from the planet, the antenna is aimed toward Earth. The rest of Magellan's orbit is spent transmitting the recorded raw data from the just-completed mapping pass, receiving telemetry instructions from Earth, and calibrating the spacecraft's altitude-control system with reference stars. Images of the surface are then constructed by computers at the Jet Propulsion Laboratory (JPL) in California.

The Magellan space probe was deployed in 1989 from the space shuttle Atlantis. *In 1990, Magellan began a series of elliptical orbits around Venus; during each orbit, the probe spends 37 minutes making radar maps of the planet. Magellan will continue its mapmaking orbits until 1995.*

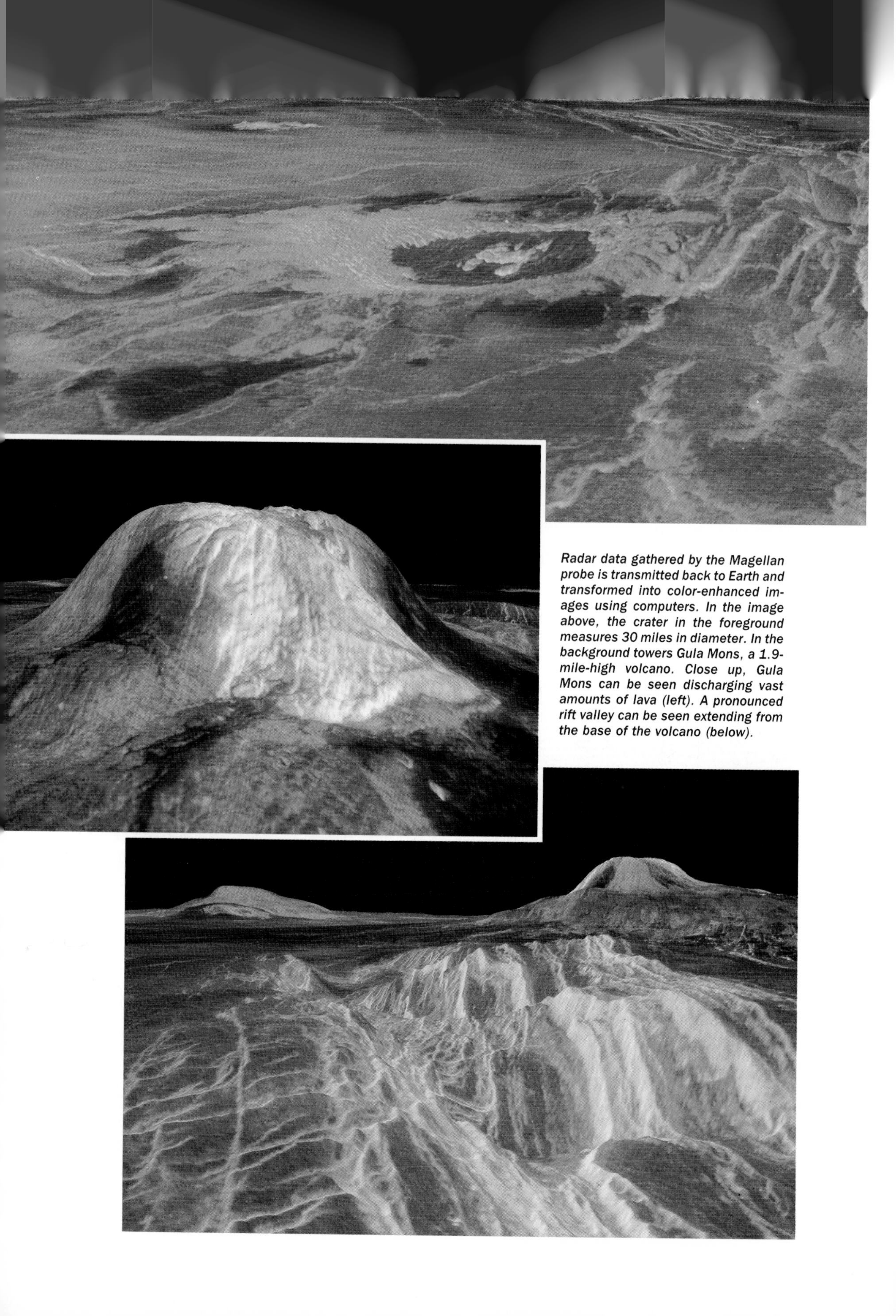

Radar data gathered by the Magellan probe is transmitted back to Earth and transformed into color-enhanced images using computers. In the image above, the crater in the foreground measures 30 miles in diameter. In the background towers Gula Mons, a 1.9-mile-high volcano. Close up, Gula Mons can be seen discharging vast amounts of lava (left). A pronounced rift valley can be seen extending from the base of the volcano (below).

After its "call home" has been completed, Magellan maneuvers back into position to begin mapping once again. Since Venus is rotating slowly beneath the orbiting spacecraft, the planet's surface is mapped in successive, slightly overlapping strips.

Each swath is about 12 miles (20 kilometers) wide, and about 9,942 miles (16,000 kilometers) long. The image strips are combined by computers on Earth into photomosaic images covering large regions of the Venusian surface.

Not without Problems

With such an ambitious mission, problems were bound to occur. Only six days after Magellan entered Venus orbit, scientists lost radio contact with the craft. After a tense 13 hours, they regained contact. But within a week, another blackout occurred. No one ever determined the cause of the problem, but engineers made adjustments to the spacecraft's software to help prevent a reoccurrence.

Then, on January 4, 1992, another problem occurred. During Magellan's 3,800th orbit around the planet, mapping came to an abrupt halt. While attempting to send its newest data harvest back to Earth, the spacecraft's main transmitter strangely began sending only a monotone carrier radio wave without the critical Venus data.

No amount of tweaking could coax the data from the transmitter, so, on January 24, scientists put a backup transmitter to work to complete the job. But the backup itself is flawed. As it warms up, its signal weakens and it begins to "whistle." While this reduces by 43 percent the rate at which Magellan beams data back to Earth, the mission at least could continue.

This was important because Magellan was about to begin its third scan of the planet—producing a remarkable three-dimensional map that would show extremely fine detail and increase the value of all data gathered to date.

Venus Unveiled

Since September 1990, Magellan's radar has been piercing the clouds enshrouding Venus. The probe has imaged 95 percent of the Venusian surface, and has produced the first high-resolution, global map of the planet. The map shows vivid images of impact craters distorted by the planet's thick atmosphere, volcanoes shaped like pancakes, and intricate networks of fractures. And, in the process, the map has shown features as small as 74 feet (23 meters) across.

Among these features, Magellan has found only 50 craters, far fewer than exist on either the Moon, Mars, or Mercury. Scientists speculate that small meteors burn up in the thick atmosphere before they can impact, while larger ones make it through unaltered to form craters. These craters may be covered by ancient lava flows or eroded in other ways. Some large, seemingly unaltered craters do appear, however, such as the 31-mile (50-kilometer)-wide Cunitz.

Volcanoes also interest Venus scientists. One of the most spectacular volcanic features is Gula Mons, named after an Assyrian goddess. Gula spans hundreds of miles across the Venusian surface, but rises to a height of barely 5 miles (8 kilometers).

Is Venus's surface still volcanically active? No one knows. Some geologists think that large parts of the planet's surface were reworked during a huge volcanic episode 500 million years ago. Others believe that Venus is turning itself inside out with volcanism, but without plate tectonics as on Earth.

And snaking some 4,216 miles (6,800 kilometers) across the planet's surface is a mysterious channel. The near-constant width of a section in the region known as Themis Regio suggests that the channel was carved by a fast-moving liquid—perhaps molten lava—eons ago.

The Mission Continues

At the start of 1992, Magellan had sent to Earth 2.8 trillion bits of data—three times more than from all previous planetary spacecraft combined. And by the time its mission is completed in 1995, that number could double. But with Magellan well into its second year of mapping, and with the majority of its data stored safely in computers on Earth, the work is far from over.

Now scientists are beginning to analyze the data in an attempt to make sense of it all. They must study the planet's features and try to understand how they interrelate. They must determine the origin and evolution of the Venusian surface. And they must begin to ponder the many questions that today we cannot even begin to imagine.

ECLIPSE!

by Dennis L. Mammana

On July 11, 1991, one of the longest total solar eclipses of the Sun in modern times occurred over western North America and the Pacific. Millions took the opportunity to watch the phenomenon.

On July 11, 1991, the people of Earth experienced the finest total solar eclipse in recent history. It occurred over some of the most densely populated regions of our planet, encouraging more than 48 million people to look skyward—many for the first time—to witness the magic and spectacle of a total eclipse of the Sun.

How an Eclipse Works

An eclipse of the Sun, ironically, is caused by the *Moon*. Interestingly, because of a remarkable coincidence found nowhere else in our solar system, the Moon and Sun appear nearly the same size in our sky. During its 28-day orbit of the Earth, the Moon regularly passes between us and the Sun. As it does, it occasionally casts its shadow onto our planet and blocks the Sun from view.

At its largest, the Moon's dark central shadow—the umbra—extends only a few hundred miles across the Earth's surface. Surrounding it is a light outer shadow—the penumbra—thousands of miles wide. An observer standing in the right place at the right time can witness an eclipse of the Sun.

If the Moon passes slightly above or below the center of the Sun's disk, and blocks only part of it from view, the event is called a partial eclipse. This is the most common type of solar eclipse, since it can be seen from large areas of the globe.

If the Moon happens to cross the Sun's center when the Moon lies farther away from Earth than normal, the Moon appears slightly smaller than the Sun, and doesn't cover the Sun's face completely. Instead, it leaves a ring (or annulus) of sunlight around its edge. This is an annular eclipse.

A total eclipse occurs when the Moon appears just the right size to completely block the Sun's disk, revealing for a precious few minutes the Sun's pearly white atmosphere—the corona.

Predicting Eclipses

Solar eclipses don't occur every month, because the Moon's orbit is tipped by 5 degrees to the Earth-Sun plane. An average of two dozen solar eclipses are visible from somewhere on Earth each decade. The most possible per year is five; the least is two. *Total* eclipses occur, on the average, less than once per year.

Eclipses can be predicted with remarkable accuracy. The interval between one eclipse and the next in a given series is called a *saros*. Each saros is about 18.03 years in length. A series of solar eclipses may run through 70 saroses and last for 1,250 years.

The first in the series is partial, with the Moon encroaching only slightly on the Sun's disk. At the next eclipse, the Moon obscures a somewhat larger area of the Sun. Each subsequent eclipse becomes more extensive, until there is an annular or total eclipse. Then a succession of partial eclipses follows, each one obscuring a smaller area of the Sun's disk than the one before. As a result, every few months a solar eclipse can be seen from somewhere on Earth. One of the best occurred on July 11, 1991.

The Great Eclipse of '91

The eclipse of July 11 was one of the longest on record; it took the Moon nearly seven minutes to cross the Sun's face.

When a total solar eclipse reaches five minutes in length, it is considered exceptional. For the period between July 18, 1898, and March 11, 2510, 1,499 solar eclipses have been tabulated. Of these, 936 are either total or annular. Only 66 of the total eclipses attain or exceed a five-minute period of totality.

The eclipse of July 11, 1991, ranks number 12 on this list. It provided the longest view of totality anyone alive today will experience from the Earth's surface—nearly seven minutes in length—and will not be exceeded until the year 2132. North Americans will not experience such a favorable eclipse until the year 2510.

For this reason, people from around the world traveled thousands of miles to experience this rare alignment of Earth, Sun, and Moon, and the awesome spectacle it produced in the sky.

On the morning of Thursday, July 11, 1991, with excitement beginning to mount, the Moon's umbral shadow first struck Earth at dawn over the North Pacific Ocean, 300 miles (480 kilometers) east of the international date line.

Viewing conditions varied along the path of totality during the July 1991 eclipse. In places where a thin layer of clouds floated high in the atmosphere, the eclipsed Sun appeared to be encircled by a spectacular ring.

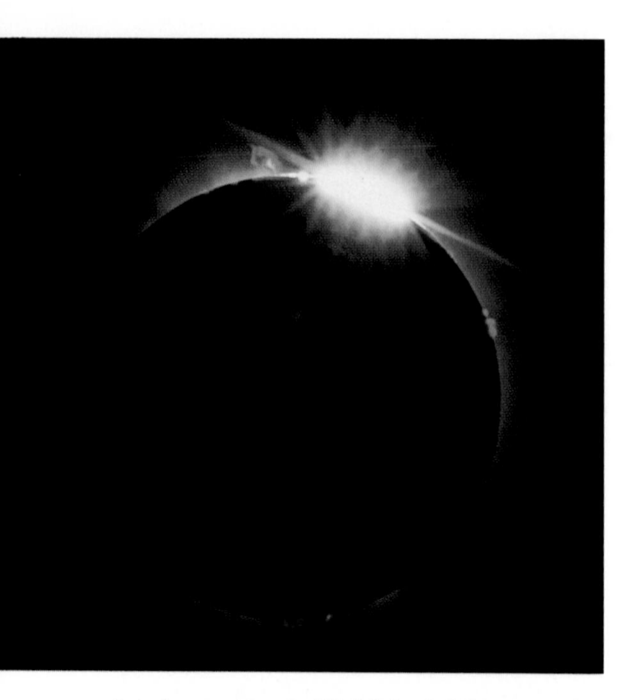

In a phenomenon called "Baily's beads," the eclipsed Sun briefly shines through the valleys at the edge of the Moon's disk. The effect is often likened to that of a brilliant diamond ring.

Racing eastward at 16,000 miles (25,750 kilometers) per hour, the shadow first encountered land four and a half minutes later on the Big Island of Hawaii. Within one minute the entire island was engulfed in the Moon's 139-mile (223-kilometer)-wide umbra. For four minutes and 10 seconds, the sky was darkened, even though the Sun stood almost 22 degrees up in the east-northeastern sky.

Flying at 5,600 miles (9,000 kilometers) per hour, the shadow continued eastward across the Pacific Ocean, heading toward the southern tip of Baja California. Slowing to a "mere" 1,450 miles (2,330 kilometers) per hour, the shadow had grown to 160 miles (257 kilometers) in width. In Baja, totality swelled to six minutes and 48 seconds, with the Sun appearing 78 degrees up in the sky.

Soon the shadow tore across the Gulf of California toward mainland Mexico and the town of Boca Teacapan, 50 miles (80 kilometers) south of Mazatlán. Observers at the tiny town of Ruiz experienced the maximum six minutes, 53 seconds of totality with the Sun directly overhead.

The shadow continued southeastward toward Central America. Within minutes, five Latin American capitals fell into darkness. At the first—Mexico City, Mexico—more than 22 million people experienced the lunar shadow at once—more than ever before in the Earth's history.

After sweeping through Central America, the umbra crossed the rain forests of the Amazon in South America. Soon it left Earth behind and drifted off into space once again. The greatest eclipse in modern times was history, leaving behind glorious memories for all in its path.

The View from Hawaii

Some of the best views of the spectacle were predicted for the Big Island of Hawaii, where an estimated 40,000 people converged. Most of the island's 9,000 hotel rooms had been booked two years in advance, and lodging was nearly impossible to find. Some people rented their houses for up to $4,000 on eclipse day. Flights to Hawaii were doubled that day, and nearly all the island's 150 highway personnel were called in to help control the heavy amount of traffic.

Eclipse souvenirs were not hard to find. And not just the usual T-shirts, coffee mugs, videos, and books, either. Entrepreneurs were offering such unusual items as etched wine bottles, earrings, golf balls, and eclipse pastries. Those with a bent for the offbeat could spend $340 to swim with dolphins during totality, watch a comic opera about the eclipse of 1706, or get a black-and-yellow eclipse haircut, with such variations as the Solar Flare and the Sun Spot.

Then, at dawn on the big day, sky watchers took their positions on beaches and rooftops. And just as predicted, the Moon began its long-heralded trek across the face of the Sun. But as the lunar shadow approached from the west, heavy clouds began to form, and spoiled the view for thousands lining the island's west coast.

The view from Mexico, however, was quite different.

The View from Mexico

Mexican officials had prepared for eclipse trekkers more than a year in advance, and in doing so, prevented the feared shortages of gasoline, water, and food that everyone had expected would occur.

For weeks before the event, eclipse fever ran high. The news media offered plenty of advice on how and when to look, and what to expect from the view. Many of the less-educated residents, however, were afraid. Some set water bottles out on the roadside to be turned into medicine by the eclipse. Villagers put red flags in treetops as protection, and others hid under blankets or in basements. One traffic policeman asked an astronomer if it was true that "during totality, one could see the face of Jesus."

On eclipse day, more than 35,000 tourists flocked to the west coast of Mexico and the Baja Peninsula. Mexican television showed hardly anything but flashy graphics of the event, minute-by-minute on-site reports, and commercials. And those who traveled long distances to see the event were understandably excited. They had heard the reports of cloudy skies over Hawaii earlier that morning. But here in Mexico, the sky was mostly clear.

Many observers set up on beaches; others waited in small towns. Some had taken cruise ships into the ocean to be closer to the path of totality. One of those was this author, whose observing station was aboard the M.S. *Jubilee*, just off the coast of Mazatlán.

The morning of July 11 began normally, with the hot summer Sun beating down from the eastern sky. As the ship headed out to sea and telescopes were being erected on deck, a brief shower caused some concern. And off in the distance swirled a different sort of nature phenomenon: an immense waterspout that resembled the tornado in *The Wizard of Oz.*

Darkness at Noon

Soon the ship reached its destination. Then, at 10:33 A.M., the southwestern edge of the Sun began to vanish as the Moon's disk appeared in silhouette against it. Shouts of "first contact!" erupted around the ship, cameras began clicking, and 1,500 pairs of eyes turned skyward. The Great Eclipse of '91 had begun at long last.

Over the next hour, the Sun's precious light slowly diminished as the Moon continued its trek across the solar face. Then, around 11:45 A.M., with more than 80 percent of the Sun's disk covered, the sky began

Over the centuries, people have interpreted an eclipse as a bad omen or as a sign of divine displeasure. Perhaps as a means of sustaining age-old traditions, hundreds of people gathered at the site of an ancient Mayan ruin in Mexico to participate in a special ritual during the sensational July 1991 eclipse.

Some Famous Eclipses in History

Superstition, public fascination, and scientific research all make up the rich lore of solar eclipses. All three of these themes were part of the Great Eclipse of '91.

2137 B.C.
The earliest written record of a total solar eclipse comes from China. According to legend, two royal astronomers named Hsi and Ho were so pleased with themselves for predicting the eclipse that they became inebriated with rice wine and forgot to alert the royal court. When darkness overtook the land unannounced, the emperor ordered them beheaded for endangering the kingdom.

1900 B.C.
In southern England stands Stonehenge, an awesome arrangement of stones that has been the subject of countless studies, poems, and legends. Evidence indicates that Stonehenge was a brilliantly conceived astronomical observatory. Certain holes were apparently used as an eclipse predictor.

1397 B.C.
First recorded observation of the corona during a solar eclipse, inscribed on oracle bones discovered in China.

585 B.C.
According to the Greek historian Herodotus, the most famous solar eclipse of classical times occurred in the midst of a battle between the Medes and the Lydians. Both sides regarded the eclipse as an omen and immediately ceased hostilities, thereby ending a six-year war.

In 1504 the inhabitants of Jamaica were astonished when an eclipse predicted by Christopher Columbus occurred right on schedule.

130 B.C.
Greek astronomer Hipparchus used the position of the Moon's shadow during this solar eclipse to estimate the distance from the Earth to the Moon.

A.D. 840
Emperor Louis of Bavaria supposedly died of fright during a five-minute solar eclipse. His three sons fought over succession, resulting in the division of his empire into what are today France, Germany, and Italy.

to darken, and the scenery glowed in muted, unearthly tones. The air cooled, and the winds became calm. A still and eerie darkness fell over the ocean. Only seconds before noon, the last piece of sunlight shone over the Moon's edge—the diamond ring.

And then, totality!

Cheers and applause erupted throughout the ship as camera shutters clicked madly. In the sky, spreading outward from where the Sun once shone, was the amazingly intricate and glowing corona—the outer atmosphere of our star. From its inner edges, where the Sun's disk was being eclipsed, a pair of huge prominences glowed with an almost "neon" pink color.

In the midday darkness, the stars of Orion, accompanied by the brilliant and unmistakable planets Venus and Jupiter, shone in the dark-blue sky. Around the entire horizon were the familiar colors of sunset—red and yellow and orange. And encircling the eclipsed Sun appeared a spectacular ring, courtesy of a thin layer of cirrus clouds high in our atmosphere.

It was this unearthly sight that thousands had traveled so far to see. Yet all the preparation and all the expectations could

A.D. 1560
The announcement of a forthcoming eclipse caused many French people to panic. They fought one another to be next in line for the confessional. One beleaguered parish priest tried to calm the crowd by announcing that, since there were so many waiting to confess, the eclipse would be postponed!

1780
During the Revolutionary War, the first American eclipse expedition was sent out from Harvard College. A special immunity agreement was negotiated with the British so that the scientists could study the eclipse unharmed by the enemy. After all their efforts, the team never saw the eclipse because their site was outside the path of totality!

1868
During this eclipse, Sir Joseph Norman Lockyer of the United Kingdom found traces of a previously unknown element—helium—on the Sun. Helium was not found on Earth for another 27 years. The new element was named for the Greek word for Sun: "helios."

1878
To avoid the prairie winds, Thomas Edison set up his instruments in a Wyoming chicken coop to view this total solar eclipse. When the Sun dimmed, the chickens returned to roost. Edison spent so much time fighting chickens that he saw only a few seconds of the more than three minutes of totality.

1919
During this eclipse, Sir Arthur S. Eddington measured star positions and confirmed Einstein's prediction that light could be bent by a massive body such as the Sun.

1925
The path of totality split Manhattan. Those above 96th Street saw a total eclipse, while those below saw only a partial eclipse.

1948
National elections in Korea were postponed because a total eclipse was to occur on the date originally set for the balloting.

1966
While thousands of Hindu pilgrims bathed in sacred tanks in northern India to protect themselves from demons, the Gemini 12 astronauts snapped the first photo of a solar eclipse from space.

1972
The world's first "floating eclipse expedition" set sail from New York City with 834 passengers to rendezvous with totality hundreds of miles at sea.

1973
The French supersonic Concorde 001 carried scientists at 1,250 miles (2,000 kilometers) per hour along with the Moon's umbral shadow, enabling them to experience a world-record 74-minute total eclipse!

never have matched the reality of the moment. We watched as our star vanished, and the world around us changed completely. Some chattered excitedly about the incredible experience. Others stood in silent awe as tears welled in their eyes. But all experienced a flood of emotions unlike any they had ever felt. And never before had seven minutes passed so quickly.

Then, just as mysteriously, the Sun's light burst into view from behind the Moon's edge. Slowly, the darkness receded, and the ship and ocean again were bathed in the Sun's warm, golden rays.

The Great Eclipse of '91 was history—now only a memory etched forever on the minds of those who had watched. But as the Sun's light returned, and emotions ran high, a school of dolphins began leaping from the water and playing in the ship's wake as if to celebrate with us humans the return of the star we share on planet Earth.

Telescopes Were Poised
Not only is a total solar eclipse an experience of intense emotion, but it is a valuable scientific event as well. Rarely does the Moon's shadow fall onto the telescope of a major

observatory. But on the morning of July 11, 1991, the Moon's shadow passed over some of the largest telescopes on our planet—the Mauna Loa Solar Observatory in Hawaii, and the cluster of instruments on Mauna Kea.

From Mauna Loa the view was spectacular. Ash from the Philippine volcano Mount Pinatubo made the mid-eclipse sky much brighter than normal. But scientists were able to obtain the biggest coronal image ever, showing features as small as 1 arc second across—about the size of a U.S. quarter as viewed from more than 3 miles (5 kilometers).

As dawn broke over the summit of Mauna Kea, clouds and fog began to build. Fortunately, as totality began, the air cooled, and the clouds dropped below the summit. And scientists used telescopes as large as 49 feet (15 meters) across to study the solar atmosphere in ways never before possible.

Astronomer Serge Koutchmy used the huge 11.8-foot (3.6-meter) Canada-France-Hawaii Telescope—the largest ever to look at the Sun—to take photos and video images of the eclipse. Many of his 6,000 video frames showed detail smaller than 1 arc second across. Images of the inner corona showed places that had been "cleared" by explosive flare activity in the previous few weeks.

Observers Donald Hall and Giovanni Fazio used a cryogenically cooled light detector on the University of Hawaii's 24-inch (61-centimeter) telescope to search the Sun's outer corona for rings of hot dust previously reported. By the end of 1991, they had found none, but they continued their search through the mounds of data they gathered.

Another 24-inch (61-centimeter) telescope was used by astronomers Jean Arnaud and Matt Penn. They used a sensitive elec-

Experiments and Discoveries from Previous Eclipse Observations

DISCOVERY OF THE ELEMENT HELIUM
Observations during the 1868 total eclipse showed spectral lines from a new element that had not yet been detected on Earth. Now known to be the second-most-abundant element, helium was first found on Earth in 1895 from gases obtained from a uranium mineral.

DISCOVERY OF "FORBIDDEN" SPECTRAL LINES
Low densities and extremely high temperatures in the solar atmosphere (the corona) allow atoms to emit light at wavelengths not normally observed in the laboratory. These "forbidden" spectral lines come from elements present in the atmosphere near the Sun's surface.

CONFIRMATION OF THE THEORY OF GENERAL RELATIVITY
Measurements of star positions near the Sun during the 1919 solar eclipse showed that starlight is bent slightly toward the Sun as it passes nearby. This was consistent with Einstein's 1917 theory of relativity.

DOCUMENTATION OF CORONA SHAPES WITH SUNSPOT CYCLE
The solar corona changes with the sunspot cycle. When sunspots are numerous, the corona is evenly distributed around the Sun. When there are few sunspots, the corona is far denser near the solar equator. These coronal gases form a solar atmosphere that reaches as far as Earth, and can change the Earth's magnetic field.

DOCUMENTATION OF RAPID CORONAL MOTIONS
Observations during eclipses showed that the solar corona could change markedly during a two- or three-hour eclipse. This evolution of coronal structures paved the way for attempts to study these motions by use of spaceborne coronagraphs. Coronal depletions and mass ejections were first regularly observed in 1972 and 1973.

This dramatic sequence shows how the Moon gradually passes before the Sun during an eclipse. The Sun's corona is visible only during totality.

UPCOMING TOTAL AND ANNULAR SOLAR ECLIPSES THIS CENTURY

DATE	TYPE OF ECLIPSE	AREA VISIBLE FROM
May 10, 1994	Annular	New Mexico through Missouri to Maine
November 3, 1994	Total	South America
April 29, 1995	Annular	South America
October 24, 1995	Total	India and Southeast Asia
March 10, 1997	Total	Siberia
February 26, 1998	Total	Northwestern South America
August 22, 1998	Annular	Islands of South Pacific
February 16, 1999	Annular	Australia
August 11, 1999	Total	Europe through India

tronic light detector to take eight 30-second exposures of a prominence at the Sun's southeastern edge. Their photos showed detail as small as 1.2 arc seconds across, and should yield new information on the corona.

An infrared-imaging experiment by Philippe Lami and Jeff Kuhn used equipment similar to military night-vision goggles to record in detail how the light of the solar corona is polarized. The photos show that the corona is elliptical and flattened along the plane of our solar system, and that the inner corona is some 5,000 times more intense than the outermost regions.

Drake Deming of the National Aeronautics and Space Administration's (NASA's) Goddard Space Flight Center covered the 9.8-foot (3-meter) NASA Infrared Telescope Facility with a huge sheet of polypropylene plastic to carefully observe features in the solar spectrum. As the Moon passed in front of the Sun, the scientists were able to learn the exact origin of these features.

Observers even used a space telescope to help out. As totality began for observers in Hawaii, the Incidence X-Ray Telescope recorded the uneclipsed Sun in X radiation, for comparison with the pictures taken from the ground in Hawaii.

When's the Next One?

Not since Halley's comet swung through our skies in 1986 has so much public attention been focused on one celestial event. For those who missed it, there will be other opportunities. Travel companies are now putting tours together for the South American eclipse of November 3, 1994.

But not until August 21, 2017, will such an event occur over the United States. On that day the Moon's shadow will sweep diagonally across the country, from Oregon through Missouri to South Carolina, inspiring a whole new generation to look skyward and experience the most majestic of all nature's wonders—a total eclipse of the Sun.

THE NEW PLANETARIUM

by Dennis L. Mammana

Say the word "planetarium," and a striking image comes to mind. Most of us think of a domed room with a large, dumbbell-shaped "bug" in its center. We envision a place where the stars, Moon, and planets can be viewed even in the daytime.

But today the planetarium has become much more. It has become a remarkable theater of illusion—a theater in which we can journey to places impossible to visit—places on our world or on others.

After experiencing such a theater, we might be tempted to think of the planetarium as a new invention. Its origin actually dates back more than 2,000 years. The first planetarium was a moving model of the sky that would hardly be recognized today.

Early Planetariums

Perhaps the first to try out such a concept was the ancient Greek astronomer Eratosthenes around 250 B.C. His model was a metal

The modern planetarium offers an engulfing experience, one that captures our senses and our imagination and carries us off to places we thought impossible.

sphere, which represented the vault of the heavens, surrounded by movable rings representing the paths of the Sun and planetary bodies across the sky. The famous mathematician Archimedes built one powered by water. It was so precise that it could accurately reproduce eclipses of the Sun and Moon.

Our modern concept of a planetarium—a theater in which audiences could sit and watch a "sky show" on an overhead screen—didn't appear until 1657. In that year, Andreas Busch constructed a large globe into which 12 people could climb. The stars were fixed on the inside of the globe, and there were rings along which the planets moved. Planetarium visitors could sit and look up at the artificial stars and planets—just as though they were under the real sky.

The planetarium "projector" didn't develop until the early part of this century. The idea came from Drs. Max Wolf of the Heidelberg Observatory and Walter Bauersfeld of the Carl Zeiss Company of Jena in Germany. Their idea revolutionized our ability to "create" the universe.

Finally, in 1923, the first planetarium projector was demonstrated. Within a makeshift plaster dome on the roof of the Carl Zeiss Company in Germany, the Zeiss Model I projector showed in stunning detail the stars, Sun, Moon, and planets as visible from Munich. Two months later, it was installed in the Deutsches Museum, and opened to the public on October 21, 1923. "The Wonder of Jena," as it soon became known, captured the public's imagination like never before.

The projector—a dumbbell-shaped instrument from which the heavens could be simulated with remarkable precision—was an amazing device. At the center of each end of the dumbbell was a high-powered lamp. Arranged around each lamp were numerous lens systems, each containing a metal slide. Tiny holes in the slides represented individual stars, and as light passed through each hole, a point of light was formed on the dome—a "star." Small projectors were attached to project images of the brightest stars and planets, while larger projectors created the Sun and the Moon.

The United States didn't open its first planetarium—the Adler Planetarium in Chicago—until 1930. At the time, it was one of only half a dozen around the world, since the projectors were tremendously expensive, and only the largest of museums could afford them. In the ensuing decade and a half, fewer than one planetarium per year was built.

Then, in 1947, a revolution began. Armand Spitz, an amateur astronomer from Philadelphia, designed and built a smaller, simpler, and less-expensive planetarium projector. Spitz planetariums began popping up in many small museums, libraries, and schools—places that could not afford the money or space for a larger instrument.

Since that time, planetariums have appeared at an average rate of 31 per year. Today nearly 1,400 exist around the world. Not all were made by the Zeiss or Spitz companies. In fact, some 165 different types of planetariums are now in operation in 64 countries on six continents.

How It Works

A planetarium instrument is not just a static projector. It is, instead, a complicated machine of optical and mechanical parts that moves images of the stars, Sun, Moon, and planets. The gear systems that perform the work have been developed through detailed mathematical analysis of the paths of the heavenly bodies. A planetarium projector can show four basic motions: daily motion, annual motion, latitude motion, and precession.

The sky's daily motion—the sunrise in the east and sunset in the west, for example—is really a result of the Earth's daily rotation. But while it takes 24 hours for the Earth to turn once, most planetarium projectors can perform the same task in less than a minute. Using this daily motion, the planetarium operator can select the sky for any time of day or night.

Annual motion is the motion of the planets (and the Earth) as they orbit the Sun. The planetarium operator can take an audience through a year in less than a minute, and select the sky of any day in any given year to project onto the dome.

Latitude motion allows the planetarium operator to simulate the changing sky from different latitudes on Earth. This enables the audience to view the sky as it would appear from the North Pole, from New York City, Rio de Janeiro, or Australia, for example. In this way, one trip around our planet can be accomplished in only seconds.

Precession is the slow wobbling of the Earth's axis. This wobble is small and slow—it takes some 26,000 years to complete one cycle. Yet in a planetarium, audiences can

Most planetariums have a dome-shaped ceiling onto which the program is projected. Programs can feature a variety of scientific topics, although most generally have an astronomical theme; the planetarium display below illustrates the relative position of the planets in the solar system.

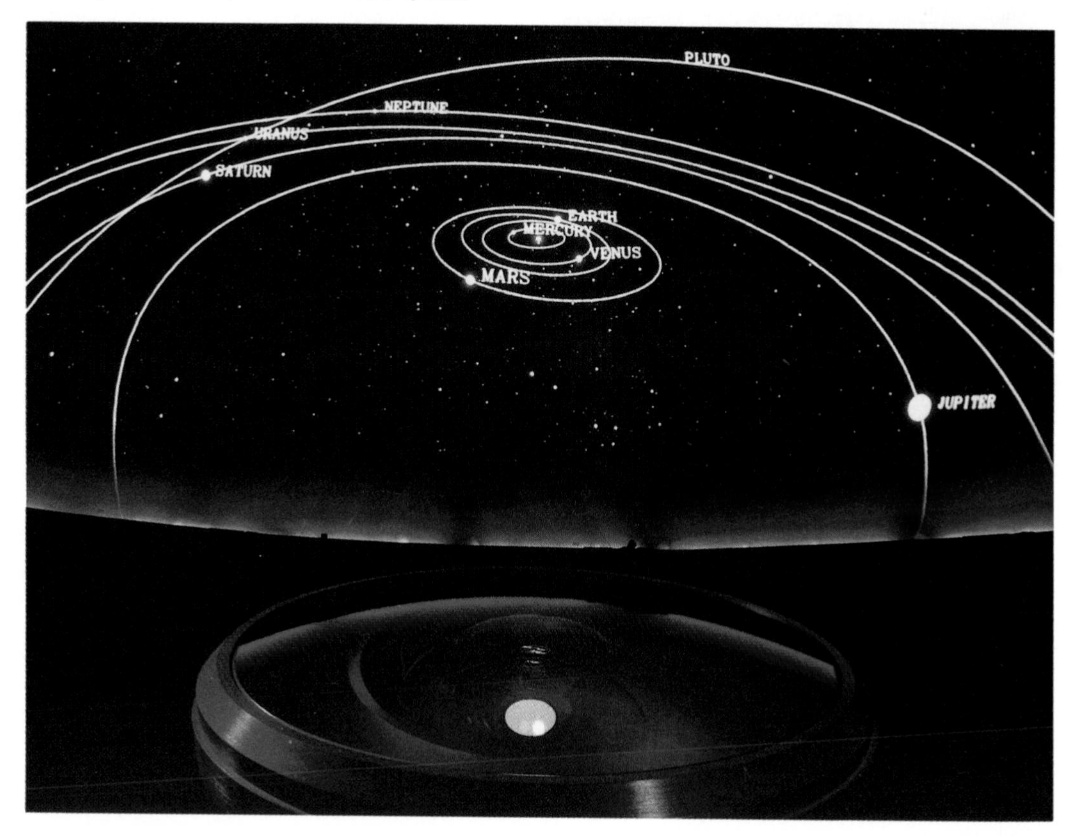

experience the changing sky, in its entirety, in less than a minute.

By moving it along any or all of its separate axes, a planetarium becomes a space-and-time machine. Not only can it show what the stars, Sun, Moon, and planets look like from our hometown, but from any other part of planet Earth as well. And, with only minor adjustments, it can let us view into the past or future—allowing us to see the sky as it was on the night we were born, or how it will appear 50 or 100 years from now—all with remarkable accuracy.

Planetariums for the 21st Century

As amazing as a planetarium instrument is, it could not remain the same over the years. If it had, the public's fascination with it would have long ago waned. Instead, planetariums now enjoy higher attendance and more-excited audiences than ever before.

One of the most exciting new technologies to burst onto the scene appeared in 1983. Introduced by Evans and Sutherland Computer Corporation of Salt Lake City, Utah, "Digistar" uses computer-graphics technology to calculate images and display them on a small monitor. A precision, high-resolution optical system then projects these computer-generated images onto the dome.

This totally new, high-tech planetarium instrument now replaces the large, dumbbell-shaped projector in the center of many domes. It can re-create the sky as seen from any location on Earth at any time of the day or night, as well as the motions of the Earth and sky—just like a traditional planetarium projector. But it can do even more.

It can display on the dome virtually any image composed of lines or dots. It now becomes possible to move at will within any three-dimensional data field. In other words, audiences can now experience simulated spaceflights to stars within a few hundred light-years of Earth, and watch how the solar neighborhood changes—now, or a million years in the past or future.

If that isn't enough, animation of such three-dimensional models as atoms, molecules, or galaxies can be projected with the same device. These instruments may one day replace entirely the mechanical projectors we often think of as a "planetarium."

Just as technology has changed, so has the idea of a "planetarium show." Some continue to be simply live lectures by an astronomer describing the night sky. Others have become multimedia extravaganzas.

Modern planetariums can create remarkably realistic three-dimensional illusions of space travel. Shows now combine planetarium projection with flashy special effects to simulate virtually any object or phenomenon in the universe. With these devices, journeys across our galaxy, onto the surface of Mars, or into a black hole have become as common as driving to the grocery store. Colorful lasers, rotating planets, swirling clouds, and whizzing spacecraft fill planetarium domes and create realistic illusions surrounding the audience.

At Chicago's Adler Planetarium, the Space Transporter booths provide visitors with a simulated look at what they could expect if they were to visit Mars, Saturn, or any other planet in our solar system.

Video projectors now create large, bright, and high-resolution animation on the giant domes to produce imagery impossible in any other medium. They are rapidly becoming common in even the smallest of planetariums. Some companies now produce computer-generated special effects that can be shown on video, and are beginning to

Space theaters differ from traditional planetariums in that their domes and floors are tilted in such a way that the audience experiences the illusion of being surrounded by space in every direction.

make obsolete the mechanical special-effects devices that have been used in planetariums for many years.

Some planetariums utilize their environments even more realistically. For example, the Reuben H. Fleet Space Theater in San Diego features an elaborate plumbing system behind its dome, connected to several tiny tubes that pierce the skin of the dome. During shows, purified water is pumped through the system and sprinkled on audiences—an ultrarealistic experience during simulated rainstorms on our planet and others.

Modern planetarium shows are seldom lectures by an astronomer. Often the lectures are prerecorded tapes narrated by well-known personalities such as Leonard Nimoy, Patrick Stewart, or James Earl Jones. When accompanied by elaborate musical sound tracks, programs come alive with an excitement impossible to capture in any other entertainment medium.

These multimedia extravaganzas are often far too complex to be operated by one person. Instead, sophisticated automation systems sometimes run the shows. Many facilities now have the capability to preprogram their shows, and store the data in a computer. With the push of a single button, extremely complicated theatrics can be created time after time.

A Portable Universe

As exciting and innovative as these new technologies are, not everyone can visit a planetarium. Many schools cannot afford to take students on field trips to planetariums, and people in rural areas often have none nearby. For this reason, Learning Technologies, Inc., developed a portable planetarium called Starlab back in 1977.

Starlab is a room-sized inflatable bubble that can be transported anywhere by one person, and set up in less than ten minutes. Inside Starlab's dome, a small projector recreates an accurate night sky that can be used with a live lecturer or combined with movies, slides, or video projection.

Space Theaters

Some state-of-the-art planetariums feature domes tilted at a 20- or 25-degree angle. These are often called "space theaters." Space theaters are identical to traditional planetariums except that their floors are tilted at the same angle as the dome. This type of configuration causes the audience to experience the illusion of being surrounded by space in every direction—above, to the left and right, and below.

Many of these space theaters are used as motion-picture houses featuring immense film formats to produce unbelievably large and crisp images. Omnimax projectors use 70-millimeter-wide film projected onto the giant domes. The Omnimax experience can be a whirling, rollicking, roller-coaster ride through some of the most remarkable terrain anywhere—from the Grand Canyon of Arizona to the inside of the human body. The Fleet Space Theater in San Diego, which opened in 1973, was the first such theater, and set the standards by which other such theaters are measured today.

Smaller facilities may use 35-millimeter film projectors—most notably Cine-360. The expense of these projectors and films is significantly less than that for Omnimax film, but their image quality is not quite as high. Nevertheless, audiences can experience similar illusions.

It's Just an Illusion

Through the advance of technology, the planetarium, which has long been used to educate the public about astronomy, can now serve one more function: to inspire. Audiences can now experience simulated space travel like never before, journey through time to appreciate evolution of our planet or the entire universe; and enjoy the magic of a star-filled night—not just from Earth, but from anywhere in the universe. And it's all done with illusions.

Even the planetarium environment itself is an illusion. The dome is often made of sheet aluminum, perforated with tiny holes. These holes, invisible to the casual visitor, make the dome virtually translucent. They help maintain ventilation of air in the planetarium theater, make the dome lighter, and allow the placement of immense audio speakers to provide spectacular sound tracks for shows. Modern planetariums also place elaborate special effects behind the dome—effects like exploding stars and spacewalking astronauts—invisible until their power is turned on.

Uses of Domed Space

The planetarium's "space" can be used for much more than astronomy shows. It also makes an excellent environment for research. For example, local and national law-

Planetariums have developed unique ways of presenting information. Adler's "Hall of Planets" features three-dimensional scale models of the Sun, planets, and largest moons of the solar system.

enforcement agencies occasionally use planetariums to re-create lighting conditions and the positions of the Sun and Moon during crimes. By using the positions of planetarium stars, ornithologists have studied how birds navigate the sky. Planetariums have helped to train astronauts to recognize star patterns as well as to navigate through space. And when the late film director Cecil B. De Mille was planning a large set in the desert for the movie *The Ten Commandments,* his staff consulted the Griffith Planetarium in Los Angeles to learn the proper Sun angles for different scenes.

Planetariums are often used for their environments as much as for their ability to simulate a realistic sky. For example, spectacular laser-animation shows are often accompanied by rock and roll, jazz, or classical music. Such performances draw audiences who might never otherwise experience the magic of a planetarium. Concerts are frequently performed under the magical spell of a planetarium starry night.

Some planetariums perform live plays under the special effects created around the audience. One of the leaders in this activity continues to be the Strasenburgh Planetarium in Rochester, New York. In 1969 it co-produced a local production of Brecht's play *Galileo*, using its realistic star field as a backdrop. Since then, it has staged dozens of other special events, including a remarkable production of *Dracula* under the stars.

A Unique Profession

There once was a time when only astronomers and astronomy teachers worked in planetariums. Not anymore. Today technicians, artists, musicians, writers, photographers, producers, actors, and computer programmers all contribute to the medium.

Anyone interested in communicating the wonder of the heavens to the general public and children can often find work in a planetarium—as a part-time planetarium operator, a volunteer, or in a number of other capacities. Never before has the planetarium profession been so accessible to so many.

Educational programs exist around the United States to help train those with an eye on a planetarium career. Major facilities, such as Rochester's Strasenburgh Planetarium and the Hayden Planetarium in New York City, offer one-year paid internships to aspiring planetarians. Others, like those associated with colleges and universities, may offer bachelor's- or master's-degree programs in planetarium education.

To foster communication among planetarium professionals, the International Planetarium Society, along with its many regional associations, publishes newsletters and journals, and hosts conferences around the world in which technical papers and open discussion can take place. This exchange of information and new ideas contributes to the general improvement of the planetarium medium and its service to the community.

Since 1977 the Starlab "portable" planetarium has brought the wonders of space and astronomy to schools and groups located in rural areas.

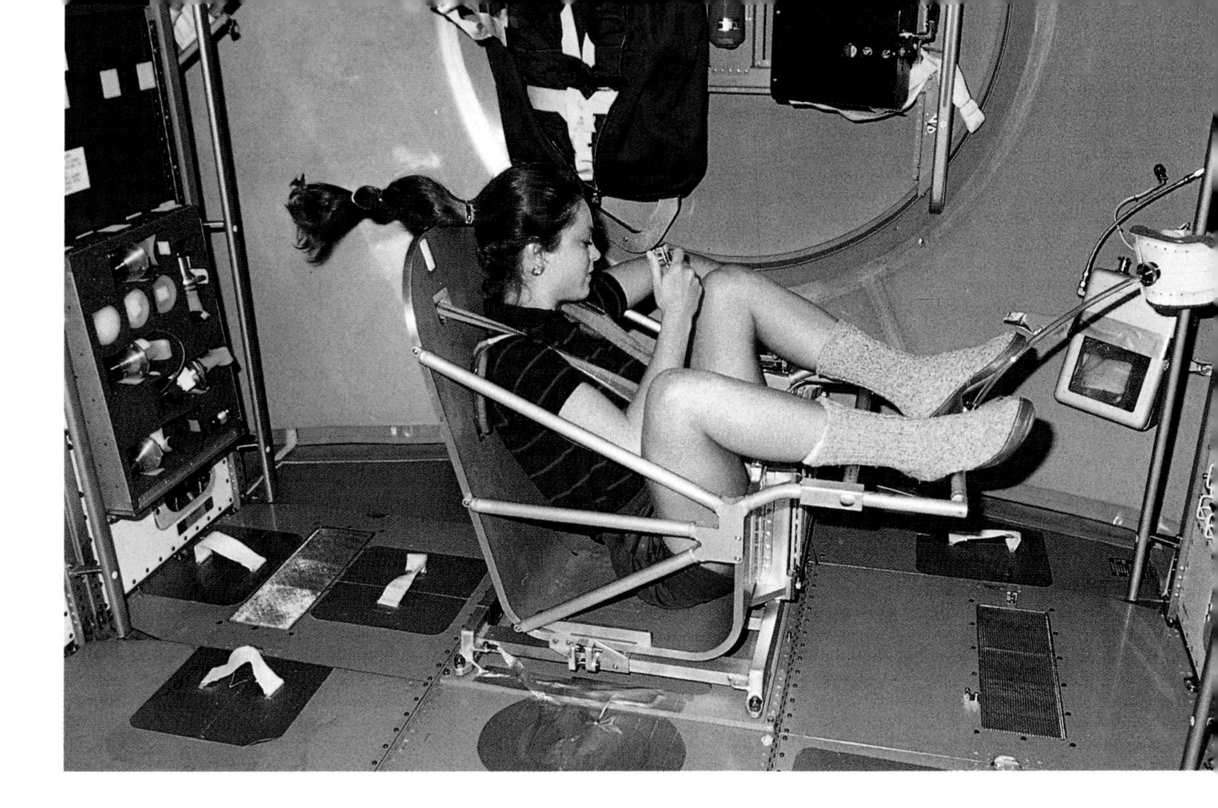

The Floating World At ZERO G

by Doug Stewart

After taking off on the last Mercury mission, Gordon Cooper settled in for a good night's sleep halfway through his journey. Compared with most of the duties of spaceflight, it seemed an easy enough undertaking. But Cooper ended up having to wedge his hands beneath his safety harness to keep his arms from floating around and striking switches on the instrument panel.

Since Cooper's flight, sleeping in space has become routine—maybe too routine. When carrying out an especially boring or tiring task, some astronauts have nodded off—only, they didn't really nod; they simply closed their eyes and stopped moving. "There are none of the waking mechanisms we're used to on Earth," says Joseph Allen, who flew on the shuttle in 1984. "Your head doesn't fall over. Pencils don't drop from your fingers and hit the floor to wake you up."

Pencils can, however, waft away. Allen discovered that if you pull open a drawer full of items that haven't been strapped down, everything drifts away, and soon "you have a real slow-motion slapstick on your hands, because you just can't get to all of them and tuck them under your arms and between your legs fast enough."

Such oddities are details that the National Aeronautics and Space Administration (NASA) didn't always foresee when spaceflight began. Now, after studying how astronauts lived and worked aboard Skylab, which hosted three-man crews on long-term missions from 1973 to 1974, the agency is pushing to build Space Station Freedom, a permanent human abode in low-Earth orbit to study how humans can thrive in weightlessness. While congressional budgetary pressures have threatened Freedom's future, the space community still hopes to launch some kind of manned orbiting laboratory. "If we're ever to colonize the solar system, we need to know more about living in microgravity," says Harvey Willenberg, chief scientist for civil space programs at Boeing Aerospace.

In the gravity-free environment of space, astronauts need a special contraption in order to determine their weight—or, more correctly, their mass.

"Can you survive a year of weightlessness? Yes, you can. Can you spend a year getting to Mars and be productive the moment you get there? That's the real question." A space station, says Willenberg, would help answer it. So while funds remain, NASA continues its research to develop a safe, familiar, even homey habitation for Earthlings who venture where they don't naturally belong.

Of course, all the research in Earthbound laboratories can't make a space station truly familiar. "Zero gravity is so bizarre," says Allen, "that even if [a spacecraft] looked like your mother's house, it doesn't feel like your mother's house, because you float, and the floating is so extraordinary. At no moment do you think, 'I'm in a nice office on the ground.' Never."

Will It Look Like Home?

So far, Skylab has been the only U.S. spacecraft worthy of the designation "space station." An enormous can that had originally been designed as a Saturn rocket's fuel tank, Skylab had over 12,000 cubic feet (340 cubic meters) of habitable space. Equal to the volume of a decent three-bedroom house, Skylab was over 50 times roomier than the Apollo command module. The crews settled in for long stays, the final one an 84-day marathon. "Unless you got on Skylab," says Pete Conrad, its first commander, "you don't have a real appreciation for zero G." If you're an American, that is.

Soviet cosmonauts have spent far more hours in space than U.S. astronauts, and since the age of *glasnost,* the space communities of the U.S. and former Soviet Union have engaged in a relatively open exchange of human-factors-engineering data. According to Jack Stokes, a human-factors engineer at NASA's Marshall Space Flight Center in Huntsville, Alabama, cosmonaut experience aboard the Soviet space station Mir supports the ideas of U.S. space station designers. "It helps to have somebody up there who's trying it and proving that we're probably on the right track," he says.

Of the lessons NASA has learned from its own astronauts, one of the biggest is that it must provide for their psychological well-being. In the days of Skylab, astronauts were largely a band of macho fighter pilots, and they were expected to perform accordingly.

Many of the scientific experiments and evaluations carried out by shuttle astronauts (above) involve the effect of weightlessness on various chemical and biological processes. With no gravity, a bubble of strawberry drink (right) accidentally released by an astronaut floats rather than splats.

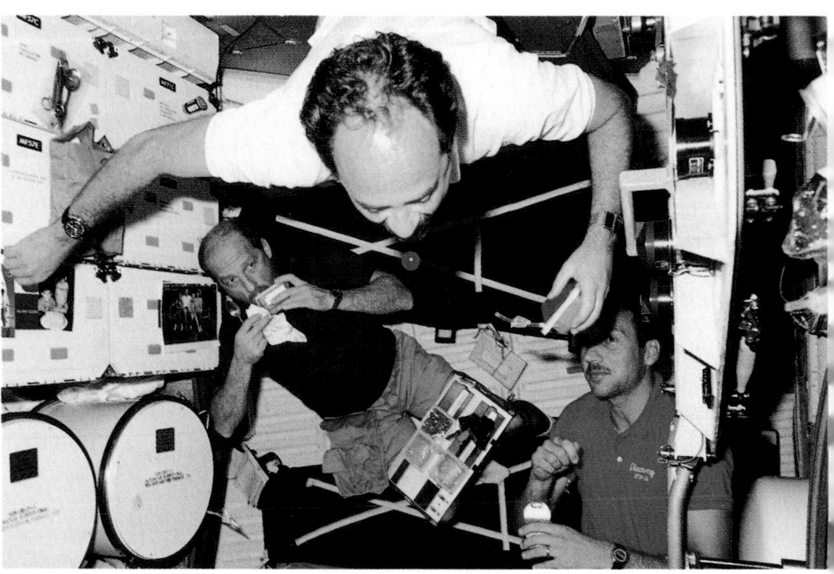

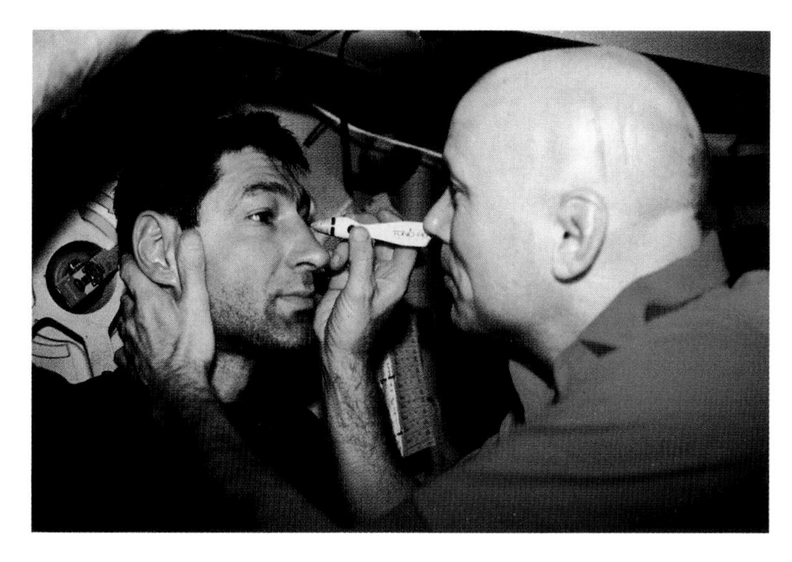

Astronauts are continuously monitored for any signs of physiological changes that might occur as a result of weightlessness.

"NASA didn't downplay psychological factors. It ignored them," says Bill Pogue, a former fighter pilot and Skylab astronaut who now advises Boeing on how *not* to lay out a spacecraft. He is still grumpy about the subject. Skylab wouldn't have had its one window if renowned industrial designer Raymond Loewy, creator of the Lucky Strike bull's-eye, and a onetime NASA consultant, hadn't insisted. In 1973, after Skylab's final crew demanded a day off after six weeks of escalating work orders, Pogue spent several hours doing nothing more than gazing out that window.

Future Designs

Any future space station will certainly have at least one window, as well as a few other comforts of home. Or at least an aerospace engineer's approximation of the comforts of home. Freedom's final layout is uncertain, but the design currently calls for a series of large metal cans linked at their ends like sausages. Work areas and eating-sleeping-leisure quarters will be housed in separate links, the "lab module" and "hab module," in NASA-speak. Plans call for a washer and dryer (to avoid wobbling the craft, the washer will squeeze clothes instead of spinning them), a Nautilus-style exerciser, and a refrigerator-freezer and trash compactor. Budget constraints, however, have already doomed plans for private, closet-size sleeping quarters, including a "togetherness suite," a pair of sleeping compartments with a central partition that married occupants could remove to engage in zero-G intimacies.

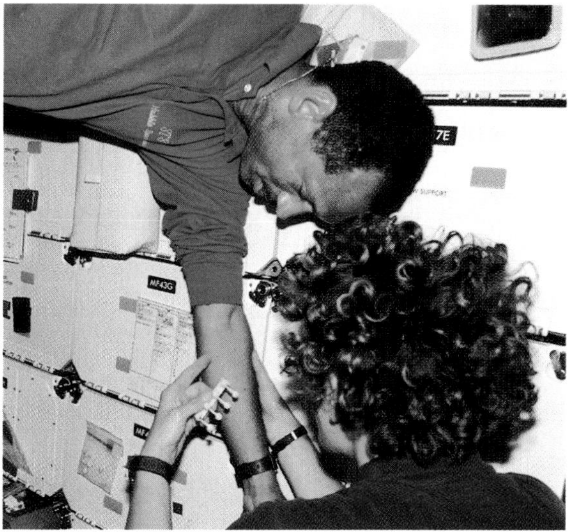

NASA's inclusion of washers and trash compactors isn't an attempt to pamper space-station crews, but rather, a sign of its current awareness that a happy, comfortable astronaut is a productive astronaut. The agency has even hired consultants to determine the wall colors that would be most simpatico with orbital life. Pink is said to be restful, though it's hard to imagine astronauts at ease in an all-pink decor. "I don't live in a pink house," says Allen. "I can't imagine why I would want to be in a pink ship."

Whatever color NASA chooses, the space-station interior will still resemble a not especially luxurious mobile home: narrow, rectangular corridors with surprisingly smooth and almost featureless walls. The barrenness is the first hint that the space station is to be parked, not in someone's backyard, but in a perpetual free-fall around Earth. Accordingly, its walls will form a sort of squash court off which humans and other loose objects will gently carom.

When Down Is Up

In zero G, one's eyes have to do what one's helpless inner ear no longer can, so NASA now takes pains to give space travelers a strong visual sense of up and down. It wasn't always so. Aboard Skylab, labels and electrical switches ran in different directions. One workstation was positioned sideways to the rest of the spacecraft. The toilet seat was set midway up a wall, like a window. "It was really squirrelly," says Pogue. Even the space shuttle, on crowded flights, forces some astronauts to sleep strapped against the underside of a crewmate's bunk. Without gravity, you can theoretically sleep anywhere, but as shuttle payload specialist Byron Lichtenberg points out, "It's hard trying to sleep like a bat on the ceiling." Awakening his first morning in space, Lichtenberg looked around at his upside-down environs and felt a pronounced wave of nausea sweep through him.

"You don't *have* to have tables on the floor. They can be on the walls or the ceiling," says Brand Griffin, Boeing's project manager for advanced civil space systems. "But in space, you like things to be pretty terrestrial." To keep crews oriented, Freedom's designers plan to have fluorescent lights running the length of the "ceiling," tables attached to the floor, and windows—not toilets—appearing on the walls.

Getting Around; Staying Put

Skylab's crews found they didn't need handholds and ladders to get around; they quickly learned to push off with one hand and float directly to their destinations. Says Pete Conrad: "You could look across the tank and say, 'I want to do three 360s and land on my feet over in that corner,' and do it." His crewmate Joe Kerwin never went anywhere in Skylab without a few slow rolls in transit.

Though the space shuttle is a bit more crowded, there's still room to maneuver. After heading aloft for the first time aboard *Columbia* in 1983, Lichtenberg remembers unbuckling himself from his seat and trying to stand up. "I immediately shot right up to

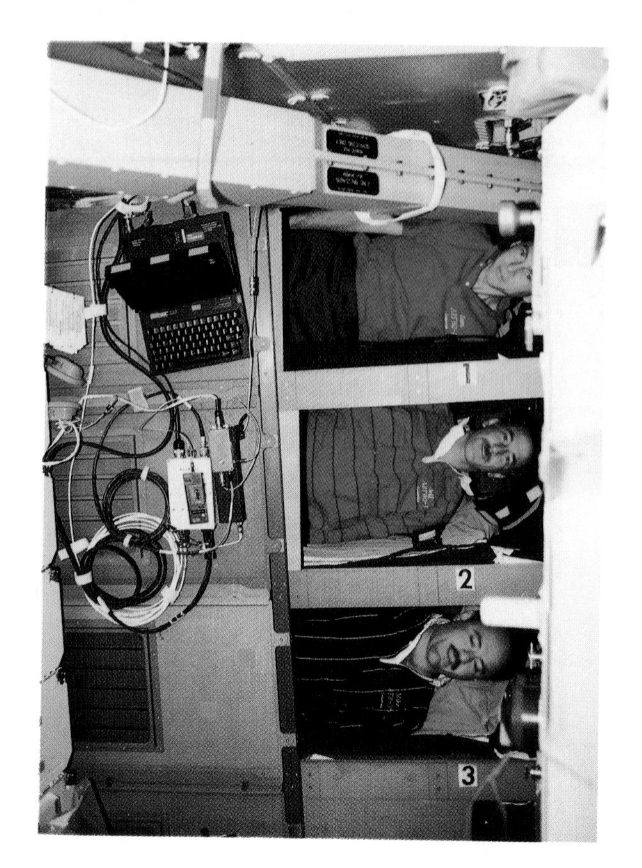

the ceiling," he says. "I said to Owen Garriott, 'Hey, we're not in zero gravity—we're in *negative* gravity!' " By the time they're back on Earth, astronauts have acquired the habit of trying to get out of their chairs by pressing lightly on the armrests. (Similarly, after returning to Earth, one astronaut let go of his coffee mug in midair and was surprised when it crashed to the floor.)

Using the lessons of Skylab, space-station designers have kept waist straps and handholds to a minimum. "We found out from Skylab that if you can nail a guy's feet down well enough, he can basically do anything he does on Earth," says Jack Stokes. He doesn't actually recommend using nails; to eat, use a computer, or do some other stationary task, astronauts now slip their feet into loops or wedges attached to the floor. Similarly, a single Velcro head strap suffices to keep sleeping astronauts from drifting out toward the ventilation ducts. For some tasks, however, a firmer anchor is required. When Pogue first tried to use a screwdriver in space, "the screw didn't twist; I did." He had to strap down both legs and his free arm before he could get the screw to turn. Spacecraft now pack electric screwdrivers.

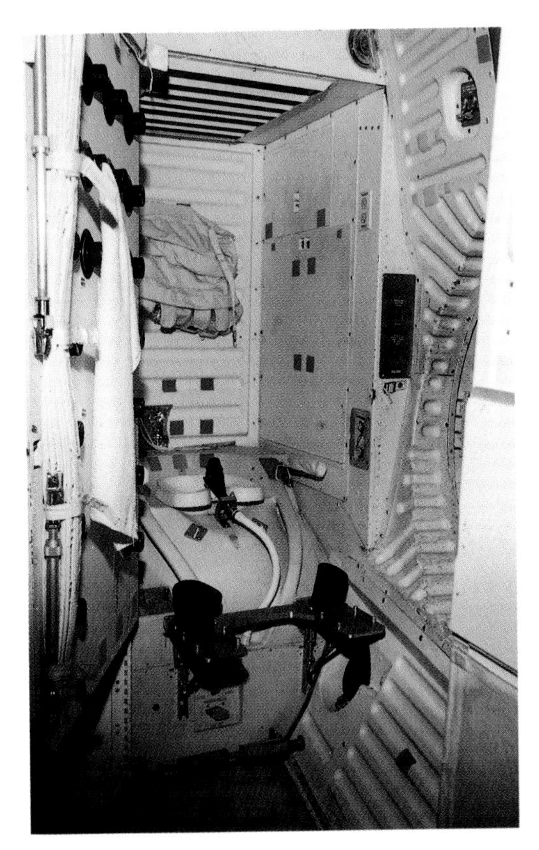

Astronauts must be confined during sleep (facing page) to prevent them from floating away. The space station will include a special "whole body shower" (above left) in which a footbar keeps the bather in one position. Lavatory functions will be less of an ordeal when astronauts have a high-tech "space toilet" at their disposal (above).

One item you won't find on any space station is a chair. In the weightlessness of space, the human body assumes what NASA calls the space-neutral body posture: shoulders hunch; arms float forward; legs and knees bend slightly. "You don't want to design things that pull the astronauts out of that position," says Boeing's Griffin. A workstation chair that flew on Skylab required astronauts to keep their abdominal muscles flexed; the chair was soon removed.

The Bathroom

Hygiene in space has come a long way since Gemini and Apollo astronauts defecated into plastic bags with adhesive-lined tops. Any space station will certainly have what NASA refers to as the "waste-management compartment"—a space toilet. It will be similar to the one that shuttle astronauts use, which is surprisingly Earth-like in appearance and function. Astronauts sit down on the shuttle toilet to perform what those in the industry call "normal bowel evacuation," while air suction guides the waste into a tank. (Because everything in microgravity floats, it is crucial to make a good seal between buttocks and toilet seat; cushioned bars across the thigh help keep the astronaut in place.) For urination, astronauts use a funnel-equipped hose attached to the toilet. On the space station, solids will be compressed and freeze-dried for return to Earth, and since water will be in short supply, urine will be purified and used for laundry and bathing.

Next to the toilet is the full-body cleansing compartment, or shower stall. This, too, will rely on airflow to make its drain work. In use, water will emerge in spherical blobs from a hand-held shower head.

Those Little Irritations

Although astronauts almost universally agree with Pete Conrad's assessment of zero G as "better than Disneyland," it has its irritations. Many have to do with the absence of convection currents, the eddies of air that gravity keeps in motion on Earth. In a spacecraft, hot air doesn't rise. Neither do bubbles, which remain in beverages and cause

A payload specialist aboard the shuttle Columbia *keeps close tabs on the animal passengers and monitors their reaction to zero gravity.*

the drinker to belch. And since sweat doesn't drip off the body, it tends to collect on the skin in ever-thickening sheets. After a session on Skylab's treadmill one day, Pogue washed all but his head before starting his work. He soon noticed a putrid odor and looked around. "Then I realized I was smelling the cocoon of smelly air around my head," he says. A more dangerous effect is the buildup of exhaled carbon dioxide around an astronaut's face; fortunately, onboard ventilation systems keep air safely circulating.

The effects of microgravity, however, aren't always irritating. A favorite recreation is playing with one's food. Surprisingly, ordinary utensils work fine in space, although food sticks to both sides. Instead of carrying food all the way to their mouths, Earth-style, some experienced astronauts use their spoons as slow-motion food catapults.

Although drinking a cup of coffee seems like the most natural thing on Earth, in space, it won't work. If you were to tip the cup back to take a drink, the weightless coffee would not roll out, so astronauts suck their beverages out of plastic bags with straws. Joe Allen enjoyed studying the behavior of orange juice after it leaked out of its container and sat quivering in a perfect golden sphere at the end of his straw. It resembled, he says, "the world's weakest, wimpiest Jell-O." Allen offers the following advisory to fellow zero-G investigators: "Don't let your curiosity tempt you into exploring a larger clump of liquid than you're prepared to drink later." If you don't catch up to your blob with a straw, it eventually attaches itself to the nearest wall or window.

Without gravity, crumbs, lint, fingernail clippings, bookmarks, and the occasional sleeping astronaut drift freely throughout a spacecraft. On Skylab the air moved up almost imperceptibly toward a grate over a ventilation intake near the top of the tank. Two or three days after it was lost, a pencil or set of reading glasses would appear on the grate. "We called it the Lost and Found," recalls Gerald Carr, Skylab 4's commander. Crewmate Pogue adds that debris from mealtime tended to nest in the metal-grid ceiling above the galley, where it was hard to wipe up. After a few months, the galley ceiling "looked like the bottom of a birdcage." "It's those little things that drive them berserk," says human-factors engineer Stokes.

It's Stokes's job to predict the problems that crop up once the tug of gravity disappears. Though he has never been to space, Stokes has flown two and a half Earth orbits' worth of parabolic arcs aboard converted KC-135 tanker aircraft, airborne roller coasters that offer zero G in hell-raising 30-second snatches. Many pieces of equipment destined for the space station are tried out by human subjects in a KC-135's padded cargo bay. Including the solid-waste toilet, Stokes recalls. Yes, in 30 seconds.

Such embarrassments are necessary if astronauts are to soar, of course. The whole idea of airborne testing is to make living and working in weightlessness easy and unremarkable for ordinary folk. If the designers do their job, says Boeing's Brand Griffin, the space station will be a boring place to live. "You'll get your Nobel Prize for what you do in the laboratory module, but the rest will be as exciting as a refrigerator," he says. Stokes agrees: "[Freedom] is just a laboratory now, and a very well equipped laboratory. It just happens to be located in a strange place."

ASTEROID THREAT!

by *Clark R. Chapman and David Morrison*

Some 130 Americans die in airline disasters in an average year. Assuming you are an average American, neither a couch potato nor a frequent flier, your own chance of meeting such a gruesome fate in any given year is 130 divided by the total U.S. population of 250 million—about one in 2 million. Assuming further that the average American has another 50 years to live, the likelihood of your eventually acquiring the epitaph KILLED IN AN AIRLINE DISASTER is roughly one in 40,000.

As best as we astronomers can tell, an asteroid about 1 mile (1.6 kilometers) across strikes Earth once every 300,000 years. A rock of that size would explode with the force of 100,000 megatons of TNT—10 times the force of the world's entire nuclear arsenal, and 5 million times the force of the bomb that destroyed Hiroshima. The asteroid would leave a crater over 10 miles (16 kilometers) across. Nobody really knows what effect it would have on the planet as a whole, but the pall of dust kicked up by the impact might block the Sun long enough to bring on a nuclear winter. Agriculture and commerce would probably cease, and most people in the world would die.

In any given year, there is a one-in-300,000 chance of such an impact occurring. That works out to a one-in-6,000 chance over the next 50 years. Thus, your chance of dying in an asteroid impact may be as high as one in 6,000—more than six times greater than your chance of dying in a plane crash.

Does that worry you? Probably not. Not many people worry about asteroid impacts—

Amazingly, Americans stand a significantly greater chance of being killed in the aftermath of an Earth-asteroid collision than they do from dying in an airplane crash.

certainly not as many people as worry about getting on an airplane. And of course there are good reasons for not becoming unhinged by the scenario we just sketched. It is probably a worst-case scenario. Many impact researchers would question our assumption that most people on the planet could be killed by the impact of a 1-mile-wide asteroid. Such an asteroid, they say, might merely wipe out the city or perhaps the country it struck.

Invisible Swarms

But such informed skepticism is probably not what keeps most people from being concerned about the danger of an asteroid impact. On the contrary, it is ignorance. Asteroid impacts happen so rarely that it is difficult for us to grasp the risk. What's more, our space-age image of Earth may be the enemy of true understanding. We have become accustomed to thinking of Earth as a lonely world, suspended in the vast emptiness of interplanetary space, millions of miles from its nearest neighbors. The reality is quite different.

Our planet resides in an almost invisible swarm of meteoroids and asteroids. The vast majority are dust grains, pebbles, or small rocks. When they meet up with Earth, they either burn in the atmosphere or fall harmlessly to the ground. But there are hundreds of rocks out there large enough to do real and possibly catastrophic damage. And someday one of them is going to hit us. We have yet to face up to that sobering reality.

In recent years, however, some of us have at least become aware of it. One person who deserves a lot of credit for that is Eugene Shoemaker of the U.S. Geological Survey in Flagstaff, Arizona. Shoemaker, a geologist turned astronomer, was an early proponent of the idea that stones from the heavens make holes in a geologist's turf. During the 1950s his turf was Meteor Crater, a 0.75-mile (1.2-kilometer)-wide, 600-foot (183-meter)-deep hole in northern Arizona. He studied the minerals inside it, the overturned rock layers on its rim, and the bits of iron strewn about the countryside. Geologists long believed the crater was volcanic, like the ones in nearby Sunset Crater National Monument. Shoemaker proved instead that it had been excavated 50,000 years ago by a tiny nickel-iron asteroid just 150 feet (45 meters) across.

For the past two decades, Shoemaker (joined in recent years by his wife, Carolyn) has been using a wide-field telescope on California's Mount Palomar to scan the skies for near-Earth asteroids—objects whose orbits approach ours, and some of which will one day poke holes in Earth or the Moon. Before 1970, only a couple of dozen of these objects had been discovered. Astronomers regarded them as curiosities rather than hazards. Through the diligent work of the Shoemakers and other observers, nearly 170 near-Earth asteroids have now been cataloged. A new one is discovered every few weeks.

We still know little about them. For starters, we don't really know how they got near Earth. Asteroids are relics of the early solar system, bits of debris that never coalesced into a planet. Most of them reside in a belt between Mars and Jupiter, in orbits that more or less parallel those of the planets. The Earth-crossing asteroids, however, zing around the Sun in orbits of all kinds, some highly elongated, some steeply inclined to the plane of the solar system. It's thought that they were deflected out of the asteroid belt and into their present orbits by close encounters with Jupiter, but the process isn't entirely clear.

Odd-shaped Chunks

We're also not sure what a near-Earth asteroid looks like. No spacecraft has yet visited one (although Jupiter-bound Galileo is scheduled to fly by a main-belt asteroid in October 1992), and through a telescope, near-Earth asteroids look like nothing more than fast-moving, faint stars. Yet by analyzing the spectrum of the light they reflect and by bouncing radar beams off these asteroids, astronomers have learned a few things about them. Near-Earth asteroids tend to be odd-shaped chunks that rotate with a period of a few hours to a few days. Most are rocky, but a few seem to be solid metal. Some may be extinct comets, consisting perhaps of a mass of carbon-rich dirt over a remnant of ice. They are probably all cratered and fractured by collisions with other asteroids.

Recently Steve Ostro of the Jet Propulsion Laboratory (JPL) made the first good radar image of a near-Earth asteroid. Astonishingly, it seems to be two separate pieces of rock, held together by the force of their very weak mutual gravity. Ostro has made

Gigantic bodies of space matter have slammed into the Earth's surface several times in the past. Arizona's 600-foot-deep Meteor Crater was created by the impact of a 150-foot-wide hunk of metal.

less-sharp images of a few other asteroids, and he suspects that many Earth-approachers may be double or compound objects. Instead of individual, solid rocks, some may be gravitationally bound rock piles.

Of the asteroids that are large enough to be readily detected, only a fraction actually have been. Astronomers engaged in the search are a small posse, and they can photograph only a tiny section of the sky each week. Near-Earth asteroids could be lurking anywhere in the sky, and most of the time they are nowhere near Earth; they are on distant parts of their orbits and are too faint to register on film. By comparing the number of asteroids that have been discovered with the fraction of the sky that has been carefully observed, Shoemaker has estimated that we have seen fewer than 10 percent of the ones larger than a kilometer, or three-fifths of a mile, across. There are 1,700 of those, he says, give or take 50 percent. There must be thousands more that are smaller than a kilometer, but still larger than the 150-foot progenitor of Meteor Crater.

From the orbital distribution of the known near-Earth asteroids, Shoemaker has calculated that one-third of them will hit Earth or the Moon sometime within the next 100 million years. (The near-Earth stock will be replenished by the asteroid belt and dying comets.) In the case of asteroids a mile wide or more, that works out to an average of one every 300,000 years. The orbits of the known asteroids have been projected into the future, and they, at least, do not look as if they will strike Earth during our lifetime. The danger, if any, to the planet's current inhabitants, will come from asteroids not yet discovered.

In January of 1991, a small asteroid, roughly 20 feet (6 meters) across, passed within 100,000 miles (160,000 kilometers) of Earth, less than half the distance to the Moon. Just two years ago, a larger rock, as wide as the Empire State Building is tall, missed Earth by twice the distance to the Moon—still a close shave by astronomical standards. Had that rock struck Earth, it would have carried the force of 1,000 megatons of TNT, the equivalent of 50,000 Hiroshima bombs. And yet we didn't even know it was coming. The asteroid was photographed by Hank Holt, Shoemaker's colleague, several nights after it had made its closest approach to Earth, as it was already zooming away. Equally large bodies must whiz by even closer every few years without being seen.

No Recent Impacts

No large asteroids have struck Earth within recorded history. But over geologic time, they surely have, and with considerable effect. Despite pockets of resistance, there is widespread consensus that asteroid impacts were responsible for the extinction of the dinosaurs and other species 65 million years ago, and perhaps for mass extinctions at other times as well. If a single asteroid killed the dinosaurs, it would have to have been at least 5 or 6 miles (8 or 9 kilometers) across. It is very unlikely, though not impossible, that an object that large will hit Earth anytime soon; we would almost certainly have seen it by now. An asteroid just 1 mile (1.6 kilometers) wide, though, would wreak havoc enough. It would be a disaster almost beyond our ability to comprehend.

The chance of its happening is small, but the consequences are potentially enormous. That is the dilemma. As a rule, our society tends to be quite cautious about such hazards. Although commercial air travel is much safer than travel by car, there is strong public pressure to reduce even further the small but real risk of plane crashes. Far less common even than airline disasters are catastrophic accidents involving nuclear-power plants. Yet the Chernobyl event actually happened, and Three Mile Island came close, so most governments have imposed very strict controls on nuclear-plant operations. Many people want even stricter controls.

The threat of an asteroid impact is at the far end of the spectrum of hazards: it is even less likely to happen than a reactor meltdown, but if it did happen, it could mean the end of civilization. So far, we have decided, by default really, that the risk is so low that we should do nothing about it.

What Can We Do?

Could we do anything about it? The first thing we could do is identify the asteroids that might threaten us. That is a simple matter of devoting more astronomers and more telescopes to the search—in other words, it is a matter of more money. It wouldn't require a huge amount: perhaps $20 million to set up four new telescopes to survey the sky, and $2 million a year to operate each one. Shoemak-

The Earth (blue) is buzzed by a swarm of asteroids that travel in orbits that do not parallel those of the planets. Seven of the 170 known near-Earth asteroids are shown here; thousands more exist.

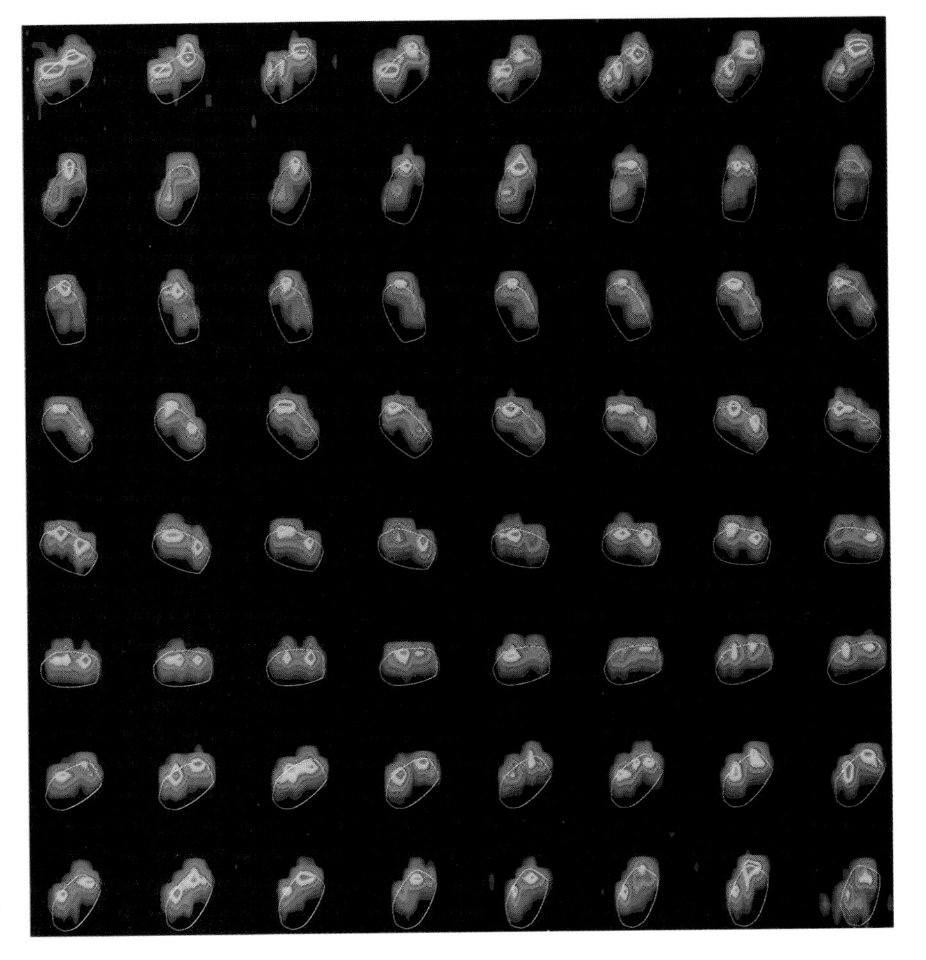

Scientists use radar to track near-Earth asteroids. This sequence (read left to right) suggests that asteroid 1989PB consists of at least two distinct pieces, each about a half-mile across. The asteroid rotates counterclockwise once every four hours.

er and other researchers have long had telescopes on their drawing boards that would be more efficient at scanning the sky for asteroids. But those plans have remained in limbo for lack of funding.

The next thing would be to figure out a way to protect ourselves. If an asteroid were due to strike within the next few weeks, nothing could be done to stop it. But if our lead time were years or decades, there is a good chance that a spacecraft could be launched to intercept the asteroid. One strategy would be to explode a series of small bombs near but not directly on the asteroid. The idea would be to deflect it gently from its path rather than blow it to pieces; a shower of asteroid fragments raining down on Earth might do more global damage than a single intact rock. If the asteroid turned out to be a loose agglomeration of rocks, like the one Ostro has studied, the job of deflecting it without disrupting it would be trickier. Obviously, more research is needed before this kind of scheme could be made to work.

There are signs that interest in the asteroid threat is increasing. The National Aeronautics and Space Administration (NASA) is studying the possibility of a reconnaissance mission to a near-Earth asteroid; the American Institute of Aeronautics and Astronautics has advocated expanded telescopic searches and more-detailed studies of deflection methods. A conference on near-Earth asteroids scheduled for summer 1992 will focus in part on the hazard they present.

The threat of an asteroid striking Earth is one that is hard for people to take seriously, and hard to talk about without sounding like Chicken Little. But the threat is real, and perhaps it is time to stop ignoring it. When you get on an airplane, you expect that the airline and government agencies have done their best to reduce the risk of a crash, and that your future is thus reasonably secure. So ask yourself: should we let the future of our whole civilization be determined by a game of cosmic roulette, or should we develop the knowledge and ability to protect ourselves?

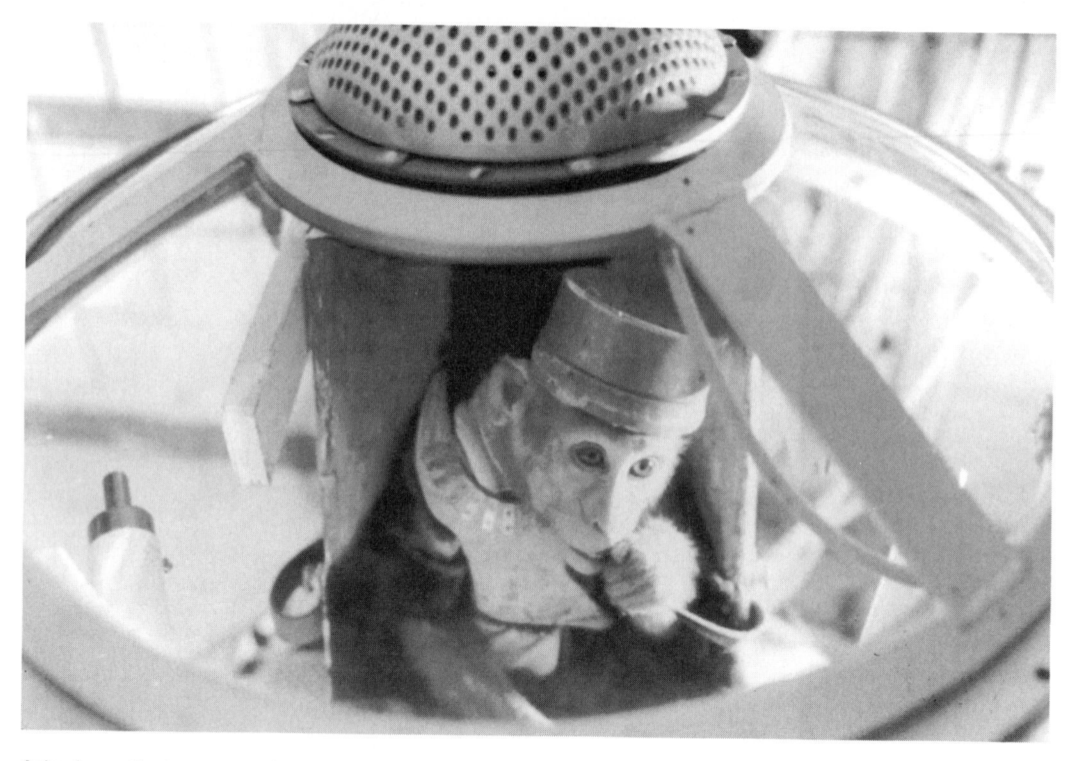

Animals are the true space pioneers, especially those launched into space in preparation for manned flights.

ANIMALS IN SPACE

by Karen Boehler

When most people think of space exploration, they probably focus on human missions. Many also think of robotic deep-space probes. But rarely do people consider the creatures who pioneered space before humans, and still serve as surrogates for future human space exploration.

Animals preceded people in space by about 15 years, and it is doubtful that space exploration would have progressed as far as it has without preliminary animal flights and experiments.

Biological experiments were regularly launched from the White Sands Proving Grounds in New Mexico. In 1946 and 1947, eight confiscated German V-2 rockets lifted off, carrying payloads of seeds, spores, and fruit flies. On June 11, 1948, the first primate went into space. An anesthetized rhesus monkey nicknamed Albert 1 rode as the sole passenger aboard Blossom 3. Albert reached an altitude of 38 miles (62 kilometers), but unfortunately died on impact when his capsule's parachute failed.

Through 1952 animal experiments continued on both V-2 and Aerobee rockets. The first surviving passengers were 11 mice—launched September 20, 1951, to a height of 44 miles (71 kilometers). A companion rhesus monkey died two hours after impact, but this constituted the first successful parachute recovery. On May 21, 1952, two capuchin monkeys named Pat and Mike became the first primates to survive actual spaceflight conditions.

Cosmocanines

The former Soviet Union began animal experiments—mostly with dogs—somewhat later than did Americans, and pursued a slower pace in the early years, but were better at recovering passengers alive. From about 1949 through August 27, 1958, the Soviets launched 42 dogs. On July 2, 1959, two dogs and a rabbit were sent to an altitude of more than 100 miles (160 kilometers), setting a weight-into-space record for that time (4,400 pounds—2,000 kilograms). One of those dogs, Otvazhnaya, had ridden on two earlier flights. In fact, many dogs used in early Soviet space experiments flew more than once. Otvazhnaya eventually made five successful suborbital rocket flights.

During the mid-1950s, under its then-director, U.S. Army Colonel John Paul Stapp, the Aeromedical Research Laboratory in New Mexico employed 38 people and housed as many as 400 animals. According to Jerry Fineg, chief of the veterinary-science division, it is difficult to envision just how scant our knowledge was back then. Scientists had no idea whether humans could survive in space—and it was in search of that answer that most early animal-in-space research was focused.

"Until the early 1960s, we did not know exactly how many G's a man was going to pull on a lift-off, and we were very concerned," Fineg says. (One G equals the normal force of gravity at sea level on Earth.) William Britz, a clinical veterinarian, worried about the other end of a mission: "The biggest question was the impact—reentry," he says.

Cats, dogs, fish, frogs, rats, and mice—all lived at the lab. Chimps, pigs, bears, and men were all subjects of choice for the Sled Track, a rapid-fire carnival ride used to study spacecraft takeoff and reentry conditions. Many also rode the Bopper, a giant slingshot for studying postreentry shock and land or water impacts, and the Daisy Track, a shorter version of the Sled Track with a gimbal-mounted seat used to analyze aerodynamic forces. During this period, ground-based experiments covered a range of space biology and biodynamics, including unusual environmental conditions and the effects of mechanical stresses on living tissues.

In 1961, Ham (left) beat out 60 contenders to become the first chimpanzee in space, spending 16 minutes and 39 seconds in suborbital space.

In 1959 the Soviets sent up Belko the dog, whose health and behavior were monitored continuously via television images relayed to Earth (right).

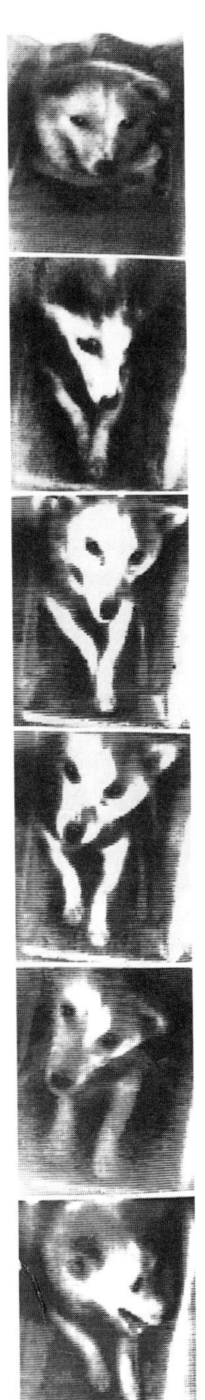

On November 3, 1957, the Soviets launched Sputnik 2 with Laika, a two-year-old female dog. She became the first living subject in orbit, and, although she died after six days (when she ran out of air and food), she galvanized the U.S. space program. Fineg says the National Aeronautics and Space Administration (NASA) came to the Aeromedical Research Laboratory. "When Laika went up, [NASA] wanted a chimp. We were the only organization capable of setting that in motion."

Space Chimps

Test launches were eventually switched to Cape Canaveral in Florida instead of White Sands. The rockets were larger and more powerful than before. In 1958 there were three mouse launches aboard Thor/Aerojet rockets. In late 1958 a joint U.S. Army/Navy project sent Old Reliable, a squirrel monkey, 288 miles (465 kilometers) above the Earth's surface on a Jupiter rocket. On May 28, 1959, the same team sent the rhesus and squirrel monkeys Able and Baker on a slightly longer flight up 300 miles (480 kilometers). The last two monkey flights in that series took place on December 4, 1959, and January 21, 1960.

Back in New Mexico, work focused on putting a chimpanzee in space. From the time Fineg and his colleagues were told their lab would be the lead agency, he recalls, they had only one year lead time. "The basic reason for sending the chimp to space was to test the reliability of the launch vehicle and the space capsule for humans."

In the early Soviet space program, a variety of animals were sent into space. Above, a rabbit and a dog are dressed in special space attire prior to lift off.

The two chimps who eventually flew, Ham and Enos, trained with 60 other chimps at the Aeromedical Research Laboratory, with eight finalists qualifying physiologically and psychologically.

Three years to the day after Explorer 1 became the first U.S. satellite in orbit, technicians strapped Ham inside a Mercury capsule like the one Alan B. Shepard would use three months later. The chimp spent 16 minutes and 39 seconds in suborbital space, and returned to what launch veterinarian Britz calls "the biggest celebration he'd ever seen." Ham's flight had made it clear humans could survive in space.

Enos became the second—and, to date, the last—chimp in space on November 29, 1961, making three orbits of Earth. Less than three months later, John H. Glenn made three orbits in *Friendship 7.*

Later Experiments

The three decades since Glenn's historic flight have seen occasional animal experiments, but nothing like the early ones that propelled humans into space. Although Gemini missions carried sea urchins, microorganisms, frog eggs, and fungus, once the critical question of human survivability in space was more or less answered, animal experiments became a back-burner issue in the United States.

Once the Soviet Union and the United States began space-station programs in the 1970s, research into animal life sciences resumed at a stronger pace. Scientists used the space environment to answer dozens of questions, from the impact of gravity on a variety of everyday animal activities to the long-term effects of microgravity on biological functions.

Skylab 3 carried as a student experiment the spiders Arabella and Anita, to see if they could spin webs in weightlessness. They could.

In the summer of 1975, the Apollo-Soyuz Test Project carried fish. Water is always compared to the microgravity environment of space, yet gravity obviously plays some role in marine habitats. Astronauts found that away from Earth's oceans, the fish didn't know which way was up.

In the 1980s, as part of the space-shuttle program, most animal experiments have been ferried aloft under the aegis of either the Get Away Special small-payload project or the Shuttle Student Involvement Program. Shuttle mission STS-4 carried fruit flies and brine shrimp. STS-5 studied the growth of sponges in microgravity. The astronauts aboard STS-7 watched a live ant colony, a student experiment to see how ants would burrow without gravitational cues.

STS-8 flew two rats to test a new animal-enclosure module. This was one of the few animal experiments to be an integral part of a mission rather than an add-on.

STS-41C was popular with the public—a student experiment had astronauts observing honeybees to see how they would construct a honeycomb in a weightless environment. Spacelab 3 on STS-51 carried two monkeys (whose floating feces created quite a mess for the human crew to cope with) and 24 rodents, and STS-29 carried four rats and 32 chicken eggs.

In one student shuttle experiment, several rats had bits of their leg bones removed to study how bones would heal in space. Plans to break the rats' bones were scrapped after animal-rights supporters objected.

Meanwhile, the Soviets pursued studies with various species, both with their continuing biosatellite series (with international participation) and aboard the Mir space station. Cosmonauts monitored quail eggs hatching in space, but the chicks could not grasp the wire mesh in their cages, and wound up dying of malnutrition or being killed.

The future of animal experimentation, now an open question, is probably neither as grim as some predict nor as hopeful as others want it to be.

A *Columbia* space-shuttle flight in June 1991 included animal experiments. The Space Life Sciences-1 (SLS-1) experiment used animals to investigate the physiological changes that occur during spaceflight, and investigated how people adapt to both microgravity and the return to Earth. The experiments involved rats and, for the first time, spacefaring jellyfish. The crew also tested new animal-enclosure facilities. (For a fuller discussion of the mission, see the table on pages 54–55.)

On a shuttle flight scheduled for the near future, a secondary payload will include eight rats as part of an experiment to study microgravity's effects on skeletal-muscle metabolism. And Spacelab-J, a joint mission with Japan set for September 1992 on *Endeavour,* will include two female African clawed frogs.

Many experiments have been designed but not yet tried. Donald Farrer, formerly of the Aeromedical Field Laboratory at Holloman Air Force Base, enthusiastically endorses an experiment called the "Mouse House," developed by animal-in-space expert James Henry in 1962. Farrer calls the experiment "absolutely fantastic, the next leap, as important a step forward as Mercury was from Aerobee." It would entail putting an adult female mouse in space with an adult male mouse to raise a generation of animals in microgravity. "It would make animals that had never experienced gravity," he explains. "Allow those animals to reproduce, and now we have a group of animals that has not only not experienced gravity, but were conceived in the weightless environment."

Fineg, who also serves as director of the Animal Research Center at the University of Texas-Austin, says he hopes such an experiment will be conducted someday. Working with the NASA Ames Research Center in California on the life-sciences portion of Space Station Freedom, he is helping design an

Space-shuttle astronaut Jack R. Lousma conducts a student-designed experiment to determine how moths, flies, and bees behave in microgravity.

8.2-foot (2.5-meter) centrifuge facility and habitat for both plants and animals. And despite recent calls for a scaled-down Freedom space station, Fineg says money seems to be available. "Life sciences has a priority based on what President Bush has indicated, and though the budget has been cut severely, the intent to go forward is there."

So while the glory days of space animal research may be long in the past, the future offers a glimmer of hope. Dr. Frederick Coulston, founder of the primate-oriented White Sands Research Center, said it best at a recent New Mexico reunion marking the 30th anniversary of Ham's flight: "Things look very, very promising for the future, and we are all very grateful to the fellows who were pioneers in all this."

REVIEW

Astronomy

THE SOLAR SYSTEM

In June 1991, bright flares on the Sun disrupted radio, artificial satellite, and utilities communications, and may have damaged solar cells or computer chips on Earth-orbiting satellites. The flares broke brightness records and may bring down some orbiting satellites, as they brought down Skylab some years ago. On Earth, expected solar neutrinos have not been found; solar physicists suggest that such an absence may mean that they need to develop a new theory of the Sun's internal operations, and perhaps a new physics in which neutrinos have mass.

Radar observations of Mercury appear to have found frozen ice patches near its poles, and a large crater, perhaps 800 miles (1,290 kilometers) wide, centered near its equator. New analyses of the chemistry of Venus suggest it may have had surface water billions of years ago. Spectacular radar images of Venus have shown us a landscape dotted with volcanoes, lava flows, and lightning.

Heat left over from Jupiter's formation appears to drive thermal variations in the atmosphere along the planet's equator. The variations occur in four- to six-year cycles, causing warm and cold periods on the planet. The variations are not connected with seasonal changes in sunlight. The Hubble Telescope observed an unusual tentlike feature created on Jupiter when the Great Red Spot—a massive hurricane larger than the entire Earth—shouldered aside a smaller hurricane.

Saturn's largest moon, Titan, has an atmosphere with both greenhouse-warming and antigreenhouse-cooling effects. Saturn itself has ribbons of infrared hot spots that circle the planet about 15 degrees north and south of the equator. The hot spots may represent holes in Saturn's thick lower atmosphere.

Large circular features seen by Voyager 2 on Neptune's largest moon, Triton, appear to be the remains of icy volcanic eruptions. The largest is 580 miles (935 kilometers) across, with a dark center and a bright periphery.

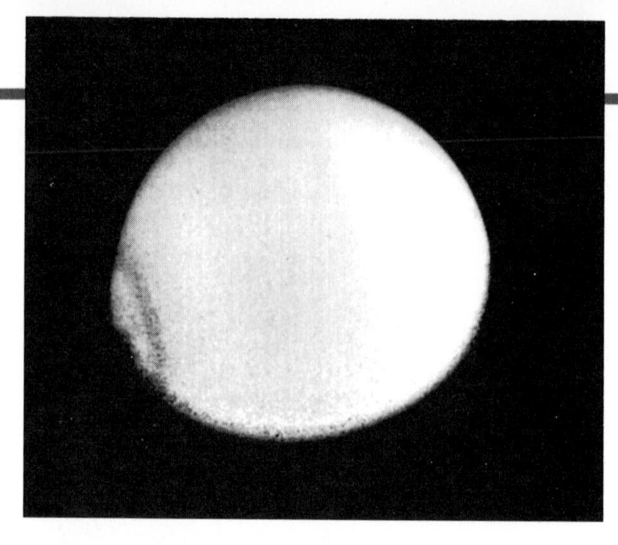

Astronomers have discovered that Titan, the largest moon of Saturn, has an atmosphere that exhibits an Earth-like greenhouse effect.

A new candidate for the crater caused when a massive object crashed into the Earth 65 million years ago has been suggested. The crater is on the Yucatán Peninsula in Mexico, and was pinpointed by tiny glass tektite fragments from Haiti.

The Galileo spacecraft, in a masterful performance of precision sighting, sent back the first close-up of an asteroid, Gaspra. The asteroid is a lumpy rock measuring about 12.4 by 7.5 by 6.8 miles (20 by 12 by 11 kilometers).

Comet Yanaka (1988r) may represent a new class of comets—comets that are interstellar travelers. Halley's comet, now speeding away from the Sun, threw off a cloud of dust and brightened briefly.

The Galileo space probe took the first close-up pictures of an asteroid. The asteroid, called Gaspra (right), is little more than a small, lumpy mass of rock.

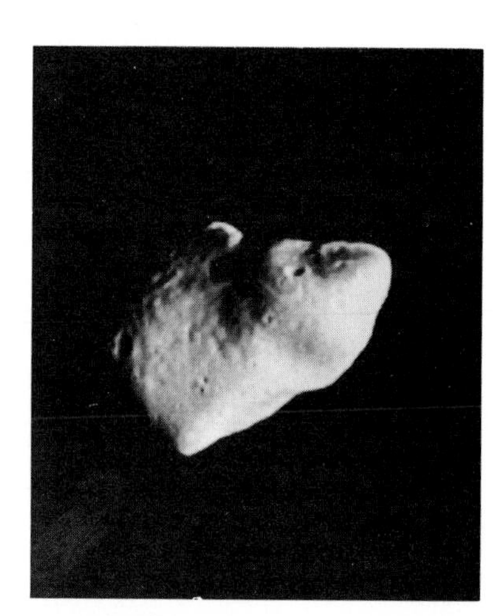

THE MILKY WAY

The first naked-eye nova found since 1975 was discovered February 18, 1992; the nova remained visibly bright for three days, dimmed, brightened slightly, and then continued to dim again.

Six brown dwarfs have been found in a photographic-plate investigation of the Hyades cluster about 150 light years from Earth. The six stars have masses about 70 times that of Jupiter, and brightnesses about one four-thousandth that of the Sun. Gamma-ray bursters—cosmic flash-bulbs—are distributed with surprising uniformity in the sky; thus, they are probably not members of our own galaxy, and they are so far unexplained.

The rotational discontinuity of stars of spectral class F5 may be caused by the formation of planetary systems.

Photographs suggest that the star Beta Pictoris may be in the process of evolving its own solar system.

Seen by the Hubble Telescope, Beta Pictoris, a star in the southern sky about 56 light years away, appears to be encircled by a large outer ring of small, solid particles and a small inner ring of diffuse gas. The inner gas ring may contain clumps of material spiraling downward toward the star. This may be a planetary system in formation, although many more observations are needed to understand this system more fully.

Hubble Telescope observations of HD 140283, about 100 light years from Earth, show it may be almost 15 billion years old, one of the oldest objects in the universe. The star contains more boron and beryllium than predicted by today's Big Bang theory.

Observations of globular clusters, perhaps the most beautiful objects visible in the heavens, have revealed a surprising number of *young* stars, (globular clusters were thought to be very old) and many millisecond pulsars. A beautiful cluster at the outskirts of our galaxy, 47 Tucanae, contains 21 massive blue (young) stars and several pulsars; M15 has many pulsars, as does Terzan 5. Surprised that these globular clusters are teeming with such pulsars, astronomers now suggest that the pulsars are vaporizing their companion stars.

THE UNIVERSE

Outside our own galaxy, the Hubble Telescope has observed blue (young) globular clusters in NGC 1275, a bright galaxy in Cassiopeia at the heart of the Perseus galaxies. Astronomers surmise that the blue globulars may have been formed in the collision of two galaxies.

Astronomers suggest that the unusual concentration of stars at the center of M87, a giant elliptical galaxy about 53 million light years away, may represent a black hole; they hope that more data to support this finding will be obtained from the Hubble Telescope.

The most-distant object, with a redshift of 4.9, has been found. The most-brilliant object ever seen is the quasar BR1202-07, as bright as 10^{15} suns (10 followed by 15 zeros). This quasar is also one of the three most-distant objects known, 12 billion light-years away. It must have existed when the universe was very young, only 7 percent of its current age. Today's theories do not adequately explain these observations. The search for dark matter and reconciliation of current cosmological theories with the observed bubbles, walls, and sheets of galaxies and matter in the universe continue. These questions generate debate and encourage the search for observations and theories for adequate explanation.

Katherine Haramundanis

REVIEW

Space Science

In May 1992, space shuttle Endeavour *crew members conducted an unprecedented three-astronaut space walk in order to save the Intelsat 6 satellite.*

MANNED SPACE PROGRAMS

▶ **United States Space Program.** During the year a committee of aerospace-industry leaders appointed by President George Bush strongly recommended that the National Aeronautics and Space Administration (NASA) refocus its program by reducing its reliance on the space shuttle and scaling back plans for the space station Freedom. With these positive recommendations, the administration proposed spending $14.7 billion on various space programs during the next fiscal year—an increase of 11.25 percent.

▶ **Spacelab.** One of the most exciting space-shuttle missions of the year was the 41st flight of *Columbia.* Tucked into its cargo bay was the 10-ton, $140 million medical laboratory known as Spacelab. During the mission, seven astronauts, 29 rats, and 2,478 jellyfish shared a nine-day biomedical experiment to study how people (and animals) adapt to spaceflight and weightlessness, and how they readjust once back on Earth.

▶ **Space Station Freedom.** In March 1992, NASA revealed plans for a streamlined space station that would be smaller, cheaper, and easier to build—one that could be ready for occupancy by a four-person crew toward the end of the century. It would use pieces assembled and tested on Earth instead of in space, and would thus require fewer shuttle flights and space walks by astronauts to build and maintain the structure. Its cost would be reduced from $38 billion to $30 billion.

▶ **European Space Program.** Under severe financial pressure, the 13 nations of the European Space Agency (ESA) agreed in November 1991 to postpone for one year their decision to begin building their Hermes space shuttle and the Columbus module for the U.S. space station. ESA ministers also decided to seek financial help from other countries to keep the program active.

▶ **Soviet Space Program.** Political fragmentation and economic upheaval in what was once the Soviet Union threw the once-powerful space program into shambles. Russian space-industry officials and scientists began to market their skills and services to other nations, including the United States. They also began to sell much of their equipment, including the massive Energia rockets. And they even offered their Soyuz ships as a possible rescue vehicle for U.S. shuttle astronauts. A new Russian Space Agency was created in March 1992, perhaps signaling an end of the chaos.

▶ **Cosmonaut Woes.** In May 1991, 34-year-old Sergei Krikalev was launched into space from the Baikonur Cosmodrome on a five-month mission as a flight engineer aboard the Mir space station. His replacement was to come in October.

When October rolled around, however, Krikalev got a surprise. Communism had fallen, and the Soviet Union had disintegrated. A new commonwealth of independent nations was being forged. The former Soviet space program was in disarray. And Krikalev, without much choice, continued to

Cosmonaut Sergei Krikalev returned to Earth on March 26, 1992—five months behind schedule.

orbit the globe every 90 minutes—a man without a country.

Finally, after 331 days in orbit, a weary yet happy Krikalev returned to Earth. By then, even his hometown of Leningrad had been renamed St. Petersburg.

In March 1992, NASA unveiled plans for a streamlined version of the space station Freedom. The facility should be ready for use before the year 2000.

THROUGH ROBOT EYES

▶ **The Hubble Space Telescope.** Scientists continued to plan a $20 million repair mission scheduled for early 1994. The mission will install corrective mirrors on Hubble, replace a camera with a more state-of-the-art unit, replace the solar panels—which now shudder when the craft moves from darkness to daylight every 90 minutes—and install a new set of gyroscopes to replace several that have failed in recent months.

Despite its problems, Hubble has made major discoveries. It photographed a remarkable storm on Saturn, demonstrated that some fuzzy points of light are actually complexes of distinct stars, and discovered the best evidence of supermassive black holes to date.

▶ **Gamma Ray Observatory.** Since its launch in April 1991, NASA's $557 million Gamma Ray Observatory (GRO) has been scanning the skies in search of gamma rays—the highest-energy wavelength of electromagnetic radiation, believed to be emitted by the hottest objects in the universe.

During the year, the GRO made a discovery so revolutionary that it could overturn major theories of the universe. It found more than 200 split-second gamma-ray bursts coming from all around the sky, not just from the plane of our Milky Way galaxy. This suggests that the gamma-ray bursts originate either from unexplained exotic sources relatively nearby, or from hugely powerful, unknown objects deep in the universe.

▶ **Galileo.** The Galileo mission was deployed from the space shuttle *Atlantis* on October 18, 1989, toward a 1995 rendezvous with Jupiter. It was designed to swing around Venus and Earth, using their gravitational fields to help it reach the giant planet.

On April 11, the craft's 16-foot (4.8-meter)-wide, umbrella-shaped antenna failed to unfurl. Engineers have tried a number of creative tricks to open it, but, to date, none have worked. If the 16-foot antenna is not repaired, a smaller antenna will be used, but this will greatly reduce the amount of data the craft can collect.

On October 29, 1991, Galileo swung past and photographed the asteroid Gaspra, an 11-mile (17.6-kilometer)-long rocky leftover from the birth of our solar system. It marked the first asteroid flyby of any spacecraft from Earth.

▶ **Magellan at Venus.** With its powerful radar-echo system, Magellan continued to probe the Venusian clouds. It has recorded features on the planet's surface as small as 325 feet (100 meters) across, and has seen huge, solidified lava flows stretching hundreds of miles across the rugged terrain. One big surprise: no surface feature seems to be older than a billion years. Since Venus suffers little erosion, and there appears to be no plate tectonics, the Venusian surface seems well preserved, allowing scientists to study how volcanism there may have changed over time. (See also *Magellan at Venus,* page 14.)

▶ **Ulysses.** The European-American solar explorer Ulysses continued its mission to the Sun. While on its way to its final destination, Ulysses has been used to monitor the region of space through which the probe travels. Scientists hope to make the first three-dimensional map of the magnetic web and particle cloud inside of which our solar system is embedded.

Dennis L. Mammana

UNITED STATES MANNED SPACE FLIGHT

	MISSION	LAUNCH/LANDING	ORBITER
1991	STS-37	April 5/April 11	*Atlantis*
	STS-39	April 28/May 6	*Discovery*
	STS-40	June 5/June14	*Columbia*
	STS-43	August 2/August 11	*Atlantis*
	STS-48	Sept. 12/Sept. 18	*Discovery*
	STS-44	Nov. 24/Dec. 1	*Atlantis*
1992	STS-42	Jan. 22/Jan. 30	*Discovery*
	STS-45	March 24/April 2	*Atlantis*

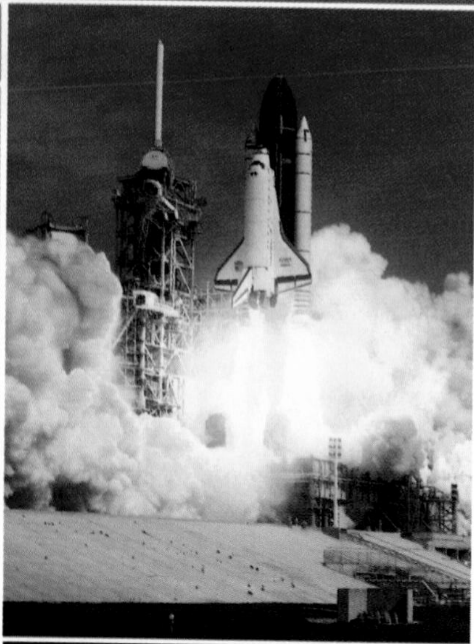

◄ *The space shuttle* Atlantis *thunders into orbit on April 5, 1991, carrying the Gamma-Ray Observatory.*

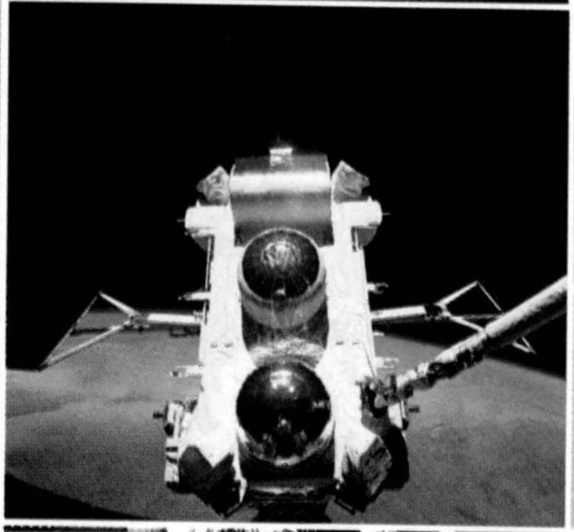

◄ *The Gamma-Ray Observatory required an unplanned space walk by crew members to unsnag an antenna.*

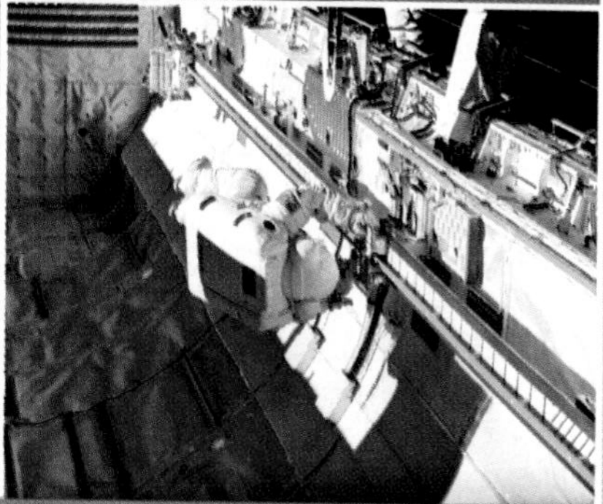

◄ *During a space walk, an astronaut moves along the shuttle's cargo bay on a device called a tether shuttle.*

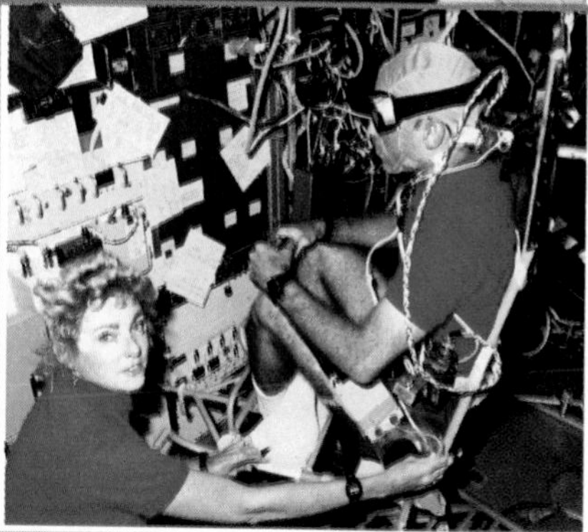

◄ *An astronaut is rotated while his head and eye motions are measured to study body orientation in microgravity.*

UNITED STATES MANNED SPACE FLIGHT

PRIMARY PAYLOAD	REMARKS
Gamma-Ray Observatory (GRO): Collects data about the gamma-ray universe.	• First extravehicular activity by U.S. astronauts since 1985. Two crew members disengaged snagged antennae of GRO. • Stayed aloft an extra day due to high winds at Edwards Air Force Base.
AFP-675: A variety of instruments to measure infrared, ultraviolet, and X-ray emissions. **Infrared Background Signature Survey (IBSS):** Collects spectral data on shuttle's rocket firings in space to aid the design of the Strategic Defense Initiative.	• Initial takeoff delayed for 2 months to repair cracks in fuel-door hinges. • Military missions declassified for the first time. • Heavy winds at Edwards Air Force Base forced landing at Kennedy Space Center.
Space Life Sciences-1 Experiments: Investigated the human body's adaptation to microgravity; 29 rats and 2,478 jellyfish studied for musculoskeletal and neurological changes that occur in zero gravity.	• Unusual postlanding procedures—crew members underwent 7 days of biomedical tests that mirrored tests performed while in space.
Tracking and Data Satellite (TDRS): Replacement satellite for 2 ailing satellites monitoring the western half of the Pacific Ocean.	• Planned landing at Kennedy Space Center. First time since 1985 that this runway considered "coequal" to Edwards Air Force Base as landing site.
Upper Atmosphere Research Satellite (UARS): Monitors chemistry, wind patterns, and heat distributions of the upper atmosphere.	• Shifted orbit to evade space debris from a discarded rocket.
Defense Support Program Early Warning Satellite: Observes rocket launches and nuclear explosions.	• Mission shortened by 3 days due to breakdown of 3 guidance computers.
International Microgravity Laboratory: 54 experiments from 16 countries to study how cells, plants, insects, and humans adapt to microgravity.	• Stayed aloft an additional day because crew had consumed less water and electricity than expected.
Atmospheric Laboratory for Applications and Science (ATLAS-01): Researches the long-term variability in the energy radiated by the Sun and the variability in the solar spectrum.	• One-day lift-off delay due to fuel leaks.

EARTH and the ENVIRONMENT

CONTENTS

JCA

COOLNESS UNDER FIRE

by Tom Flynn

The devastation to Kuwait is both real and dramatically exaggerated. There is no doubt that Saddam Hussein savaged the country. He unleashed death, suffering, and terror on its people. His soldiers vandalized Kuwait. Hotels, businesses, and apartment buildings trashed. Museums and hospitals looted. Government records burned. Zoo animals eaten or shot for sport.

But the sudden end to the war, coupled with the restraint of the Allied Coalition, left Kuwait generally intact. Basic services have now been restored; schools, banks, and ports are reopening; Kuwait Airlines has resumed service; and general commerce is getting back on its feet. Kuwait City stands, and restoration estimates have dropped sharply from the original $100 billion-plus figures.

Panorama of Destruction

At the end of the 100-hour blitz that blew the army of Iraq out of Kuwait, the most spectacular and cataclysmic evidence of Saddam Hussein's occupation was to be found in the liberated country's oil fields. Saddam Hussein torched them. Within days of Desert Storm's end, the panorama of destruction was breathtaking.

The valve heads had been blasted from some 750 high-pressure wells. Nearly 650 were ablaze, with the others gushing thousands of barrels of oil onto the desert floor, forming eerie dark lakes. (It could have been even worse; during the occupation, some courageous Kuwaitis cut lines to the explosive charges at the wellheads.)

Fire roared out of the ground from virtually every compass point. No two fires were alike, but each shared a subterranean heartbeat pumping a volatile mixture of oil and gases up to its wellhead.

Iraqis fleeing Kuwait in the last days of the Persian Gulf War opened hundreds of oil wells. Some ignited into huge fires; others simply gushed oil into the surrounding desert, forming giant lakes that impeded fire fighters in their efforts to extinguish the blazes.

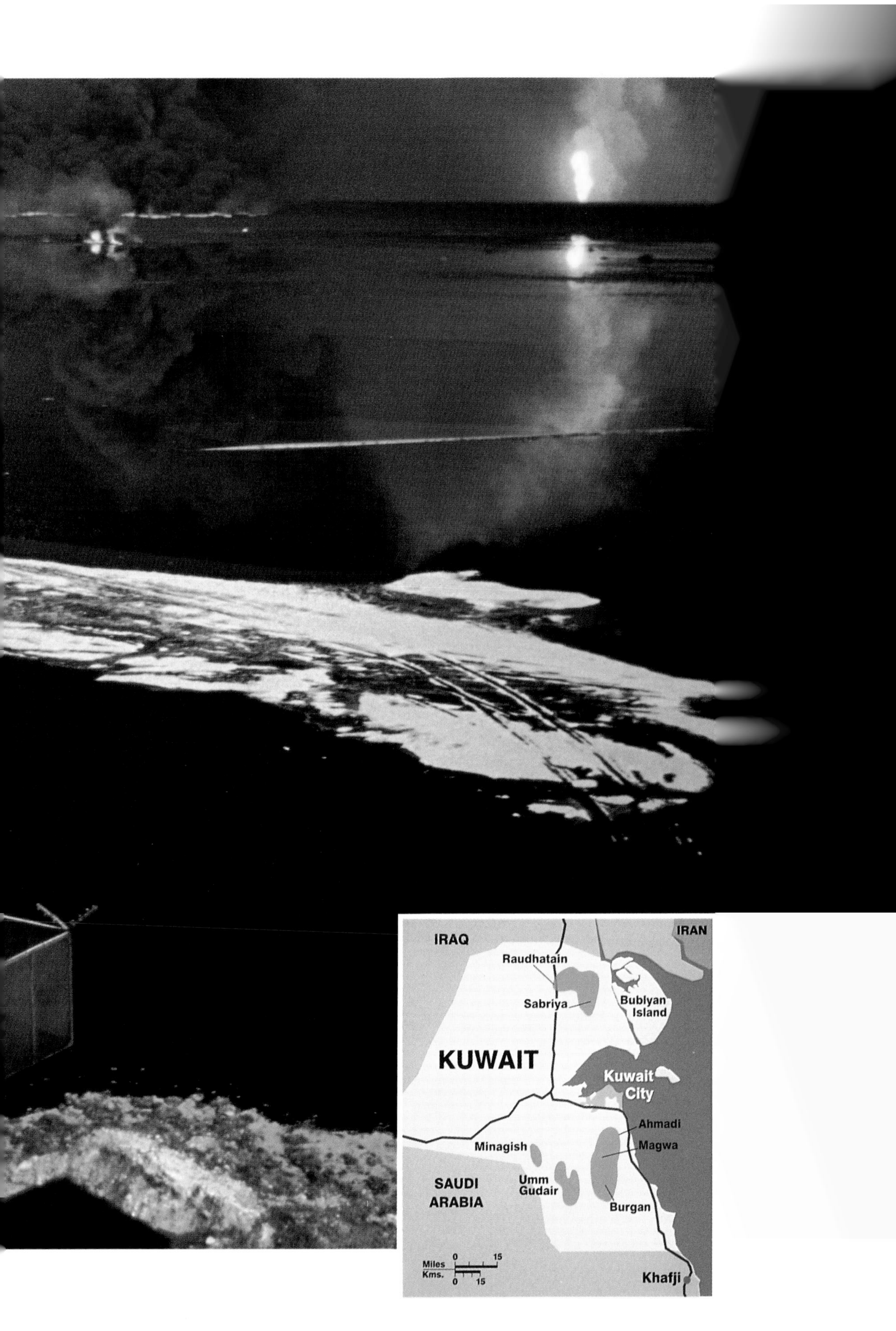
IRAQ
IRAN
Raudhatain
Sabriya
Bubiyan Island
KUWAIT
Kuwait City
Ahmadi
Magwa
Minagish
SAUDI ARABIA
Umm Gudair
Burgan
Miles
Kms.
0
15
0
15
Khafji

Some wellheads spewed waterfalls of red fire and black oil interspersed with showers of cascading white sparks. Others pumped straight up with a thunderous rhythm as gases under great pressure detonated half a dozen times above ground before drifting away in smoke.

Smoke and lakes of oil were everywhere. One spectacular lake engulfed the walled compound of a farm, creating an eddy of petroleum that moved in and out of buildings and through a grove of tall palms sagging under the buildup of airborne oil on blackened, sticky fronds. Overhead hung clouds of white and black and gray smoke, the variation in color indicative of the volume of water that was being sucked from below ground. Occasionally wind and ground-level heat sculpted the smoke into dark tunnels and canyons. Sometimes it all massed in impenetrable blackness.

Mobilizing the Restoration Effort

It was clear immediately that handling the oil fields of Kuwait was a monumental, one-of-a-kind project.

Kuwait Oil Company (KOC), the Coalition governments, corporations such as the Bechtel group of companies, and others had anticipated great destruction, and they had spent thousands of hours on advance planning. Before anyone could get back into Kuwait after the war, planning specialists had anticipated finding at least 200 sabotaged wells and forecast the need for an overall force of about 4,500 workers. Their foresight would pay off, but not even these experienced hands could have predicted the magnitude of the challenge, the complexity of the problems, or the abruptness with which the war would end and all resources would be plunged into the restoration effort.

First, there was no water, electricity, food, or facilities. Second, the country was littered with unexploded mines, bombs, grenades, artillery shells, and every variety of ordnance the opposing armies had thrown at one another. Finally, a good portion of the fires were already inaccessible because oil lakes covered the roads, and fire had spread to the ground surrounding many of the wells.

On the scene and ready for work were the fire fighters hired by KOC. They were all experienced, world-renowned hands: Wild Well Control from Texas, Safety Boss from Canada, and Boots & Coots and Red Adair, also from Texas.

First up was the issue of how to feed, house, and equip a work force that would grow into the thousands by the time the effort was fully mobilized. Virtually no resources existed in-country.

Project director Tom Heischman took charge on the ground in Kuwait. Responsible for meeting all support requirements for the fire fighting, Bechtel launched an intensive on-site assessment and planning effort, paired with a global procurement program.

About 500 miles (900 kilometers) to the southeast, Bechtel established a lay-down yard, docking facilities, and warehousing at the free port of Jebel Ali in Dubai. As a central transshipment point into Kuwait, the port receives and processes hundreds of shipments from chartered seacraft and aircraft, including the Soviet super cargo plane, the Antonov.

Accommodations in Kuwait were arranged in apartment buildings near the Gulf, which the Iraqi soldiers had used as redoubts while waiting for the Marine Corps amphibious assault that never came. Each unit had to be stripped to bare walls to remove the vestiges of the vandalism and the caches of ammunition, weapons, and poison gas.

Through all of this, a gigantic blanket of smoke lay over most of Kuwait. At high noon, people frequently used flashlights to see the street curbs. Outdoors, their clothing quickly became speckled with fine droplets of misting oil. Windshield wipers proved useless, and drivers resorted to repeated wipedowns with paint thinner.

Against this backdrop, officials of KOC initiated the fire-fighting assault with two missions: stop the smoke clouds choking most of the country, and restore petroleum production as quickly as possible.

Fighting an Inferno

Fighting a wellhead inferno is demanding and very dangerous. Yet it is a straightforward and rather fundamental exercise. The first level of strategy is gaining access. The inventory of required resources begins with nature's own—wind, water, and earth. The last burning well in the Ahmadi field (see map), which was a key source of the smoke shroud over Kuwait City, is illustrative.

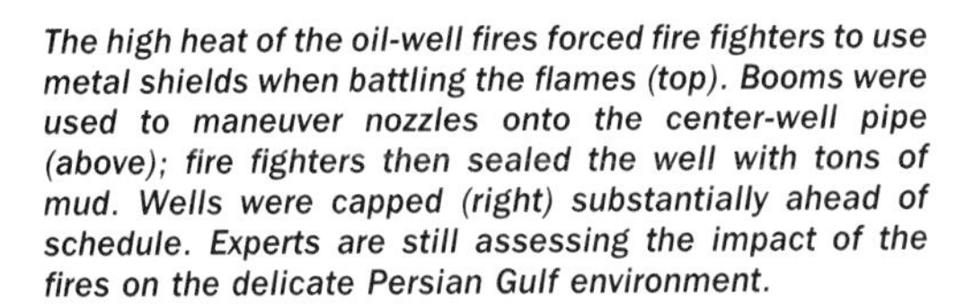

The high heat of the oil-well fires forced fire fighters to use metal shields when battling the flames (top). Booms were used to maneuver nozzles onto the center-well pipe (above); fire fighters then sealed the well with tons of mud. Wells were capped (right) substantially ahead of schedule. Experts are still assessing the impact of the fires on the delicate Persian Gulf environment.

Officially designated AH-74 by Kuwait Oil Company, it was a 10,000-barrel-a-day gusher spewing a pair of infernos from two belowground levels. Pat Campbell, co-owner of Houston-based Wild Well Control, says the fire was as complex a challenge as any his team had faced. "This was one of the biggest and best of the wells anywhere in Kuwait," Campbell explains. "When the Iraqis blew it, they set off two fires, and both of them were shooting off in all sorts of directions."

Explosive-ordnance-disposal crews first cleared the area of live munitions to secure access to the wellhead. "The Allies fired millions of tons of bombs, rockets, and other ordnance, at least 20 percent of which never exploded," says Terry Farley, president of Bechtel Construction and the company's senior officer in Kuwait. "We deal with that every day. The toughest job is to hold back our people from the work until we clear out each area. Kuwait is not a place where you say, 'Heads up!' You have to be looking down, around, and ahead of you all the time. We cleared one area with British and French demolition teams, and three weeks later hit an antiarmor/antipersonnel mine with a pickup truck and ended up with six folks in the hospital."

Once the ordnance was removed or otherwise rendered harmless, project personnel dammed the oil flowing through the area and prepared a new road to the wellhead, bulldozing some 1,640 cubic yards (1,500 cubic meters) of compact landfill over the burning ground around the well.

A million-gallon seawater lagoon was excavated at the site, then lined with heavy plastic sheeting and filled from water lines connected to new and existing pipelines to the Persian Gulf. Pumps and hose lines capable of delivering 6,000 gallons (22,700 liters) of water a minute to the blaze were installed before the final attack began.

A huge hardened mass of coke—solids produced by partially combusted oil mixed with sand—had built up around the base of the wellhead, hiding it from view and complicating the control effort. Preparing for the final assault, Campbell's crew took care of the coke pile with 200 pounds (90 kilograms) of dynamite, exposing the already damaged valve system on the wellhead.

Then, under a constant spray of water from four separate hoses, they ripped off the valve wreckage with a long boom device and dragged the area clean. Stripping the area straightened out the first fire into a jet-engine column burning straight up at a temperature in excess of 2,000° F (1,100° C). The second fire sprayed out in a quarter-circle arc immediately behind.

Under Campbell's direction, and with the wind at their backs, the Wild Well crew maneuvered a boom-mounted pipe nozzle over the 7-inch (17.7-centimeter) centerwell pipe and jammed it into the gusher. Through connecting hoses, tons of a heavy mudlike material were pumped in, sealing off the pipe. The second fire fell within hours to the same technique.

By August 1991, the pace of the fire fighters' work was dramatically accelerating, and predictions of an effort that might take up to five years were dismissed as wildly exaggerated—like much of what has been reported about the restoration of Kuwait.

In early September, there were 17 crews in the fields, about the maximum that could be supported efficiently. They reached the halfway mark September 8, with 375 wells brought under control. Kuwait mobilized a new national fire-fighting crew to join the original four teams. Iran and the People's Republic of China also mustered teams, and seven more crews, from Great Britain, France, Canada, Hungary, Romania, the Soviet Union, and Argentina, began joining the fire-fighting force in September.

Two of Kuwait's seven major oil fields were completely contained, and the pace was now averaging more than four wells a day. There was cautious talk that all of the wells could be brought under control before the end of the year.

At the same time, the project team addressed the issue of what to do with the lakes of oil. By most estimates, at least 70 million barrels had spread into the desert. Richard Hindson, a mechanical field engineer from Bechtel's London office, directed the development of an experimental recovery pond with skimming devices and a rudimentary system to clear contaminants from the oil. It showed that commercial-grade petroleum could be recovered in substantial quantities. An innovative pilot program designed by Travis Brown reached 10,000 barrels a day. Despite the economic prospects, as well as environmental mitigation, Hindson notes the substantial challenge in dealing with the oil-impregnated sand beneath the lakes and areas where evaporation has left thin layers of oil the consistency of asphalt.

Mind-boggling Logistics

From any perspective, the huge effort of massing nearly 1,000 engineering and construction professionals within six months was an extraordinary accomplishment.

Concurrently, KOC, with Bechtel's management assistance, mobilized an international force of nearly 8,000 manual workers. They came from Afghanistan, Australia, Bahrain, Bangladesh, Belgium, Brazil, Canada, Colombia, Djibouti, Egypt, France, Germany,

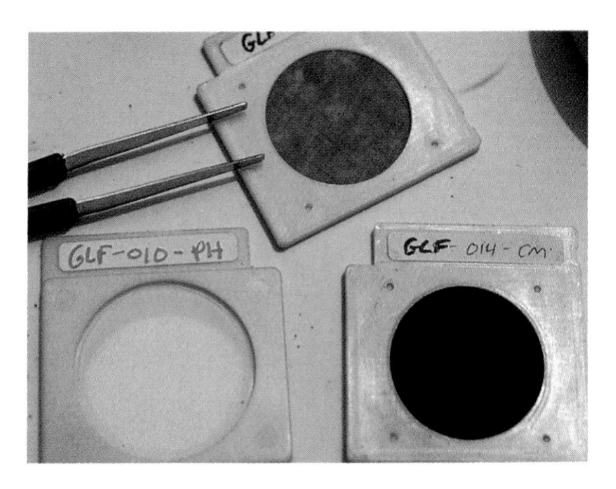

Each burning oil well (right) increased the air-pollution levels over Kuwait. Harvard University researchers used filters to monitor the air in Kuwait. A new white filter (above, left) turned gray after two hours exposure (top) and black when exposed overnight.

Great Britain, India, Indonesia, Iran, Ireland, Kuwait, Lebanon, Mexico, the Netherlands, New Zealand, Nigeria, Pakistan, the Philippines, Saudi Arabia, Somalia, Sri Lanka, Syria, Tanzania, Thailand, Tunisia, the United States, and Yugoslavia.

Six full-service dining halls fed 9,000 workers three meals a day. A 24-hour-a-day safety program included two helicopter-evacuation teams, a 40-bed field hospital, and a team of 100 professional medical personnel, paramedics, and other staffers on duty at seven medical stations.

The achievements of the international work force are impressive. Together, they have laid more than 90 miles (150 kilometers) of pipeline, a system capable of delivering 20 million gallons (75 million liters) of water a day to the fire sites. The team excavated 400 lagoons, each filled with 1 million gallons (3.7 million liters) of water.

This kind of progress does not come easily. The work routine in Kuwait was generally six weeks on and two off. "The six weeks does not include any days off, and the workdays run 12 to 16 hours," says Heischman.

The project team has massed and deployed 125,000 tons of equipment and supplies, including some 4,000 pieces of operating equipment ranging from bulldozers, cranes, and front loaders to ambulances and other support vehicles. This equipment has come from 12 nations, most of these members of the Allied Coalition.

Clearing Skies

Meanwhile, Kuwait is advancing toward full restoration of oil production, as well as refining. "Unfortunately, many of the wells are beyond repair, and new ones will have to be drilled," says engineering manager John Newman. "We have restored partial production to the refineries. In addition, new gathering stations and loading facilities will have to be designed and built. The final solution to the full restoration of oil production is still some distance down the road."

Nevertheless, the first postwar pumping of oil to two of the original 26 gathering centers began in late May 1991, and by early September, KOC was pumping well over 200,000 barrels of oil per day.

By the end of summer, the skies over Kuwait City were clearing rapidly. Only occasionally, as the desert winds came out of the north where fire fighters were still engaged in the Raudhatain field, would a pall of smoke settle on the capital. Satellites, whose eyes had guided Allied bombing, sent back photographs of the reemergence of Kuwait from months of darkness. By early November the last fire was capped.

SECRETS OF SNOW

by John Carey

In recent years, scientists have discovered many secrets harbored in nature's white quilt. Spying through satellite eyes winter after winter, researchers have watched the snow cover grow and shrink, providing tantalizing evidence that the Earth may now be warming up. They have learned that the size and thickness of the winter blanket of Asian snow may affect weather and climate thousands of miles away. And by digging through layer after layer of past snowfalls, scientists are beginning to piece together the intricate atmospheric dance that causes ice ages or searing heat. Once all the data are analyzed, says University of New Hampshire glaciologist Paul A. Mayewski, "there will be a fantastic tale to tell."

After snow piles up on the ground, its ice crystals begin to change. Water vapor migrates upward from the warmer earth to the colder surface, causing the snow near the ground to form brittle, loosely arranged crystals known as depth hoar. And as mice and voles scamper through this easily breakable layer, they fashion tunnels and chambers—a protected habitat deep under the snow. Since the temperature remains an almost constant 30 to 31° F (−1° C) in this so-called subnivean space, even on bitterly cold nights, "it is a very nice place to live," says Bill Schmid, a University of Minnesota biologist. Not surprisingly, he says, snow cover is full of animal and insect life.

The icy cover, however, does not always protect these creatures from predators. Measurements by Schmid show that snow is remarkably permeable to gases. Voles and mice don't have to dig ventilation shafts to get oxygen to their snowy chambers, as naturalists had once thought. But this also means a fox can smell prey far under the snow's surface, and dive through the crust to grab it.

Pollution Record

Tasty voles are not the only tidbits hidden under snow cover. On the glaciers of Greenland, built up by the snowfalls of thousands of years, scientists are digging for a different kind of prey. Drilling deep into the ice, then analyzing the resulting ice cores for every-

The stillness of a snowscape belies nature's winter activity. The lynx above has a sense of smell acute enough to detect a mouse burrowing through the snowpack.

thing from carbon dioxide to pollutants, researchers hope to learn what makes the Earth warm or cold—information that might offer a glimpse of what the future holds.

Take, for example, the sulfate record in snow. When tossed up into the atmosphere by both volcanoes and fossil-fuel burning, sulfates block some of the Sun's light and heat from reaching the Earth's surface, thus cooling the globe. Paul Mayewski and other scientists are analyzing ice cores collected by drilling deep into the layers of Greenland ice. By measuring everything from carbon dioxide levels to particles in each layer of ice core, much as a forester might study tree rings, Mayewski has been able to show that sulfate levels from pollution increased dramatically between the 1940s and 1980s, then stabilized as pollution controls went into effect.

Such discoveries may help explain a great puzzle. Once carbon dioxide from industry, power plants, and motor vehicles began pouring into the air, the planet should have begun to warm from the so-called greenhouse effect, especially during the second half of this century. But actual temperature records show that there was little warming between 1940 and 1980. The sulfates offer a solution. "They can explain why we get cooling from the 1940s to the 1970s, and warming in the 1980s," says Mayewski. Ironically, one implication is that global warming might be slowed down by spewing out more pollution.

Mayewski is also after bigger game. In order to predict what will happen as levels of carbon dioxide and other greenhouse gases increase in the atmosphere, scientists want to know what caused climate changes in the past. Were dropping levels of carbon dioxide the trigger for the ice ages, for example? The answer is crucial, because it may tell us what will happen now with rising carbon dioxide levels. Mayewski and colleagues are now poring over all information collected from the snows of the past thousand years, and preliminary results are encouraging.

Measures of key gases show that levels of these substances in the atmosphere changed during the Medieval Warm Epoch, a balmy period about 1,000 years ago. "We have some very believable trends," says Mayewski. The next—and crucial—step is figuring out which changes actually caused the Earth to either warm or cool.

But if snows of centuries past can uncover the inner workings of the Earth's thermostat, today's snow cover offers telling evidence of actual changes in our own climate. Geographer David A. Robinson of Rutgers University has been collecting satellite data on the extent of the white snowy blanket over North America. And recently he made a startling finding: the size of the snow cover shrank some 10 percent in the late 1980s. In fact, says Robinson, "1990 is the least-snowy year in the entire 19-year satellite record." The average snow cover for the year was about 3.8 million square miles (9.8 million square kilometers) compared with about 4.4 million square miles (11.3 million square kilometers) in 1978. What's more, most of the change comes from earlier spring snowmelt. "It fits very well with many of the theoretical predictions of how snow will change with global warming," says Robinson.

Does that mean that the Earth is warming up from the greenhouse effect? Not necessarily. Nineteen years is too short a time to tell. That's why Robinson has been examining records kept by backyard weather watchers, searching for clues to the amount of

Scientists called glaciologists analyze samples of ice extracted from the lower reaches of a glacier for traces of elements that may provide evidence of major climate changes in the past.

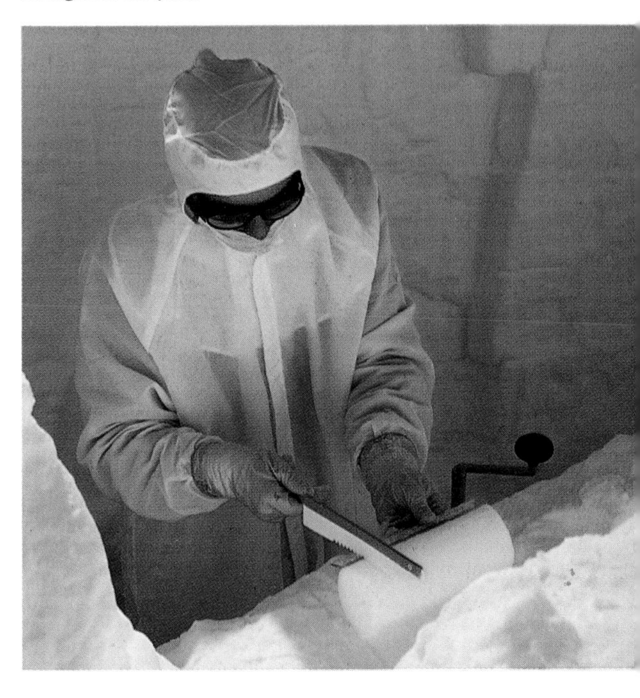

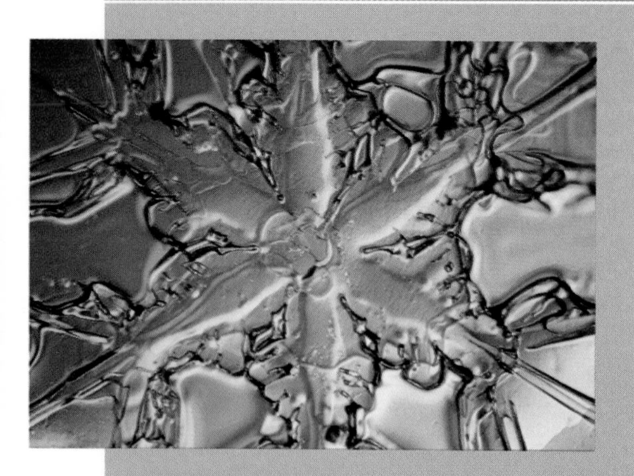

SNOWFLAKE!

Snow consists of beautiful hexagonal accumulations of ice crystals formed from water vapor. Each individual snowflake is said to be unique, but is this really so?

Maybe. In 1989 Nancy Knight, a researcher at the National Center for Atmospheric Research (NCAR), claimed that she saw two *identical* snowflakes. Other meteorologists have since disputed her claim, arguing that while two snowflakes could appear to be alike, they aren't, because subtle differences probably exist at each snowflake's molecular level.

Snowflakes are distinct from sleet (frozen raindrops in the United States, mixed rain and snow elsewhere) or hail (a frozen raindrop coated with successive layers of ice) or even graupel (a kind of soft hail that *looks* like miniature snowballs as it falls). A snowflake begins as submicroscopic ice crystals. Its crystalline shape is a product of the below-freezing air temperatures spawning it. Once formed, each crystal attracts a microscopic dust or salt particle as its core. Continuing to attract more particles and to grow as they fall, the crystals tumble and strike each other until perhaps a million crystals unite to form an ordinary flake.

Snowflakes aren't really white, either. Instead, each one contains millions of tiny, clear, light-reactive prisms. While the prisms disintegrate the light that strikes them into all colors, our eyes receive a sensory overload and are unable to perceive "rainbow" snow, and so change the snowflakes we see back to white.

Gode Davis

snow cover in the past. He has found that the current decline in snow cover may seem larger than it really is, because the 1970s were unusually well endowed with snow. "When we looked at the central and northern Great Plains, we recognized that the past several decades have been the snowiest of the last century," he says. "So the diminished snow cover in the late 1980s might just be a return to what we saw earlier in the century."

While shrinking snow cover doesn't yet offer proof of greenhouse warming, Robinson's results to date "do bode well for the potential of snow as a monitor of climate," he says. If the shrinking trend continues for another decade or so, scientists may well proclaim that global warming is occurring. And changes in the amount of snow can have far-reaching effects, both for animals and for the world's weather. With low snow cover, the populations of some animals, such as mice, could be seriously hurt. Because that affects the food chain, lack of snow could have detrimental effects on the ecosystem.

Predicting El Niño

When it comes to weather, the amount of snow may be vitally important. New results from computer climate simulations suggest Asian snowpacks help determine whether or not the world experiences the El Niño phenomenon, where abnormally warm temperatures in the Pacific wreak havoc with fish stocks and bird colonies. The mechanism proposed by climatologist Tim P. Barnett of Scripps Institution of Oceanography goes like this: More snow on the Asian continent in winter causes the land to warm up slower in spring. Since prevailing winds are driven by the temperature difference between land and sea, cooler land, in turn, means weaker winds. That affects the whole circulation of the trade winds over the Pacific, generating an El Niño. The idea is still speculative, awaiting better snow-cover data and more-sophisticated computer models. But if it holds up, and if it is possible to accurately measure the amount of snow across Asia, says Barnett, "we might be able to better predict the coming El Niños."

For scientists, nature's white quilt is a tool for learning. As they dig deeper into the Earth's mantle of snow, researchers may eventually solve the great climate mysteries that confront us today.

CHERNOBYL'S WORSENING AFTEREFFECTS

by Glenn Alan Cheney

On April 26, 1986, at 1:23:40 A.M., one of the four reactors at Chernobyl in the Soviet Union exploded. As the power plant burned for the next 10 days, an estimated 50 tons of nuclear fuel and other radioactive materials gushed into the atmosphere. The wind blew the clouds of fallout north and west, then south and east, dispersing deadly radionuclides across the farmlands and forests of the western republics of the Soviet Union.

By April 28, Sweden was detecting the presence of radioactivity in the air. The isotopes were of such a variety that they could have come only from a nuclear reactor gone haywire. Wind currents pointed the blame at the Soviet Union. Only then did the Soviet government admit that it had a reactor burning out of control.

Though the clouds of dwindling radioactivity have circled the Earth several times, the consequences of the disaster have not

Scientists are only beginning to assess the long-term effects of the 1986 explosion at the Chernobyl nuclear power plant. Radiation has caused untold damage to the local environment (above); many areas remain off-limits (top) to all but official personnel.

Following the explosion, the damaged reactor was entombed in concrete. Radioactivity levels at the crater formed by the explosion still remain dangerously high (left). The other three reactors at Chernobyl operate as usual (above).

blown away. In fact, the seriousness of the accident is only now coming into focus. That focus, however, is blurred by conflicting politics and a dearth of reliable information.

A lot has happened since 1986. The Union of Soviet Socialist Republics, once the secretive suppressor of all bad news about itself, has broken up into a number of independent states. Among them are the three most affected by the Chernobyl disaster: Russia, Byelorus, and Ukraine.

No longer under Soviet domination, these three new countries are gradually uncovering the depth and extent of the disaster. Much information remains hidden within the bureaucracy that the Soviet system left behind, and much has been lost, altered, or destroyed in attempts to ward off blame.

How It Happened

The original Soviet explanations lay the blame for the accident on a handful of operators at the plant. Subsequent information, however, indicates not only incompetence among the technicians at Chernobyl, but also very poor design engineering and an inept and corrupt bureaucracy.

The four reactors at the Chernobyl complex (and a fifth that was under construction) were graphite-modulated, channel-tube reactors known by their Russian acronym, RBMK. The graphite modulator is a set of graphite bars that, when lowered into channels among the nuclear fuel, absorb loose neutrons. Unabsorbed neutrons collide with other molecules of nuclear fuel, cracking them into isotopes and releasing energy. The channel tubing circulates water through the reactor. As the water turns to steam, it cools the reactor and powers the turbines that generate electricity.

The accident occurred during a poorly planned experiment to see how well the reactor would function during a major power outage. To emulate a worst-case scenario, the plan called for the blocking or disconnecting of almost all safety systems. The supervisor of the plant, who had only superficial understanding of nuclear-energy production and safety, also approved the disconnection of the emergency core cooling system.

The experiment went awry as the plant attempted to shut itself down by using electricity from its own generators. According to plan, the generators had been disconnected from the reactor, but were still generating electricity as they gradually slowed. The operators had not correctly calculated changes in power supply, steam pressure, and other parameters, however, and they quickly began to lose control of the reactor.

Then, to their surprise, they discovered that only 18 graphite rods had been lowered into the nuclear fuel. There should have been at least 28 in position. The operators attempted to shut the reactor down immediately by lowering all rods into the fuel. But the rods jammed, probably due to the buildup of excessive heat. Due to the design of the rods, the partial insertion caused a power surge that may have been 100 times more than normal. The result was a series of steam explosions that quite literally blew the lid off the reactor, destroyed most of the roof above it, and sent 50 million curies into the sky. (The curie is a unit of radioactivity; the accident at Three Mile Island released just one-thirteen-millionth of one curie.)

Only recently has the world come to know what happened over the next few days. The 45,000 inhabitants of Pripyat, a town built for power-plant workers just 2 miles (3 kilometers) away, were at first told nothing of the accident. Though they could see a pillar of fire "redder than red" stabbing into the sky above the plant, they were told to continue life as usual.

As radioactive graphite and isotopes showered down, children walked to school, and parents went about their business. The town health official denied any danger, though he himself kept his children home and gave them iodine pills to minimize danger to their thyroid glands. Doctors passed out iodine pills in school, but then allowed the children to walk home. Special trucks washed the streets with a foam meant to clean up radiation. People who walked in the foam soon found their ankles blackened by radiation.

Evacuation began only on the afternoon of the next day. By then, however, nearly everyone in town had received far more radiation than should be allowed in a lifetime. They were ordered to bring only their documents and enough clothes for two days. Officials locked apartment buildings as the residents left. Families then stood outside under intense radiation for up to an hour as they waited for buses to take them to Kiev, the Ukrainian capital, about 60 miles (96 kilometers) to the south.

Meanwhile, helicopters were dropping loads of boron, lead, and sand into the burning reactor. The boron was meant to absorb neutrons and prevent the fuel from reaching a critical state and exploding. The lead was to absorb heat and shield the atmosphere from subatomic particles. The sand gradually buried the fuel and kept oxygen from the graphite fire.

Unfortunately, the lead evaporated as soon as it hit the white-hot nuclear fuel, which may have reached temperatures of 9,000° F (5,000° C). The vaporized metal shot into the air and rained down across hundreds of square miles. This poisonous element, long ago banned from gasoline in the United States for health reasons, still lies over vast areas of Ukraine and turns up in food and drinking water.

Fire fighters sent onto the roof of the plant had little protection from the radiation. Each person would pick up a piece of nuclear fuel and toss it back into the remains of the reactor, then go to a hospital. The careers of these first "liquidators" were 90 seconds long. In many cases, that was enough to kill them. Likewise, helicopter pilots who hovered momentarily over the reactor were doomed to death or years of serious illness.

How many died of radiation sickness? For years the Soviet government admitted to the deaths of only 31 fire fighters. Recent investigations put that number at 256, and it is very likely the true total is much higher.

Some 600,000 liquidators battled the fire and, over the course of several months, built a "sarcophagus" around the wreckage of the plant. Most of them had little idea what danger they faced. Due to the intense radiation, construction was extremely difficult. Steel battlements had to be moved into position by bulldozers, then filled with concrete. Miners had to tunnel under the plant to reinforce it. Often their only protection was a simple gauze mask. The liquidators were not supposed to be exposed to more than 35 rems—the maximum permissible lifetime radiation dose—but often they arrived at the hospital after being exposed to 200 rems.

No one knows how many of these liquidators have died or how their health has suffered. Most were soldiers who have been dispersed across the country, preventing any meaningful statistics. The others are left wondering about their fate. Often they do not know how much radiation they absorbed. They see themselves and their fellow liquidators suffering from fatigue, illnesses more serious than normal, and what seems to be a higher incidence of death by cancer and heart disease at an early age.

The governments of Russia, Byelorus, and Ukraine have only recently begun to keep records of the fate of liquidators. So far, the governments lack sufficient evidence to

At the time of the explosion, local people were not informed of the dire health risks they faced. Many people still refuse to leave their homes (below), despite persistent dangerous radiation readings.

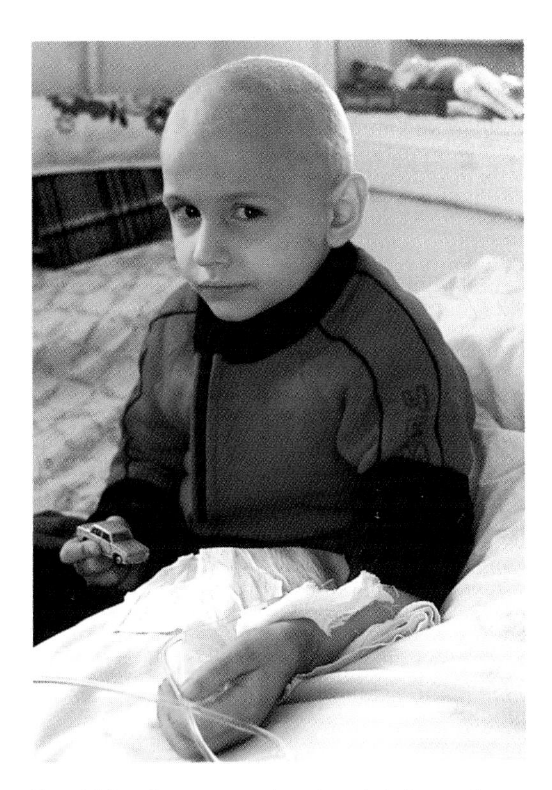

The childhood cancer rate has skyrocketed among area children in the aftermath of Chernobyl. Nearby hospitals generally lack the technology and the facilities to provide adequate treatment.

clearly link radiation to long-term health problems. For this reason, it is impossible to attribute specific deaths to exposure to radiation. Nongovernment organizations estimate the actual number of deaths to be between 3,000 and 10,000.

The director of the morgue in Kiev says that he has never been asked to keep special records on the deaths of liquidators. He reports two significant facts, however. One is that the death rate of people not currently under the care of a doctor has risen 2,000 percent. The other is that the incidence of brain hemorrhage among young men has soared in the past few years, while it was rarely observed before the Chernobyl disaster. Such hemorrhages are often the result of weakened blood vessels, which in turn are often the result of exposure to radiation.

"Chernobyl AIDS"

But it isn't that simple. Alcoholism can have the same effect on the cardiovascular system. It can also lead to cancer. Although there are no statistics to verify a correlation, alcoholism is relatively high among the liquidators. One reason may be that alcohol is rumored to reduce the effects of radiation. Liquidators at the reactor site were issued doses of vodka. Some doctors say that alcohol in the body during exposure to radiation may have a slightly beneficial effect. But they warn that the alcohol yields more-harmful effects than does the radiation. Nonetheless, many liquidators, telling themselves they're going to die anyway, continue to drink heavily in hopes of cleansing their bodies of radiation.

Liquidators are not the only people with health problems. Doctors report an unprecedented health situation among children. Their immune systems are weaker than normal. Their white-blood-cell and red-blood-cell counts are far below the norm. Doctors say the condition, which is at epidemic proportions, qualifies as an acquired immune deficiency. Informally, they call it "Chernobyl AIDS."

Due to their weakened immune systems, young people heal more slowly and contract childhood illnesses more easily than they used to. Often a vaccination touches off the disease it was supposed to prevent. In Ukraine, only 30 percent of school-age children are of normal health.

Not only are children sick more often, but their symptoms are more complex and sometimes even baffling. Treatments can have unpredictable effects. A medicine meant to reduce fever, for example, might increase blood pressure instead. In effect, each patient has a new and unique disease—a medical nightmare in which traditional treatments are useless or even dangerous to implement in patients.

Doctors in Ukraine suspect that the Moscow Scientific Institute has information on radiation-related health problems that could be of use to victims of Chernobyl. The information came from studies made in 1968 after a large population near Cherbinsk, Russia, suffered exposure to high doses of radioactive wastes. So far, however, the institute has been unwilling to share what it knows.

At a pediatric hospital in Kiev, doctors have developed a software program that makes it possible to treat an overall medical condition rather than a specific disease. The program analyzes data from a wide variety of medical tests and calculates the appropriate treatment. Ironically, the hospital lacks the equipment needed to perform all the necessary tests, and there are no ribbons for the computer printer.

It is suspected that radiation has had negative effects on fetuses and human reproductive systems, too. The rate of birth defects has increased 180 percent. The birthrate has plummeted, and miscarriages and premature births have increased. Today the death rate in Ukraine exceeds its birthrate. Ukrainians joke bitterly that they have become an endangered species.

Cancer incidence has also risen dramatically. Cancer of the thyroid has increased by 400 percent. This gland tends to collect deposits of iodine. People who ingested the radioactive iodine 131 isotope soon had a concentrated source of radioactivity in their bodies. This was especially true of children. While there had been only one case of thyroid cancer in Ukraine per year before 1986, there were six in 1989, and 20 in 1990. Doctors expect the numbers to keep rising.

Doctors also expect the number of cases of leukemia to begin rising over the next two or three years. According to experiences in Hiroshima and Nagasaki, leukemia tends to develop five or six years after initial exposure to radiation.

The Synergy Effect

All of these health problems are complicated by various other factors. Chemical pollution may be having more-negative effects than is radiological pollution. The use of chemical fertilizers introduces a wide variety of dangerous elements and compounds into the environment. The combustion of gasoline puts lead into the air, a problem compounded by the lead thrown into the burning ruins of the Chernobyl reactor. Unregulated industrial processes have poisoned air and water. Many of these chemicals produce the same physiological problems as radiation.

On top of that, the people of the former Soviet Union are suffering from malnutrition. Shortages of fruit, vegetables, and milk leave children without essential vitamins. When milk does become available, it is suspected of containing strontium 97, a radioactive substance that cows have ingested from contaminated pastures. In fact, after the disaster, the "acceptable" level of strontium in milk, as established by the Soviet government, qualified it as nuclear waste that would need special disposal. Consequently, many children's immune systems either lack calcium or suffer radiological attack from the strontium that has accumulated in their bones.

All of these problems combine to produce what doctors call the *synergy effect*—the phenomenon of the whole equaling more than the sum of its parts. When the effects of radiation, malnutrition, and pollution combine, the results are worse than could be expected by merely totaling the three factors.

Lack of data also makes it hard to determine the causes of health problems. Before 1986, statistics gathered by the Soviet government were often altered to suppress fears

Farmers continue to report a high incidence of animals born with grotesque deformities, apparently from radiation. In humans, freakish birth defects have contributed to a rising infant-mortality rate.

Contaminated soil has led some plants to grow to gigantic size. Cucumbers from the giant plant at right would be considered unfit for human consumption.

of widespread cancer and other diseases. After the Chernobyl disaster, doctors were ordered, under threat of arrest, not to tell their patients that their illnesses might have been caused by radiation. Unknown numbers of people supposedly died of sore throats and bronchitis that were probably thyroid problems or cancer. Consequently, doctors cannot compare numbers from before and after the disaster, and thus cannot attribute health problems to their probable cause.

The Soviet government also attempted to prevent the public from knowing how much radiation surrounded them at the time of the accident. All Geiger counters and dosimeters used to measure radiation were confiscated. The government claimed to be monitoring food supplies, but it is believed that due to the shortages of food, many contaminated crops and animal products found their way to the market. The Soviets also shipped contaminated food throughout the country, often mixing clean and contaminated products to lower the average level of radiation.

Ironically, the new government of Ukraine may also be trying to hide the truth about the radiological situation. After dividing the country into four zones of decreasing levels of contamination—a total of 25 percent of Ukrainian territory—the legislature passed a law granting inhabitants of each zone certain privileges. Kiev was considered uncontaminated. Recent tests, however, have revealed that all of Kiev is contaminated. The Ukrainian government, however, refuses to recognize the test results, possibly because it would exempt 2.5 million people from paying income taxes—a fiscal catastrophe that the all-but-bankrupt nation cannot readily afford.

The current radiological situation is unique in history. Never before have so many people been exposed to long-term, low-dose radiation. Recent studies have indicated that no dose of radiation is safe, and that continued exposure to low levels of radiation may be more dangerous than short, intense exposure. With 4 million people throughout Europe and the former Soviet Union suffering various levels of ongoing exposure, estimates of "extra" deaths due to cancer—that is, cases that would not have developed in the absence of radiation—have been adjusted from a few thousand to a few hundred thousand. Only the next decade or two will yield the final results of what some scientists have called the largest human experiment ever performed.

NATURE BLOOMS IN THE INNER CITY

by Roger Cohn

On a glorious spring Sunday in New York City, two scenes are unfolding. Each represents a small victory in the frustrating and sometimes futile battle to save nature—or at least a few patches of nature—in the heart of the nation's largest and most densely populated megalopolis.

Deep in the South Bronx, amid ghostly blocks of abandoned buildings and rubble-pocked lots, the urban tillers of the Garden of Happiness are emerging from a winter of hibernating behind barred windows, and are reclaiming their piece of the Earth. On a city lot that was once littered with the stripped-down hulks of stolen cars, they are again beginning a ritual of faith in the restorative powers of nature and of their community. Armed with hoes and rakes and pitchforks, more than a dozen neighborhood residents are busily preparing their garden plots for the new season's first planting.

"This garden has brought the community together," says Karen Washington, who helped lead the fight to organize the aptly named Garden of Happiness three years ago. "Here everyone has a shared bond, and that's coming together and making something beautiful out of nothing."

Twenty miles (32 kilometers) away on Staten Island, in a prairielike meadow dwarfed by the looming towers of the Verrazano-Narrows Bridge, Marc Matsil of the New York City Parks and Recreation Department is proudly surveying the grassland and marshes surrounding Eib's Pond, the city's most recently acquired sliver of parkland. In what must be regarded as a mini-urban miracle, the 20-acre (8-hectare) pond site was saved from a planned townhouse development that would have filled in the pond's shoreline and cut it off from the public.

Urban Triumphs

Saving a piece of nature—whether it be an endangered bit of open space like Eib's Pond or a manicured neighborhood garden amid the hard-bitten pavement of the South Bronx—is no easy task in New York City. Even so, along with the incessant barrage of media stories of oil spills off Staten Island and neighborhoods polluted by illegal dumpers and insidious drug pushers, there are some real triumphs to be reported from the trenches of the city's environmental wars.

Indeed, the battle is being waged successfully on a number of fronts.

In many of New York's most troubled inner-city neighborhoods, from Bedford-Stuyvesant in Brooklyn to Manhattan's Lower East Side and Harlem, residents have taken over blighted lots and transformed them into

For many people, an urban garden is their closest connection to nature. The young boy below revels in the scent of a homegrown tulip.

Flowers and vegetables thrive in a garden wedged between the tenement buildings of New York City's Lower East Side. Such gardens have helped nurture a sense of community pride among local residents.

thriving community gardens where everything from collard greens to primroses now grows. By the city's estimate at least 700 community gardens now bloom throughout New York, serving both as pockets of green and as symbols of the residents' determination to regain control of their turf.

"Once people see what they can make out of the ground here, it gives them a different perspective on what their neighborhood can be," explains Sophie W. Johnson, executive director of the Magnolia Tree Earth Center, which operates a thriving community garden and an environmental-education program for youngsters in Bedford-Stuyvesant. "People need to be able to touch the earth and feel a connection with nature."

For many New Yorkers, particularly those who spend their weekends in the inner city rather than at the beaches of Long Island, a neighborhood garden—where they can touch the soil and grow the calabaza pumpkins of their native Puerto Rico or the okra and sunflowers they remember from the family farm down South—is probably their closest connection with the natural world.

Yet within the city's five boroughs, remnants of an even wilder, largely unspoiled New York also survive, such as the Bronx woods where red fox can be spotted, and the Staten Island hillsides where endangered plants like whorled mountain mint thrive. Some of these areas are included in the 7,200 acres (2,900 hectares) of parkland that the city Parks Department has designated as natural areas and protected from development. Other environmentally fragile sites, however, from Eib's Pond to the bird-rich wetlands of Jamaica Bay and the Arthur Kill, are hanging on only tenuously, despite the extraordinary efforts of conservationists and parks officials to preserve them.

Perhaps New York City's toughest environmental battle is being fought in its classrooms, where a small but determined group of educators is struggling to instill in city kids a sense that the urban environment is worth saving. At Public School 21 in Bedford-Stuyvesant, Sheila Dunston teaches her students that removing graffiti and litter from the neighborhood are real contributions to the environment, as significant as any distant efforts to save South American rain forests. And at the Brooklyn Center for the Urban Environment, kids from throughout the city are exposed to innovative outdoor lessons on topics ranging from habitat preservation to eastern bird life without ever leaving Brooklyn's Prospect Park.

"We shouldn't be spending our environmental-education dollars to hire buses to take kids out to the country," says John C. Muir, the center's executive director. "The lessons of nature can be learned right here."

Gardens of Paradise

"Maybe we should call this the Lower East Side Flyway," Sandi Andersen says as she leads a visitor on a tour of several gardens that are flourishing amid the decaying tenements of New York's Lower East Side. In the

John Muir (below), director of the Brooklyn Center for the Urban Environment, teaches innovative lessons on topics ranging from habitat preservation to bird life. At Eib's Pond on Staten Island, protecting the environment also provides novel recreational opportunities for urban kids.

past decade, 40 community gardens—roughly one for every square block—have sprouted here in what is still one of the city's toughest neighborhoods. For Andersen, who tends her own plot in the place her Puerto Rican neighbors have named El Jardín del Paraíso (the Garden of Paradise), this gardening renaissance has meant not only a bit of greenery, but also the welcome arrival of migrating birds that used to steer clear of places like the Lower East Side.

"We've created these little pockets of green that together make up a mini-wildlife preserve," says Andersen, who also works as an organizer for the Green Guerillas, a non-profit group that promotes community gardens in the city.

The Green Guerillas, back in the 1970s, were instrumental in pioneering community gardening in New York. As businesses and the middle class fled the financially strapped city, a group of young activists took over an abandoned lot on the Bowery and started gardening. They called themselves the Green Guerillas, and soon, on a corner previously known only as a hangout for panhandlers, a handsome garden was blossoming, with lush planting beds and flowering fruit trees. "People would come by and see what we were doing," says Barbara Earnest, the group's executive director, "and they'd say, 'Hey, we've got a vacant lot in our neighborhood. Why can't we do this, too?' "

The Green Guerillas promoted the use of "seed grenades"—balloons or Christmas tree ornaments packed with wildflower seeds and fertilizer that could be tossed over a fence to "liberate" an inaccessible lot. Soon the organization was providing gardening advice and supplies to block associations throughout the city that were taking over their own neighborhood eyesores. In the late 1970s, city officials recognized this grass-roots movement and established Operation Green Thumb, a program that provides vacant, city-owned lots to nonprofit groups for use as community gardens. Currently, there are nearly 500 gardens operating under the Green Thumb program alone, enabling New York to begin to rival such longtime urban gardening centers as Philadelphia, Milwaukee, and Seattle.

For many New Yorkers, organizing a community garden serves as a catalyst for regaining control of a troubled neighborhood.

Community gardens often serve as a catalyst for regaining control of a troubled neighborhood. Vacant lots that were hotbeds of crime or trashy eyesores have been reclaimed.

"A garden can be a very empowering thing," says Jane Weissman, the director of Operation Green Thumb. "It makes people feel like there's hope. Starting a garden is saying that things can get better, that we're going to be here next year."

Ethnic Diversity

A tour of New York's community gardens provides a telling portrait of the city's ethnic diversity. At El Jardín de la Esperanza (the Garden of Hope) in Spanish Harlem, the Ortiz and Colon families regularly compete to see who can grow peppers as hot as the ones they remember from their native Puerto Rico. In Chinatown, Asian immigrants grow Chinese cabbages and bok choy. At the Garden Beautiful on 153rd Street in Harlem, okra and mustard greens are among the most common crops, and one elderly gentleman is even reputed to have successfully harvested cotton a few years ago.

The diverse assortment of New Yorkers tending community gardens was also evident at the annual "spring giveaway" sponsored by the Green Guerillas at their home base on the Bowery. Neighborhood groups, invited to

Sisters from the Missionaries of Charity in the South Bronx take shrubs and seeds provided by the Green Guerillas to their garden lot in the most desolate section of the Bronx.

pick up free plants and materials, came from all over the city—some by subway or bus, some in cars and borrowed station wagons, some on foot with shopping bags and carts. Two elderly ladies from the Young Senior Citizens Garden in Coney Island carefully selected their quota of rosebushes, proudly pointing out that they will also be planting peanuts and tobacco this year. Nearby, a burly dockworker from Brooklyn hauled yew bushes and mulch to his pickup truck, noting that his block association is starting a garden on what had been the neighborhood eyesore: a waterfront dump site.

Perhaps the most unlikely urban gardeners to show up were Sister Françoise and Sister Christina of the Missionaries of Charity in the South Bronx. They arrived by subway from their convent in what is perhaps the South Bronx's most desolate section, a site selected by Mother Teresa herself as eminently suitable for serving "the poorest of the poor." There the sisters have taken an adjacent lot and, with the help of neighborhood kids, converted it into a garden so lush with flowers and fruit trees that it now rivals any garden found on posh Park Avenue.

As they consulted with Barbara Earnest about how to use chopped banana peels to fertilize their rosebushes, the sisters sounded like two rather savvy horticulturists. The youths in the neighborhood have become protectors of the garden, Sister Françoise says, declaring it off-limits to vandalism. Indeed, she adds, the only real problem is that kids sometimes pluck peaches and pears from the trees before the fruit is ripe. "They get a little excited," Sister Françoise explains. "We have to teach them when it's time to pick. You see, they've never seen fruit growing on a tree before."

Environmental Education

In New York City's public schools, as in many hard-pressed big-city school systems, learning about the natural world and environmental issues is still considered something of a frill. According to Michael Zamm, director of environmental education for the Council on the Environment of New York City, only about 15 percent of the city's schoolchildren receive environmental education on an ongoing basis. A recent report prepared for the city Board of Education urges that topics ranging from water conservation to recycling be included in the standard curriculum.

"Environmental education is not treated like geometry or American history," says Zamm, who was a member of the task force that prepared the report. "It's just never been a priority."

Indeed, the challenge of educating urban kids about the environment has so far been tackled primarily by individual teachers who take the initiative and develop their own programs. At Public School 175 in Harlem, science teacher Lettie Hartwell spurs her students to develop a Success Garden, where they plant everything from apple trees to native rhododendrons and maintain their own composting bin. At Brooklyn's Public School 21, Sheila Dunston makes the environment the primary topic of her weekly science-club sessions.

Piecing together an environmental curriculum requires considerable ingenuity on Dunston's part. In her spare time, she attends educational seminars sponsored by the National Audubon Society and the Council on the Environment, and this summer hopes to attend Audubon's ecology workshop for teachers. Still, her lessons retain a decidedly urban emphasis. At one recycling class, Dunston tried to get a group of first graders to understand that their garbage does not just disappear when it is dumped into the apartment-house incinerator. "It makes smoke, that's right," she told her students. "And nobody likes to breathe that smoke. So what we need to do is produce less garbage."

Some of the most progressive environmental education takes place, not in city classrooms, but at the Brooklyn Center for the Urban Environment. John C. Muir, the center's executive director, believes the traditional approach to teaching city kids about nature and ecology displays an inherently antiurban bias. "The idea always was to take the kids out somewhere beautiful in the country for the day and have them walk in the woods," says Muir. "And then they'd come back home and think, 'Well, I've been to the environment—and it's out there where it's all lovely and green. I must live in the antienvironment.' "

In contrast, nearly all of the center's two dozen or so field programs take place within the 500 acres (200 hectares) of Prospect Park in Brooklyn, one of the crown jewels of the city's still-impressive park system. Here young children are introduced to the concept of natural habitats on a tour of the park's meadows, woods, and streams. Junior-high-school kids take the Trail of the Waters tour, tracing Prospect Lake water through swamps, ravines, and waterfalls to the source—a pipe connected to the city's water-supply system.

In one of the center's most innovative classes, Endangered Species-Endangered Spaces, students are told of a fictional developer's plan to replace one of the park's prime ponds with an amusement park. The students are asked to play the roles of neighborhood property and business owners, conservationists, construction workers, and legislators, and to decide the project's fate. "It becomes a real moral dilemma for them," says Tom Oppecker, director of the center's school programs. "Most times, they want both the amusement park and the pond. We want them to realize they can't have both."

This novel exercise in role-playing sometimes has rather unpredictable results, as was evident recently when Oppecker presented it to a group of eighth graders from the Canarsie section of Brooklyn. First, he took the students down to the pond for a field primer on just what could be lost. He pointed out the birds along the pond's edge and, using a net and a bucket, scooped out water and leaves from the bottom to display the

Gardens flourish even in city squalor. Some neighborhood garden groups bring video cameras to their weekly meetings to scare away drug dealers. Others sponsor evening barbecues to bring gardeners together.

pond's teeming aquatic and insect life—tiny snails, leeches, a toad beetle, damselfly larvae.

When it came time to begin the debate on the amusement park, the students played their roles perhaps a bit too zealously. "Look," urged 14-year-old Tom Mennel, who brought a frightening realism to the role of business owner. "We gotta have this amusement park. We can make a lot of money out of it. And we can give some of our profits to the park to keep the conservationists happy."

The construction workers agreed, as did the homeowners, represented by young Adam Cooper. "It'll increase the value of my property," he said. "I'll sell my house and move to Florida."

Ultimately, the conservationists won at least a partial victory, as the legislators, after much wheeling and dealing, settled on a compromise plan: the pond would be saved—but a recreation center complete with video arcades and pinball machines would be built elsewhere in the park.

A Daunting Task

"Just look at the mess they've made up here," grumbles Marc Matsil of the Parks Department as he tromps along a muddy hilltop a few hundred yards from Eib's Pond. The ground here, just beyond the boundaries of the city's newest parkland, has been torn apart by heavy construction equipment being used to build a new townhouse development nearby. A disgusted Matsil, who has been monitoring the construction work with a hawk's eye, quickly makes note of several violations he plans to report to environmental officials.

As he starts downhill, Matsil's mood quickly improves. Surrounding the pond is a peaceful, prairielike meadow of native grasses—slender stalks of little bluestem and clumpy, dense switchgrass—plants more commonly found on Nantucket Island than on Staten Island. Preserving this meadow, one of the few remaining in the New York area, was a major reason why the Parks Department considered the acquisition of this site so critical.

Only four decades ago, the meadows surrounding the pond were part of the Eib family dairy farm, which provided fresh milk to a growing Staten Island neighborhood. But the pond was gradually boxed in by suburban-style houses on one side and a subsidized-housing project on the other. By 1981 developers had acquired the 20-acre (8-hectare) site for two proposed townhouse developments. They planned to fill in portions of the pond and leave the remnant as little more than a backyard amenity for the new homeowners.

With the support of a fledgling group of Staten Island environmentalists, the Trust for Public Land (TPL) launched a spirited campaign to save the pond. After lengthy and sometimes testy negotiations, TPL succeeded in acquiring it from the developers and donating it to the city. "This place may not be everybody's idea of Walden Pond," says Richard Mudgett, a retired teacher who lives in the neighborhood and helped lead the preservation fight. "But for a lot of people around here, it's the closest they'll get to a real piece of nature."

Surely, Eib's Pond is not pristine wilderness. The skeletons of an old Ford and a Datsun lie half-submerged in its marshes. Strewn along the paths that surround it are the rusted chunks of a Frigidaire and a depressing assortment of oil cans, snack-food wrappers, and beer bottles. The Parks Department, now in its first full season of managing the pond, has scheduled an intensive cleanup of the site for this summer, and plans to install guardrails around the perimeter to deter dumping.

At an Earth Day celebration, the residents of the local Staten Island community made their own contribution to the future of Eib's Pond. Grabbing pitchforks and shovels, they gathered behind the Park Hill Apartments housing project, in a hillside meadow overlooking the pond, and planted an assortment of native trees and shrubs provided by the Parks Department. For some, like Andrea Davis, a Park Hill tenant, it was the first time they had ever planted anything. "How deep should I dig?" Davis kept asking. "What should I do with the rocks?"

Marc Matsil, handing out blueberry bushes and red-stemmed dogwoods, found it a most gratifying scene. "If people don't see ball fields or playgrounds, they tend to think of a place like this as worthless land, a dumping site," he said. "But if they can get out and do something positive, they begin to appreciate what's here. It gives them a stake in the future of the land."

Cheap Power Doesn't Come CHEAP

The James Bay Project will provide the energy of over 25 nuclear-power plants, but quite possibly at the cost of a damaged environment and a permanent disruption of Indian traditions.

by Theodore A. Rees Cheney

Energy experts are searching frantically for a nonpolluting, renewable, and inexpensive source of power. Nuclear power, coal-fired energy, petroleum, and solar energy each fail in one or more of these requirements. But many believe that hydroelectricity may fit the bill. What could be cheaper, cleaner, and more renewable than electricity generated by falling water?

In Quebec, not far from the densely populated and highly industrialized northeastern provinces and states of North America, flows all the water one could possibly want, passing through a pristine wilderness.

Project James Bay

Canada's Hudson Bay thrusts itself deep into the heart of North America. This deep, nearly Texas-sized body of salt water connects with the Atlantic via Hudson Strait at the northern end of the bay. The strait and the bay itself freeze solid by mid-fall, and do not open for

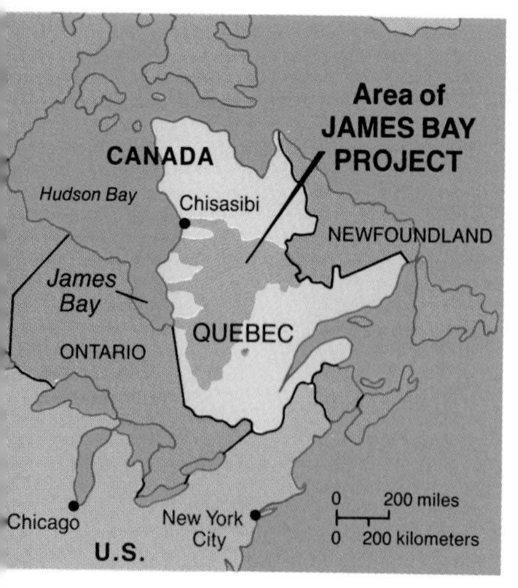

When completed, the James Bay Project will redirect the flows of three major rivers. Some 650 dams and dikes will be built in Quebec (above). Already, the resulting ecological havoc has seriously endangered the caribou and beluga whale populations (far right).

sea traffic again until mid-July. Most of the traffic heads for Churchill, Manitoba, halfway down the western shore.

James Bay, the shallow southerly extension of Hudson Bay, washes Quebec on its eastern shore, and Ontario on the south and west. Most of James Bay freezes over by January, and thaws out by mid-June. James Bay is significant because of all the freshwater rivers that enter it.

Quebec Premier Robert Bourassa saw this vast runoff as a key solution to his province's economic future, especially if it should secede from Canada, as the province threatens periodically to do. Back in 1971 Bourassa decided to dam all major Quebec streams that fall toward James Bay.

Although one of the most ambitious hydroelectric projects ever attempted, James Bay I came on-line ahead of schedule and under budget. By the 1980s Montreal's nightclubs glowed gaily with electricity from some of the 14,800 megawatts generated in the "wilderness" only 620 miles (1,000 kilometers) north of Montreal. Today Hydro-Quebec, a huge, province-owned utility, generates electricity at La Grande Complex, tapping the energy from many streams that flow from Quebec into James Bay.

We call it wilderness where all the required dams, dikes, powerhouses, and transmission lines scarred the land, but some 15,000 Cree Indians and Inuit call this particular wilderness "home," and depend on its resources for survival. These indigenous people, along with thousands of citizens worldwide, are desperately concerned about the project's effects on the wilderness.

And no wonder. Project James Bay I, a world-class hydroelectric complex, has required the excavation of 342 million cubic yards (262 million cubic meters) of earth, the pouring of 600,000 tons (544,000 metric tons) of concrete, and the creation of five reservoirs covering an area more than four times larger than Rhode Island. The main dam (La Grande-2), towering 53 stories tall and almost 2 miles (3 kilometers) long, generates over 10,000 megawatts of electricity (seven times that of Hoover Dam).

Project James Bay II, scheduled for completion within the next 10 years, will

involve a number of dams and dikes within the Great Whale River watershed to divert rivers and create hydroelectric reservoirs. On its way to Hudson Bay, the water will fall through turbines to generate 3,000 megawatts of electricity at an overall construction cost of $6 billion.

The second half of James Bay II will involve the Nottaway, Rupert, and Broadback rivers near the southern end of James Bay. Their reservoirs and dams will generate another 8,400 megawatts for sale to the industrial Northeast.

The total James Bay Project will build 650 dams and dikes, cost $47 billion, affect in various ways an area larger than France, generate 28,000 megawatts (the equivalent of 25 or 30 nuclear-power plants), require 3,300 miles (5,300 kilometers) of high-tension lines (carried by 12,000 towers), and will raise ecological havoc, especially along the rivers.

Repercussions

Historically, the expected result of such a technological invasion on an area would be for the more primitive peoples to take on the culture and value systems of the invader. The Cree and Inuit, already seeing signs of their cultural system eroding, are not at all happy about this possibility.

Initially the huge hydroelectric project didn't worry the Cree Indians. After all, it provided thousands of high-paying jobs for Cree men and women more characteristically underemployed in the southern cities or working for themselves in the wilderness as trappers, hunters, and fishermen. And the James Bay and Northern Agreement provided the Cree and Inuit with what will eventually cost Canada and Quebec about $2 billion, of which $136 million went as cash payments directly to the Cree and Inuit peoples. The same agreement also gave the indigenous peoples hunting, fishing, and trapping rights on many thousands of square miles.

But then some environmental repercussions became evident. Some rivers had their directions reversed to make them flow through the project's great turbines. The Caniapiscau River, for example, which once flowed northeast into Ungava Bay, now has been made to flow southwest through La Grande's turbines and into James Bay.

Such environmental manipulation resulted in unanticipated tragedies. Migrating herds of caribou had been crossing the Caniapiscau at a shallow ford for thousands of generations, but now they found the river running not only in reverse, but rushing in flood stage. Ten thousand caribou drowned trying to use the only crossing they knew.

Other rivers will have their flows manipulated. The Broadback River, for instance, flows into James Bay at a natural rate of about 14,000 cubic feet (396 cubic meters)

The Cree and Inuit were relocated when their hunting grounds and villages were flooded. While they now have electricity and television, they face a rapid deterioration of their traditional heritage.

per second, normally dropping to a quarter of that during the winter. Once it is part of the James Bay hydroelectric system, it will flow most of the year into James Bay at an astounding rate of 80,500 cubic feet (2,280 cubic meters) per second.

During the winter, when everyone down in civilization wants heat, the river's rate will have to increase to 140,000 cubic feet (3,964 cubic meters)—40 times that of its natural winter flow. This will create wide mud flats around the reservoirs. Fish, mammals, and other fauna and flora may find it impossible to adapt to such sudden and drastic changes in the natural scheme of things.

Methyl Mercury

When Hydro-Quebec began changing La Grande River, Fort George, a bustling center for Inuit trade, was moved to the mainland upstream, where the utility company created the village of Chisasibi (Big River) for the displaced people. La Grande now flows with a volume perhaps three times what it carried when the new, carefully laid-out village was named. The increase resulted from Hydro-Quebec's diversion of several rivers into La Grande to generate even more power.

Chisasibi, itself now served by electricity from the power complex, seems to some a great improvement over Fort George's lamplit shacks and tents. Families now live in reasonable comfort. On their VCRs, they watch "Wheel of Fortune," "The Simpsons," and sports programming—the cultural fare of the civilized south. Adults and children also receive educational programs never before available. But not all advances have improved Cree lives.

Beginning in 1984 some older Cree in the newly built village of Chisasibi developed strange symptoms. Their bodies shook, their limbs grew numb, and they suffered neurological damage. Doctors found that two out of three residents had unacceptably high concentrations of methyl mercury in their bodies. They called the disease *nimass aksiwin* (fish disease), as the mercury came in the main component of the Cree diet: fish.

This unanticipated environmental impact came from a most unlikely combination of events. When all forested acres drowned beneath the rising reservoirs, bacteria arose from the decomposition of so much organic matter. The bacteria interacted with the non-toxic mercury in the soils and rocks to form toxic methyl mercury. The fish breathed the water, fish ate fish, people ate fish—and the rest is history, the history of not making careful environmental-impact studies.

In a wilderness where people depend on plentiful and healthy caribou and fish, life begins to seem tenuous. The white doctors believed they have solved the problem of *nimass aksiwin* by telling everyone, especially women of childbearing age, to no longer eat fish (perhaps only for about 25 years).

Road to Ruin?

At first the new road that Hydro-Quebec pushed through the wilderness from Montreal seemed a blessing, but residents have since learned that a road goes two ways.

The same road that provides food and clothing at much lower prices also brings in alcohol and drugs, substances almost never seen before. People with more salary than they've ever seen before find it hard to resist

The diverted La Grand River flows through a narrow sluiceway at a recently constructed dam site (top). The water flowing through the dam turns turbines like the ones under construction above.

buying what television implies "everyone" in the civilized world already has.

Native families now suffer greatly from the abuse of these substances, and traditionally strong family ties have come untied. These issues combine with other new problems. Some traditional traplines lie under 500 feet (150 meters) of reservoir water; the Big River, now running so powerfully, erodes the banks and dumps the mud out into James Bay, changing the food-production ecology of the estuary and the bay; and the rapidly running river no longer freezes so early in the season, and breaks up too early in the spring, thus greatly inhibiting the seasonal movements of people and animals across a river of few bridges. *Endoohoo* (the Cree concept of living off the land), a concept around which an entire culture molded itself, seems certain to disappear.

James Bay II

Some 165 miles (265 kilometers) north of Chisasibi, the dual communities of Kuutjuaraapik (Inuit) and Whapmagoostui (Cree), usually referred to jointly as Great Whale River or simply Great Whale (or, in French, *Poste de la Baleine*), have sat in tenuous friendship on the sandy terrace above the Great Whale River for 50 years. Many of the Inuit people who used to live in the extreme north were lured southward by employment opportunities—first by various mining operations, then by the Mid-Canada Early Warning radar line during the 1950s.

During the early 1970s, about 700 Inuit, 700 Cree, and 200 whites lived at Great Whale, the gravel airstrip informally separating the Cree from the Inuit and white communities. A 737 jet arrived from civilization like clockwork every Thursday. The James Bay and Northern Quebec Agreement later formalized the separation.

The village of Great Whale sits astride the North American tree line's northern limit. You can see many trees to the south of the Great Whale River, but almost none to the north. Many of the village's people still depend on wood for fuel, and because of all the years they've cut wood near the village, they must now travel far up the river valley to find enough wood for fuel. Hydro-Quebec plans now to build three powerhouses along Great Whale River to produce 3,000 megawatts of electric power.

To produce that much power, it will have to divert other rivers, such as the Nastapoka, into Great Whale River, which may then be sent through a tunnel into Hudson Bay. Gone will be the wood supply, gone the drinking water the river now supplies the village, and probably gone will be the white beluga whales the village people hunt in the mouth of the river.

The region's first road will soon connect the village with Montreal. Prices of food and other civilized goods should go down, but will this community suffer the same cultural shocks as Chisasibi?

With the arrival of television and VCRs, will other Inuit and Cree cultural traditions disappear? Will the next generation know how to hunt the whale, the caribou, the beaver, the fox? After all these years living in the shadow (and beneficence) of the Canadian government, they've already forgotten how to build an igloo or handle a dog team.

Forewarned by the experiences of Chisasibi, and even by those of the Alaskan natives under the pressures of the petroleum dollar, will they avoid the deadly lures of alcohol and drugs? Will their families fall apart? And what will happen when this present construction employment stops? Will the youth of the north take the new road south?

Cloudy Outlook

James Bay Projects I and II will have positive economic impacts on the indigenous peoples in addition to the negative social and environmental impacts. Although the positive economic impact upon the Cree and Inuit may be short-lived, the positive impact upon Quebec's and Canada's economies may be substantial and long-lived.

Quebec's economy has improved so much in recent years that economists say it should be able to withstand the stresses of separation or secession from Canada, should that actually occur. Quebec, the size of Europe, has about 25 percent of Canada's population, and creates about 25 percent of its wealth. Part of this wealth comes from the sale of its natural resources, especially hydroelectric power. The availability of great amounts of cheap power also lures manufacturers that require huge amounts of electricity, such as plants that convert bauxite to aluminum. Three aluminum plants are under construction in Quebec for over C$2 billion. A British aerospace company plans to build a C$90 million plant at Mirabel Airport outside Montreal. The plant will build landing gears for the European Aerobus passenger aircraft. A Norwegian firm is building a C$550 million magnesium-processing plant at Becancourt, not far east of Montreal. And they all depend upon cheap hydroelectricity.

Environmentalists point out, however, that despite what Hydro-Quebec claims, James Bay will produce much more electricity than Quebec can use itself, and that to pay the project's immense costs (about C$2,300 per capita), Quebec must export the rest of the electricity to other provinces and to the northeastern states of the United States. Without these export sales, the James Bay Project will have been a waste of money and an unnecessary destruction of a wilderness and indigenous cultures.

Protests by U.S. and Canadian environmental groups have caused several of the U.S. states to rethink their promised contracts. The citizens of Maine, for example, have already voted down a statewide referendum that would have allowed transmission lines to cross the Maine wilderness. Vermont protesters have even conducted street theater in the streets of Burlington to protest that state's signing of a $4 billion contract to buy 340 megawatts of power from Hydro-Quebec annually. In April 1992, New York Governor Mario Cuomo announced that the state will cancel its $17 billion, 21-year contract with Hydro-Quebec.

The future of James Bay II, at least, looks cloudy. As a result the future of the Great Whale River environment and the indigenous cultures looks a bit brighter—for the moment.

STORM CHASERS

by Nick Ravo

To Jim Leonard, a hurricane hobbyist, winter here is wimpy, what with travel-poster skies, slow-motion surf, and nary a tropical disturbance in sight. "BORing," he says. "Not even a thunderstorm."

Last autumn, as soon as the hurricane season ended, Mr. Leonard did what only the most fanatical storm chaser would do: he quit his job, that of a dispatcher for the Florida Power and Light Company, packed his rain gear, and moved 8,000 miles (13,000 kilometers) to Guam. "Nonstop action: typhoons year-round, maybe 20, 25," he says in an interview. "Super typhoons, too."

Mr. Leonard, a 42-year-old bachelor, is one of a growing number of thrill seekers who pursue dust devils, hurricanes, typhoons, thunderstorms, tornadoes, waterspouts, and other extreme weather, usually by car or pick-up truck. (Vans have rolled over in high winds.)

"It has become a real big sport," Mr. Leonard says.

Many of these storm chasers—the hard core probably number no more than a few hundred—are amateurs who are fascinated by thunder and lightning. "Some people do it to relive a childhood experience or work out their childhood fears," says William Hauptman, a playwright, short-story writer, and tornado tracker from Texas whose book *The Storm Season* (a novel about storm chasing) was recently published.

A few storm chasers like Jack (Thunderhead) Corso, a postal supervisor who lives in Harrison, New York, have turned their hobby into lucrative small businesses, selling videotapes and photographs of nature on the rampage.

"I'm number one for quality of close hits, hits within 100 yards [100 meters]," says Mr. Corso, who has been photographing lightning since 1976. "I average one real close call a year. You've got to take that risk to get that special shot."

Others have a more academic mission. Tim Marshall of Lewisville, Texas, a civil engineer and meteorologist, has worked at the Institute for Disaster Research at Texas Tech University in Lubbock, studying storms. He also publishes a bimonthly newsletter for 500 subscribers, *Stormtrack*. "We get calls from people," Mr. Marshall says, "especially from Desert Storm. They didn't get enough action over there, so they want to be in a ditch and have a tornado go over them."

A tornado (below) provides an especially tempting target for the dedicated storm chaser. Some storm lovers who find American weather phenomena too mundane make frequent visits to tropical islands prone to hurricanes.

Mr. Marshall has also produced a 19-page storm-chasing manual (page 15, Safety Rule E: "After sighting a tornado, maintain a safe distance at all times"). The manual is not available to the public. "It has a disclaimer," he says. "We don't want people to go out and be kamikaze pilots."

Patrick Hughes, the managing editor of *Weatherwise* magazine, says that the presence of storm chasers at disaster sites has grown to the point that the National Weather Service would like them to organize and set safety standards. Movement in that direction has been slow, however.

He also says there might be a need to educate some storm chasers about ethics and disaster etiquette. "There have been situations when you have people who have lost their homes or their families in a tornado, and then some guy with a camera comes in like they have a trophy," he says.

But he adds that trophy hunters are "an extreme minority," and that the number of storm chasers has increased because of the spread of small video cameras. "It is a primal thing," he says.

He says television-news teams in helicopters are now among the most ardent storm chasers. Many local stations sell other stations what they film. "There is a Minnesota television station that has the tape to beat," Mr. Hughes says. "It's from a helicopter shooting around and down into a tornado. Everyone wants to duplicate that. It is spectacular stuff."

Close Calls

Only one person is believed to have been killed chasing a storm. Mr. Marshall says it happened in the mid-1980s when a chaser driving back from a tornado swerved off the road when he saw a rabbit. "The rabbit got him, not the tornado," says Marshall.

Richard Horodner, a former accountant who is now a hurricane chaser in Miami, Florida, says he came close to death once or twice, in two hurricanes, first in 1979, then 10 years later. "In Frederic," he says, referring to the 1979 hurricane, "we were just sitting there in Pascagoula, then all of a sudden, half a gas station flies by. The worst, though, was in [Hurricane] Hugo, where all the glass was breaking around us, and furniture was flying. For about 15 minutes, I thought we were in trouble."

Neil L. Frank, a weather forecaster for KHOU-TV in Houston who is a former director of the National Hurricane Center, is critical of most storm chasers, including Mr. Horodner. "I've been concerned that these guys are going to end up losing their lives," he says. "And is there any scientific value in what they are doing? No, not really."

Mr. Horodner's interest in hurricanes is singular. The walls of his home are covered with photographs of hurricanes and hurricane-tracking maps. During the off-season, he says he limits his viewing of The Weather Channel to only about an hour a day.

Mr. Horodner became enthralled by storms while he was growing up in south Florida. He graduated to chasing hurricanes in 1965, and has since followed more than 30 tropical storms up the Atlantic Coast or along the Gulf Coast. Occasionally he has ventured to a Caribbean island, or to Baja California; there he has rented cars to chase storms. "For obvious reasons, I don't tell the rental-car people what I'm there for," he says.

For most of his chases along the Atlantic seaboard or Gulf seaboards, Mr. Horodner, who is 44, uses a pickup truck owned by a fellow hurricane chaser in Daytona Beach, Florida. To measure wind velocity, the truck has an anemometer rigged on the back. He says the highest wind that he has recorded is 135 miles (215 kilometers) per hour.

"There is no better feeling than being in the eye of a hurricane and looking up," he says. "You get the experience of being in a black hole without being crushed to death."

Connecting with Reality

Mr. Horodner says he has earned $30,000 to $50,000 a year selling videotapes of his adventures. In May 1992, Mr. Horodner was a featured guest at the Hurricane Survival Trade Show in Miami, a first of its kind.

"Why do I do it?" he asks. "I do it because I get off on it, the feeling of the power in the middle of a storm. It is kind of a therapy for me. You connect with reality."

Mr. Hauptman agrees wholeheartedly. "Storms are beautiful, awe-inspiring," he says. "It is a natural urge to chase them. The problem is when they chase you. My brother said that seeing a tornado is like seeing a rattlesnake. There is something paralyzing about them."

Some chasers may work as official storm spotters for the National Weather Service. Working with the aid of ham radio operators, they report the precise location of a storm and warn local officials whose municipalities lie in the storm's projected path.

Mr. Horodner, however, says that storm chasing can attract its share of "crazies," a category from which he excludes himself. He gives this example:

"It was Gilbert, 1988, South Padre Island. I was stuck in the sand right near the coastline, and all of a sudden two guys came walking in from the surf. They were firemen from Dallas. They wanted to see a big storm. It was exciting. They were having a good time. One of them had a beer in his hand."

Mr. Corso chases tornadoes as well as lightning. Each May, he spends his vacation time prowling "Tornado Alley," a strip extending from Dallas to Iowa in which scores of twisters have been spawned. "We sleep in hotels, make day-to-day forecasts, and try to get to the heated areas," he says.

Mr. Leonard considered moving to that part of the country last fall, when he was thinking of relocating to where the physical atmosphere would be more unstable; he also contemplated Corpus Christi, Texas. But he chose the Pacific island of Guam at the behest of Barbara White, an old friend and fellow hurricane chaser who once lived in Miami. "Chasing typhoons is our main form of entertainment," Ms. White says.

Mr. Leonard plans to fly to Pacific islands and atolls that might be in a typhoon's path. Since he arrived on Guam three months ago, he says, he has captured pieces of storms without having to leave home. One was Super Typhoon Yuri, whose eye came within 50 miles (80 kilometers) of the island. (A super typhoon develops winds over 150 miles—240 kilometers—per hour.) "It was great: 170 mile-per-hour winds, gusts to 200," he says.

Man, of course, does not live by typhoons alone, not even on Guam. Mr. Leonard is mowing lawns on the island until he can land a government job, some kind of job, to help pay his way.

"I know it sounds strange," Mr. Leonard says wistfully. "But some people climb mountains; some jump out of planes. Hurricanes and tornadoes are my big thing. I like to be in them."

OFFSHORE OIL: How Deep Can We Go?

by Peter Britton

Twenty-four years ago, the *D/V Glomar Challenger* hovered over the Sigsbee Knolls—a group of subsea promontories nearly 500 miles (800 kilometers) due south of Louisiana. Some 11,600 feet (3,500 meters) below, its drill bit entered the soft mud of Challenger Knoll and penetrated 470 feet (140 meters) into late-Miocene sand. Soon the National Science Foundation's (NSF's) Deep Sea Drilling Project had a fine core sample on deck—and what a core sample it was.

The project's aim was to determine what was going on under ocean floors and to investigate the continental-drift theory of Earth's land formations. Researchers found many answers, but at longitude 92°35.2′ west, latitude 23°27.3′ north, they found more than anyone had bargained for.

Global Marine engineering development director Sherman Wetmore, whose company operated the drilling vessels, recalls how "one minute, paleontologists and geologists aboard were enthralled just by the concept of drilling thousands of feet into the Earth beneath 10,000 feet [3,000 meters] of water. The next minute, they were flabbergasted as the 6.5-foot [2-meter]-long core sample came up. The tube was oozing something no one thought would be there—that 'something' was crude oil."

The drill bit had penetrated a perfectly normal geological formation in which hydrocarbons—oil and gas—could collect and languish under heat and pressure. But the best that *Glomar Challenger* could do was plug the hole, mark the spot, and have the core analyzed by a team of geologists, who pronounced it "bona fide crude oil." There was little else anyone could do: recovery of oil at that depth was unthinkable, for in 1968, 600 feet (180 meters) of water was about the limit for offshore oil production.

Last Great Adventures

Now oil companies are starting to do the unthinkable. Although offshore-oil technology proceeds at a slow pace, the talk has turned dead serious about oil drilling and recovery in water depths approaching 10,000 feet (3,000 meters).

Says Shell Offshore president R. L. Howard: "Then, years ago, the oil industry took a dim view of deepwater possibilities. But now we see it as a major opportunity. It's the economics that are the biggest hurdle for a technology ready to go." Conoco-Du Pont engineer William Dietrich calls deep offshore oil "one of the last great adventures on the face of this Earth." The message is that recovery of these deepwater deposits can and will be done. The question: When?

And why now for deep oil? The answer is a convergence of national need, threatening geopolitics, domestic moratoriums on other offshore-drilling locations, the willingness of banks to finance such ventures, and ever-improving technology, where a minor revolution is taking place, including:

- Rig automation that replaces the roustabout and increases safety and efficiency while decreasing deck load.
- The replacement of divers by remotely operated or autonomous vehicles, another move to get people out of the most dangerous zones.
- Three-dimensional seismic exploratory techniques that lend depth and sense to arcane underground visual information.
- Horizontal drilling techniques that can reach out in many directions and levels from a platform that services as many as 44 wells.
- The advent of expert systems in interpreting the flood of information collected during exploration, drilling, and production.
- The resurgence of steel- and/or composites-reinforced concrete for use in floating-platform construction. Low cost, strength, and resistance to the powerfully corrosive effects of salt water are the attractions.
- New long-life drill bits with internal cameras that send back real-time bit's-eye-view images.

The Ocean El Dorado is a hybrid rig that could drill for and produce oil in waters 10,000 feet deep. The huge structure, with a deck the size of two football fields, would be moored by a complicated system of chains and steel cables that would keep the rig stable in winds over 120 miles per hour. Ocean El Dorado would take 4½ years to build, and could ultimately accommodate as many as 64 wells.

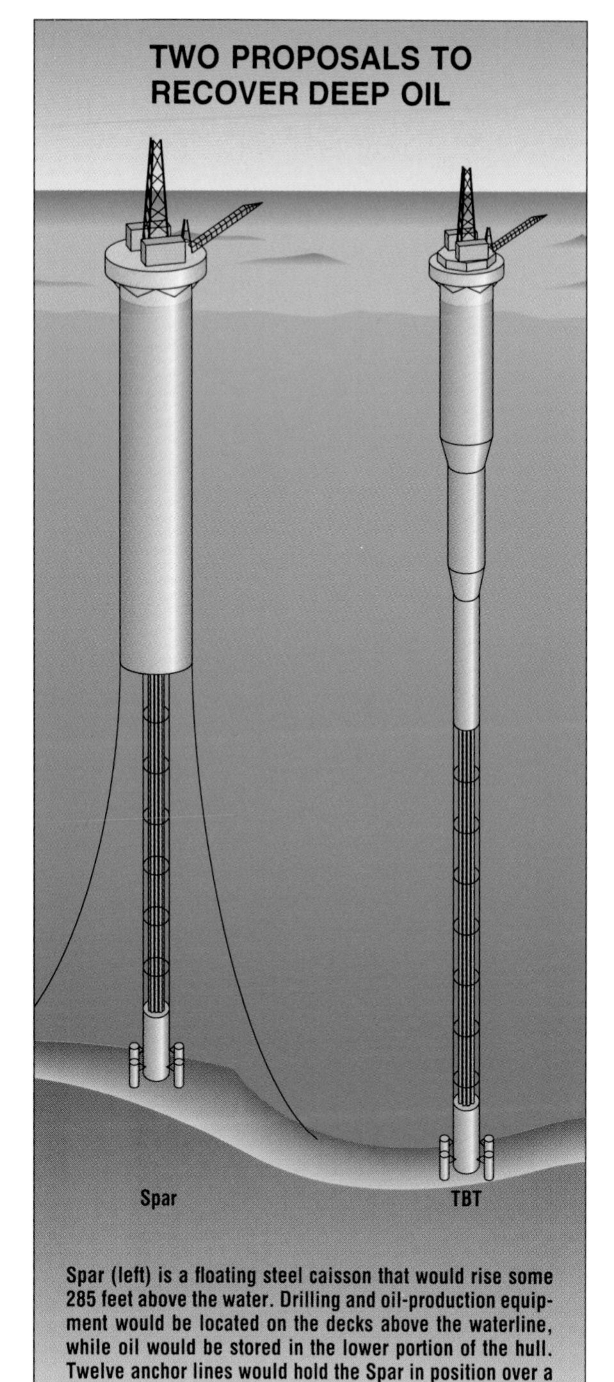

Spar (left) is a floating steel caisson that would rise some 285 feet above the water. Drilling and oil-production equipment would be located on the decks above the waterline, while oil would be stored in the lower portion of the hull. Twelve anchor lines would hold the Spar in position over a multi-well sea-floor template. A Spar proposed for the Gulf of Mexico would operate in water 2,700 feet deep. Another way to tap oil deep beneath the sea would be with the tension buoyant tower (TBT). The TBT consists of tapering, cylindrical buoyancy modules that would extend hundreds of feet beneath the surface. Tendon lines extending to the sea floor would moor the structure in place. The TBT could be used in water depths ranging from 1,000 to 10,000 feet.

- Continued refinement in measurement-while-drilling techniques that use down-hole motors and electronics to send information to the surface by a coded system of pulses.
- The widespread use of composite materials like Kevlar aramid fiber both below and above the surface of the water.
- New safety measures, including an intumescent spray used on the outer side of 18-inch (45-centimeter)-thick walls to block the hottest of hydrocarbon fires (about 2,000° F, or 1,100° C) for up to four hours.
- Detection by airborne lasers of faint oil slicks and gas traces on the water surface from subsea seeps.

Beyond these advances, there is yet another compelling reason for deep-oil recovery: a growing industry belief that the remaining "elephant" fields are only in deep water. So new is this thought that the industry itself seems confused about what "deep" and "ultradeep" actually mean. When pressed, Shell's Deepwater Projects manager Carl Wickizer allows that, for now at least, deep begins at 1,500 feet (460 meters), and ultradeep at 7,500 feet (2,286 meters). Wickizer should know. His company holds the current world record of 7,520 feet (2,290 meters) of water (henceforth, "feet of water" is abbreviated as "fow") for drilling an exploratory well.

Confusing Crossroads

Shell, the world's largest oil company in terms of equipment, has long been the leader in the search for ever-deeper offshore oil. The company currently produces about 150,000 barrels of crude oil and almost a billion cubic feet (28 million cubic meters) of natural gas daily in the Gulf of Mexico. In addition, Shell owns 35 percent of the total leased acreage in the Gulf in water more than 1,500 feet (460 meters) deep. The company's Bullwinkle platform, 1,615 feet (492 meters) tall and operating nicely in 1,350 fow in Green Canyon Block 65, holds the water-depth record for fixed platforms.

In 1989 Shell Offshore announced the discovery of 220 million barrels of oil (some experts claim considerably more) in Garden Banks Block 426, 214 miles (344 kilometers) southwest of Morgan City, Louisiana. Named Auger, this field is an elephant by Gulf standards, some 10 times the size of the typical Gulf shelf project. Located in 2,860

The Auger tension leg platform (TLP), now under construction in two shipyards, is scheduled to start oil production in 1993 in 2,860 feet of water. The steel cables that hold the TLP in position are anchored to the sea floor and pull down against the buoyancy of the semisubmersible platform. The platform is expected to produce 40,000 barrels of oil and 150 million cubic feet of natural gas each day.

fow, Auger is scheduled to start production in 1993. While this depth may sound only a tad deeper than normal, it is roughly the depth at which a whole new regime begins. To get the oil from Garden Banks, Shell Offshore has had to decide which of several available technologies could best be scaled up and reconfigured to handle the many challenges associated with water this deep.

It's a confusing crossroads. Most in the industry expect fixed platforms like Bullwinkle and concrete gravity-based platforms to become too expensive and susceptible to wave-induced fatigue at 1,500 fow. Guyed and compliant towers may work to 3,000 feet (915 meters), while jack-up rigs are good only to about 500 feet (150 meters). Moored semisubmersibles are thought to be effective to about 4,000 feet (1,200 meters) and when dynamically positioned, to greater depths—although at great cost for thrusters and fuel. Drill ships operate well to more than 7,000 feet (2,100 meters), but only for exploratory drilling, not production. Some say tension-leg platforms (TLPs) may top out their usefulness at 4,000 fow; others put that figure closer to 7,000 feet (2,100 meters).

Shell finally committed to an unprecedented in-house program of enormous cost, complexity, and challenge to design and build the Auger TLP, a $1.3 billion investment that signals the beginning of oil recovery at these depths. About two-thirds of the investment will be for platforms, facilities, and pipelines, with about one-third planned for drilling and completion of wells.

Says Howard: "The Auger project represents a major step for the industry in really deep water. Deepwater oil is worth about 40 percent less than a barrel of shelf oil. The midpoint of production for deep water is projected to average about 17 years after discovery; that's a delay of about 10 years over the shelf project. So, on a present-value basis, the average deepwater barrel is worth a lot less than a barrel on the shelf."

The folks at Shell are nothing if not optimistic, especially after the company's recent discovery of rich new oil fields in the Gulf of Mexico. Already, Shell is drilling a well in 4,340 fow. Wickizer predicts that "we'll be producing at 6,000 feet [1,830 meters] by the end of this decade. There are more elephants out there, and they're all deep—or ultradeep."

MEASURING UP

AUGER
3,275 FEET TALL
IN 2,860 FEET
OF WATER

BULLWINKLE
1,615 FEET TALL
IN 1,353 FEET
OF WATER

SEARS TOWER
1,454 FEET TALL

Platforms and Towers

But the challenges of deep-oil recovery remain open to entrepreneurial engineers, and the place to shop one's wares is at Houston's annual Offshore Technology Conference

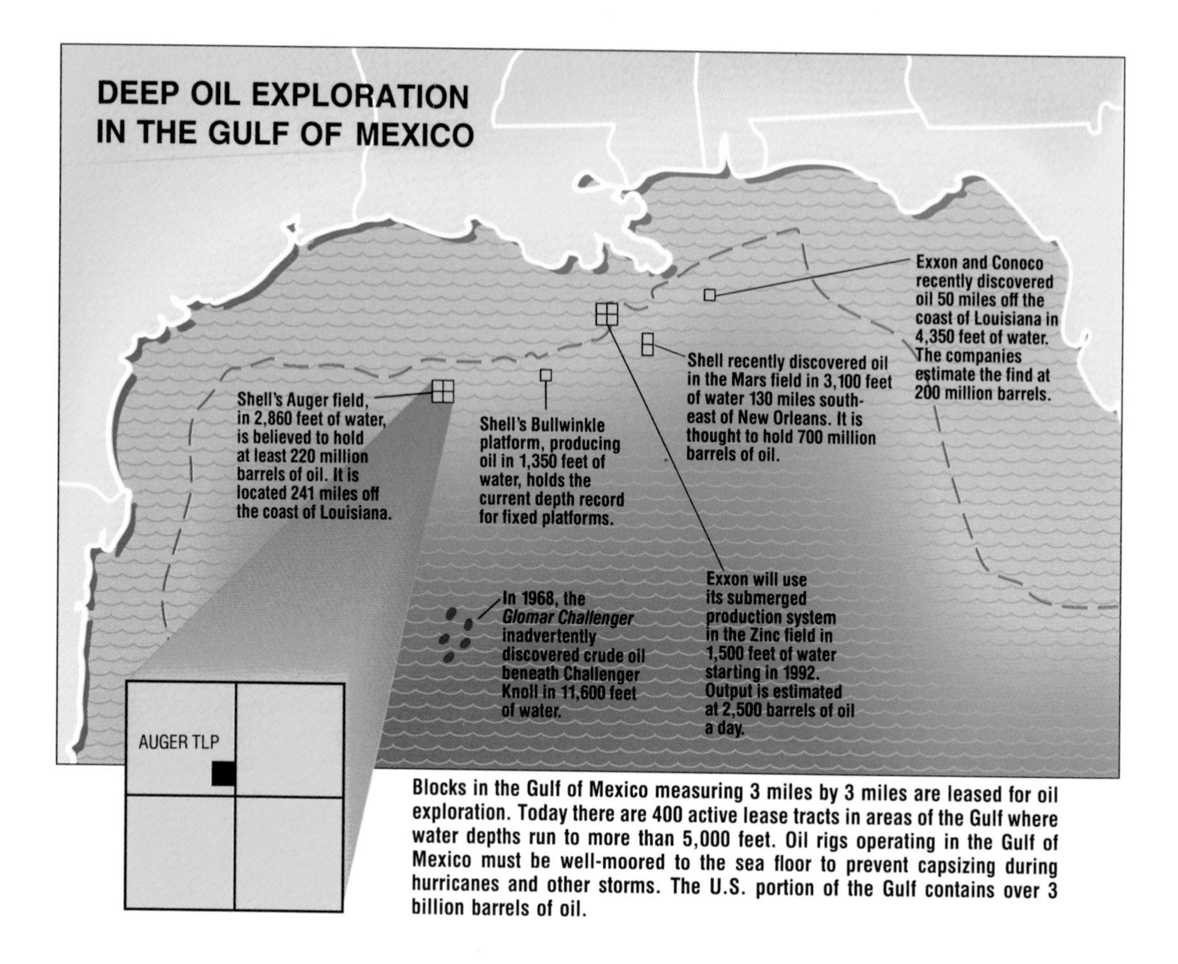

Blocks in the Gulf of Mexico measuring 3 miles by 3 miles are leased for oil exploration. Today there are 400 active lease tracts in areas of the Gulf where water depths run to more than 5,000 feet. Oil rigs operating in the Gulf of Mexico must be well-moored to the sea floor to prevent capsizing during hurricanes and other storms. The U.S. portion of the Gulf contains over 3 billion barrels of oil.

(OTC). In 1991 some 34,000 industry folk visited the four-day feast of seminars.

At OTC, one can find concepts that range from informed hunches to the results of years of costly research. One of the companies represented was Deep Oil Technology (DOT), supported in the early 1960s by a consortium of oil companies to study methods of drilling and production in the then-deep waters of 600 feet (180 meters).

DOT's earliest concept was the original tension-leg platform, wherein a floating deck of steel or concrete is winched into the water by cables attached to the ocean bottom. The platform stays under tension, remaining stationary and stable in the worst of weather conditions. This technology was borrowed from a 1940s idea to build mid-ocean airports for refueling stops in the Atlantic; Shell's Auger TLP is its latest descendant.

Meanwhile, DOT is currently designing advanced concepts for extremely deep water. Says DOT's John Halkyard: "We feel that the TLP has to be modified for use in really deep water. The Spar floating drilling production and storage structure is one such derivation. It's more than 700 feet [210 meters] high and is floated to a location like a log or spar, then tilted upright and moored in place."

In addition, DOT engineers envision the tension buoyant tower (TBT) as another promising concept. This is, in effect, a full drilling-and-production deck sitting atop a huge elongated buoyant steel barrel that tapers to a thick tube. The 750-foot (225-meter) tube houses both the telescoping marine riser and the tendons that moor the TBT to the ocean floor. According to Halkyard, both elegant systems will operate to 10,000 feet (3,000 meters)—and beyond.

Submarine Drilling?

In the category of informed hunches at the OTC, U.S. Navy and human-powered submariner Ted Haselton showed his patented, novel proposal: a 300-foot (91-meter)-long submarine that can travel at 20 knots. Once on location, it tilts to the vertical position and rides below the surface, save for a helicopter landing pad and an access tube.

The submarine features a variable-pitch, six-bladed bow propeller, which supplies power for both travel and station keeping. All drilling equipment is self-contained, and the marine riser exits through the propeller, which is at the deep end of the vessel. The chief advantages are very low heave and much less steel.

Wetmore of Global Marine doesn't see a future for Haselton's submarine, summing up offshore state-of-the-art possibilities: "I don't think we'll ever see drilling from submarines. But never say never." The concept, says Wetmore, is feasible, but not practical, calling for high-pressure and saturation divers.

"But," he adds, "unmanned and remotely controlled subsea production would be feasible." Wetmore explains that a three-phase pumping system is needed to pump rocks, sand, oil, and gas together 100 miles (160 kilometers) to shallow water for production on- or offshore.

Seafloor Operations

Of the methods down the road many experts believe that the seafloor-type operation is most likely. There has been a trend in recent years to move workers off the drill floor and out of the ocean; this means automation and remote operations. The logical extension is, indeed, full subsea operation, and Exxon has been the leader in this area. Some believe that this represents the ultimate future of offshore oil recovery.

Exxon, the world's largest oil company in terms of assets, has been developing its submerged production system (SPS) since the 1970s. The first application will be in 1992 in the Zinc field of the Mississippi Canyon in 1,500 fow. The SPS will produce 2,500 barrels of crude and condensate (condensed natural gas) per day, along with 115 million cubic feet (10 million cubic meters) of gas. The oil and gas will be transferred via pipeline to a platform six miles (10 kilometers) away in 468 fow.

The next step will be the key one, about 50 miles (80 kilometers) east in Block 211. There Exxon and Conoco recently announced a find in 4,350 fow after drilling through 3,000 feet (915 meters) of salt to a depth of 13,000 feet (4,000 meters) under the seabed. The two oil companies think that 200 million barrels of oil may await them.

Monster Rigs

There are other proposals as well. Not far from Wickizer's office at One Shell Tower in New Orleans are the offices of Odeco, a company with a long history of research in semisubmersible technology. In the lobby is a handsome gray-and-black model of the Ocean El Dorado, the vessel that Odeco's Mark Childers believes will be just the thing for drilling and production in 10,000-foot (3,000-meter) waters anywhere in the world.

The Ocean El Dorado is a new breed of monster rig, designed for monster duty. A daring hybrid marriage of a steel semisubmersible and concrete tower, Ocean El Dorado could change the concept of both deepwater rigs and deepwater rig building in the Gulf. It is truly massive. At 230,720 tons displacement, it is four times the size of Odeco's dynamically positioned Ocean Deep TLP, which is designed for 7,500-plus fow and is larger than any semisubmersible in use today. Ocean El Dorado has a 350-foot (106-meter) operating draft (a standard rig at 23,000 tons has an operating draft one-fifth that figure).

Ocean El Dorado's deck would be slightly larger than two football fields side by side. It would have 64 well slots, with "Christmas trees"—the bristling clusters of valves and pipes that control the flow of oil—located at platform level.

Extensive analysis has convinced Odeco that a spread mooring system would be the best compromise to keep the rig in position. It would employ 24 lines up to 5,000 feet (1,500 meters), and 36 lines up to 10,000 feet (3,000 meters), using six-inch (15-centimeter) chain and five-inch (12-centimeter) wire rope. This, says Odeco, would hold the vessel within 1.6 percent of water depth with a 120-knot steady wind, and within 1.94 percent with a 5-knot current in 10,000 fow. It is, of course, designed to withstand the Gulf's worst weather.

Even the construction would be revolutionary. The Gulf Coast tradition is steel; Odeco's vision is the slip-form-concrete construction method. It would take up to four and a half years to build. Says Childers:

LOOKING FOR OIL WITH SOUND WAVES

The U.S. Department of the Interior is surveying the vast 700,000-square-mile Gulf of Mexico with sophisticated electronic equipment and such techniques as three-dimensional seismic surveying. In 3-D surveying, a ship tows a series of compressed-air guns, perhaps 56 of them in seven rows of cables. Thousands of acoustic receivers run on cables parallel to and on each side of the array of air guns. The entire assembly can be up to 4 miles long. The air guns send acoustic pulses through the water that penetrate the sea floor to a level where oil is suspected to lie. The reflected waves are "collected" by the receivers and relayed to the ship via fiber-optic cable. The data are later interpreted by a supercomputer, which produces a 3-D cube of data that shows the spatial relationships of salt domes, faults, and other sea floor features that can indicate oil deposits.

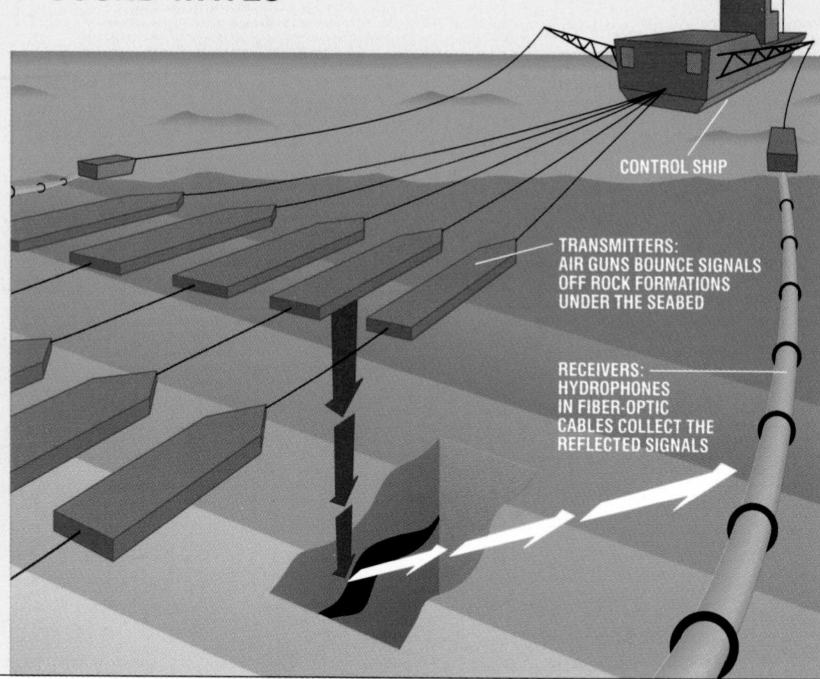

"Technically, this system looks feasible. With a large enough field, the economics appear reasonable. But with a decade needed for completion, the present industry climate—uncertainty in the price of oil, estimated project costs, actual size of the field, and flow rates—makes Ocean El Dorado questionable."

The Ocean El Dorado remains a hypothetical monster, dependent, perhaps, on the success of Shell's Auger TLP. So, for the moment, Shell Oil commands the industry's attention—and a look at Shell's Auger TLP is equally fascinating. Wickizer points out that "this TLP may be the most complex machine ever built by the offshore industry."

All Decked Out

The Auger TLP is, basically, a highly complex, computer-designed creature of 73,000 tons displacement that carries an industry's hopes on its broad decks. Auger will have 32 well slots, with wells to 20,000 feet (6,100 meters) below the seafloor. It is expected to produce 40,000 barrels of oil and 150 million cubic feet (14 million cubic meters) of natural gas daily. The five-story "safe-haven" personnel facility will house 112 people. One of the items the industry will be watching closely is the double mooring system to be used—the first for a TLP. For, in addition to the 12 regular 26-inch (66-centimeter)-diameter steel tendons anchored to four 60-foot-square templates on the sea floor, there's another wrinkle in the system.

From the base of each of four 74-foot (22-meter)-diameter columns, two mooring lines will be attached to two 30-ton fixed fluke-type anchors. The lines will be comprised of an 8,000-foot (2,400-meter) section of five-inch (12-centimeter) wire rope and an 1,800-foot (550-meter) piece of $5\frac{3}{16}$-inch (13-centimeter) chain.

The huge platform is taking shape in two major fabricating yards and, most impressively, within the intricate depths of a Cray computer in Shell Offshore's headquarters. There, each and every piece of steel, configuration, and man-machine interface is analyzed in the rigorous software database known as Plant Design Management Systems (PDMS).

This CAD system enables designers to work in three dimensions and visualize results ahead of fabricators. PDMS also has "walk-through capabilities" and a feature called "clash." This alerts Shell's engineers and designers when equipment and/or people are incompatible as presented.

As prodigious and cleverly designed as the Ocean El Dorado and the Auger TLP are, nature still boasts phenomena to be reckoned with. Offshore oilmen still recall with dread the infamous Bay of Campeche Ixtoc I blowout in the summer of 1979. In only 164 fow, this 30,000-barrel-a-day runaway was finally stopped after Red Adair's oil-fire-fighting company drilled a relief well nearby. But today in the Gulf of Mexico, there are only a handful of vessels capable of aiding a stricken well at the depths now being approached. Fortunately, however, careful mud analysis can give clear indications of the presence of high-pressure gas deposits like those found at the Ixtoc blowout. But there are other potential troublemakers lurking. Hydrostatic pressures of 1,200 pounds (545 kilograms) per square inch at 3,000 fow are new factors. Elements of drilling such as remotely operated vehicles, marine risers, and blowout preventers must be redesigned to withstand these pressures; the 35° F (1.6° C) temperatures at the seabed combine with the pressure and gas to create the climate conducive for hydrates and paraffins. These are thick, icy substances that can cause malfunctions.

The 100-Year Storm

The age-old threat of hurricanes and the prodigious waves they can spawn is another cause for concern. The rigs are built to withstand what's known as the 100-year storm, a theoretical cycle of weather in which an exceptionally severe storm occurs about every 100 years. In the Gulf of Mexico, such a storm can and does cause waves up to 76 feet (23 meters). But the main concern is not for the platform, but for the marine riser. This is the steel casing that houses the drill pipe, mud, and whatever else runs from the rig deck to below the ocean floor. Even though the riser is given added buoyancy by air cans and syntactic foam (a highly buoyant mix of plastic and glass microballoons), its 1 million-pound (450,000-kilogram)-plus weight and bulk make it a big problem under certain circumstances.

The 2-mile (3.2-kilometer)-long, 21-inch (53-centimeter)-diameter pipe can, in fact, snap or buckle in severe weather. This is avoided by "pulling the string"—taking up the entire riser—and placing it on deck. The deeper the water, the bigger, longer, and heavier the pipe—and the bigger the platform needed to accommodate it. To ease these problems, a lightweight composite riser is currently being tested.

Fast-Eddy Threat

Then there's the natural phenomenon called the Gulf of Mexico loop current. This is an influx of warm tropical water that periodically enters through the Yucatán Strait, meanders about the Gulf as far north as the Mississippi Canyon, and then exits with the Gulf Stream around the tip of Florida.

Every now and then, a massive anticyclonic gyre will break off from the deep-running loop current when it gets about halfway to U.S. shores. Recent eddies have been drolly dubbed Nelson Eddy, Fast Eddy, and Eddy Murphy, and they can cause big mischief. They are capricious, strong, and deep, with speeds above 3 knots at the surface, and extending as deep as 2,000 feet (600 meters).

The potential for harm to the weighty and massive marine drilling riser is frightening to offshore oil operators. The current can embrace the rounded riser, setting up vortices on the other side. These will appear first on one side, then the other, within a period of four to six seconds. This can set up a fatal oscillation in the riser that can last for weeks, all the while weakening the giant tube.

Once in the grip of, say, Fast Eddy, nothing can be done. But satellites can be used in winter to detect the arrival of a loop current by changes in surface-water temperature. And an oil-company consortium is setting up a surface detection system. The only other defense is applying strakes—spirals of small wings that trip the turbulent flow and interfere with the vortices. Shell drilling engineering adviser Gary Marsh has perhaps the best remedy: "If a big one comes, we pull the riser." Permanent installations, such as the Auger production riser, are designed to prevent such vibration.

But where there's a drill, there's a way, and Shell's Marsh puts the deepwater movement in perspective. "Can we drill at 10,000 feet [3,000 meters]? Of course. Can we make money? That's harder to answer."

He's not overly concerned, though. "I fully expect that one of these years, the word is going to come down from the boardroom to go and do it. And we will."

REVIEW

Energy

NATIONAL ENERGY POLICY

After years of pressure to establish a national energy policy that would include strategies for reducing the country's dependence on oil, and spurred on by the Persian Gulf War's potential effect on availability of oil, the administration finally proposed a National Energy Policy early in 1991. Environmentalists were outraged that a major provision of the policy was to expand domestic oil exploration—including in the Arctic National Wildlife Refuge—instead of promoting energy conservation and development of alternative energy sources. There also was much disagreement about the proposal to streamline licensing for new nuclear-power plants. Controversy continued throughout the year, with many proposals and counterproposals being offered and argued in Congress, before the Senate, after a lengthy filibuster in November 1991, postponed further action.

FOSSIL FUELS

Crude-oil output in the United States dropped in 1991 to its lowest level since 1950 as exploration and production declined rapidly throughout the year. Some industry observers stated that fewer oil and gas rigs were in operation in 1991 than at any other time since 1901. This decline was felt in the labor market: the number of exploration workers dropped in the past decade from 692,000 to 392,000 in 1991. Imports also dropped in 1991, to 45.5 percent of U.S. usage, down from 47 percent in 1990.

Prices responded to the Persian Gulf War by rising to as high as $40 a barrel on January 1, 1991, but by March, they had dropped again to the $18–$20 range, where they remained throughout the year. The temporarily higher prices early in the year combined with nationwide cutbacks in demand as the recession lingered to bring about the first decline in demand since 1983. Demand fell approximately 2 percent, to 16,648,000 barrels a day from the 1990 level of 16,989,000 barrels. Oil's contribution to U.S. energy consumption declined in 1991 to 40.3 percent, its lowest level since 1950, and down significantly from its 1977 peak of 48.7 percent.

Natural-gas consumption remained steady, supplying about a quarter of the nation's energy needs, down from the 1970 record of 33 percent. Prices for producers, though, continued falling rapidly, particularly during the winter, to about $1.05 per 1,000 cubic feet (28 cubic meters) in early 1992, down from $2.30 in October, and much lower than the record $3.50 in February 1982. Lower prices benefited only the middlemen, however, as prices to consumers remained constant at about $6 or more per 1,000 cubic feet.

Lyondell Petrochemical, one of the nation's largest refiners, started processing used motor oil into gasoline, and planned to reach an output level of 30 million gallons a year. Although this would account for only a little over 2 percent of the 1.35 billion gallons of used motor oil produced every year, it was considered a significant step in dealing with this serious environmental issue while recycling a valuable fuel.

Coal production dropped slightly in 1991, to 998 million tons from 1990's record 1.029 billion tons. Utilities used coal to produce 55.1 percent of the nation's electricity in 1991, and coal supplied about a quarter of the country's overall energy consumption. With many European countries increasing their coal usage, exports also increased to well over 110 million tons. As the Clean Air Act amendments begin to take effect in mid-1992, operators of many coal-burning plants faced the prospect of having to spend as much as $38 billion to reduce sulfur dioxide pollution that results from burning coal.

HYDROELECTRIC POWER

Canadian plans for greatly expanding its hydroelectric-power output and exports were dealt a serious setback in early 1992, when the state of New York backed out of a 21-year contract to buy as much as 18,000 megawatts of electricity—10 percent of the state's consumption—beginning in 1995. The hydro project had become increasingly controversial over the past few years, with environmentalists protesting the flooding of more than 1,700 square miles (4,400 square kilometers) of pristine wilderness in the James Bay area, Cree Indians protesting the loss of their traditional hunting and fishing grounds, energy demand in the United States falling far short of projected levels, and numerous other legal difficulties.

NUCLEAR POWER

Nuclear power produced a record 22 percent of the country's electricity in 1991, enough electricity to light 64 million homes for a year. In spite of the record output from the nation's 111 licensed plants, 1991 was a troubled year, with a fire at the Maine Yankee plant, a power failure at a New York plant, the decision to permanently shut down a plant in California and one in Massachusetts, and the increasing national opposition to low-level radioactive-waste-disposal sites as deadlines mandated by federal law approached, and potential locations were being picked in more than a dozen states.

With the breakup of the Soviet Union, some scientists expressed concern for the safety of the former country's nuclear-power plants. Some experts even warned of another Chernobyl, as similarly designed plants were said to be receiving less maintenance, there was a shortage of parts and equipment, sabotage was on the rise, and supervision and standards were slipping. An accident at a Chernobyl-like Russian nuclear plant near St. Petersburg in March 1992 raised fears internationally, but later reviews confirmed that its problems were far short of those experienced more than five years earlier at its sister plant.

Using cleaner-burning gasoline may help reduce the levels of automotive pollutants in the atmosphere.

ALTERNATIVE-ENERGY SOURCES

Although there was much disappointment that the proposed National Energy Policy continued the trend of reducing funding for research, there was progress on several fronts. Fuel cells, which combine hydrogen and oxygen to produce electricity, developed into a usable technology in the 1960s. They have become more efficient and more cost-effective, particularly when this clean energy source is compared to other sources with their expensive pollution controls.

In October the U.S. Department of Energy (DOE) announced a four-year, $260 million project—half from DOE and half from industry—to develop new batteries for electric vehicles. A study in Cologne, Germany, however, warned that if electric cars were to replace current internal-combustion vehicles, pollution could worsen, as many new electricity-generating plants would have to be built.

In August, three California utilities announced plans to build a $39 million, 10-megawatt solar plant that will use mirrors to focus the Sun's rays on molten salt, which in turn will be used to heat water to steam to drive a turbine generator.

In the absence of significant federal support, many electric utilities and industry groups took the lead in innovative projects aimed at developing existing and new alternative-energy technologies. Utilities also stepped up ongoing efforts to encourage and provide incentives for conservation, both as a means of reducing pollution and as a means to avoid having to build expensive new plants.

Anthony J. Castagno

REVIEW

Environment

Now that the use of drift nets has been internationally renounced, many fishermen have returned to more traditional types of nets to catch fish.

U.N. CONFERENCE ON THE ENVIRONMENT

"Can the Whole Planet Get Its Act Together?" A more appropriate keynote for the environmental situation in 1992 could hardly be found than that April *Los Angeles Times* headline. Twenty years after some 160 of the world's nations scrutinized global environmental problems at a historic United Nations conference in Sweden, they prepared to gather again in Rio de Janeiro (Brazil) in June for more-positive action: international commitments to reforms aimed at averting global catastrophes and fortifying the Earth's quality of life. President Bush headed up the U.S. delegation.

Two far-flung problems dominated months of preconference deliberations: the possibility of planet-wide warming because of accumulation of waste gases; and the continued shrinking of the stratospheric ozone layer that protects the Earth from excessive ultraviolet radiation. Action to stem ozone depletion, now observed from the Antarctic to the North Pole, is already under way as major nations commit themselves to stopping the emission of chlorofluorocarbons, the synthetic industrial chemicals believed mainly responsible for the disintegration of the ozone layer.

The "global-warming" situation poses more complexities because this phenomenon is attributed mainly to accumulated carbon dioxide, the universal residue of combustion. Still to be threshed out was international agreement on limiting emissions of carbon dioxide; while some countries had endorsed controls, the United States, the largest source of such emissions, had balked at subscribing to limits so as not to antagonize industry.

Further impeding global environmental accord was the insistence by the smaller Third World nations that they deserve compensation from the industrialized "North" for any constraints on their development occasioned by pollution controls. A draft declaration endorsed by prospective participants in April approved financial assistance to the Third World. But it left unresolved the financing of worldwide environmental rehabilitation, the extent of technical assistance to underdeveloped countries, and the protection of the world's forests.

INTERNATIONAL ACCORDS

But the lack of global advances did not preclude progress on many individual fronts. The 12-nation European Economic Community (EEC) ordered, subject to parliamentary ratification, sharp reductions in the sulfur content of all diesel and heating fuels within 30 months. Sulfur-combustion residues have been blamed for endemic regional corrosion and for damaging as much as 75 percent of the forests in Europe.

In a major policy shift, the United States dropped its opposition to a 38-nation treaty that would ban mining in the Antarctic for 50 years.

A series of international accords was reached among both European and Pacific nations to ban or restrict excessive commercial exploitation of marine life, including dolphins, whales, bluefin tuna, sea turtles, and salmon. Several nations renounced the widespread use of drift nets, which extend for as much as 30 miles (48 kilometers) and indiscriminately kill all the creatures they envelop.

THE BIOSPHERE

In a unique ecological experiment, four men and four women sealed themselves off from the outside world in a 3.15-acre (1.2-hectare) glass-and-steel "greenhouse" in the desert at Oracle, Arizona. The chamber was outfitted with 3,800 species of animals and plants, along with water and other basic chemicals designed to produce a self-sustaining natural environment. The "Biosphere II" experiment was projected to go on for two years at a cost of as much as $150 million; major initial funding came from the Texas Bass family, but the project was expected to possibly produce a wide variety of commercial spin-offs, along with revenues from sightseers allowed to peer in at the experiment from the outside.

DOMESTIC ISSUES

In 1991 and 1992, the United States erupted in a welter of controversy on environmental issues ranging from towering forests to tiny birds and plants. Combative partisans by turns urged placing human interests above those of biota, preserving humanity's natural heritage, or working out effective compromises. The diverse views were debated endlessly in the courts and administrative tribunals, with few definitive solutions. In 16 states, voters were asked to decide a total of 28 environmental initiatives proposing regulations on subjects ranging from surface mining to glass recycling.

While communities from coast to coast instituted systems for recycling household refuse, the overall problem of disposing of society's wastes seemed as formidable as ever. Local disputes proliferated over the siting of landfills and incinerators. Surveys showed that trash from at least 43 states was being shipped to other states. Some materials intended for recycling, particularly newspapers and glass, simply piled up in small mountains, with no takers. One waste load from Long Island was traced through a five-day journey to Illinois. Nor was the waste problem by any means limited to the United States. Shiploads of old newspapers moved to the Netherlands; Argentina found itself the recipient of plastic wastes from as far away as Germany. Solid waste was going through the same phase once seen with air and water: simply moving pollution from one place to another.

Meanwhile, no progress at all was registered in developing a viable repository for high-level radioactive wastes from nuclear plants, which need to be isolated for eons. The federal government has designated Yucca Mountain in Nevada as the likeliest site, but scientists said its geological suitability might not be determined for several years yet.

CALIFORNIA DROUGHT

Extraordinary rainfalls finally dropped a moist curtain on the five-year drought that had beset California and some other sections of the West. But authorities cautioned that it would take several more years of exceptional precipitation to bring long-parched reservoirs up to historic reserve levels.

Even as the rains sluiced down, the seaside city of Santa Barbara, severely afflicted by the extended drought, completed construction of a $30 million desalination plant to convert seawater to fresh water. It will be used as a backup source for water in future shortages.

Gladwin Hill

Bedecked in futuristic uniforms, eight "Biospherians" sealed themselves off from the outside world in what is hoped to be a self-sustaining environment.

REVIEW

Geology

PLATE TECTONICS

Two geologists announced a controversial new theory about Earth's wandering continents. They proposed that North America was connected to Antarctica for more than 1 billion years—almost a quarter of the planet's history. The two researchers developed the idea after noticing striking similarities between rocks on the two continents that now sit on almost opposite ends of the globe. The rock evidence suggests that North America collided with the ancestral Antarctic continent sometime before 1.6 billion years ago, a time when only single-celled organisms populated the Earth. The two continents remained side by side for eons, and then split apart around 600 million years ago, about the time when the first multicellular animals appeared. Geologists have long suspected that North America connected to many other continents as part of a giant landmass that existed from 600 million to 800 million years ago. But most believed that the western end of North America had attached to some portion of Siberia. The new theory holds that western North America actually bonded to Australia and Antarctica.

Before it split away 600 million years ago, Antarctica may have been part of North America's western frontier.

VOLCANIC CHAMBER

Scientists drilled the deepest hole ever in the ocean floor, attempting to reach into the lowermost section of the crust, a region that represents the fossilized remains of ancient volcanic chambers. Scientists want to reach these chambers because they form part of the seafloor spreading system that creates the ocean crust covering two-thirds of the planet. This hole, located near the Galápagos Islands, has been drilled on and off for 12 years. But scientists had considered abandoning this effort because drilling equipment had become stuck in the hole. The recent expedition cleared the clogged equipment and deepened the hole to within a few hundred yards of the volcanic chambers. Oceanographers are planning future missions to reach that goal.

ARGENTINE CRATERS

Geologists working with an Argentine fighter pilot identified a chain of craters formed during a rare type of meteorite impact in Argentina sometime in the past 2,000 years. The chain of oblong craters apparently formed when a meteorite hit the ground at a shallow angle, breaking into chunks that ricocheted forward. Planetary scientists have seen craters from this type of impact on other planets, but they have never before found an example of a glancing impact on Earth. The researchers who identified the Argentine craters estimate the meteorite was 490 feet (150 meters) in diameter and was traveling at roughly 15.5 miles (25 kilometers) per second, equivalent to 55,000 miles (88,500 kilometers) per hour. Because the craters have not yet eroded significantly, the crash apparently occurred in the recent geologic past, during a time when people inhabited this region and could have witnessed the fireball and its cataclysmic impact.

Richard Monastersky

Oceanography

HEARD ISLAND EXPERIMENT

On January 26, 1991, oceanographic history was made when an underwater acoustic signal sent from the southern Indian Ocean was received on the Atlantic and Pacific coasts of the United States, as well as at many receiving stations around the world. These initial results were just the beginning of the Heard Island experiment—a prelude for the annual measurement of oceanic sound waves that could eventually document both the existence and extent of actual global warming.

The selection of the Indian Ocean's Heard Island as the site of the acoustic-transmission experiment was a deliberate one. Not only does the island's vicinity provide an underwater "view" of both U.S. coasts, but in the deep environs surrounding the island, the sound channel through which the experimental signals are sent is at an ocean depth of 3,000 feet (915 meters)—where water temperatures remain relatively constant.

The constant temperatures are crucial to the experiment's long-range success. Because the speed of sound in the ocean is greatly affected by water temperatures along the sound path, and because sound travels faster in warmer water, even infinitesimal changes of ocean temperatures caused by global warming could be detected if continuous measurements are taken over a period of years.

The specific measuring technique being used in the experiment is acoustic tomography, where sound waves are generated, then detected, and their motion is computed to yield a picture of the vast ocean spaces between the sound waves.

DEEP DIVES

A unique deep-sea video camera developed by Woods Hole Oceanographic Institute in Massachusetts was able to take the clearest pictures yet of the sunken luxury liner *R.M.S. Titanic.* On March 8, 1992, another prized Woods Hole research tool, the 25-foot (7.6-meter) *Alvin* (one of only six deep-diving manned submersibles in the world) made its deepest dive yet: 8,300 feet (2,536 meters) down to the Pacific Ocean's floor along the East Pacific Rise (about 9 degrees north latitude) off the Mexican coast. The East Pacific Rise is part of the 46,000-mile (74,000-kilometer) Mid-Ocean Ridge, an underwater mountain range that circles the planet like the seam on a baseball—and is the largest geologic feature on Earth.

THE OCEANS AS REGULATORS

Significant 1991 results also revealed that variations in the Earth's rotational speed are caused by ocean currents—along with melting ice sheets, winds, and disturbances in the planet's liquid-metal core. According to a report published in *Science* (August 9, 1991), the length of a day has diminished by an average of 1.4 milliseconds each century, although it was found that short-term fluctuations could shorten a day by as much as 5 milliseconds.

More controversial was an article published in the British journal *Nature,* presenting corroborating oceanic evidence that the Earth's climate, through its complex ocean-atmosphere system, actually regulates itself, enabling global temperatures to remain within certain broad limits, thus diminishing the threat of a runaway greenhouse effect.

Dr. Veerabhadran Ramananthan, a climatologist at the Scripps Institute of Oceanography at La Jolla, California, showed that as sea temperatures grew warmer, huge, towering cumulonimbus clouds were formed. As they penetrated cold atmospheric regions, the thunderclouds' tops became so thick that they blocked sunlight from reaching the ocean—thereby creating a natural regulatory thermostat.

Gode Davis

Seismology

A particularly deadly earthquake struck northern India near the border with Nepal on October 20, 1991. With a measured magnitude of 7.1, the quake shook the foothills of the Himalayas, killing over 2,000 people. This region of the world experiences frequent earthquakes because it sits in the middle of a geologic collision zone. The Indian plate is crashing into the Asian plate, moving northward at a rate of 2 inches (5 centimeters) per year. That compaction generates stress in the crust, which builds the Himalayan Mountains and causes earthquakes.

A strong jolt, measured at magnitude 7.0, hit the Republic of Georgia in the former Soviet Union on April 29, 1991. The seismic shock killed at least 114 and left 67,000 people homeless. Like the Himalayas, the Caucasus Mountains are being raised by a collision between two great plates. Northern Peru suffered a magnitude-6.7 earthquake on April 5, 1991, which killed at least 100 people. At least 47 people died in a large earthquake that shook Costa Rica and neighboring Panama on April 22, 1991. The quake measured a magnitude of 7.6.

The largest quakes in the vicinity of North America struck off the coast of northern California and southern Oregon. Three shocks ranging in size from 6.2 to 7.1 occurred about 620 miles (1,000 kilometers) off the coast near Crescent City, California, during July and August of 1991. On August 17, a magnitude-6.0 jolt hit onshore near Cape Mendocino.

A magnitude-5.8 quake awoke residents of the Los Angeles area on June 28, 1991. This shock, centered under the San Gabriel Mountains, was the sixth earthquake to strike the region since 1987. Over that period, seismic activity has moved northward across the San Gabriel Valley toward the mountains. Seismologists are concerned this trend could foreshadow a larger earthquake in the area.

Volcanology

The Hudson Volcano in southern Chile erupted in mid-August. The volcanic cloud reached over 11 miles (18 kilometers) into the atmosphere and deposited ash up to 620 miles (1,000 kilometers) southeast of the volcano. Debris from the eruption caused respiratory ailments in people and damaged automobile engines and other mechanical devices.

The eruption of Mount Pinatubo in the Philippines last year spewed millions of tons of sulfur into the stratosphere, where it formed tiny droplets of sulfuric acid. The droplets have accelerated the destruction of the protective ozone layer. Because the droplets also reflect sunlight, they have cooled the globe slightly. The volcanic sulfur should slowly fall out of the atmosphere in the next year or two.

The eruption of Mount Pinatubo strewed large regions of the Philippines with its fallout, causing considerable hardship for the people and great economic damage.

Geologists studying ancient lava flows in Siberia believe that these massive volcanoes may have caused the greatest set of extinctions in Earth's history. The Siberian lava deposits date to the time, 248 million years ago, when 96 percent of all ocean species and most dominant land animals vanished—clearing the way for the world of the dinosaurs.

Richard Monastersky

Weather

Once again in 1991, Americans saw more of what they have come to expect in terms of weird and wacky weather across the country. Over the past few years, it seems that the weather simply can't make up its mind; this past year was no exception.

The year ushered in some colder-than-normal temperatures, particularly across the Pacific Northwest. The California drought continued unabated, making headlines as it went into its fifth year. The central part of the United States remained dry, despite the unusually powerful thunderstorms and tornadoes that swept through the area in mid-January. Vacationers in Florida enjoyed record heat, while the Northeast remained tranquil. The Midwest had an abrupt taste of summer during February, with International Falls, Minnesota, heating up to 55° F (12.7° C) on February 2. Three days later, in the East, the Big Apple reached 70° F (21° C).

SPRING 1991

Even the most avid weather watcher had trouble telling when spring began in the East, especially with places like Boston having just experienced their warmest winter in 121 years. Early spring often brings healthy doses of Arctic air, but such was not the case in 1991, as the northern jet stream stayed unusually far north. This trend would prove to be a harbinger of things to come for the Northeast. Elsewhere, welcome rains graced California during the spring, helping to temporarily quench the very parched earth. Of course, all was not warmth and rains: parts of Wyoming saw 80 inches (200 centimeters) of snow during April; that same month, Yankton, South Dakota, broke records with a high temperature of 93° F (33.8° C).

The end of spring turned out to be the warmest on record for many places throughout the Ohio Valley and into the Northeast. A deep trough across the West and a strong ridge in the East produced a warm spring in the East and cool weather across the intermountain West.

SUMMER 1991

The heat and humidity that dried things out across much of the West during the early summer gradually spread into Kansas and Oklahoma, producing the most arid conditions since the devastating drought of 1988. The Far West, meanwhile, remained cool while the East sweltered. The hot, rainless weather produced scattered areas of drought across southern New England, where Hartford, Connecticut, saw temperatures soar over the century mark three days in a row—the first time ever. Later in July the weather seesawed, with the East enjoying cooler temperatures while the West soared above 90° F (32° C). Unusual as it seemed, July 1991 brought rain to Southern California; Los Angeles received over 0.1 inch (0.25 centimeter) of rain, a record for that city.

As summer drew to a close, a major hurricane slammed into New England, the worst such storm since 1938 in some places (see page 107). The heat continued in the mid-Atlantic area, bringing the hottest August ever to that region; ultimately the steamy weather spread to the Plains, where residents sweated out temperatures some 12° F (8° C) above normal.

Occasional thunderstorms offered only temporary relief from the heat that enveloped much of the East during the summer of 1991. Some of the storms caused much personal property damage.

A powerful coastal storm created very high tides along the Northeast coast on Halloween 1991, causing much shoreline erosion. The crashing waves nearly demolished the seaside home of President Bush (shown at left with First Lady Barbara Bush inspecting the damage) in Kennebunkport, Maine.

FALL 1991

Although it was generally warm during September, alternating weather patterns across the lower 48 states gave Americans a taste of things to come, while not letting anyone forget just how hot the summer had been.

On September 16, 30 cities set record high temperatures; three days later, 50 cities set record lows. By month's end, snow had fallen across northern Maine, a sure sign that the cooler weather had started to win its battle against summer's heat. October was another warm month. Temperatures in Phoenix ended up reaching 100° F (37.7° C) or higher during the first 11 days of the month; to the north, in Montana, temperatures fell below 0° F (−17.7° C) on numerous occasions.

The end of October brought the Northeast a great coastal storm on Halloween. The large ocean storm developed across the Canadian Maritimes and then moved southwest, prompting surf advisories to be issued from Maine to Florida. Atlantic City recorded its second-highest tide ever from this storm. President Bush's summer home in Kennebunkport, Maine, sustained heavy damage.

November continued the stormy trend, with snow and cold across the Midwest. The Minneapolis–St. Paul area had a remarkable 55 inches (140 centimeters) of snow during the month of November alone! From the last three days of October through November 11, 640 record lows entered the record books over the eastern three-quarters of the United States.

WINTER 1991–92

The year ended with drenching rains across Texas, where flooding caused millions of dollars in damage. Rain was measured in feet across portions of the state, where the deluge forced people from their homes for the holidays. Pacific storms brought much-needed rain to Southern California, although the heavy precipitation after years of drought produced flooding and even destroyed some homes. Arctic air made some minor incursions south before Christmas, then retreated north for part of January. The East thus continued its string of mild months. For southern New England and parts of New York, the winter was virtually snowless.

El Niño, a periodic warming of the waters off the coast of Ecuador, made an appearance in late 1991 and continued into 1992. Many experts feel that this phenomenon, thought to be a contributor to global-weather fluctuations, may have led to the storminess across California and the mild winter in the East.

WORLD WEATHER

Elsewhere in the world, extreme weather made news on virtually every continent. In April 1991, a devastating cyclone hit the poverty-stricken country of Bangladesh. Winds of 146 miles (235 kilometers) per hour buffeted the port of Chittagong. Flooding and winds claimed the lives of over 140,000 people and 1 million cattle, and destroyed 1.4 million houses.

Residents of Pakistan and western India endured an early-June heat wave, with temperatures climbing as high as 122° F (50° C). Early summer brought torrential rains to China, where many homes were destroyed and millions of people had to be evacuated. Over 2,000 people lost their lives.

The eruption of Mount Pinatubo in the Philippines led meteorologists to speculate as to what influence on climate the dust that was spewed into the atmosphere would have. One thing for sure: the rains that followed the eruption caused numerous mud slides and hundreds of deaths.

Australian flooding, European snowstorms, and "black rain" caused by Kuwait's oil fires were just some of the other bizarre weather phenomena reported around the world in 1991.

THE 1991 HURRICANE SEASON

According to Dr. Richard Pasche at the National Hurricane Center in Miami, Florida, 1991 was a quieter-than-normal season for hurricanes. When all was said and done, the season produced eight storms, four of which evolved into hurricanes.

Hurricane Bob was the first and most talked-about storm of the season. It developed near the Bahamas in mid-August, then rode north on a powerful jet stream, striking land near the Buzzards Bay area in Massachusetts. Winds of up to 120 miles (193 kilometers) per hour struck the Rhode Island coastline; widespread damage was reported. Portions of Cape Cod were also devastated. Shoreline homes as far north as Maine were affected, prompting officials to declare states of emergency in many areas. Bob would prove to be the only hurricane to make landfall in 1991 (usually two hurricanes a year hit the U.S. mainland). Claudette in September and Grace in late October threatened only ships at sea. Grace would eventually become absorbed by the large "extratropical" storm in the Canadian Maritimes that went on to hit the Northeast coast. Interestingly, as that storm moved away from the coast early in November, it developed into a small, compact "no-name" hurricane before dying in the Atlantic.

WEATHER HIGHLIGHTS 1991

FEBRUARY
- **2:** International Falls, Minnesota hits 55° F.
- **5:** New York City hits 70° F.
- **27:** Los Angeles receives 2.17 inches of rain, most in four years.

APRIL
- Parts of Wyoming receive 80 inches of snow.

JULY
- **19–21:** Hartford, Connecticut, tops 100° F three days in a row.

AUGUST
- **18–19:** Hurricane Bob hits New England.
- Hottest month ever in mid-Atlantic region of U.S.

SEPTEMBER
- **30:** Northern Maine receives snow.

OCTOBER
- **1–11:** Phoenix tops 100° F each day.

NOVEMBER
- Minneapolis receives 55 inches of snow.

All of the storms in 1991 formed north of 25 degrees latitude. This is highly unusual, as it meant that no hurricanes developed in the "true" tropics. Furthermore, no named storms developed in the Gulf of Mexico. The last time such a season occurred was in 1962; before that, you'd have to refer to weather records from the 1920s to find such a year.

Some scientists feel that the abnormally quiet hurricane season may be attributed to the El Niño phenomenon and abnormally strong westerlies that tend to rip the storms apart before they get a chance to develop.

David S. Epstein

HUMAN SCIENCES

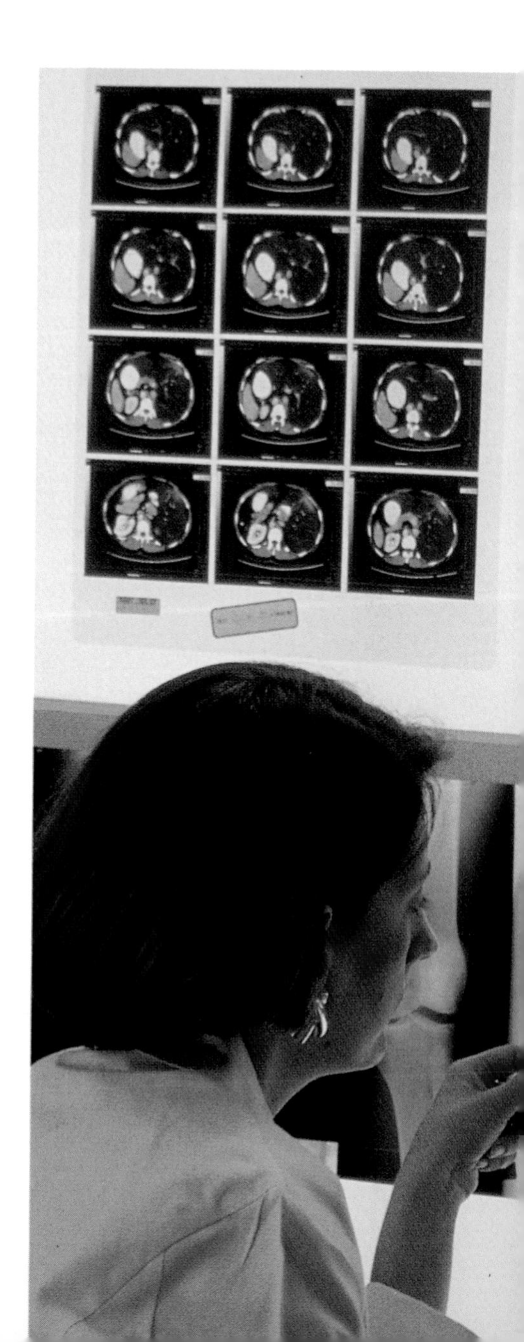

CONTENTS

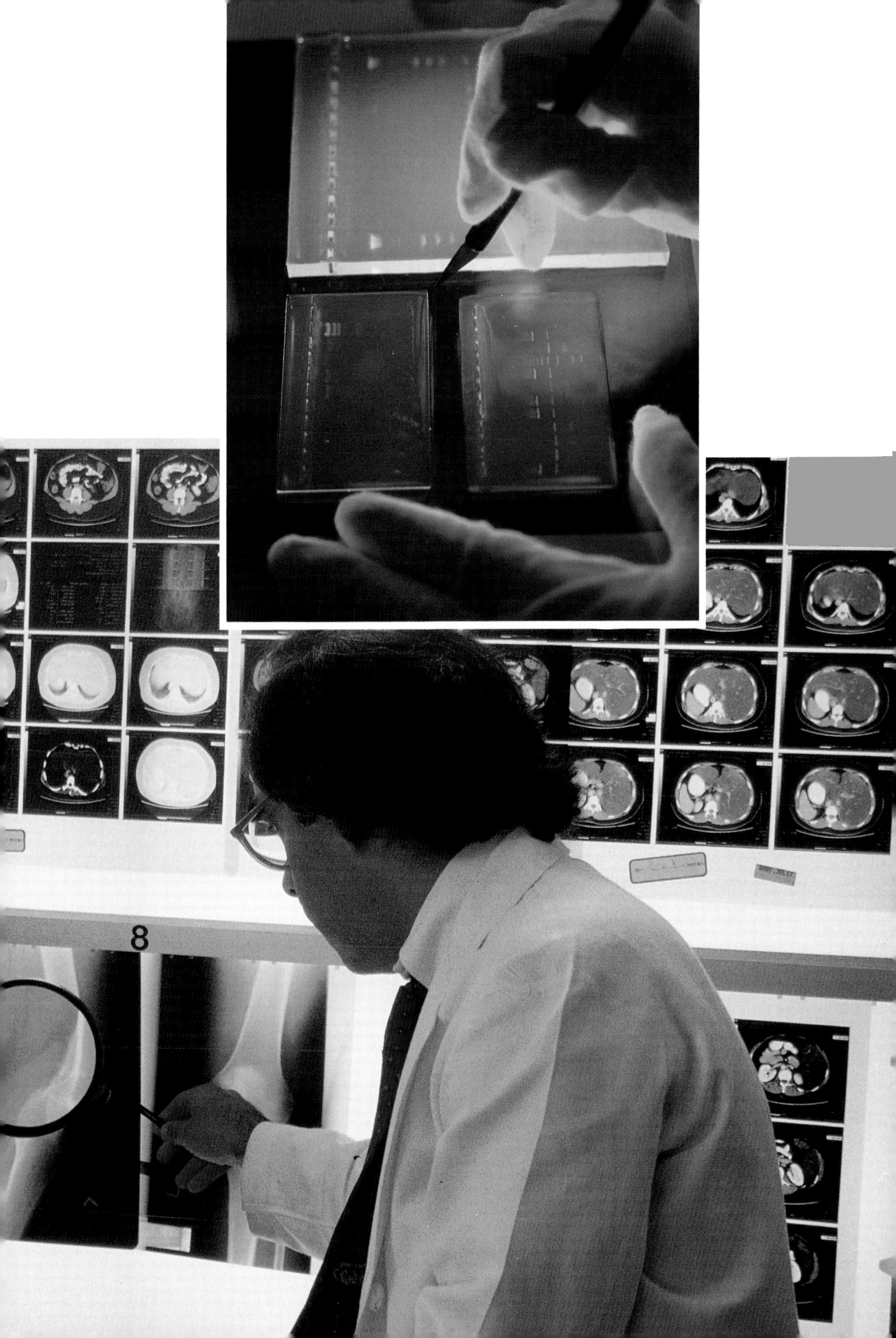
8

THE GOLDEN HOUR:
Advances in Emergency Medicine

by Wendy J. Meyeroff

- *A 16-month-old child dies of cyanide poisoning after swallowing the solvent his mother used to remove her artificial fingernails.*
- *A boy punctures his thigh when his wood-carving chisel slips; his older brother, upon seeing the blood, faints backward, splitting open his scalp on the doorjamb.*
- *A 20-year-old comes into an emergency room with a stab wound and collapses. Opening up the patient's chest, the physician finds the wound has penetrated the heart, but the physician manages to plug the hole with a finger until a surgeon arrives.*
- *An Amtrak train carrying more than 600 people crashes into the rear of three Conrail freight locomotives in Chase, Maryland, strewing wreckage for miles.*

The above scenarios—all true—don't even begin to represent the variety of cases that fill America's emergency facilities each day. Approximately 92 million people seek care in the country's emergency rooms every year for problems ranging from simple bruises to life-threatening multiple injuries.

Yet most people are astonished to learn that emergency medicine has been available for treating civilians for less than 30 years, though military efforts (albeit often primitive) go back much further. For example, Napoleon's chief surgeons provided "ambulance service," which consisted of throwing soldiers injured at the front into a covered cart and dragging them back to doctors in the rear. Florence Nightingale brought some order to trauma nursing back in 1854 by insisting on cleaner conditions and beginning an accurate record-keeping system of supply inventory, admissions, and deaths. The first medical air evacuation was reportedly the removal of 160 wounded soldiers by hot-air balloon during a siege on Paris in 1870.

In fact, most of today's practices in emergency medicine began on (or at least close to) the battlefield. The process of *triage*, for example, which comes from a French word meaning "to sift," began in wartime. Army field medics and surgeons, working under frenzied conditions, had to separate the patients who were most badly wounded and who needed treatment immediately from those who could wait until battlefield conditions were more stable.

In medical emergencies, patients have the best shot at survival if they receive proper treatment within 60 minutes—the so-called golden hour.

Another principle adapted from the battlefield was getting aid to the patient as quickly as possible. Solutions ranged from sending trained medics into the field so as to stabilize the wounded at the scene, to improved methods of transport, like the helicopter pickup and delivery depicted so well in the movie and television show *M*A*S*H*. Between World War I and Vietnam, the lag time between injury and surgery dropped from 18 hours to a mere 65 minutes.

The Golden Hour

Yet while the military was making great advances, the civilian side of emergency medicine was in poor shape. In the 1960s more people were dying on U.S. streets than in wars. Finally, in 1966 Dr. R. Adams Cowley wrote a paper called "Accidental Death and Disability: The Neglected Disease of Modern Society." That report included a concept that has since become a standard in emergency medicine in general—*the golden hour*.

Just as Army medics had realized the importance of decreasing the time it took for a patient to receive treatment, Dr. Cowley felt that for a patient to have the best chance of survival, he or she must receive definitive care within 60 minutes. More specifically, doctors define the golden hour as the *total* amount of time that should elapse from initial examination (whether it's out in the field by a paramedic or in the hospital's emergency room, or E.R.) to actually wheeling the patient into surgery. Today some experts in emergency medicine talk about the "platinum half hour," generally agreeing that the patient whose condition is *immediately* life-threatening—approximately 5 to 7 percent of emergency patients—should be seen by *experts* within *20 to 30 minutes* of injury.

Many advances in emergency medicine were first developed on the battlefield. Between World War I and the Vietnam war, the lag time between injury and surgery dropped from 18 hours to just 65 minutes.

It is critical to emphasize the word "experts." A secondary principle of the golden hour is not just getting the patient to the hospital but rather getting the patient to the *right* facility with the specific experts and equipment needed. At the time of Dr. Cowley's report, this was a revolutionary idea, since it wasn't unusual to find a heart-attack victim being treated by a dermatologist who just happened to be pulling E.R. duty that night.

To help ensure better patient care at the emergency-room level, the American College of Emergency Physicians (ACEP) was founded in 1968 (though emergency medicine wasn't officially recognized as a major medical specialty until 1979). The ACEP established protocols for training doctors specifically in the area of emergency medicine. These specialists are taught to properly evaluate, stabilize, and initiate treatment for emergency-room patients with injuries that threaten life and limb.

Emergency Situations

So what are the conditions most likely to require emergency help? Each year, there are numerous drownings, poisonings, amputations, and other emergencies, but in general the two problems that most commonly result in a visit to the E.R. are trauma and heart attack. Between them, these two emergency situations annually cause the premature deaths of 1 million Americans, and the disability of another million.

Dr. John Barrett, director of the Trauma Center at Chicago's Cook County Hospital, notes that trauma is "the leading cause of death in Americans, particularly American males, under 47 years of age." One in every five Americans suffers from some sort of traumatic injury, with falls being the leading cause of nonfatal injury in the United States (close to 800,000 hospitalizations, and many more cases that didn't require hospitalization). Motor-vehicle accidents are the leading cause of fatal traumas, resulting in approximately 46,300 deaths annually.

The American Trauma Society defines trauma as "any physical injury caused by violence or other forces," and it includes car crashes, falls, strangulation, and knife or gun wounds. Trauma cases are classified as either *blunt trauma* or *penetrating trauma*. In the latter, some type of foreign object penetrates the body, most commonly a knife or a bullet. Not surprisingly, urban facilities tend to see more instances of penetrating trauma; for example, at Cook County, 50 percent of the 450,000 patients seen each year have been either shot or stabbed.

Blunt trauma occurs when a person hits, or is hit by, something. Motor-vehicle crashes are blunt-trauma cases. The odds of survival of victims in rural versus urban crashes underscores the advantages of quick access to trained care: although numerically, more car crashes occur on urban roads, approximately two-thirds of the deaths occur on rural highways, simply because the victims tend to be farther from help, or, more precisely, from specialized trauma care.

Heart Attack

The second-most-devastating emergency situation is a heart attack. Heart attacks claim 1.5 million victims every year, half of whom die. Of those who die, about half are victims of a phenomenon called *Sudden Death Syndrome*, the unexpected short-circuiting of the heart's electrical system. Although this syndrome is lethal within four or five minutes, timely intervention can markedly increase the chances of survival.

Two of the greatest advances in increasing survival rates in cardiac patients are CPR (cardiopulmonary resuscitation) and defibrillation. CPR is a method of manually compressing the chest to keep oxygen-rich blood flowing to the heart muscle, combined with mouth-to-mouth resuscitation, until more sophisticated methods of intervention can be applied. Defibrillation uses an electrical current to shock the heart and, in doing so, restore its normal rhythm.

Emergency Treatment in the Field

The idea of providing in-the-field care is another idea adapted from the military. Since the late 1960s, the concept of the "prehospital provider" (generally based on the old Army field medic) has become an integral part of American emergency care. Nationally, the Emergency Medical Service (EMS) receives a call every other second. Approximately 20 million to 25 million people annually receive care from EMS personnel.

There are essentially three levels of emergency-service technicians: *first responders*, the people who tend to be the first on the scene, such as fire fighters and police officers, and then two levels of *emergency medical technicians* (EMTs). EMT-B's are trained in *Basic Life Support* (BLS), which is generally defined as being able to provide CPR and mouth-to-mouth resuscitation. EMT-P's (paramedics, also called EMT-A's) have more-extensive training known as *Advanced Life Support* (ALS), which permits them to administer certain drugs and even use special equipment such as defibrillators.

In June 1980, another major breakthrough in emergency care occurred: the American College of Surgeons adopted a program known as Advanced Trauma Life Support (ATLS). The two-day course provides physicians with basic guidelines to follow when evaluating an emergency patient's injuries. These guidelines remind them to first check the "ABCs." Dr. Carlos Flores, assistant director of Emergency Medical Services at New York City's Bellevue/NYU Medical Center, explains that these are the three most important steps to follow in treating any medical emergency: *(a)* to make sure that the patient has a viable *Airway*, *(b)* that he or she is *Breathing* through that airway, and *(c)* that his or her *Circulation* is good.

Once these steps are performed, ATLS provides guidelines for a quick, primary observation of the patient, and then a more

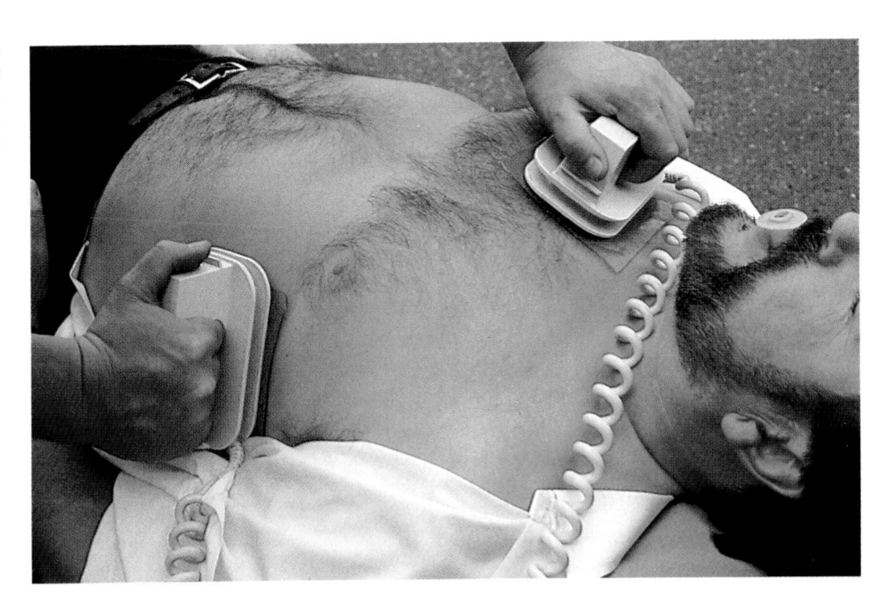

Defibrillation, an emergency procedure in which a brief electric shock is administered to the heart, has greatly reduced deaths due to heart attack.

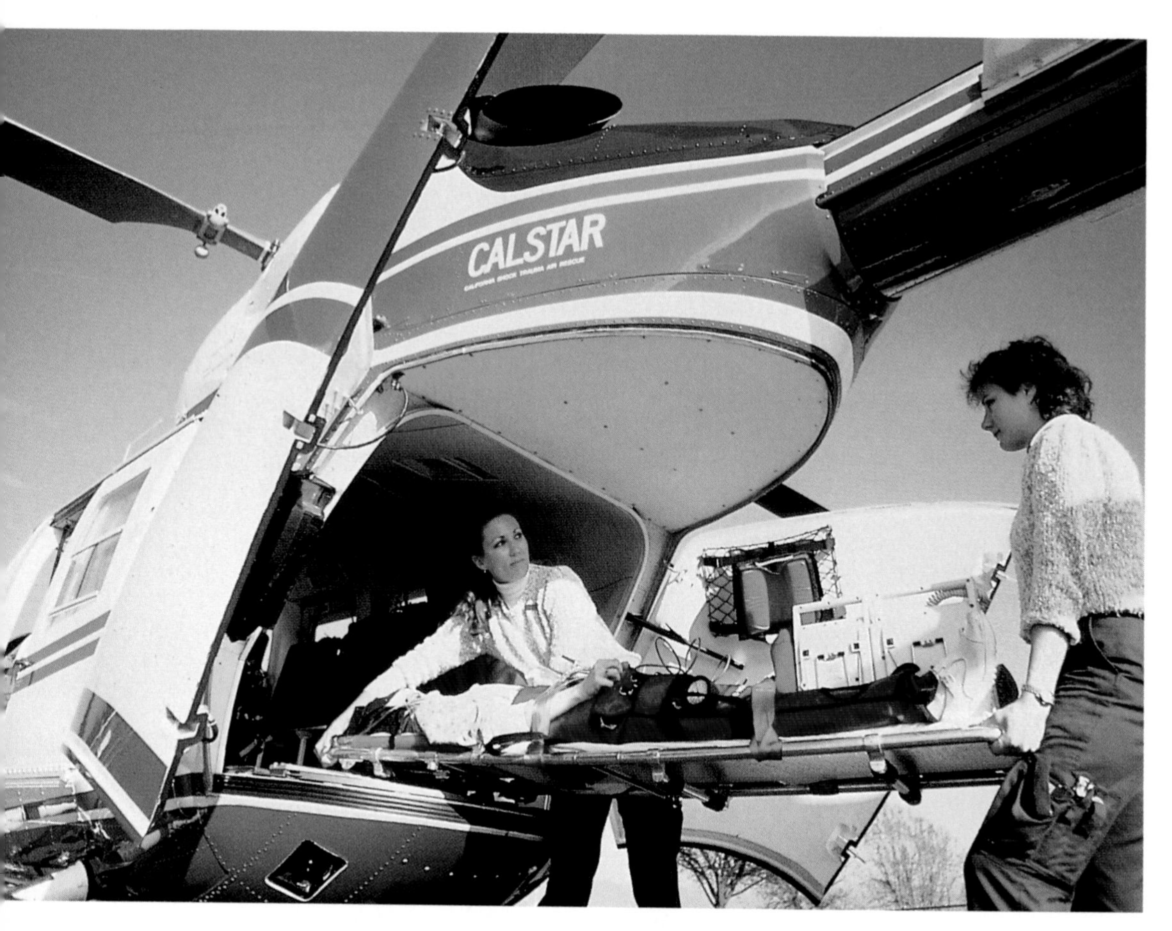

Helicopters are used to whisk the critically injured from an accident in an isolated area to the nearest hospital, or to transport a patient from a local emergency room to a burn center or other specialized facility.

thorough, secondary observation. For example, in a preliminary observation, it might be immediately apparent that the patient has a fractured pelvis, but it might take a secondary look before it is noted that the fractured bone is causing abdominal bleeding, which requires immediate treatment.

Getting a patient to the correct facility also presents certain transportation challenges. To increase a patient's survival odds, trauma centers use two methods of transport: the "Y-Model" is typically used in a rural situation. The patient is sent first to the nearest medical facility (such as the local E.R.) for stabilization and evaluation, and then transferred to the specific type of care facility needed, such as a burn center or pediatric-trauma facility. In an "X-Model" of transport, the patient is delivered directly from the injury site to his final destination.

Emergency-Medicine Advances

Improvements in technology, from equipment to medicines, are constantly increasing the chances of survival for patients in emergency situations. Among these advances:

• *Rapid transfusers* can quickly push fluid into the blood, and thus help stabilize the patient more quickly.

• *Blood warmers* help prevent patients from going into shock when refrigerated blood is pumped into the body.

• *Enhanced 911* lines, including those that record the phone number of the caller in case the victim becomes incoherent or unconscious.

• *End-tidal CO_2 ($ETCO_2$) monitors* provide the ability to ventilate a patient—to make sure he or she is getting enough oxygen and exhaling carbon dioxide (CO_2). In a hospital, one method is to draw *arterial blood gases*

(ABGs) and check the levels of each gas in the blood. $ETCO_2$ monitors provide technicians with a noninvasive (and more accurate) way of determining how well a patient is being ventilated. In early 1992 Samaritan AirEvac in Phoenix, Arizona, started testing the feasibility of using such equipment in airplanes, where there is usually no room to draw ABGs.

• *High-frequency ventilators* increase the survival rates of patients with *adult respiratory-distress syndrome* (ARDS), a common postoperative complication of trauma patients. Of those patients who develop ARDS, 50 to 70 percent die when the lungs swell and stiffen, impeding the flow of oxygen. Dr. James Hurst, director of trauma and critical care in the University of South Florida's Department of Surgery, notes that conventional ventilators work at too high an air pressure for sick lungs to handle; the new high-frequency ventilators deliver smaller volumes of air at less pressure.

• *Thrombolytics*. Some of the most-talked-about advances relating to the emergency treatment of heart problems are in the area of drugs, particularly those that dissolve the blood clots that cause heart attacks. The oldest (introduced in Germany in 1979) is called streptokinase, which is believed to be most effective when administered in the first three hours after attack. In the late 1980s, a drug called tissue plasminogen activator (t-PA) started gaining acceptance. If injected four or five hours after a heart attack, t-PA may save more heart muscle than other drugs. But it is expensive—at about $2,000 a dose, it is 10 times more costly than streptokinase.

Some doctors are beginning to wonder if t-PA is worth the cost, particularly in light of overseas studies that showed a 25 percent reduction in deaths when streptokinase and aspirin are given within a few hours of heart-attack symptoms appearing. A third drug, eminase, is now being tested in the field: a test program in Cleveland, Ohio has equipped EMTs with eminase plus sophisticated cardiac monitors. Experts hope that if these drugs can be administered in the field, mortality rates from heart attacks can be reduced by up to 50 percent.

Emergency medical technicians (EMTs) provide first aid at the scene of an accident. Their efforts often make the difference between life and death for an accident victim.

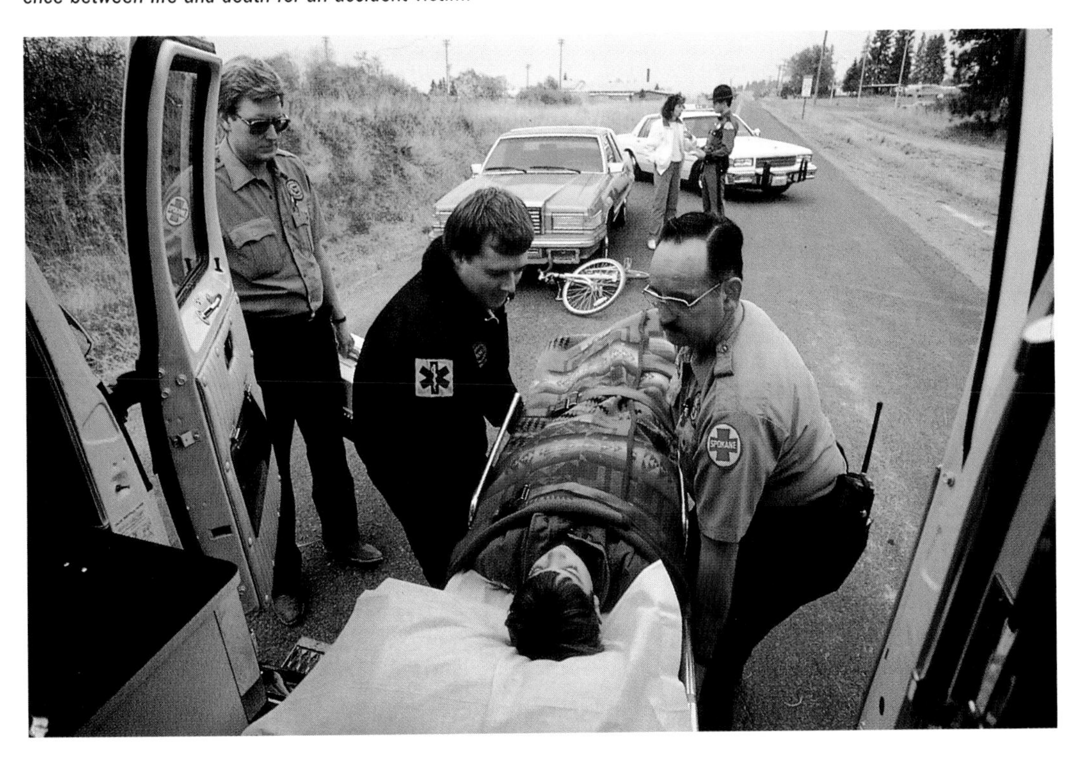

The Crisis in Emergency Medicine

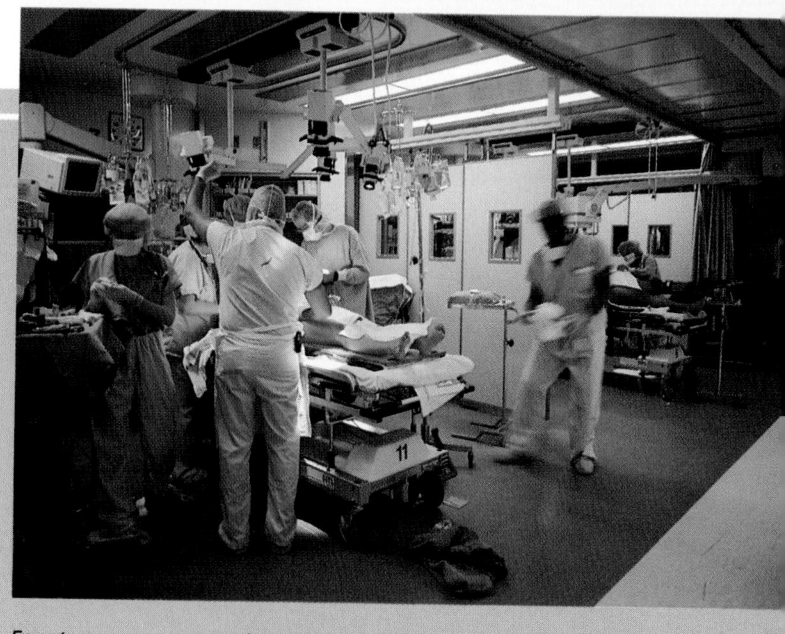

Emergency rooms across the nation are burdened by budgetary restrictions, understaffing, and overcrowding. In some emergency rooms, patients must wait hours before seeing a doctor or receiving treatment.

Emergency medicine has made bold advances in the last decade. But it would be remiss to cite all the marvels of emergency medicine in 1992 without briefly noting the problems. Like health care in general, emergency medicine faces a major crisis as it approaches its 30th anniversary. Emergency departments—from local community rooms to specialized trauma centers—are generally underbudgeted, understaffed, and overworked. In 1965 there were 30 million visits to the short-term hospitals that operated such departments; by 1990 that number had more than tripled to 92 million. This, says the American Hospital Association, occurred at a time when the overall number of hospitals dropped from 7,123 to 6,649. Trauma centers have been hit particularly hard and are showing the strain; since 1983, 66 centers across the country have closed.

• *Exosurf Neonatal* is a new drug helping to fight the pediatric form of respiratory-distress syndrome (RDS). Experts estimate that about 50,000 infants develop RDS each year and 5,000 die. Exosurf supplies a synthetic lung surfactant, the substance that maintains the shape of the small lung sacs. Without surfactant (which the body normally produces), the lungs stiffen and can ultimately collapse.

• *Anti-TNF* is a monoclonal antibody that is currently being investigated for its ability to block the production of *tumor necrosis factor* (TNF). The University of South Florida is one of the institutions investigating anti-TNF. Dr. Hurst explains that after a traumatic injury, the body's white blood cells release TNF to fight infection. Unfortunately, those substances sometimes go haywire and overwhelm the system. It is hoped that anti-TNF will block this abnormal immune response.

• *Epidermal growth factor* is an experimental substance that may speed up the time it takes certain injuries to heal. It might eventually be a great help in treating burn victims and in other patients needing skin grafts.

Nationwide Emergency Systems

Although most emergency centers are likely to see only one or two truly life-threatening emergencies a night, there are two situations that can severely test the system to its limit. The first, and more common, is the Multiple Casualty Incident (MCI). It has been defined as a "situation in which the local system is [temporarily] under-resourced at the scene," but which can eventually be managed. Train wrecks, floods, and car crashes are a few of the situations that are classified as MCIs.

The Maryland Institute for Emergency Medical Services Systems (MIEMSS) is the nation's most sophisticated system for dealing with such incidents on a statewide basis. Begun in 1961, the system now includes 10 trauma centers (including those specializing in shock and pediatric trauma), a Police

The reasons reflect the crisis in America as a whole. Approximately 37 million Americans without health insurance crowd emergency rooms with even simple ailments, using E.R.s as their primary source of all medical care. Even those with insurance—faced with skyrocketing costs for preventive medicine—often wait until a real health crisis forces them in. The facilities are further crowded by the large number of the homeless, some of whom are truly sick, while others pretend to need help so as to find shelter for the night. And the problem is not just an urban one: in a small town of 44,000 people in North Carolina, the chief of emergency service reports 48,000 emergency-room visits a year.

Of course, most of these patients can't afford to pay for the services they need. And the problems are even worse in trauma care—even those with insurance coverage often cannot afford the prohibitive cost of major trauma treatments. Thus, faced with the choice of allowing the trauma center or emergency department to bleed the rest of the facility dry, or closing the "unprofitable" department, many hospitals—albeit reluctantly—choose the latter. That leads to a "Catch-22" situation in which the remaining facilities are further overburdened (patient crowding is so severe that waits of up to 12 hours for a bed are not uncommon).

Prehospital providers, who have made a major impact on survival rates, have also been devastated. When Congress passed the Emergency Medical Services Systems Act in 1973, it was hoped that 304 EMS regions across the country would be established. That dream never materialized, and was eventually decimated by the concept that the federal government should throw the fiscal ball back to the local level. Though a few (about 50) of the original departments still survive, and some—like Houston, Texas, and Seattle, Washington—have actually flourished, by 1981 the system was essentially dead. Today local communities each provide their own emergency response teams.

Unfortunately, it may not be enough. As Dr. John Barrett of Cook County Hospital in Chicago notes, "[Today] 30 to 40 percent of our patients die of preventable causes, like bleeding to death from a ruptured spleen." We can help those patients, he points out; we have the knowledge to save them—but not unless the mechanisms are in place.

Med-Evac program, a statewide emergency-medical-communications system, and programs to train and certify prehospital providers. MIEMSS played a crucial role in patient survival during the Amtrak train crash mentioned earlier; of the 600 people involved, more than 400 were evaluated and processed at the crash scene (at a treatment center set up on-site), and another 175 were transported to hospitals. There were 15 fatalities on-site, and one person died at the MIEMSS Shock Trauma Center.

The second program is one that (fortunately) few Americans have needed to become aware of—the National Disaster Medical System (NDMS). Originally based on military needs, NDMS was converted to a civilian program in 1984. The system has yet to be activated in the United States, since it is designed to be used only in the type of mass casualty in which local medical resources are destroyed or totally overwhelmed. Thomas P. Reutershan, NDMS Task Force director, notes that the system wasn't needed during the 1989 San Francisco earthquake because existing facilities stayed intact and were able to handle the crisis. But if, for example, a recurrence of the 1857 quake in Fort Tejon, California, were to take place, the consequences could be devastating. Back in 1857 the area was essentially unpopulated, so no injuries or fatalities were reported in the quake, which perhaps registered at greater than 8.0 on the Richter scale. But today more than 10 million people live in that area. Estimates indicate that a quake of equal intensity would cause between 3,000 and 14,000 deaths and between 12,000 and 55,000 injuries requiring hospitalization.

All of these components—trained field personnel, new medications and equipment, programs for treating multiple and even mass casualties—constitute one of medicine's giant steps: a true system of emergency medicine.

HAIR: Personal Statement to Personal Problem

by Devera Pine

From the shaved heads of medieval monks, to the long-haired hippies of the 1960s, to the spiked hairdos of today's punk rockers, hair has always made a personal statement.

"It's one of the leading ways people can establish their individuality and express their style," says Jerome Shupack, M.D., professor of clinical dermatology at New York University Medical Center in New York City. "Hair has had sociological importance throughout the ages."

Because of its importance, anything that happens to our hair that we can't control—falling out or turning gray, for instance—can be the source of much anxiety.

In the United States, some 35 million men are losing or have lost their hair from male-pattern baldness, according to the

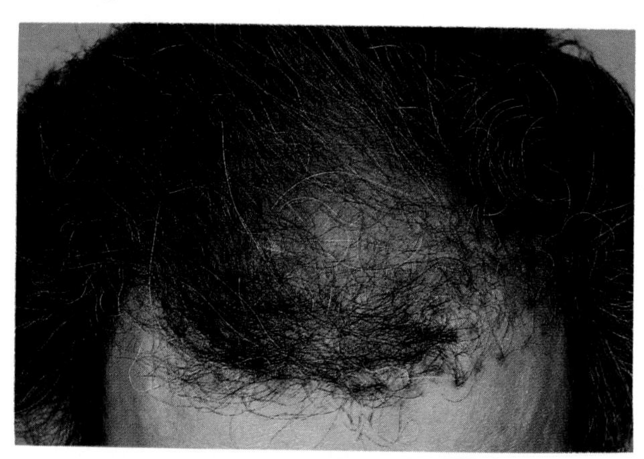

Using hair to express individuality (top) becomes more difficult when baldness sets in. Some men remedy the situation through hair transplants, in which plugs of "donor" follicles are used to fill in bald spots (above).

American Hair Loss Council. Approximately 20 million women have experienced a similar loss of hair (from female-pattern hair loss), and an estimated 2.5 million Americans have lost their hair due to other causes.

The Basics

Hair is produced by hair follicles—indentations of the epidermis (outer skin layer) that contain the hair root, the muscle attached to it, and sebaceous, or oil, glands. Hair is made up of dead cells filled with proteins. The cells are woven together like a rope to form the hair fiber. The hair fiber, in turn, has three layers: the outer cuticle, with its fish-scale-like structure; the cortex, which contains the bulk of the fiber; and the center, or medulla. Hair color is determined by melanocytes, cells that produce pigment. When these cells stop producing pigment, hair turns gray.

Although it seems as if the hair on your head is always in the process of growing, hair actually has active and rest phases. The growth, or *anagenetic,* phase lasts two to six years. At any given time, about 90 percent of scalp hair is in the growth stage. The remainder is in the rest, or *telogenetic,* phase; this lasts two to three months.

Once the rest phase is over, the hair strand falls out, and a new one begins to grow. As a result, it's considered normal to lose from 20 to 100 hairs a day, says Diana Bihova, M.D., a dermatologist in private practice in New York City. Only a change in your regular pattern of loss is considered abnormal, but many things—including genetic factors, diet, stress, and medications—can change that pattern.

In many cases of male-pattern baldness, minoxidil, an FDA-approved drug, can slow down or even reverse hair loss.

Baldness: Manifest Destiny?

The most common cause of hair loss in both men and women is rooted in genetic predisposition. Called *androgenic alopecia,* it is known as male-pattern baldness in men and female-pattern hair loss in women. (Alopecia is the scientific term for baldness.) According to the American Hair Loss Council, genetics accounts for 95 percent of all cases of hair loss in this country.

Baldness results from a combination of genetic factors and levels of testosterone (a hormone produced by the adrenal gland in both sexes and also by the testes in men). If hormone levels are right, then the hair follicles will express their genetic destiny by growing for shorter periods and producing finer hairs. In men, who have higher levels of testosterone than women do, this eventually results in a bald scalp at the crown of the head, and a horseshoe-shaped fringe of hair remaining on the sides. In women the hair thins all over the scalp; the hairline does not recede. This type of hair loss doesn't usually show up in women until menopause; until then, estrogen tends to counteract the effects of testosterone.

One Approved Drug

The only drug approved by the Food and Drug Administration (FDA) to treat pattern baldness or hair loss is minoxidil topical solution (Rogaine), which is rubbed into the scalp. Originally approved for hereditary male-pattern baldness in 1988, it was also approved for treating female-pattern hair loss in August 1991. However, it should not be used by pregnant or nursing women.

In his dermatological practice, Arthur P. Bertolino, M.D., Ph.D., director of the hair consultation unit at New York University, says that this lotion helps hair grow in 10 to 14 percent of the people who try it. He estimates that approximately 90 percent of the time, Rogaine at least slows down hair loss. (Minoxidil is also available in tablet form to treat severe high blood pressure. Oral minoxidil has a potential for serious side effects and is not approved by the FDA for treatment of baldness.)

No one is certain yet just how topical minoxidil works to promote hair growth. "One theory is that it dilates the blood vessels, so it

The uncertainty of toupees is eliminated through hair-bonding. In this technique, bonding material is placed along the hairline (below left); a hairpiece is then attached to the bonding material (below). The effect is remarkably realistic.

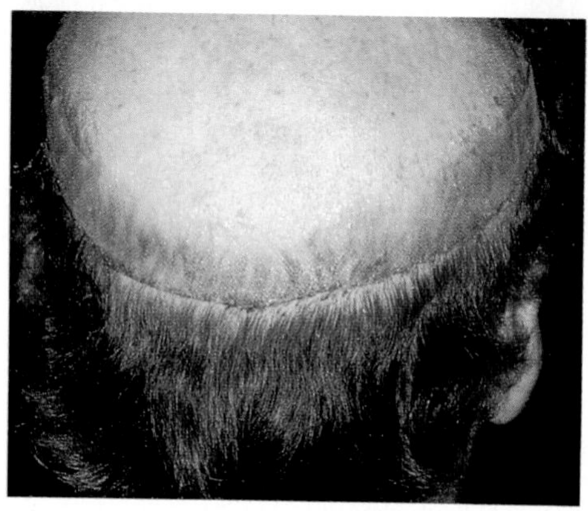

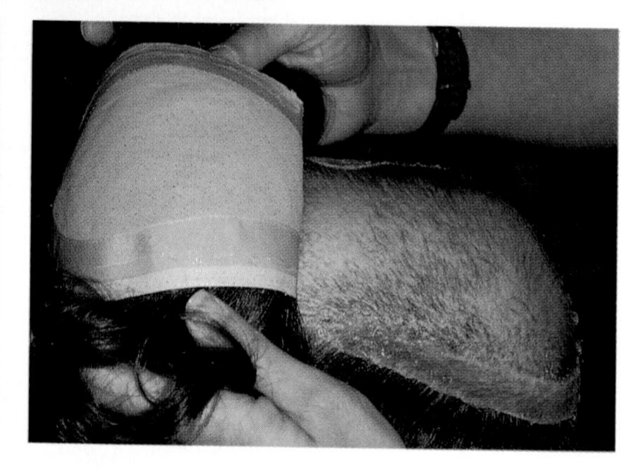

may stimulate nourishment of follicles," says Bihova. Alternatively, Rogaine may take tiny, peach-fuzz-producing hair follicles and convert them into large hair follicles that produce normal-size hairs. Again, no one knows for sure.

What is certain is that, at least in men, Rogaine works better on patients who fit a certain profile: they've generally been bald for less than 10 years, have top-of-the-head bald spots that are less than 4 inches (10 centimeters) in diameter, and they still have fine hairs in their balding areas. "The process begins very early," says Bihova. "I see 19-, 20-year-old males who have it."

The most common side effects with this medication are itching and skin irritation. Also, according to Bertolino, once you stop using Rogaine, any hair that grew as a result will fall out. Finally, the drug is expensive: in 1990 it cost about $600 a year to use it twice a day.

Transplants

Baldness can also be treated with hair transplants, in which plugs of "donor" follicles from the patient's scalp are used to fill the hairline. Although hair transplants work well in both men and women, the treatment tends to have a more dramatic effect on appearance in men with bald spots than it does on women with thinning hair.

"The less hair you have, the more drama in the change," says Robert Auerbach, M.D., associate professor of clinical dermatology at New York University School of Medicine. However, the American Hair Loss Council warns against attempting to replace lost hair with hairpieces sutured to the scalp. The FDA has not approved any products specifically intended for this purpose; however, this does not preclude a physician from using sutures, which are approved devices, for this purpose. Although the procedure is legal, it can result in scars and infections.

In the much-advertised hair-weaving technique, a base on which to attach a hairpiece is woven into the hairline (photos below). The front section of the hairpiece is secured by means of an adhesive or two-way tape. Clients attest to the results.

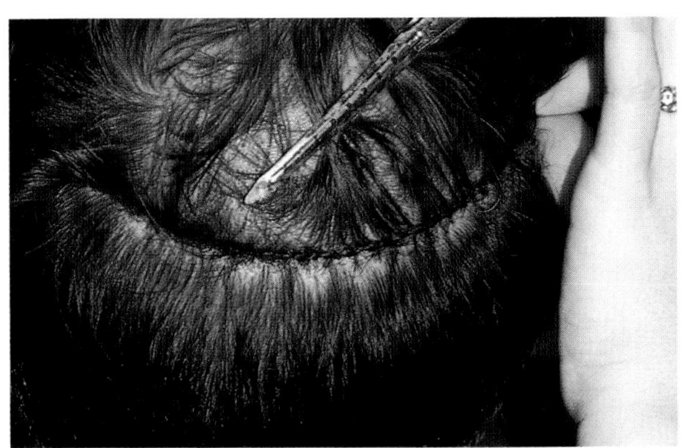

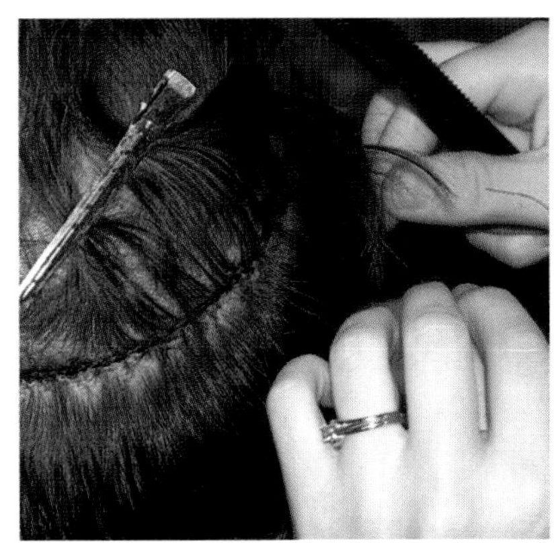

Another treatment for male-pattern baldness, hair implants made of high-density artificial fibers surgically implanted in the scalp, was banned by the FDA in 1984 because the fibers cause infection. These artificial implants are the only baldness-treating device the FDA has ever banned.

Products That Don't Work

The so-called "thinning-hair supplements," "hair-farming products," and "vasodilators" for the scalp will not promote hair growth, says Mike Mahoney, a spokesperson for the American Hair Loss Council.

Thinning-hair supplements are nothing more than hair conditioners that temporarily make your hair feel or look a little thicker. The main ingredient in these products—polysorbate—is also found in many shampoos. Promotional materials for hair-farming products claim that they will release hairs that are "trapped" in a bald scalp. Mahoney says these products, many of which are herbal preparations, can do no such thing. And so-called vasodilators do not increase the blood supply to the scalp, and do not promote hair growth.

Everyday Hazards

While male- and female-pattern baldness results in permanent hair loss, other factors can cause temporary loss of hair. For instance, the drop in the level of estrogen at the end of pregnancy can cause a woman's hair to shed more readily. Two or three months after a woman stops taking birth-control pills, she may experience the same effect, since birth-control pills produce hormone changes that mimic pregnancy. Physical stress, such as surgery, or major emotional stress can also cause hair loss.

"I've seen women start losing their hair before getting married," says Bihova. Even jet lag can have a similar effect.

In most of these cases, the hormonal imbalance or stressful situation will correct itself, and the scalp will soon begin growing hair again. But, says Bihova, since most women are extremely upset by even a temporary hair loss, many dermatologists treat these conditions with either topical steroid preparations or localized injections of low doses of steroids. Bihova emphasizes that these are local, not systemic, injections of steroids; therefore, the shots do not have the same risk of dangerous side effects as systemic steroids. However, only a board-certified dermatologist should administer this treatment, she says.

The list of causes of temporary hair loss goes on: pressure on the scalp from wigs or hairdos that pull too tightly can cause it; a fever of 103° F (39° C) or higher more often causes hair loss six weeks to three months later; and some medications can cause a temporary loss. Examples include vitamin-A derivatives such as Accutane, cough medicines with iodides, antiulcer drugs, some antibiotics, beta-blockers, antidepressants and amphetamines, antiarthritis drugs, blood thinners, some cholesterol-lowering agents, aspirin taken over long periods, some thyroid medications, and chemotherapy.

You Hair What You Eat?

Although nutrition does play a role in hair loss and in the overall health of your hair, only extreme nutritional deficiencies or excesses will cause hair loss. For instance, people with anorexia and bulimia may temporarily lose hair. So will others suffering from malnutrition.

"It's pretty rare in the U.S." says Bertolino. "If someone was on a real strange, restrictive diet, it could happen to them."

Megadoses of some vitamins—particularly A and E—and an iron deficiency may lead to hair loss. People who claim they can determine which vitamins are lacking in your diet by analyzing your hair, however, are not speaking from a scientifically sound basis. The test used with this type of hair analysis—atomic-absorption spectrophotometry—is a legitimate analytical-chemistry method; however, used on hair, the results of this test do not correlate with nutritional status, says NYU's Shupack. "Because of the sociological importance of hair, a lot of people try to cash in on it," he says. "[Hair analysis] is all witchcraft as far as I'm concerned."

There are, however, a few legitimate hair tests for poisonous substances such as arsenic and lead.

In the common process called male-pattern baldness, hair recedes in the front (top left) and thins at the crown of the scalp. Ultimately, the hairless areas can merge, forming "horseshoe baldness."

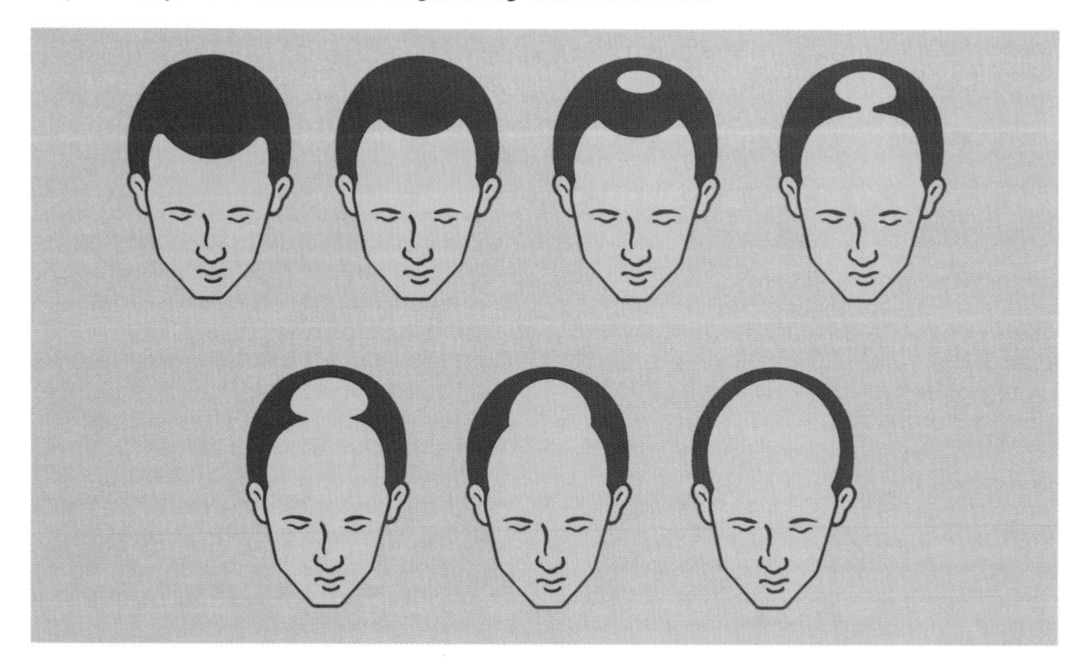

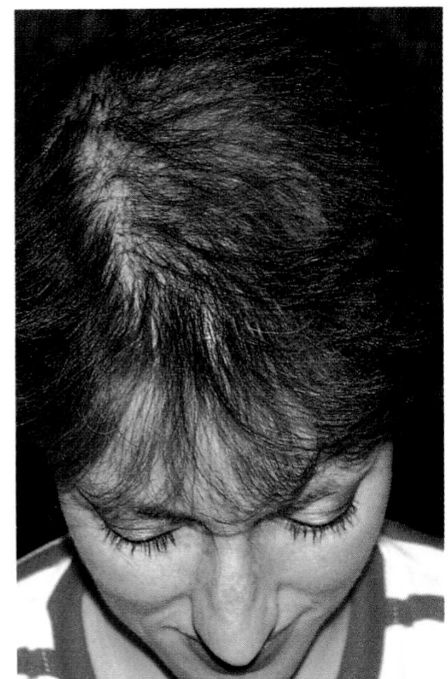

Although excessive shampooing, perming (near right), and dyeing can damage the hair, such treatments rarely result in hair loss. Only a small number of women are subject to female-pattern baldness (far right), which, like the more common male variety, is thought to have a genetic origin.

For Beauty's Sake

Every time you shampoo, blow-dry, perm, straighten, or dye your hair, you damage it slightly, says Bertolino. For the most part, hair can withstand this type of treatment. But overzealous beautifying can damage the hair fiber, resulting in many broken strands and a frizzy, split-end look. For instance, if you bleach your hair and then have a bunch of perms done in a short time, you're heading for trouble.

Misuse of hair cosmetics can cause the hair to break as it comes out of the scalp, says Frances Storrs, M.D., professor of dermatology at the Oregon Health Sciences University. Permanent-wave solutions break the bonds that hold hair together, and then reform them. But with a perm that is not diluted right or not rinsed off properly, for instance, those bonds may not re-form, and the hair would soon fall out as a result. Fortunately, most professional hairdressers know how to use perms correctly, says Storrs.

Most hair dyes are not as irritating as permanent solutions, mostly because they do not break the bonds between hair fibers, and are therefore not likely to cause a hair loss, she says. However, a severe allergic reaction to hair dye could cause hair loss. "The allergy is pretty common, actually," says Storrs. Permanent solutions can also cause allergic reactions, though that's a rare side effect.

Other beauty-related manipulations of the hair can cause problems, too: hot irons, cornrows, and braids may bring on temporary or permanent hair loss. If the hair breaks often enough, the follicles may eventually not be able to produce normal hair, says Bihova. "If someone has a problem with thinning and excessive loss, we advise being gentle," she says. "Don't use rollers; don't use blow-dryers on a hot setting; don't wear tight hairstyles." Rough shampooing may accelerate any loss, though it's usually not a problem in people with healthy hair.

The Medical Side

Some hair loss is the result of a type of immune disorder known as *alopecia areata*—some 2.5 million people suffer from this condition in which antibodies attack the hair follicle, causing the hair to fall out. Alopecia areata often causes small oval or circular areas of hair loss. However, in some forms of the condition, all the scalp hair falls out; in other forms, all body hair is lost. Although the loss is usually temporary, the condition can recur. Treatments include topical steroids or the use of chemicals to produce an allergic reaction to start the hair growing again.

Finally, chronic, systemic conditions—including one form of lupus, abnormal kidney and liver function, and hypothyroidism or hyperthyroidism—can affect the hair. If you're experiencing hair loss, see a doctor. He or she will want to order some basic blood tests to rule out any medical cause.

THE AGE OF GENES

by Shannon Brownlee and Joanne Silberner

The winsome, sable-haired four-year-old didn't even know she was making history, or care. By the time of the injection last year, she had been poked and prodded so often that she could not be bothered to take her eyes off the cartoon she was watching on TV. But to everyone else in her hospital room at the National Institutes of Health (NIH), this particular shot was ushering in the dawn of genetic medicine.

The injection marked the first human trial of gene therapy, a revolutionary means of treating a disease by giving a patient new genes. The girl suffers from an extremely rare inherited disorder in which faulty genes have crippled her immune system, leaving her vulnerable to the slightest infection or illness. To treat this disorder, her doctors removed immune cells from her blood, fitted them with a new gene, and reinjected them into her body. Today, four months after her last dose of the ground-breaking therapy, the girl's immune system appears to be fending off infections normally.

In the nearly 40 years since James Watson and Francis Crick elucidated the twisting structure of DNA, scientists have probed deeply the workings of the molecule that governs all living cells. They have learned how a single, mutated gene can clog the lungs of a child suffering from cystic fibrosis, and how other lengths of DNA open a cell's portals to the AIDS virus. Just in the past year, researchers have announced the discovery of at least four new human genes responsible for ailments ranging from deafness to sterility. And while finding a new gene is only a step toward vanquishing a disease, says Nobel laureate David Baltimore of Rockefeller University in New York, "every disease we know about is either being attacked with genetics or is being illuminated through genetics."

Experiments with gene therapy represent a giant step into the medicine of the future. An advisory committee at the NIH, which keeps a tight rein on such experiments, has approved three more gene-therapy proposals to treat cancer and a deadly inherited form of high cholesterol. In other laboratories, scientists have rigged the lining of blood vessels with genes that deliver anticlotting drugs; in still others, they are exploring genes that could make cells resistant to the AIDS virus.

Yet, for all the good that molecular medicine will do, the ethical dilemmas are grave. The advances bring closer the day when parents can endow children not only with health, but also with genes for height, good balance, or lofty intelligence—uses of genes that trouble most researchers and ethicists. Of more immediate concern is the possibility that health insurers, employers, and the government will gain access to genetic information and unfairly discriminate against people on the basis of their genes.

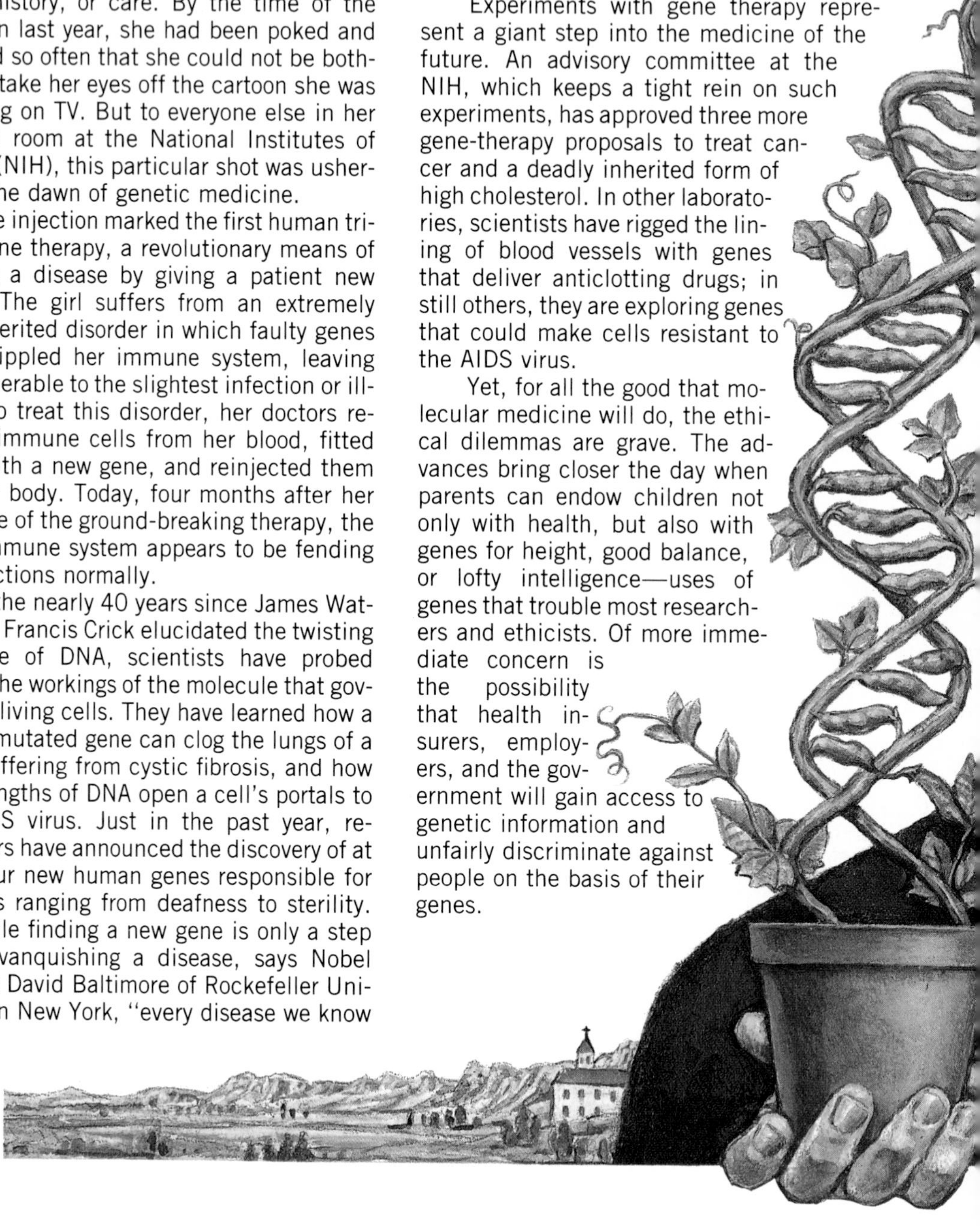

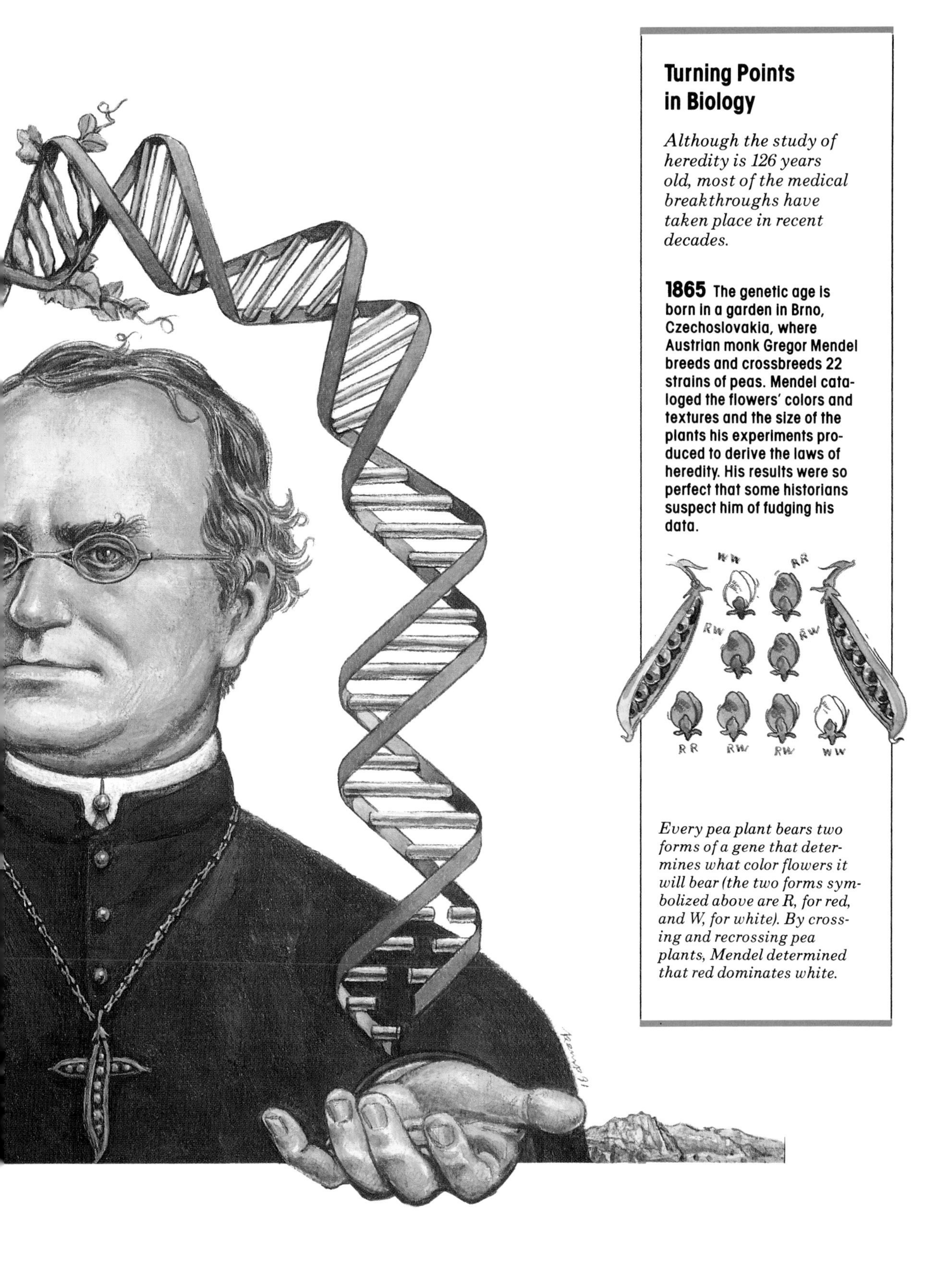

Turning Points in Biology

Although the study of heredity is 126 years old, most of the medical breakthroughs have taken place in recent decades.

1865 The genetic age is born in a garden in Brno, Czechoslovakia, where Austrian monk Gregor Mendel breeds and crossbreeds 22 strains of peas. Mendel cataloged the flowers' colors and textures and the size of the plants his experiments produced to derive the laws of heredity. His results were so perfect that some historians suspect him of fudging his data.

Every pea plant bears two forms of a gene that determines what color flowers it will bear (the two forms symbolized above are R, for red, and W, for white). By crossing and recrossing pea plants, Mendel determined that red dominates white.

Gene Hunt

While ethicists and Congress ponder who should be privy to the secrets of an individual's DNA, the pace of genetic discoveries is rapidly accelerating. Twenty years ago, researchers had identified only 15 of the human species' estimated 50,000 genes. Now, with the advent of new technologies, the count stands at more than 2,200; and over the next 15 years, an ambitious scientific endeavor known as the Human Genome Project will spend $3 billion developing even faster methods to identify the remainder. "Biology is in its golden age," says Gary Nabel, a physician and gene therapist at the University of Michigan, who has watched the advent of genetics transform his profession. "When people are practicing medicine 50 years from now, they'll be doing it in ways we would never have imagined."

Doctors have known since the time of Aristotle that certain families carry traits for diseases. But how those traits are passed between generations remained mysterious until the 1869 discovery of deoxyribonucleic acid, the immortal coil of inheritance known as DNA. This long and slender molecule spins out genetic instructions within nearly all living cells from conception until death. Inside a human cell, the strands are scrunched into 46 tiny packets, called chromosomes, too small for the eye to see. Unwound and laid end to end, a single cell's DNA would equal the height of a man.

Two key discoveries ensured that genetics would dominate the future of medicine. In 1953 Watson and Crick determined the manner in which DNA is constructed of four smaller molecules, called bases, that are strung together like letters in an alphabet. Within a decade, researchers solved the riddle of how DNA carries inherited traits from parents to child: they established that segments of DNA are, in effect, the words spelled out by

The anatomy of a gene

Packed into nearly every cell in the body is a blueprint of sorts: the long, twisting strands of DNA, which encode the instructions for making all the proteins necessary for life.

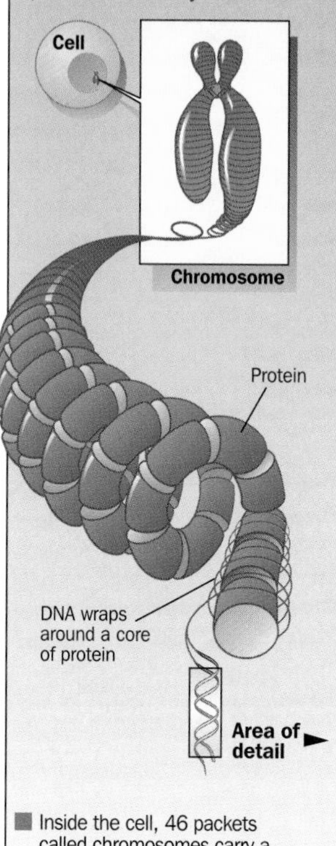

Inside the cell, 46 packets called chromosomes carry a total of 6 to 8 feet of of tightly coiled DNA. Along its length lie the human body's 50,000 genes, most of which carry the instructions for the body's proteins.

Double helix of DNA

Adenine (A) bonds with thymine (T)...

...guanine (G) bonds with cytosine (C).

Unwound from the chromosome, a gene looks like a twisted ladder. In fact, it consists of two separate strands bearing four types of compounds, called bases. These bases constitute the four-letter alphabet that makes up the genetic code.

1869 DNA, the twisting strands of inheritance often referred to as the blueprint of life, is discovered by Johann Friedrich Miescher, a Swiss chemistry student who purified it from the sperm of salmon and from white blood cells derived from surgical wounds. Miescher had no inkling of what he had stumbled upon.

1952 The "Waring Blender Experiment" proves DNA is the substance that transmits inherited characteristics from one generation to the next. At the time, scientists were arguing whether genetic information is carried by a protein or by DNA, which is an acid. In what is now considered a biochemically sloppy experiment, Martha Chase and Alfred Hershey use

the common kitchen appliance to separate the protein coats of viruses from the DNA to prove which is the real blueprint of life.

this four-letter alphabet. Just as the English language uses 26 letters in various combinations to make thousands of different words, different arrangements and repetitions of these four bases make all of the body's genes. Most genes are responsible for providing cells with instructions for making one of the thousands of proteins that form the building blocks of all cells and are essential for life.

The implications of the seminal discoveries were enormous for medicine. If scientists could ferret out the misspellings—or mutations—in genes, they could understand and perhaps cure the 3,500 known inherited diseases. Inherited disorders result when a mutation in a gene gives cells faulty instructions to produce aberrant proteins. Sickle-cell anemia, for example, is the tragic result of a single incorrect letter in the 60,000-letter gene for hemoglobin, a protein that carries oxygen and gives red blood cells their color. The tiny flaw in the gene produces a misshapen protein that cannot bind oxygen, leading to excruciating pain, weakness, and premature death.

From gene to protein

In an intricate series of events, the machinery of the cell deciphers the genetic code to manufacture proteins.

First, the DNA's double helix splits down the middle like a zipper. This exposes its bases to the cell's protein manufacturing machinery.

Each base constitutes a letter in the genetic code, and every group of three bases corresponds to a particular subunit of protein, called an amino acid. The cell's protein manufacturing machinery "reads" the bases three at a time. It then adds the corresponding amino acid to the growing chain, which will make up the final protein.

Every protein consists of a particular sequence of amino acids, strung together like beads. The different sequences give each protein a different three-dimensional shape, which determines its function in the body.

Double-stranded DNA

Single-stranded DNA

A G C T C T A G C G T T A G G T T A C C G

Amino acid

Protein strand

Folded protein molecule

By the 1970s a handful of scientists had begun to speculate that such genetic disorders could be treated by inserting good genes into a sick person's cells. Much of the impetus for trying gene therapy in clinical trials came from W. French Anderson, a physician at the National Heart, Lung and Blood Institute, who first tinkered with using genes to correct illness in 1974, and pioneered last year's first human experiment. The disease Anderson and his colleagues R. Michael Blaese and Kenneth Culver chose to investigate was adenosine deaminase (ADA) deficiency, an inherited disorder that destroys the body's immune cells.

Desperate Measures

Though rare, the disorder was targeted for two major reasons: researchers had already cloned, or

1953 **James Watson and Francis Crick unravel the double-helix structure of DNA, along with Rosalind Franklin and Maurice Wilkins, at the Cavendish Laboratory in Cambridge, Great Britain. Franklin died four years before this ground-breaking work was honored with the 1962 Nobel Prize.**

1964 **Marshall Nirenberg cracks the genetic code. Along with Har Khorana and Robert Holley, Nirenberg deciphered the "language" that enables DNA to be translated into proteins, the building blocks of all living cells. DNA's code consists of varying combinations of just four bases. The work of Nirenberg, Khorana, and Holley was recognized with the 1968 Nobel Prize.**

1972 **Recombinant DNA comes of age: Stanford biochemist Paul Berg ties together two lengths of DNA from different species using "sticky ends," which prove extraordinarily handy for gluing bits of DNA together. He shared the Nobel Prize in 1980.**

1974 **Cloning the gene requires breaching barriers that normally separate species. Stanford's Stanley Cohen, Annie Chang, and Herbert Boyer spliced DNA from a frog into *E. coli*, which ably churned out copies of the foreign gene. *E. coli* is now an indispensable cloning machine.**

1975 **The exact sequences of bases that make up a gene are determined by Walter Gilbert and Allan Maxam of Harvard University and Fred Sanger of Cambridge University, who developed two techniques for reading the genetic code one letter at a time. Gilbert and Sanger shared the 1980 Nobel Prize.**

1977 **Genetic medicine moves a step closer when Yale's Sherman Weissman and a team of researchers pinpoint the single mutation responsible for sickle-cell anemia. The defective beta globulin gene was the first human disease gene decoded and cloned.**

The first human gene is spliced into the bacterium *E. coli*, which begins pumping out the hormone somatostatin. The next year, human insulin is synthesized by genetically engineered bacteria, and four years after that, bioengineered insulin becomes commercially available.

duplicated, the gene for ADA in the lab, and their patient's condition was desperate, since the only other treatment for the disease was failing her. In healthy people, ADA breaks down a naturally produced compound that is toxic. Without ADA the compound builds up in immune cells and kills them.

In theory, gene therapy for ADA was deceptively simple. For this first experiment, the researchers removed immune cells from the girl's blood, supplied them with good copies of the gene for ADA that they lacked, and then put them back into the little girl's veins, where the cells produced ADA.

In practice, however, the procedure was not only clever, but exceedingly complex. The gene for ADA was ferried into the cells by special viruses, also the product of genetic manipulation. The viruses had been genetically crippled so that once they deposited their cargo of genes, they died without causing infection. Anderson and his colleagues incubated the girl's rejuvenated immune cells in a tepid broth to increase their numbers, and then injected about a billion of them into her blood, where they appear to be thriving.

With the spectacular success of this first experiment, the gates have swung wide to an extraordinary array of genetic therapies. The NIH's Recombinant DNA Advisory Committee, or RAC, has approved an experiment to treat three cases of familial hypercholesterolemia, a disease that can doom children to an old person's death from heart attack. In this inherited disorder, mutated genes render the liver all but incapable of removing a particular form of cholesterol from the blood, and the arteries of a child with the disease are often lethally clogged by age 10.

To correct this flaw, James Wilson of the University of Michigan and Howard Hughes Medical Institute plans to remove about 15 percent of a child's liver and add good genes to the cells. Once the revitalized cells are put back in the liver, they should sop up some of the excess cholesterol in the child's blood.

Nobody is expecting miracles from this experiment, least of all its architect. "We're looking for an effect for a couple of years," explains Wilson. The problem with this and other genetic therapies is that most cells have finite life spans—in some cases, just a few weeks—and the therapeutic effects last only as long as the cells themselves. An experiment with rabbits, for example, lowered the fat in their blood by 40 percent for about 10 days.

For more long-term improvement and certainly any true cure, researchers have been searching for what is called a "stem cell," a long-lived parent cell that serves to renew the many different types of cells in various tissues. In the liver, for instance, stem cells are thought to lie dormant until called upon to replace damaged or spent liver cells. Finding and tinkering with these hardy cells has become something of a Holy Grail for modern biology.

The existence of stem cells was first theorized in 1961, but they proved exceedingly elusive until 1991, when biologists at Stanford University and SyStemix Inc., a California biotech company, isolated what they believe is the human-bone-marrow stem cell. Assuming they are right, much still remains to be done before stem cells can actually be fitted with new genes. But when researchers succeed, they will be able to conquer many inherited disorders.

Even some diseases that are not inherited may one day suc-

CYSTIC FIBROSIS: New Hope

Scientists are identifying genes at the rate of one a day, but such feats represent only the first step on the long road to cures. Cystic fibrosis (CF) is a clear example. The 1989 discovery of the gene that causes CF buoyed the hopes of families beset by the deadly inherited disease. But the gene merely provided the game plan to the search for a cure. After the initial excitement, scientists settled down to the tasks of defining the precise makeup of the newly discovered gene, and figuring out what protein it makes and what that protein does.

Within a year of the discovery, several labs had nailed down the molecules involved in the disease. The blueprint—the gene—told researchers what the protein it manufactures must look like, and with that knowledge, they tracked it to the surface of lung cells. The normal form of the protein, called CFTR, plays a key role in pumping water into and out of lung cells. The abnormal form of CFTR can't move water around. Without that water, the lung's mucous lining thickens into a haven for bacteria, resulting in recurring lung infections and early death.

Living "Pharmacies"

Planting the protein in lung-cell membranes is one way to set things right, but to do that, scientists need enough protein to work with. In October 1991, researchers from Genzyme Corporation in Framingham, Massachusetts, reported that they had loaded up mice, called transgenics, with healthy CFTR genes, and that the mice were producing the protein in their milk. Such living "pharmacies" may supply healthy proteins to treat CF in the future.

Unlike protein delivery, gene therapy would provide a more lasting fix. Ronald Crystal of the National Heart, Lung and Blood Institute has saddled normal CF genes onto cold viruses. He has successfully ferried the gene into human cells in culture and into lung cells of living animals.

But while the discovery of the gene has opened up a broad avenue of research, it has also brought on a dilemma common to genetics: whether to test people to see if they have a defective gene. About 12 million Americans carry a single faulty CFTR gene. They won't know it until two of them have a child who inherits a defective gene from each parent—or unless their genes are tested. But since some relatively rare types of cystic fibrosis have yet to be defined, the test can identify only about 85 percent of the carriers. Because of the imprecision, the government will hold off on general screening.

Preventative Strategies

The gene has also led the way to two experimental forms of prevention. Several in vitro fertilization clinics are developing a method of testing an eight-cell embryo for CF by removing and checking the genes of one cell before it is implanted into the mother. Moving even further back in the chain of life, Yury Verlinsky of Illinois Masonic Medical Center tested an unfertilized egg from a CF carrier and determined that it did not hold the faulty gene. He reinserted it in the woman, and the resulting fetus was CF-free.

While the road from gene to cure is often bumpy and winding, in this case, progress has been rapid. Two years after the gene's discovery, researchers are "years, not decades" away from a cure, according to Cystic Fibrosis Foundation medical director Robert Beall. That will be none too soon for the 30,000 people in the United States with CF, and for the parents who pray every day that something will happen before their child runs out of breath.

Joanne Silberner

1980 A brainstorming session leads to a seminal idea by David Botstein, Ray White, Mark Skolnick, and Ronald Davis, who devise a means for pinpointing the location of genes along a stretch of DNA using RFLPs, fragments of DNA whose lengths vary from person to person.

1982 Transgenic mice are created by Richard Palmiter and Ralph Brinster by injecting mouse eggs with the gene for growth hormone. The technique allows scientists to create mice that suffer human

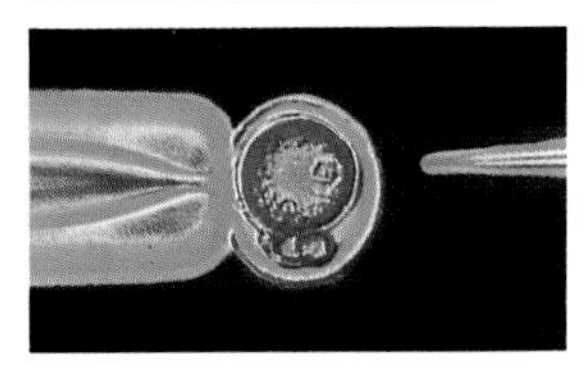

ailments. In 1988 mice with engineered susceptibility to cancer become the world's first patented life-form.

1990 A four-year-old girl with a rare genetic immune disorder is the first recipient of gene therapy. Doctors insert a gene into immune cells taken from her own blood.

The Human Genome Project begins. Its goal: to map all the genes in the human body at an estimated cost of $3 billion.

1991 International furor ensues when the U.S. Government quietly applies for patent rights to more than 300 genes whose functions are not yet known. Critics fear a "gene rush" by those hoping to cash in on biotech profits.

cumb to gene therapy. Recently, the RAC approved two experiments, in which genetically engineered tumor cells are injected directly into terminally ill cancer patients. Cancer quickly spreads through the body by evading the immune system, and the goal of the new treatment is to teach the body's immune cells to recognize and kill tumors. Steven Rosenberg, chief of surgery at the National Cancer Institute (NCI), removed tumors from his patients and endowed the cancer cells with a gene that manufactures tumor necrosis factor, a protein that regulates the body's immune system.

Immunity Lessons

Rosenberg injected less than a thimbleful of engineered cancer cells under the skin of his two patients' thighs, where small tumors should grow. In theory, these genetically engineered lumps will teach the patients' immune systems to recognize the peculiarities of their own cancer cells. Once instructed, their immune systems will seek out and destroy the rest of their tumors.

Ultimately, researchers would like to deliver genes directly to cells without removing them from the body. For example, Michigan's Nabel hopes to win approval from the RAC for a scheme to send genes into tumors by wrapping the DNA in a fat particle, called a liposome. In animal experiments, he has injected liposomes near tumors, and the cancer cells have willingly absorbed the DNA.

Gene therapy may also prove to be the most potent weapon against AIDS, even though this devastating disease is not an inherited disorder. Nava Sarver, chief of the NIH effort to find new drugs against AIDS, estimates that there are dozens of strategies under investigation for using genes to make cells resistant to the AIDS virus. Biologists have now successfully provided human cells in the laboratory with genes that can thwart the virus' attempts to enter and to reproduce itself. Eventually, doctors will be able to catch patients soon after infection and bolster their cells with resistant genes. "Without molecular biology, treating AIDS would be impossible, a shot in the dark," says Sarver. "Now we have the knowledge to home in on specific targets in the virus and specific steps in the process of infection."

Indeed, there is hardly a malady that is not being tackled with the tools of genetics and molecular biology. From infectious diseases to the chronic killers like cancer and stroke, identifying the role of genes is now considered indispensable toward understanding and conquering illness.

The flood of new gene discoveries has already begun to explain a host of diseases. At the 8th International Congress of Human Genetics, held in October 1991 in Washington, D.C., nearly 5,000 researchers from around the world described dozens of new bits of DNA. For example, a Scottish and German team disclosed that certain men who are sterile possess a tiny mutation in a gene on the Y chromosome, the first mutation ever associated with male infertility. Another gene was linked to a susceptibility to multiple sclerosis, while a cluster of genes appears to make some people especially prone to cancer if they smoke. The genes in question govern enzymes that break down and eliminate toxic substances, and a smoker who carries these genes can develop tumors as much as 50 years earlier than a smoker with a different genetic makeup.

No disease has given up more of its secrets to genetic sleuths than cancer. Thirty years ago, both the understanding of tumors and the treatment for them were exceedingly crude. Doctors employed radiation and chemotherapy, but the toxic side effects were often as bad as the cancer itself. Cancer's cause was equally murky. Scientists had observed that cigarette smoke and certain chemicals caused tumors, but they had no clue how.

Genes, it turns out, control critical steps in the terrible pathway to cancer. It is now known that some cancers erupt when the genes that tell cells to divide are mutated. The cell is permanently switched on by these so-called oncogenes, and it divides wildly out of control. Researchers have also uncovered genes that normally halt cell division, but, when mutated, fail to hold cells in check. At least five cancers owe their existence to these anti-oncogenes, including one for a childhood form of kidney cancer.

A cure for cancer has yet to emerge from this newfound knowledge, but understanding oncogenes has allowed scientists to sharpen their search. "We used to think that cancer came from something outside the cell," says John Minna, from the University of Texas Southwestern Medical Center. "The fact that

OF MICE AND MEN

Pink, hairless, and slightly larger than jelly beans, the newborn mice all look alike. But 10 weeks later, a few are twice the size of their full-grown siblings. No ordinary rodents, these ''transgenics''—part mouse, part man—are helping revolutionize the study of genetics.

Genetic engineer Ralph Brinster of the University of Pennsylvania with a normal mouse (right) and a transgenic mouse that is helping to revolutionize the study of genetics and illness.

Deep within their cells, transgenic mice harbor an alien piece of DNA—in this case the gene that produces human growth hormone (HGH). Transgenics are created by injecting a foreign gene into fertilized mouse eggs, a technique that is producing better animal models for illuminating dozens of illnesses and testing possible treatments. Says Richard Palmiter of the University of Washington, a pioneer in the field: ''The transgenic mouse is a wonderful experimental tool to study almost any question of genetics.''

Engineered Eggs

Though transgenics are increasingly commonplace in research, constructing the mice still leaves much to chance. Researchers use the same technique employed by Palmiter and Ralph Brinster of the University of Pennsylvania when they made their giant mouse, one of the first transgenics, in 1982. They began by coupling the gene for growth hormone with a bit of DNA that regulates gene activity within cells. The scientists then removed a mouse's fertilized eggs, less than 12 hours old. They injected hundreds of copies of their hybrid DNA into the microscopic embryos, which were then implanted in a foster mother.

The technique does not always work. The hybrid genetic material sometimes disrupts a gene that is vital to the mouse, and often it simply fails to insert itself. As a result, only about 3 of every 100 manipulated eggs develop into transgenic mice. Scientists have now devised a new technique, called homologous recombination, which allows them to aim a piece of DNA at a particular mouse gene and replace it.

A decade ago the word ''transgenic'' had not been invented. In 1990 alone, more than 600 scientific papers described strains of transgenics with symptoms of diseases ranging from arthritis and diabetes to obesity. Thousands of varieties of these ''guinea mice'' are now providing models for testing possible treatments. In 1991 scientists at Lawrence Berkeley Laboratory reported they had protected mice from cardiovascular disease by endowing them with genes for high-density lipoproteins, or HDLs, the ''good'' cholesterol.

A team at Harvard Medical School has patented transgenics that reliably develop cancer, and other researchers have created mice carrying the gene for beta-amyloid protein, the principal component found in people with Alzheimer's. The mice have since developed the brain deposits that characterize the degenerative disorder.

Genetic Surprise

Transgenics have proved to be such powerful tools in part because they often reveal genetic surprises. Rudolf Jaenisch of the Whitehead Institute in Cambridge, Massachusetts, unexpectedly discovered that mice can live without a certain class of molecules that are considered crucial to the immune system's ability to differentiate between the body's own cells and invading viruses and bacteria. Jaenisch used homologous recombination to eliminate the gene for the molecule. Most researchers had assumed that without it, the mice would either die before birth or be prone to deadly infection. But to Jaenisch's astonishment, the mice are healthy. In fact, they are now being used to study the role these molecules play in transplant rejection and infection. Such is serendipity in the lives of mice and men.

Charlene Crabb

cancer is caused by mutations in normal genes switched our attention to factors inside the cell. Now it's all fitting together." Instead of searching for toxic chemicals that are only slightly better at killing cancer cells than they are at killing normal cells, researchers are seeking drugs that specifically block the actions of cancer-causing genes. For instance, Georgetown University researchers are experimenting with a drug that blocks the activation of an oncogene in breast cancer.

Already cancer genes have made their way into hospitals as diagnostic tools. The presence of the oncogene called *neu* in breast-cancer cells, for instance, signals a particularly aggressive tumor, and calls for the most radical treatment.

Indeed, DNA's most significant contribution to all of medicine thus far is in the form of such diagnostic tests. Genetic probes now exist for an array of inherited disorders, such as hemophilia, muscular dystrophy, and polycystic kidney disease, all of which can be detected even before birth. Soon other tests will flag the subtle genetic tendencies toward more-common ailments such as heart disease and mental illness, and in the future, doctors will be able to identify many diseases before they happen, and will be able to do something about them.

Test Conundrums

In the meantime, however, many genetic tests grant people only the dubious power to peer into their medical futures. The test for Huntington's chorea, for example, does no more than tell those who bear the mutated gene that they will one day develop a disease that cannot be treated.

Genetic tests will force parents to make wrenching decisions. Many might choose to abort upon learning their child bears the gene for a lethal illness, such as Tay-Sachs disease, a disorder that kills long before puberty. But what if the gene confers a slight tendency toward colon cancer, which won't show up, if it does at all, until the bearer reaches the age of 50? As testing becomes more widespread, so, too, will the need for genetic counselors, specialists trained to instruct about the meaning of genetic test results. Even now the American Society of Human Genetics is pondering the wisdom of testing all adults of childbearing age for cystic fibrosis—about 100 million people.

The pressure to use genetic tests will be difficult to resist. According to a study released last year by Congress' Office of Technology Assessment (OTA), 20 percent of companies already use genetic tests on employees, in part to hold down corporate health-care costs. In addition, 15 percent of 400 employers surveyed by an insurer intend to screen the dependents of prospective employees. While such practices may make economic sense, ethicists worry that they are discriminatory, particularly since genetic traits often cannot predict with certainty if or when their bearer will fall ill. A person carrying the recently discovered genes for lung cancer, for example, may avoid the disease by not smoking.

Health insurers, too, will be using genetic tests in the future, a prospect that many find troubling. Insurers routinely deny coverage to people with expensive "preexisting conditions," a list of dozens of diseases including many genetic disorders. Widespread genetic testing will greatly expand this list, since everyone carries at least five genes that could lead to serious illness. "Policies governing the use of genetic information need to be debated and put in place early in this decade, not after problems emerge," warns Robert Weinberg, an oncogene specialist at the Whitehead Institute for Biomedical Research in Cambridge, Massachusetts, writing in *Technology Review.*

Society's knotty decisions will become even more tangled as the Human Genome Project lumbers toward its goal of mapping the location of every human gene, including those that govern such traits as intelligence and coordination. That knowledge will expand the potential of genetic engineering far beyond the correction of disease, and push it toward the realm of social engineering.

Some find such prospects so harrowing that they would halt all genetic research. But most scientists believe the world must proceed into the new genetic age. "This notion that there are things too dangerous to know is fundamentally antiscience and antiprogress," says Rockefeller University's Baltimore. "It is not a good representation of what the human spirit wants." Rather, it is a reflection of fear. But while entering the age of genes indeed may be inevitable, society has a great deal of choice in how it uses the new knowledge.

COMPANIONABLE CANINES

by Jack Fincher

Look at her. She's *so* anxious to please. And why not? It cost Canine Companions for Independence (CCI), a California-based organization that trains dogs to assist people with disabilities, more than $10,000 to get Filly ready for this. But as I nervously take my place in the wheelchair that has been assigned to me, I can sense that she's already losing interest.

"Sit up straight!" shouts instructor Laura Pintane. She is not reprimanding Filly. She is talking to me. "It's important that you project an air of authority. Filly has been taught to do what she needs to do, but before she will do it, she must know that she's working for *you*."

My assignment, Pintane goes on, is to get and keep Filly's attention. "Generally we don't believe in giving treats as incentives or rewards; otherwise, you wind up with a dog that won't work without them. Praise or a spoken correction should be all that's required. Because people confined to wheelchairs usually are restricted in movement, we prefer to command by voice alone, if possible. Now go ahead. Talk to her." Pintane waits to hear me command the dog.

"Filly?" I say. She ignores me and glances up at Pintane. Not hearing any orders, Filly gazes serenely across this shopping mall in the city of Santa Rosa to where five other novices are working with their newly assigned dogs. For the past two weeks, CCI instructors have bombarded these recruits

"Man's best friend" is a particularly appropriate description for dogs that have been specially trained to assist people with disabilities.

with demonstrations, practice sessions, lectures, and follow-up quizzes on everything about canines from health and history to psychology and the theory of learning. The stakes are far higher for them than they are for me. I'm just a journalist without any disabilities, trying to experience what they and their dogs must go through. Three of them are confined to wheelchairs. The fourth, although erect and able to walk, has had his diseased joints surgically fused. The fifth is hearing-impaired.

"Don't *ask* for Filly's attention," Pintane lectures me. "*Demand* it. Try again."

I do. Filly looks up at me, then away. Ah, progress.

"This time tell her to 'heel,' " Pintane says. "Then push off as you speak."

Together, we lurch into motion, Filly smoothly in step by my side—at first. All too soon, she is forging to the front, where she cannot be aware of my movements, and Pintane is again dissatisfied.

"Bring her in," she urges. "Let her know who's boss."

I am afraid Filly already knows.

For the next couple of hours, we "heel," "stay," and "wait." I say "Lie down," and Filly lies down. I say "Retrieve," and she brings me several small objects from the floor. I say "Pull," and she tugs the wheelchair along at a brisk clip. I feel a warm glow of accomplishment as Filly begins to respond to my commands more and more quickly. As that sense of confidence seeps into the tone of my voice, I can see that Pintane is right: an emotional bond has already begun to form between Filly and me. We're starting to feel comfortable with each other.

Then comes the terrible moment of truth known as "the transaction." I steer Filly and the wheelchair into a candy shop. There, in tandem, we slowly navigate the narrow aisles, taking care to avoid the merchandise-filled shelves that loom like treacherous canyon walls. Having rehearsed me, Pintane perversely waits by the door for me to bring her a bag of jelly beans. Finding a stack of them, I stop.

"Filly, look," I say. When I see her look, I say, "Get it." Filly gently takes one in her jaws. "Hold," I say, wheeling to the checkout as she carries it. "Up." Filly stands up with her forefeet on the counter. "Give." She gives the item to the smiling clerk, who rings it up and gives it back.

Are we on a roll, or what?

"Down." Filly gets down and stands beside me, her teeth delicately holding the bag of jelly beans by its edge. Then, for some reason, her mouth opens, and—*splat*—the bag drops to the floor.

Part of a training session includes rehearsing a "transaction" (below), in which a series of commands instructs the dog to pick up a book, bring it to the check-out desk, and give it to the librarian for processing.

With a sinking feeling, I repeat the required commands, and my spirits soar when Filly promptly obeys. "In lap," I say. She jumps up, deposits the bag of jelly beans on my chest, and accepts a heartfelt hug.

Then she gives the clerk my credit card to complete the transaction. Together, man and dog pass the test.

I'm exhausted. As we pause to rest, we are approached by a couple and their two toddlers. Having been prepared for something like this by Pintane, I go into my spiel. "Please don't pet Filly without asking me first," I tell them, managing a grin, but feeling foolish for forbidding such a friendly act. "She's a working dog, not a pet." Only afterward do I learn how important it is at the beginning of a relationship that the master of such a dog make sure that he or she is the only channel for all of its pleasures and, indeed, for everything in its life. "I know that sounds too controlling, but believe me, this can be dangerous work," says Bonita Bergin, the determined schoolteacher who, 16 years ago, founded CCI.

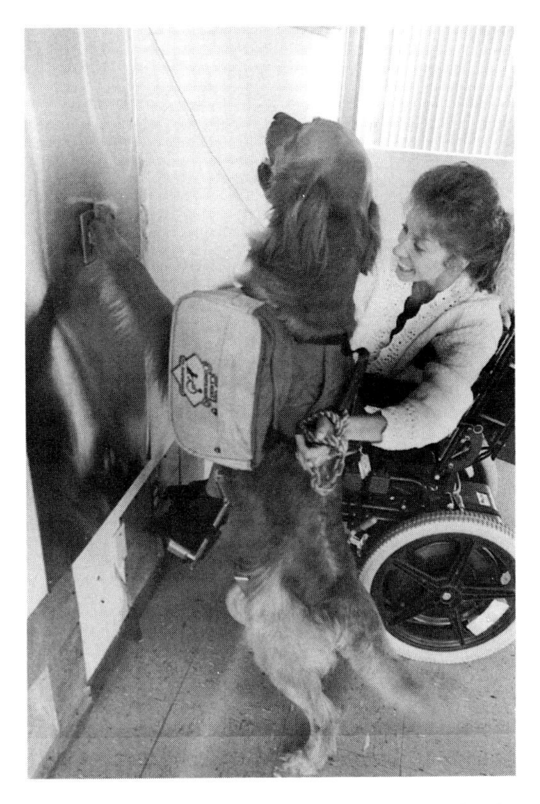

A well-trained service dog can perform a variety of tasks for its master, including pressing elevator buttons (above), opening doors, turning light switches on or off, and even fetching food from the refrigerator.

Lifesavers

For years, dogs have been used to guide blind people; using them to assist people with other disabilities is a relatively new idea. CCI is one of a few dozen nonprofit organizations that, over the past 15 years or so, have placed more than 800 "service" dogs with physically disabled people all across the country, and 2,500 "hearing" dogs with people who are hearing-impaired. "You've seen how little function and movement some of our candidates have," Bergin adds. "Can you imagine one of them stopping his wheelchair if he's leashed to his dog and doesn't have it under complete control? It could dash into traffic after a cat. Lives depend on those dogs."

That's true in more ways than one. Two years ago, Vicki White was driving her van through a rough neighborhood in Marin, California, when she was held up by three muggers. She was alone and physically helpless—except for her muscular black Labrador, Nomad. Canine companions are not taught to attack or defend, but rather, on the command "Guard," to bark in alarm. This time, however, White's anxiety must have communicated itself to Nomad. Given the command "Guard," he spontaneously assumed what his master now calls his menacing "vulture" posture. "Whoa, lady, be cool," one of the muggers muttered as they melted back into the darkness. Another time, Nomad pulled White from a rain-swept street in Oakland after her electric wheelchair shorted out. "Both times," she says, "I believe he saved me from serious injury, if not worse. I knew he was going to change my life. I had no idea how much."

Working dogs can restore lives as well as save them. Five years ago, hairstylist Debbie Walrod, an athletic young Californian, lost both legs and all of her fingers except the base of her right thumb as the result of a blood infection. She was fortunate to be alive, but don't try telling that to a lithe beauty who had been a ballerina since childhood and had modeled professionally as a teenager. "I'd lie in bed and just sleep or cry," she remembers. "I told my mother that I was so pathetic I couldn't kill myself if I wanted to, and I did want to."

Then Walrod met Oregon, a curly-haired dog trained by CCI. It was love at first sight, but there was one hitch. Oregon had been raised by CCI's regional director and was devoted to her, a fact Walrod learned when Oregon abruptly detoured her wheelchair into the director's office during a meeting. Could such a strong attachment possibly be overcome? The final test was a battery of commands that required Oregon to cross the room off-leash and come to Walrod, ignoring the presence of his first master, the director. If he was unable to do that, Walrod and Oregon wouldn't make it as a team.

"At first, he went over to her," Walrod remembers, "and nuzzled her hand. But she wouldn't speak to him, look at him, or anything. Boy, was he confused! I gathered my confidence and said, 'Oregon, no! Come here!' He looked back and forth. I kept at him, and finally he came to me. I looked at the director, and a tear was running down her cheek." Today Debbie Walrod once again leads an energetic life, styling hair, making jewelry, and, in her spare time, helping CCI raise pups. Oregon is the reason why: "He's my hands, my feet, my best friend."

Multitalented Partners

It's amazing how many things dogs like Oregon can do for people who could not manage without their help. Service dogs can turn light switches on and off, raise and lower window shades, lock and unlock wheelchairs, select books from shelves, open and shut doors and drawers. One has been taught to brace while its partially paralyzed owner grasps the handles on its "backpack" and hoists herself erect. Another picks up its owner's immobile legs, using the tongues of her shoes, and periodically crosses and uncrosses them to stimulate circulation. Yet another knows how to fetch its master a beer from the refrigerator. "Hearing" dogs learn how to use their bodies to alert their hearing-impaired masters to ringing telephones and doorbells, buzzing smoke detectors, dinging microwave timers, and bawling babies. Steve and Nancy Bock of San Jose, California, even use their canine companion, Gala, as the medium for exchanging love notes. The only problem is that sometimes the messages get punctured by tooth marks.

Unlike guide dogs for the blind, an idea that was conceived in Germany after World War I, canine companions for people with physical or developmental disabilities or hearing impairment originated here in the United States. When Bergin was taking special-education courses in northern California during the mid-1970s, she was struck by the emphasis that was placed on what professionals could do for people with disabilities, rather than what the people could do for themselves. Earlier, traveling and teaching in the Orient, she had seen disabled individuals leaning on their burros as they hobbled about hawking their pots and pans. "I wondered how we could give our people the same freedom and flexibility to fend for themselves," she remembers. "If it couldn't be burros, it had to be dogs."

While teaching in public schools, Bergin got a second job at a kennel, where she observed canine behavior and sought advice from experts. Dog trainers told her it couldn't be done. Dog-training manuals said it wasn't possible. Veterinarians said it was a nutty idea. Undeterred, Bergin began to call social agencies, explaining her plan and asking for referrals. Some social workers responded negatively, too. But professionals who worked with disabled people were desperate, and were willing to try anything. They helped spread the word that Bergin was looking for candidates who would give it a try.

Finding the people was not a problem, but, as the experts had predicted, training the dogs was. The puppies Bergin tried out were either unsuited for the work or too undisciplined to be taught. Older dogs, on the other hand, were apt to have too many bad habits. So Bergin finally decided to breed the animals herself, pick the pups that demonstrated an aptitude for the work, then figure out how to train them. And that's how CCI got off the ground.

At one point, Bergin's 18 dogs shared the plasterboard walls and bare concrete slab of a house she and her husband were building outside Santa Rosa. That was in 1977. Initially she had little idea of what to look for in a dog's manner or behavior that would translate into effective learning. While some dogs were satisfactorily placed, many others proved unsuitable. Eventually experience showed that golden retrievers and Labradors—and their crossbreeds—were the best at service; Pembroke Welsh corgis and Border collies excelled at signaling.

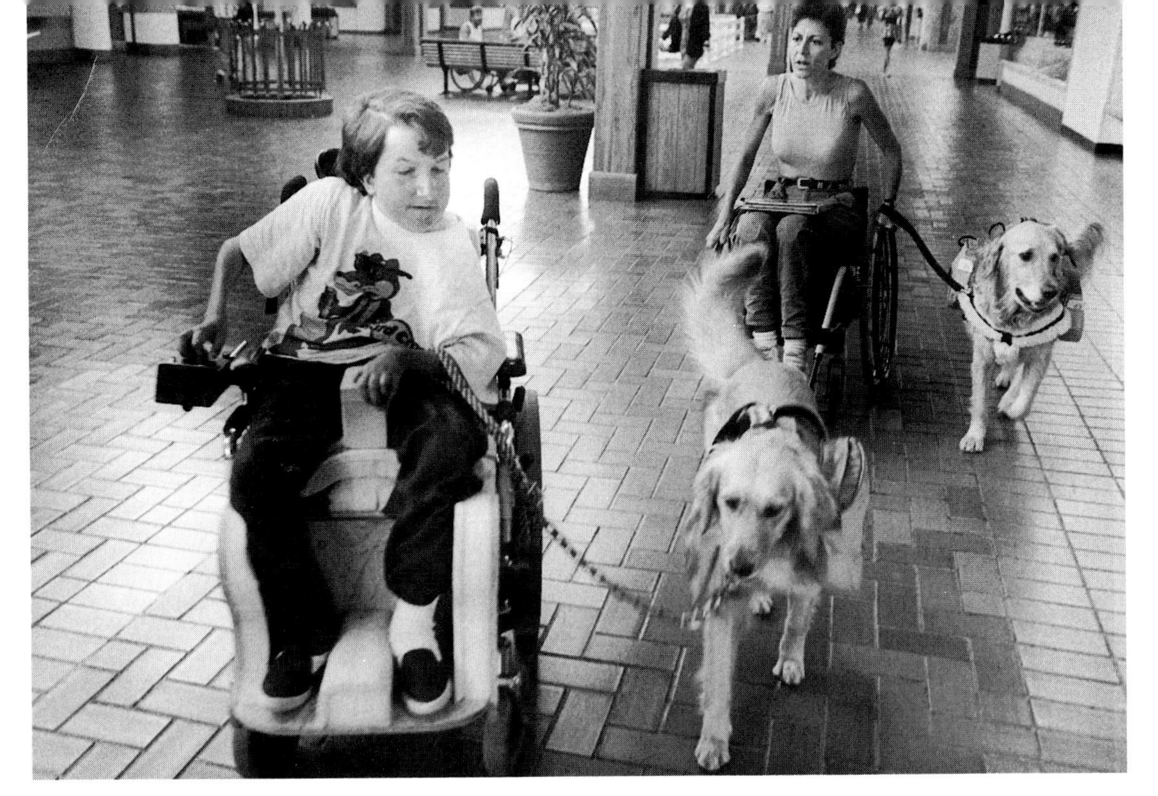

A shopping mall provides an ideal training ground for newly matched wheelchair-bound people and their dogs. The malls are a natural obstacle course, filled with elevators, tight store aisles, and distracting crowds that provide challenges for master and dog alike.

In 1981 Bergin moved her dogs into an old rented house in town. Then she transferred them to a condemned chicken coop that leaked whenever it rained. CCI limped along with a couple of minimum-wage assistants and volunteer help. Bergin did everything from training the dogs and coaching her disabled clients to cleaning out the kennels. It took her nearly two years to get an animal ready to work; then she would give it to a client. It is no exaggeration to say that the operation was not cost-effective.

Then, one evening in 1983, a man identifying himself as a foundation president called. He and his wife had been visiting San Francisco, he said, and had decided to drive to Santa Rosa. They were coming right over. Minutes later an elegant limousine pulled up in the pelting rain. Out of it issued an elderly woman in a wheelchair and several other passengers. "We phoned a couple who had one of our dogs, and asked them to rush right over and demonstrate," says CCI veterinarian Ruth Daniels, who started out working for the organization as a volunteer. "Afterwards the foundation sent us a check for $10,000. It was a miracle!" CCI razed the chicken coop and built its current facility.

Another turning point had come a year earlier, when Bergin went across town to see the man who arguably qualifies as the dog's best friend—cartoonist Charles Schulz, the creator of Snoopy. Would "Sparky" Schulz, as his friends call him, consent to front, in some dignified manner, for such a demonstrably dogworthy organization? Schulz wrote out a check for $10,000 on the spot, and later paid for and narrated a documentary to be used in promotion. His wife, Jean, joined the CCI board and is now its president. Last year, her husband chaired a drive that raised more than half a million dollars.

Bergin left CCI recently to start up another organization, the Assistance Dog Institute. Her goal is to help others get still more service-dog programs into operation. As for CCI, it now has an annual budget of more than $2 million and a staff of 65. In addition to its headquarters building in Santa Rosa, CCI operates a veterinary clinic there. It also maintains instructional centers in Ohio and New York, as well as at two locations in California. Once the dogs are bred, volunteers raise the pups and teach them basic manners and commands. When the youngsters are 17 months old, they are farmed out to the instructional centers, where college-educated trainers put them through an intensive eight-month "finishing school."

With a decade and a half of experience, CCI has a pretty good idea of what makes for a reliable canine companion. National training manager Clark Pappas explains it in this way: "We want dogs that are submissive but confident and contained, not out-front. That are malleable, not aggressive or dominant. That have asking eyes, not telling eyes." These traits are all well and good, but how do you go about breeding for such characteristics? Basically by putting together dogs that have successfully gone through the program. It doesn't always work. "The softer males we like aren't always the best studs," concedes Ruth Daniels.

You Never Know What Will Go Wrong

Currently about one out of every three puppies makes the grade. To be more cost-effective, CCI wants to see that figure improved. But when it comes to animal temperament, you never know what will go wrong. It turned out, for example, that one dog refused to accept credit cards. Or, as a trainer put it, "Mame didn't take plastic." Nobody knew why, but eventually the problem was solved.

Dogs are specially bred for the specific traits required to be acceptable service dogs. Volunteers raise the pups and teach them basic commands. At 17 months the puppies undergo an intensive eight-month "finishing" program.

Just what CCI is up against can be seen in the saga of one handsome golden retriever who sired 240 of the puppies that have been raised by the organization before a devastating Achilles' heel was detected in his makeup. At times, this dog refused to accept commands and resisted carrying anything on his back, including the small backpack that some canine companions must wear. These deficiencies were transmitted to some of his descendants, making them unfit for service. This is why dogs are now required to pass the CCI curriculum before they are used for breeding.

Hereditary roadblocks aside, CCI's track record is impressive. Currently about 600 of its graduates are at work in 38 states and four foreign countries. This is a tribute not only to the dogs, but also to the ingenious curriculum Bergin and her associates have developed. It features such clever wrinkles as a series of obstacle courses that promote physical and mental flexibility; a set of exercises in "prolonged eye contact" that heighten attention span and increase responsiveness;

and, for puppies, a number of relatively simple tasks (holding a pencil while sitting; carrying a paper bag) that help prepare them for the kind of work they will do as adults.

The sequential system of instruction involves 60 separate commands—building blocks for a repertoire of jobs that couple a person's needs to the dog's innate desire to please and its capacity to learn. Simple acts are taught first, then strung together to embody harder tasks. Thus, "Up" and "Shake," once mastered, can be combined into "Switch"—the command for turning off a light. Filly's jelly-bean transaction resulted from seven different commands she had been taught, which even I could string together.

Those who are going to receive the dogs must also be taught the skills they will need to master the animals. To qualify for a dog, applicants must endure a gauntlet of application forms, a two-year waiting list, and a series of interviews, all leading up to boot camp. A three-week pressure cooker held three times a year, boot camp annually brings together a total of 75 to 80 people and a like number of dogs for tryouts at CCI's regional training centers. The teams either bond or wash out. Approximately one in 15 human recruits fails to make it for physical or emotional reasons. Many of those who don't make it the first time try again. Only a handful have been rejected twice. John Giacomazzi of San Jose flunked his initial try primarily because his voice wasn't expressive enough. Giacomazzi practiced his inflections at home, on his mother, for a year ("Mom—Look! Get it! Good!"), came back, got another dog, and graduated.

Four of the five rookies in my boot camp faced similar frustrations. Mary Sexton of Seattle, Washington, speaks well, but during boot camp her confidence was undercut by problems with spatial orientation. Leaving a Santa Rosa college campus one afternoon, she almost drove her wheelchair off the side of a ramp when she became visually confused by the sidewalk's brickwork pattern. Tim Beidler of Portland, Oregon, who lost the use of his legs in an auto accident, suffered a setback when the dog assigned to him proved too difficult to control. He had to switch animals with Rob Roseberry of Victoria, British Columbia, a strapping 185-pounder disabled by a disease of the nerves and muscles in his legs. Roseberry had found that his upper-body strength could not compensate for the fear of failure that was getting in the way of communicating with his first dog, Xylan, a free spirit with a very short attention span.

Gene Hopkins, the man from Salinas, California, with the surgically fused joints, had an unsettling experience when his young Border collie, Drake, spied his previous owner strolling in the mall and bolted off in pursuit—still leashed to his new master. Instinctively, Hopkins, who has had plenty of hard falls, went into a protective somersault and crashed to the floor unhurt. A day later Drake saw his first basketball being dribbled in a gym. He had to be restrained. Diagnosis: an exaggerated impulse to act on his visual attraction to movement, a trait that makes his breed excel at herding. The next morning, Drake was forced to endure a ball being rolled past him time after time as the stern command "Stay" was punctuated by emphatic tugs on his choke chain whenever he failed to obey. In less than an hour, he was cured.

Eventually all five of my classmates passed. At first, Mary Sexton tried too hard and wanted to quit when her initial efforts didn't succeed. On solo day, she wept. "Don't give up on me," she pleaded with her instructor; "if you don't, I won't give up on myself." Assured that her instructor wouldn't, Sexton pulled herself—and her dog, Mame—together. "I learned more about myself than I ever expected," she recalls. "Somehow the dog picks up on emotions that I would normally hide, and that forces me to be more honest."

Tim Beidler succeeded, too—but barely. At one point, he jerked the leash angrily when Xylan sniffed the ground. "Watch that," Beidler was told in the trainers' post-mortem. "It affects how people perceive you with the dog." Even Rob Roseberry finally had to accept the fact that he couldn't simply muscle his way to success. "What a time to be all thumbs," he quipped when he had trouble attaching Daisy's backpack with his numbed fingers.

When boot camp ended and everyone was getting ready to go home, it was Gene Hopkins who came up with a fitting benediction. "From now on," he said with an affectionate look at the dog that had done everything asked of him, "we're in this thing together. Aren't we, Drake?"

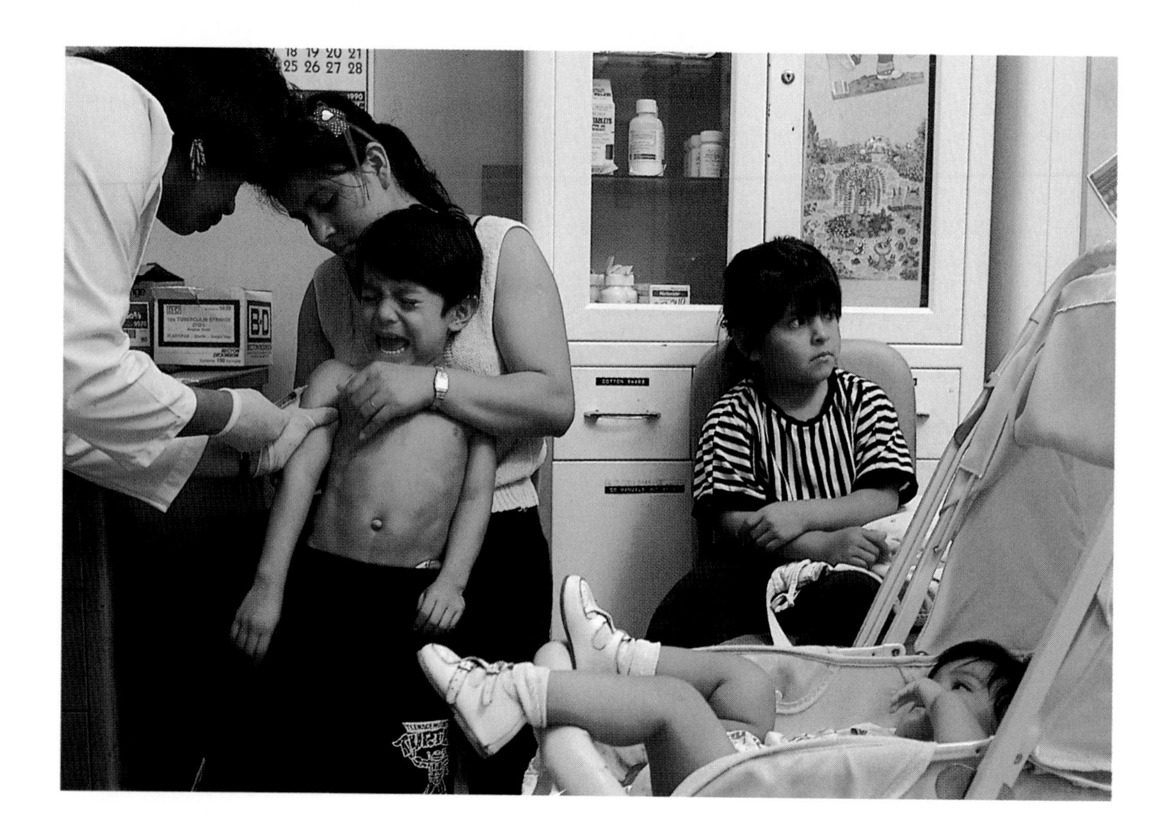

COMEBACK DISEASES

by Devera Pine

Not too long ago, measles, mumps, rheumatic fever, tuberculosis, and polio were considered illnesses of the past: they were diseases that maybe our parents had once contracted long ago; diseases that modern medicine had long since conquered. Unfortunately, though, nothing could be further from the truth: all these diseases are still with us. In fact, they are currently making a surprising and sometimes frightening comeback.

Measles

Before the development of a vaccine for measles, an estimated 4 million to 5 million people came down with the virus each year. Measles, which is transmitted by sneezes, runny noses, and the like, causes a high fever, rash, cough, sensitivity to light, and white blisters in the mouth. Although measles usually runs its course in a little more than a week, it can result in such deadly complications as pneumonia and encephalitis, particularly in older people, infants, and individuals with impaired immune systems.

The vaccine, introduced in 1963, significantly reduced the danger of catching measles: from 1980 to 1988, on average, there have been only about 3,000 reported cases each year in the United States according to William L. Atkinson, M.D., M.P.H., a medical epidemiologist with the Centers for Disease Control (CDC). However, more recently we've been in the midst of a mini-epidemic of sorts: in 1989 there were some 18,000 cases of measles, and in 1990 there were another 28,000 cases. Although the exact figures for 1991 have not been published yet, Dr. Atkinson estimates the final numbers will show about 10,000 cases of measles nationwide. What happened? Why didn't the vaccine wipe out the disease?

Outbreaks of measles and mumps still occur among children and young adults, most of whom did not follow the prescribed sequence of vaccinations.

Part of the answer lies in neglectfulness: "The U.S. has been less successful than other countries in getting the entire population immunized," says Lorry Rubin, M.D., chief of infectious diseases at Schneider Children's Hospital of Long Island Jewish Medical Center in New Hyde Park, New York.

In the United States, for instance, the major outbreaks of measles have occurred in racial- and ethnic-minority children under 5 who live in inner-city areas. According to the CDC, surveys of areas in which measles outbreaks have occurred indicate that as few as 50 percent of the children were vaccinated by their second birthday. In addition, many recent outbreaks have occurred among young adults on college campuses.

The resurgence of measles may also be due in part to vaccine failure. Vaccines don't always "take"; indeed, the vaccine for measles has a failure rate of 5 percent.

Since measles is a serious disease, it's imperative to be sure you've been immunized: if you were born before 1956, chances are you have a natural immunity to the disease. If you were born after that year, you probably need a second dose of the vaccine. Children should be immunized twice: once at 15 months and again at age 5.

Mumps

With the introduction of a mumps vaccine in the 1960s, the incidence of this disease has declined dramatically, from more than 185,000 cases in 1967 to 5,300 cases reported in 1990. While mumps has not reached the epidemic proportions of measles, small outbreaks of the disease nationwide make it clear that mumps is still very much with us today.

Mumps is a viral infection characterized by swollen glands in front of the ear, a low fever, and headache. Given rest and painkillers, mumps is not serious most of the time. However, it can result in inflamed testes or ovaries, and in adults it can cause nerve deafness, meningitis, or encephalitis.

If you were born after 1956, make sure your mumps vaccine is up-to-date. Children should be vaccinated at 15 months and again at age 5. Fortunately, these recommendations should be easy to follow, since the vaccine for mumps is generally given with the vaccine for measles in the Measles, Mumps, Rubella (MMR) shot.

Rheumatic Fever

After fading into the ranks of rare diseases in the 1970s, rheumatic fever has made a sudden return: over the past 10 years, there have been several outbreaks of the disease, including an outbreak of 200 cases in one hospital alone. In 1981 there were 264 reported cases of rheumatic fever, though that number has since dropped: in 1990 there were 108 cases.

Rheumatic fever occurs only after a strep infection. The body reacts to streptococcal bacteria by producing antibodies; these antibodies, in turn, affect other tissues throughout the body. For instance, the antibodies can inflame the joints, the heart muscle, and the heart valves; rheumatic fever can also result in nodules under the skin, and involuntary movements (known as Sydenham's chorea).

Although one of the hallmarks of rheumatic fever was the strep throat and high fever that preceded it, in the most recent outbreaks, many sufferers did not develop a throat infection, says Arthur A. Klein, M.D., professor of clinical pediatrics at Cornell Medical School and executive vice-chairman of pediatrics at the New York Hospital-Cornell Medical Center. These cases of "silent" strep throat, when left untreated, increase the risk of rheumatic fever.

Interestingly, despite the reappearance of rheumatic fever, there have not been increased outbreaks of strep infections, notes

Critically ill rheumatic fever patients from the 1950s greatly benefited from treatment with the then newly developed "wonder" drug cortisone.

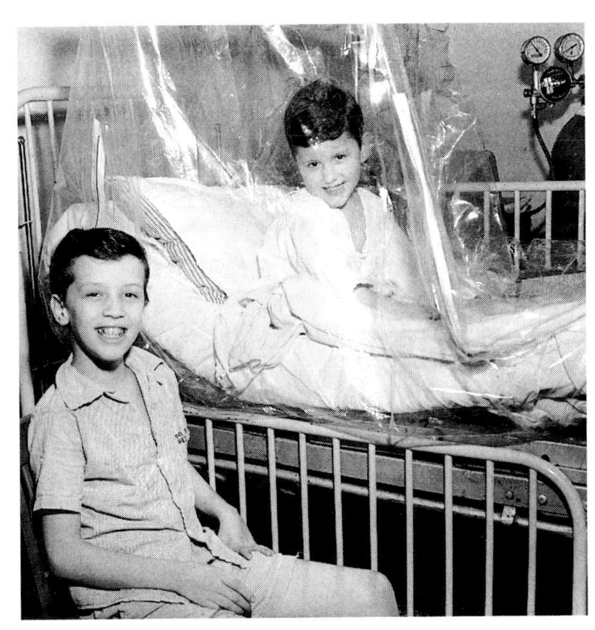

ILLNESSES REVISITED

DISEASE	MEASLES	MUMPS
CAUSE	Virus spread via airborne droplets produced by sneezing or coughing.	Virus spread via airborne droplets produced by sneezing or coughing.
SYMPTOMS	Appear 9-12 days after exposure; runny nose; cough; red, watery eyes; high fever; rash. Complications include pneumonia, ear infections, and encephalitis.	Appear 2-3 weeks after exposure; low fever; pain and swelling in neck glands; headaches. Rare complications in post-puberty patients include inflammation of testes and sperm ducts in males; inflammation of ovaries in females; nerve deafness; viral meningitis.
TREATMENT	Supportive care, including bed rest, fluids, and mild, fever-reducing medication.	Supportive care, including bed rest, fluids, and mild fever-reducing medication.
PREVENTION	Two doses of Mumps, Measles, Rubella (MMR) vaccine at 15 months of age and again at age 5.	Two doses of MMR vaccine at 15 months of age and again at age 5.

Alan L. Bisno, M.D., chief of medical service at the Miami Veterans Administration Medical Center and professor of medicine at the University of Miami School of Medicine. One possible explanation for this is that the strep bug may have undergone a change so that some forms of it are more likely to result in rheumatic fever. "We have found that there are changes in the strep population in virulence," says Dr. Bisno. "Some strains capable of causing rheumatic fever have come back with a bang." In fact, at about the same time we started seeing outbreaks of rheumatic fever, toxic strep syndrome—the syndrome that resulted in the death of Muppet creator Jim Henson, for instance—also appeared on the scene.

Fortunately, not every case of strep throat results in rheumatic fever—and not every case of rheumatic fever results in heart damage. To head off the disease, see your doctor promptly if you or your children have a significant sore throat, especially if it is accompanied by a fever. Antibiotics can cure a strep throat; aspirin and cortisone or prednisone may be used to treat the various symptoms of rheumatic fever. In addition, people who have had a bout of rheumatic fever are generally advised to take daily doses of penicillin for up to five years to prevent further strep infections.

Tuberculosis

One of the most frightening comebacks of all has been the reemergence of tuberculosis (TB) and the evolution of a deadly, drug-resistant strain of the disease. According to the American Lung Association, before 1985 TB in this country was on the wane. Since then, however, the trend has reversed: in 1990 there were 25,701 cases, and in 1991 there were 26,283 cases.

The increase is in part due to the human immunodeficiency virus (HIV): TB is one of those diseases that we all pick up, but that, normally, our immune systems are able to contain, says John L. E. Wolff, M.D., chief of infectious diseases at Beth Israel Medical Center, North Division, and clinical associate

In the early 1900s, children with tuberculosis were often placed in special hospitals called sanatoriums. The woman at far right wears the Double Barred Cross, now the American Lung Association's Chistmas Seal insignia.

RHEUMATIC FEVER	TUBERCULOSIS	POST-POLIO SYNDROME
Group A streptococcus bacterial infection of upper respiratory tract that is ineffectively treated.	Tubercle bacillus bacteria spread by airborne droplets produced by coughing; produces a latent form with no symptoms and an active form.	Uncertain, but some theories suggest that past polio sufferers now suffer from nerve and muscle overuse syndrome as adults.
Appear 3-5 weeks after strep infection; sore throat; fever; painful, swollen joints; rash. Complications may include inflammation of heart muscle, deformed heart valves.	Fatigue; loss of appetite; weight loss; headache; low fever; cough	Extreme muscle weakness; body fatigue; pain in muscles or joints; breathing difficulty.
Aspirin to treat fever and painful joints; if heart problems develop, prednisone or cortisone may be prescribed.	Exposure to TB requires a 6-month course of isoniazid to prevent active TB from developing. Active TB requires long-term program of antibiotics.	Supportive care, including anti-inflammatory medications to relieve pain; reducing activity level.
Early detection and effective antibiotic treatment of strep infections of the upper respiratory tract.	Protection from exposure to people with active TB; regular screening for TB at 3-year intervals.	None known; any past polio patient who is beginning to notice symptoms should have a baseline exam so that doctors can monitor if symptoms are worsening.

professor of medicine at New York Medical College. But if the immune system is compromised by, for instance, the HIV virus, TB becomes activated, he says. Healthy people have a 1 in 10 chance per lifetime of developing active TB after they've been infected with the bacteria; immunosuppressed people have a 1 in 10 chance *per year* of developing active TB.

The current epidemic has also developed among the homeless and among immigrants from countries with ongoing TB epidemics. However, since the bacteria that cause tuberculosis are transmitted by prolonged contact with people who have an active case of the disease, there is a danger that the epidemic will spread to the general population. Symptoms of active TB include coughing, fever, chills, and weight loss.

An even scarier prospect is the emergence of a new, drug-resistant form of TB. Normally, it takes about six months of treatment with three different antibiotics to cure a case of TB, says Lee B. Reichman, M.D., M.P.H., professor of medicine and director of the pulmonary division at the University of Medicine and Dentistry of New Jersey. However, if a person with active TB takes his or her medicine haphazardly, the TB bacteria can develop a resistance to one or more of the drugs. Furthermore, that person can now infect other people with the new, drug-resistant TB. According to the CDC, during the first quarter of 1991, 3.1 percent of new cases of TB and 7.4 percent of recurrent cases were of the drug-resistant type.

Although drug-resistant TB can be treated with antibiotics, these drugs are very toxic, expensive, and not always effective. Drug-resistant TB has a 50 percent mortality rate, according to Dr. Reichman, who is also president-elect of the American Lung Association. Furthermore, unlike ordinary TB, it is not preventable: if someone has been exposed to TB, a six-month course of a drug called isoniazid will generally prevent them from developing active TB. "If the TB is drug-resistant, we don't know what to do," admits Dr. Reichman.

To protect yourself against TB, consult your doctor about taking isoniazid if you are HIV-positive or if you live or work with someone who has TB.

Postpolio Syndrome

Once the great scourge of childhood, polio has been nearly eliminated in the U.S. In fact, the last epidemic of this paralyzing and sometimes fatal virus occurred in 1979 among 10 Amish children who had not been vaccinated for religious reasons. These figures are in sharp contrast to the epidemic of 1952, for instance, in which 58,000 people came down with the disease.

Today vaccines protect most of us against polio. Lately, though, polio has been

The close quarters and unsanitary conditions of homeless shelters contribute to the spread of TB. Homeless people may already be ill or have impaired immune systems. When in close contact with carriers of the disease, such people are especially susceptible to infection.

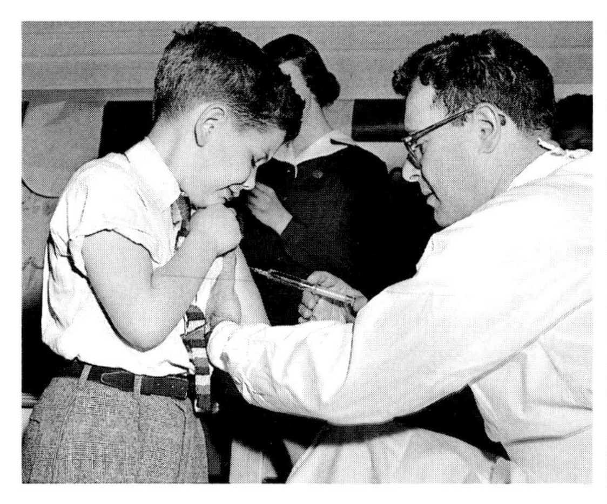

The polio vaccine (above), developed in the 1950s, has all but eradicated this disease. Some people who overcame polio as children are now finding themselves subject to nerve and muscle debilitation as adults (right).

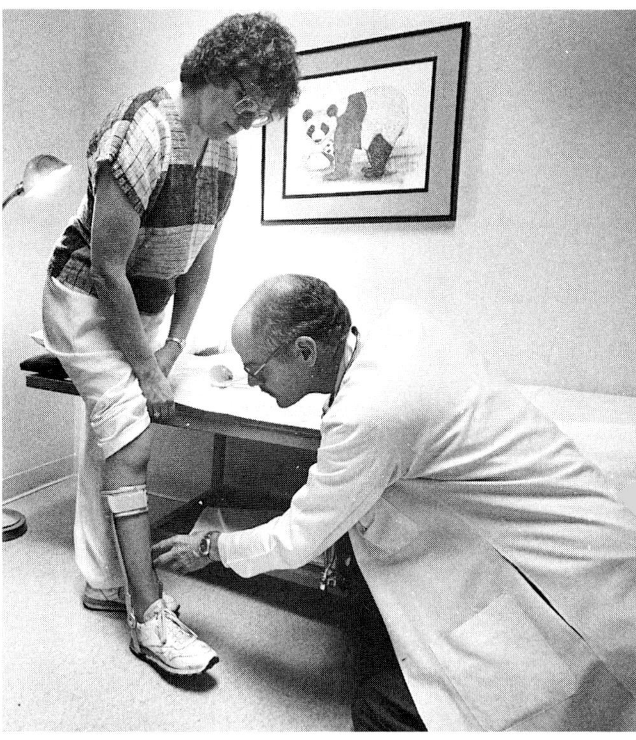

in the news again due to new symptoms related to the disease: postpolio syndrome.

Postpolio syndrome affects men and women who had paralytic polio as children and teenagers. Its major feature is unaccustomed or very intense fatigue that tends to come in the mid- to late afternoon, as well as new muscle weakness, explains Lauro Halstead, M.D., director of the postpolio program at the National Rehabilitation Hospital in Washington, D.C.

The syndrome is not a reactivation of the virus, says Beth Israel's Dr. Wolff. Rather, it is more likely the result of a kind of overuse phenomenon: the polio virus paralyzes by killing or injuring nerves that control muscles. The paralysis is not always permanent, however. In many cases the remaining healthy nerve cells grow additional axons to pick up the "orphaned" muscle fibers so that the muscles can move again.

According to one theory, postpolio syndrome occurs when, after years of doing double duty, these nerve cells begin to tire. "You have a nerve designed to drive maybe 100 motor cells driving 1,000 or perhaps 2,000," says Dr. Halstead. "I think metabolically, the nerve becomes exhausted—it can no longer make the necessary energy packets required."

Other experts speculate that some kind of immunology mechanism—perhaps an autoimmune response to the virus—may be at work, says Dr. Wolff.

Whatever the cause of postpolio syndrome, the result is that some polio survivors begin to lose their ability to move: they have to return to walking with a brace or cane, use a wheelchair, stop working, and otherwise modify their living habits. "Psychologically, it can be devastating," says Dr. Halstead. "There are a number of folks who have severe physical deformities from their polio. Or they may have had breathing problems way back, and now they're having to go back on nighttime ventilation."

So far the treatment for postpolio syndrome is experimental: researchers are investigating the use of growth hormones, steroids, and other medications (including drugs that stimulate the transmission of messages between nerves and muscles). In addition, nonsteroidal anti-inflammatory drugs (NSAIDs) can help relieve pain. One bright note on the horizon: a recent study by the Mayo Clinic in Rochester, Minnesota, suggests that postpolio syndrome may not be progressive. Five years after they were first examined, the subjects in the study had not continued to lose significant amounts of muscle strength due to postpolio syndrome.

If you had polio as a child and are beginning to experience muscle weakness, fatigue, and breathing or sleeping difficulties, see your doctor for a medical exam and evaluation of your muscle strength.

THE ATHLETE'S DILEMMA

by Jared Diamond

My dorm at Harvard was popularly known as the jock house because so many of the college's athletes chose to live there. Those jocks are responsible for some vivid images from my undergraduate days—not images of stirring athletic triumph, necessarily, but of admirable and almost superhuman attempts to train their bodies for the moment when triumph would be within reach.

Standing on line in the cafeteria, I would watch the jocks pile their trays so high with food that it began to fall off. I especially remember a discus thrower who would routinely, at each lunch and dinner, take nine dishes of ice cream for his body to burn off during long workouts. He was grimly determined to shovel down as many calories as possible, in the hope that he could thereby run his body's engine faster, train harder, and cream his competition.

Now, 35 years later and at UCLA, I no longer eat in the dorms or watch slack-jawed as college athletes consume mountains of calories. But I still watch them burning those calories. Up my street every day chug the long-distance runners in training, doing their 120 miles (190 kilometers) per week. In the UCLA pool, swimmers churn away at their 20,000 yards (18,000 meters) a day, while triathlon competitors barrel past on their 100-mile (160-kilometer) bike rides. Like their Harvard counterparts of a generation ago, today's UCLA athletes push themselves to eat and exercise. They tell themselves that if only they can find a better diet, if only they can train more hours per day, they, too, will be able to utterly defeat their competition. Their methods and diets may have changed in the past few decades, but their goal remains the same: to run the body's engine faster.

Alas, as a physiologist, I am beginning to suspect that their hopes are doomed to frustration. Certainly it pays to train hard, and certainly athletes will still be shaving seconds off record times for years to come. But ultimately the biological engineering of the body may condemn us—and all other animals as well—to remain under an unyielding ceiling that limits just how much we can usefully take in, and hence just how much we can put out. This ceiling limits all our uses of metabolic energy—not just for relatively nonessential purposes, like athletics, but also for such critical activities as staying warm or nursing babies.

A competitor in the Tour de France bicycle race must sustain an extraordinarily high metabolic rate. Yet even when consuming an astounding 9,000 calories a day, a cyclist's metabolic rate reaches a value only three times that of a deskbound worker.

What Is Metabolism?

Does such a ceiling indeed exist? And if so, why? Why should evolution have arranged things so that we are constrained in how much speed and power we can maintain, how extreme the environmental conditions in which we can live, how many babies we can nourish? Let's start by reminding ourselves what metabolism represents. To use a perhaps well-worn but nonetheless apt analogy, imagine my Harvard classmate's mouth as the opening of a car's gas tank, into which he poured his nine dishes of ice-cream fuel at every meal. Like car engines, human bodies operate by burning fuels, converting chemical energy into heat or useful work. In an engine the fuels are hydrocarbons, and the reactions go on at high temperatures—typically 4,500° F (2,500° C). In our bodies the fuels are carbohydrates, fats, and proteins, and we make specialized catalysts called enzymes that permit the reactions to proceed at a temperature of about 98° F (36° C).

The car engine will use some of the resulting energy to propel the car, but the laws of thermodynamics oblige it to waste (from the car owner's point of view) some of the energy in the form of heat. Our bodies, however, use that heat to a good purpose: to keep us warm and maintain our body temperature, even when we're surrounded by a cold environment. Besides that heat, we also extract useful energy, which we devote to five purposes, the most visible of which is muscular work. We also use that energy to keep our cells and tissues constantly in good repair, to synthesize new tissue as we grow, and, if a pregnant female, to synthesize the fetus' tissue and then to synthesize milk for the newborn infant.

During our waking hours, our metabolic rate is lower when we're lying motionless in bed in the morning, having just awakened, with our stomachs already emptied of yesterday's meals and our intestines no longer doing the work of digestion. Even under these nontaxing conditions, we continue to burn our chemical fuels at the rate of about 1,640 calories per day for the average man, 1,430 calories per day for the average woman. (One calorie as defined by nutritionists equals 1,000 calories as defined by chemists, and it is the quantity of heat required to warm 1 kilogram of water by 1° C at a pressure of one atmosphere.) That invisible expenditure of energy goes to maintaining our body temperature, repairing our molecules and cells, keeping our nerve and muscle cells ready to fire, and permitting our kidneys to keep making urine. All those expenditures, as we lie in bed seemingly motionless, but actually hard at work, constitute what's termed our basal metabolic rate.

Sustained Metabolic Rate

Any muscular exertion burns calories over and above that basal metabolic rate. So does shivering to stay warm or secreting milk to nurse a child. We can sustain a metabolic rate 100 times our basal rate during a few seconds of intense exertion, or 20 times our basal rate for many minutes while running or cycling. To do so, we have to burn some of our stored fuel supplies, which consist mainly of glycogen and lipids stored in our muscle and liver and fat cells. We gradually lose weight as we burn those stores, and we regularly do so for hours between meals. If necessary, we can starve and continue to metabolize while losing weight for many days. Usually, though, we eat several times each day, then proceed to absorb nutrients out of our intestine and to resynthesize the fuel we have stored.

Naturally, most of us use more energy during the day than at night. However, averaged day and night over many weeks, the rate at which we burn calories has to equal the rate at which we absorb calories over that time. Otherwise we would be losing or gaining weight indefinitely. That long-term average rate is what I call the sustained metabolic rate, which varies greatly depending on a person's average level of activity.

The lowest sustained metabolic rates are those of sedentary people like scientists who write articles for popular magazines, and desk-bound editors who must read the articles the scientists write. Reported values are only about 2,000 calories per day (1.4 times basal metabolic rate) for nonpregnant women scientists, 2,800 calories per day (1.7 times basal values) for men scientists. Pregnant women have higher metabolic rates (about 2,450 calories per day) than nonpregnant ones because of the energy required to make the fetus and placenta. People leading strenuous lives have still higher metabolic rates: for example, 3,800 calories per day for miners, 4,100 for soldiers in boot camp, 5,000 for polar explorers pulling heavy sleds in extremely cold weather, 6,000 for UCLA water-polo players in training, and 8,000 calories for triathlon aspirants in training. These values for sustained, time-averaged metabolic rates are respectively 2.3, 2.5, 3.0, 3.7, and 4.9 times basal rates.

Cranking It Up

All these numbers, though, are just values for what people actually metabolize. There's no proof so far that these values approach the maximum metabolic rates possible. How much higher might we crank up and sustain our metabolism?

Certainly it seems that there must be a limit. In cold weather, for example, we start eating more to stay warm; and even in Chicago in January, it's easy for us, as long as we're normally clothed, to process the few hundred extra calories we need daily. Granted, it would be more of a challenge for us to maintain our body temperature while going naked outdoors in Chicago winters. In principle, though, we could—if there were no ceiling at all on our metabolism, and if we could generate enough heat by shoveling in 30,000 calories of deep-fat-fried noodles every day.

I've never met or heard of anyone who's attempted to spend the winter naked outdoors in Chicago. However, there are fiercely determined athletes who attempt the equivalent of this stunt, with the difference that they seek to turn calories into muscular work, not heat. In doing so, they produce the highest human values of sustained metabolic

No matter how hard and how long they train, a championship-level swimmer or an Olympic gold-medalist skater still labor under a metabolic ceiling that limits exactly how high their internal engine can rev.

rates that I know of. These are the cyclists who race in the Tour de France, probably the most grueling athletic event in the world. In just over three weeks, they cover more than 2,000 miles (3,200 kilometers) and ascend and descend dozens of mountains. Every day of the race, they shovel in calories at rates inconceivable to us armchair mortals. So demanding is their task that many, despite their truly extraordinary motivation and training, still drop out before finishing.

In 1986 a group of physiologists from the Netherlands chose five particularly zealous cyclists who they thought were likely to finish the Tour de France (that year a race of 23 days, 2,377 miles [3,825 kilometers], and 34 mountains). Four of the athletes actually did. The researchers kept tabs on each cyclist's consumption of bottles of high-energy drinks, which provided much of the calorie intake, and had each man keep a daily record of all other food consumed. Metabolic rates were also measured directly by a method using ingested isotopes.

The cyclists managed to gulp food so fast that they nearly maintained their body weight despite their great exertions. On average, they lost only 3 pounds (1.4 kilograms) from their mean body weight of 153 pounds (70 kilograms). Their time-averaged metabolic rates ranged from 4.1 to 5.6 times basal values; the highest average food consumption was 9,160 calories per day.

That sounds stupendous. But before you get too excited about it, reflect that it's barely more than three times the rate of a desk-bound editor, and not even six times that of a

A yellow warbler parent, searching for food to feed its young, can speed up its calorie-burning machinery to almost three times the normal rate.

fasting man lying in bed and doing absolutely nothing. We already know that for short periods—say, the time required to run a 50-yard dash—even we flabby nerds can burn calories at many times the cyclists' daily average rate. Why couldn't these superathletes sustain higher average rates than they actually did? Why didn't they just wolf down more fried noodles, gulp more gallons of Gatorade? Why didn't they shove in, say, 30,000 calories per day and turn their wheels of metabolism three times faster? Had any cyclist succeeded in doing that, he'd have left the competition gasping in his dust. That even these most fanatic of fanatics metabolized no faster than they did says to me that they really were close to a ceiling on their sustainable metabolic rate.

Metabolic Slowpoke

Could it just be that we humans are metabolic slowpokes among animals, a result of our 5,000-year exploitation of animal and machine power rather than muscle? To make sure, my UCLA colleagues Charles Peterson and Kenneth Nagy and I gathered data on the sustained metabolic rates of 36 vertebrate species besides ourselves.

These were normal, physically active wild animals, running around and doing whatever they chose to do. The species were very diverse and included iguanas, mice, swallows, penguins, monkeys, kangaroos, and koalas. They ranged from a one-eighth-ounce lizard to a 24-pound (11-kilogram) koala. The metabolic rates varied nearly 7,000-fold among these species (more than 50,000-fold if you include humans), and, for obvious reasons, rates were higher for big animals than for small ones, and higher for warm-blooded birds and mammals than they were for cold-blooded reptiles.

But the factors by which each species cranked up its metabolic fires—that is, the ratio of its sustained metabolic rate to its basal metabolic rate—fell pretty much in the human range for all of them: the animals' ratios ranged from 1.0 to 6.9, while ours range from 1.4 (nonpregnant female scientist) to 5.6 (the most vigorous male Tour de France cyclist). Only four species topped that cyclist's value of 5.6, and they didn't do it by much: a small Australian opossum reached 5.7; a macaroni penguin reached 5.8; and a gannet (a large seabird) and a little Australian marsupial mouse both topped out at 6.9, a rate that has yet to be beaten by any observed human or animal subject.

These results become puzzling when you reflect that, like us, animals can maintain much higher metabolic rates for short periods, during which they operate off stored energy reserves. Wolves, for example, have been measured burning calories at 32 times basal levels. But, again like us, all the animals for which we had data seemed unable to sustain such metabolic rates indefinitely—or at least they "chose" not to do so.

Were all these animals in the wild really trying as hard as they could? Well, one might expect that some certainly were: namely, those birds that were flying around almost nonstop from dawn to dusk to find food for their chicks. Yet even these birds sustained metabolic rates less than seven times basal values. Apparently, they, too, like the Tour de France cyclists, had bumped up against a metabolic ceiling.

To be certain, however, we needed to see how fast metabolism could operate in animals whose motivation could be not just observed, but also manipulated. For this part of our study, our subjects were female laboratory white mice that were trying to feed their young.

Nursing Mothers

Any woman who has breast-fed her baby knows that nursing doesn't come cheap. Nursing mothers have to eat a lot, both to maintain their own bodies and to produce the milk they need for feeding their babies. Yet humans have it easy compared with many other animals, including laboratory mice: a human mother typically has only one baby at a time, and she stops nursing when the baby is still less than one-quarter her own weight. The mice, in contrast, have a normal litter of eight pups, which are nursed until each is 40 to 50 percent as large as the mother herself. Just before she weans them, the mother is providing the nourishment for a combined mass of animal perhaps three times her own weight. (To match this, a human mother would have to nurse simultaneously a dozen rapidly growing 30-pound babies.) It's not surprising, then, that among wild mammals, the individuals requiring the most food often seem to be the lactating females.

My colleague Kim Hammond set out to test whether there is any limit to the metabolic rate these mother mice can sustain. Into the cages of mice that had just given birth, Hammond placed additional newborn pups from another litter. The mothers proceeded to adopt and nurse some foster pups along with their own—but only up to a limit. Many mothers adopted half a dozen pups to reach a total litter size of 14. But when offered even more pups (Hammond tried up to 26), the mothers either adopted only some and refused others, or else tried to, but just couldn't, nourish them all. One especially motivated mother, however, managed to wean a litter totaling 16 pups, double the natural litter size.

The "supermoms" reached a daily food consumption nearly four times that of a nonpregnant mouse just sitting in her cage and not doing much of anything beyond waking up and walking around occasionally. Their calorie utilization (including the calories exported into milk) was about seven times the basal metabolic rate of a nonpregnant mouse of the same body weight. That puts their ratio of sustained to basal metabolic rate above the values for the Tour de France cyclist (5.6), and essentially the same as the ratios for gannets and marsupial mice (6.9). But it doesn't break that record.

Why not? What was it that limited these mice's ability to nurse more babies? One possible, banal explanation is that the mothers might have been held back by their having

The metabolic ceiling is well illustrated by the dilemma of the laboratory mouse: even the most motivated mother can produce only enough milk to nurse 16 young.

only 10 teats. In reality, though, most were able to rotate 14 pups among 10 teats, just as human mothers with triplets rotate three infants among two breasts. The mouse supermom who nursed 16 pups hit on the efficient solution of dividing the pups into two piles and commuting between them, ensuring that there was a teat for each pup in each pile.

Instead, something less obvious must have been limiting the metabolism of mother mice, just as something limits gannets and cyclists. If not even the urgings of motherhood could motivate these mice to crank up their metabolism any higher, it suggests to me that they really couldn't go any higher. Apparently, some physiological bottleneck sets a ceiling on their, and our, sustained metabolic performance.

Metabolic Bottleneck

Just where in the body is that bottleneck? And what principles of evolutionary design prevented natural selection from producing animals that overcame this limitation?

To even begin to answer such questions, we must leave the secure ground of established fact and turn to speculation. First of all, there are lots of places where the bottleneck could lie. It might, for example, reside somewhere in the body's systems for acquiring energy: it could be some limit on the intestine's ability to absorb food, or on the liver's ability to handle the absorbed nutrients, or on the kidneys' ability to excrete the resulting wastes. Or the bottleneck might reside in the body's systems for burning energy: a limit on muscles' ability to do work, say, or on mammary glands' ability to secrete milk, or on the combined ability of all tissues to produce heat.

While we don't yet know the answer, there are some clues. In the case of the intestine, it's easy to decide whether that organ is imposing a bottleneck on food absorption. All we need to do is calculate how much sugar a mouse's intestine can absorb, and compare that with how much sugar mice actually eat. It turns out that sugar-absorptive capacity exceeds sugar intake by only a modest safety margin in nonpregnant mice. Remember, however, that in supermoms, food consumption was nearly four times normal; clearly, if there were not some accompanying change in these mice's intestines, they couldn't have absorbed all the sugar they were taking in.

In fact, the supermoms did quickly grow bigger intestines, but the resulting increase in absorptive capacity wasn't quite as large as their increase in food intake. Hence, their safety margin became even narrower, and they came closer to the point at which their intestines couldn't keep pace with the ingested food. That calculation suggests that metabolic rate might be limited by the intestine's ability to absorb food.

However, it's useless for your intestine to absorb lots of food if your liver isn't going to be able to keep getting rid of the resulting wastes. Another colleague of mine, Eric Toloza, studied mice living in a room maintained at cool temperatures. The chilliness promptly stimulated the animals to double their food consumption to keep warm. Those mice grew not only bigger intestines, but also bigger livers and kidneys, suggesting that those organs also had been operating with little reserve capacity. When the metabolic rates of these animals increased to maintain their body temperature, all those organs had to grow to increase their capacities and to avoid becoming a bottleneck on food consumption and heat output.

If that's true, then many organs may simultaneously impose a ceiling on metabolic rate. I don't know of similar studies in humans, but triathlon competitors are known to have enlarged hearts. I'll bet anything that they, and Tour de France cyclists as well, also have an enlarged intestine, liver, and kidneys. My own guess is that it will turn out that no single organ imposes the bottleneck on metabolic rate. That would be a silly way for evolution to have designed an animal. If by merely enlarging the intestine, say, the whole animal could have metabolized faster, then why hadn't natural selection made that organ larger in the first place?

The Best That Nature Permits

I'd guess instead that evolution has efficiently designed all our organs to be adapted to the same maximal metabolic rate, and all our organs to impose the same bottleneck. No one organ would bottleneck the whole animal, but conversely, no oversize organ would be wasting space on a uselessly large metabolic capacity that other organs couldn't keep up with. Whether nature really has followed such rational design principles, I hope we'll discover in the next few years.

Even an animal at rest needs to sustain a certain metabolic rate to maintain and repair its body. The land iguana above gnaws on a cactus to satisfy its body's most basic energy needs.

Why is there any limit on how far above basal metabolic levels we can crank up our sustained rate? Is there any way to increase our sustained rate without increasing basal rate as well? Probably not, because any metabolic machinery in our bodies incurs idling costs: the costs that I already mentioned as going into the basal metabolic rate. We keep spending energy, even as we lie motionless in bed, to maintain our bodies on the chance that we might have to lift a finger or nurse a baby or step outdoors into the cold later that day. Sure, the marathon runner can exercise and build his muscles, but every bit of extra muscle adds to those idling costs. Furthermore, the runner can't just add some muscle and nothing else: the extra muscle needs more intestine, liver, and kidney to fuel it and clear the wastes. Hence, there's no way to increase sustained metabolic rate without also increasing basal metabolic rate.

Does this mean that there is no point in training if you're a competitive athlete? Of course not. First of all, you have to train to approach your metabolic ceiling, and only a few are going to do that. Second, you can come closer to your ceiling by figuring out better ways to train: that's how Ironman Triathlon times have fallen from just over nine hours to just over eight in the past six years.

Finally, I'm talking about a ceiling on sustained metabolic rate averaged over many days, not about a ceiling on metabolic rate during a few minutes or hours. Yes, sustained metabolic rate limits your training schedule over the course of many days. Yes, it directly limits your performance in very long events, like the Tour de France. But in comparatively brief events like a marathon, it limits your performance only indirectly, through its effect on your training. During the marathon, you power yourself by stored energy reserves, and you don't have to sustain your effort by food consumed during the race itself.

What all this means is that I doubt any average-size athlete will ever maintain a training program of much more than 10,000 calories a day. Most of us—and athletes most of all—don't like to be told, "You'll never be able to do it." We drive ourselves to achieve the impossible. We live by the belief that the human spirit, armed with willpower, can triumph over all obstacles.

But the human spirit lives within a body that nature designed wisely for us. In particular, we are designed such that all our parts work in harmony. That harmony doesn't permit one part to go on burning calories at a rate faster than all the other parts can deliver. Marathon runners can train to push at 17 miles (27 kilometers), but not through the ceiling at 10,000 calories.

Still, that's a more rigorous training program than most marathon runners have time to put in. So I urge you to get out there, train hard, and not get discouraged when you can't get past that barrier. You'll be doing the very best that nature permits.

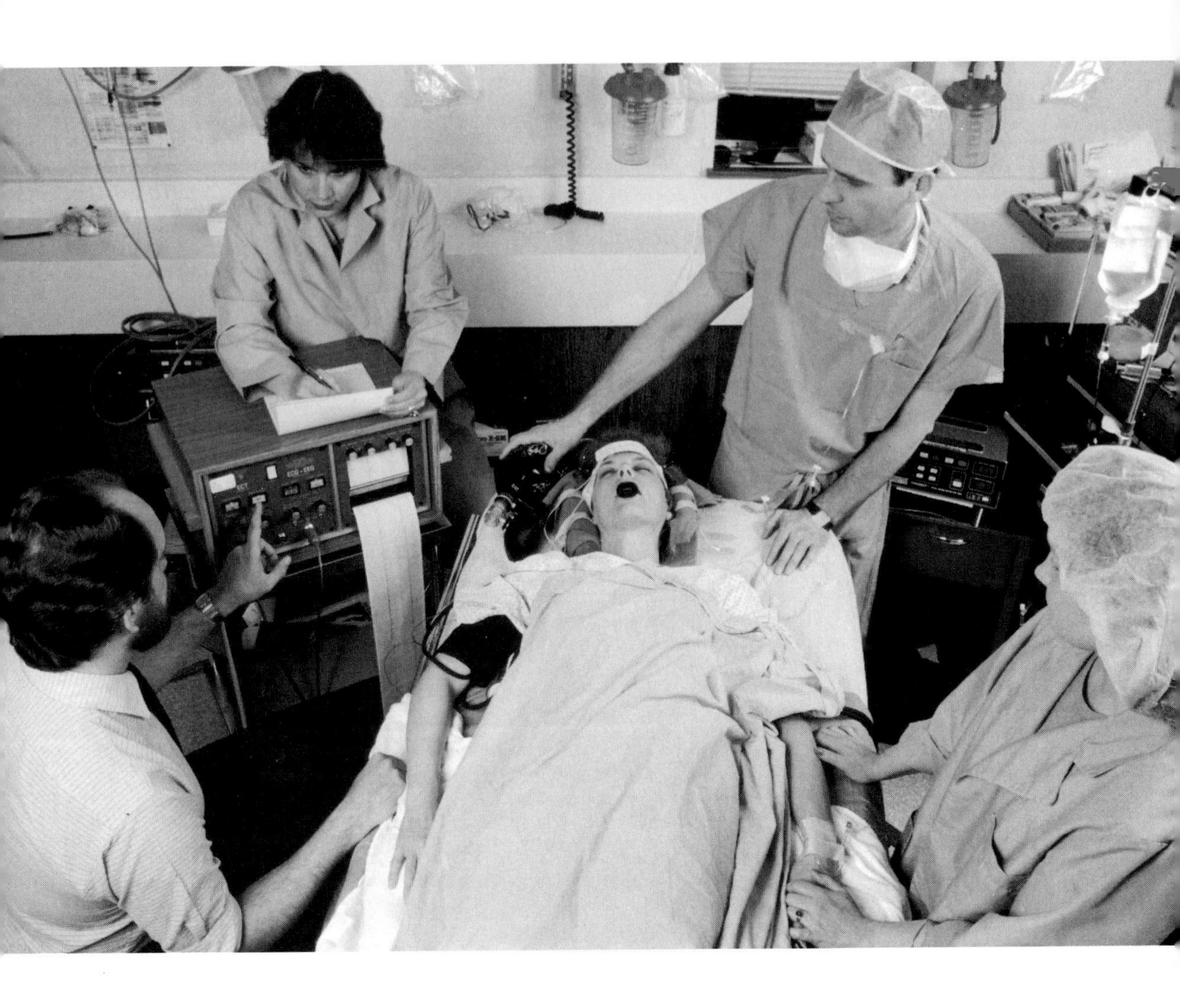

THE QUIET COMEBACK OF ELECTROSHOCK THERAPY

By Daniel Goleman

For Big Nurse in *One Flew Over the Cuckoo's Nest*, it was a tool of terror, and in the public mind, "shock therapy" has retained the tarnished image given it by Ken Kesey's novel: dangerous, inhumane, and overused.

Despite its poor public image, electroconvulsive therapy, or ECT, is in the midst of a quiet revival. In improved forms, electroshock therapy has gained renewed credibility in psychiatry, emerging as the treatment of choice for the most severe depression when drugs and other therapy fail to help.

While there are no nationwide figures on how often electroshock therapy is used, one indicator is the number of treatments paid for by Medicare, the federal health-insurance program for 33 million elderly and disabled people. In 1986 there were 88,847 sessions that were reimbursed, and in 1988 there were 96,276, an increase of 10 percent.

Among psychiatrists, electroconvulsive therapy is viewed as a safe, legitimate way to treat severely depressed patients. The procedure has been greatly refined to minimize the patient's discomfort.

Another sign of renewed vitality is the amount of scientific research being conducted on the treatment. "In the early 1970s, there were only about 125 scientific articles published each year on the topic," says Dr. Max Fink, a psychiatrist at the State University of New York at Stony Brook who is the editor of *Convulsive Therapy*, the main journal on the topic. "By 1980 there were 150, and by 1988 it had climbed to 250."

A Pattern of Expansion

Within the past four years in the New York City metropolitan area, five hospitals have established new units for electroshock treatments, a pattern experts say is typical of the country as a whole. But manufacturers of the devices used in such treatments declined to give sales figures.

Earlier this year the American Psychiatric Association recommended electroshock therapy for severe depression and specified precisely when and how it should be used. In May 1991, at the association's convention in New York City, the report giving the recommendations, "The Practice of Electroconvulsive Therapy," was the number-two seller among about 100 titles offered by the publisher, the American Psychiatric Press. The best seller, as it is every year, was the annual review of new developments in psychiatry.

The patients for whom electroshock therapy is recommended by the report are among the most troubling in psychiatry: they are beyond the reach of drugs or other treatment, and many are so severely depressed they do not eat, sleep very little, and are suicidal. Many suffer from delusions. But in about 80 percent of cases, the report said, electroshock therapy can lift their depression within a few weeks.

"There's no question that ECT is making a comeback, despite its terrible public image," says Dr. Richard Weiner, a psychiatrist at Duke University and head of an American Psychiatric Association task force on the treatment. "It's safer and more effective than ever."

Treatment Induces a Convulsion

Patients receiving shock therapy are given muscle relaxants and a short-acting general anesthetic. Then a brief pulse of electric current is passed through the brain. The current induces a convulsion, or short seizure. For reasons that are still unclear, the seizure seems to relieve the symptoms of depression in many people.

A typical course of electroshock therapy involves 6 to 10 sessions, usually given three times a week.

Dr. Weiner and other proponents say the side effects of electroshock therapy, which in its earlier forms included broken bones, heart attacks, and memory loss, have largely been eliminated. New methods use lower levels of electric current, which proponents say reduces memory loss. Muscle relaxants and anesthesia prevent the muscle spasms that in the past sometimes led to broken bones.

In the 1940s and '50s, electroshock therapy was widely used in mental hospitals, often indiscriminately: patients with problems as diverse as schizophrenia and obsessive-compulsive disorder were treated with it. With the introduction of psychiatric drugs in the 1960s and '70s, and the acknowledgment that electroshock had been abused, it fell into disuse; many hospitals stopped using it altogether.

Within the past decade, though, psychiatrists have recognized that medications do not work with many of the most severely depressed patients, while electroshock does, leading to the recent increase in its use.

Despite the changes, critics say, patients can still suffer long-term memory problems as a result of the procedure.

"Those psychiatrists who use electroshock don't admit it can cause permanent intellectual deficits, not just short-term memory loss," says Dr. Lee Coleman, a psychiatrist in Berkeley, California. "I don't think patients ever really know what risks they are taking when they give their consent for shock therapy. It can leave patients as confused and amnesiac as a brain injury."

Other critics attribute the recent revival of electroshock therapy in large part to its quickness, a virtue that fits well with new efforts by health insurers to save money on hospital costs.

"The use of ECT has increased primarily because insurance companies are pressing to get patients out of the hospital sooner, and they don't want to pay for drugs or therapy that take longer to clear up depression, but have fewer relapses than ECT," says Richard S. Abrams, a psychiatrist at Northwestern University.

Level of Current Varies

Advocates and critics alike agree that the practice of electroshock therapy appears to have changed markedly from the days when its abuse and its sometimes devastating side effects shaped its very negative public image. "There's been a sea of change in our understanding of ECT in the last decade," says Harold Sackeim, deputy chief of biological psychiatry at the New York State Psychiatric Institute.

One refinement is in tailoring the level of electric current to each patient. "Until about 10 years ago, patients were commonly given the maximal charge, with machines set at the upper limit," says Dr. Sackeim, a psychologist who has guided a federally financed program of research on electroshock therapy. "But patients vary greatly in the level of current at which they have a seizure."

For example, patients with thinner skulls need less current, since more electricity passes through the skull to the brain, Dr. Sackeim says.

"On average, a woman needs about 80 percent as much current as a man, and the older you are, the more you need," he says. "You need less current if it is applied only to one side of the head instead of both. A woman of 35 might need half the current necessary for a man of 70."

The level of current is important, Dr. Sackeim notes. His research and that of others showed that confusion and memory loss, the main psychological side effects, were caused by too high a level of current.

"When the dosage is too great or too frequent, patients can have grave memory loss and acute confusion," Dr. Sackeim says. "But if you do it properly, two or three weeks later, patients have no sign of any cognitive problems except for an inability to remember the several hours before the treatment."

One way to reduce memory loss, the task force report recommends, is for the shock to be administered only to the right side of the head. Among most patients, that bypasses the brain's verbal center, which is usually in the left hemisphere, thus minimizing confusion and memory loss.

In a study now under way, Dr. Weiner says that electroshock-therapy patients surveyed six months after their treatment were highly satisfied with the results. "Compared to their attitudes before the treatment, they have a much higher belief that ECT works, and much less fear that it is dangerous," he concludes.

The reason for some of the renewed interest in using electroshock therapy is that there is a growing realization of the limits of medications to treat depression, psychiatrists say.

"After all these years, there's a bit of disillusionment about drugs being the answer to depression," Dr. Weiner says. "We used to think that you could just keep trying one antidepressant after another until you found the right one, but there just is not always one that will work, or that works without having serious side effects."

For the Severely Depressed

About 80 percent of the patients who receive electroshock therapy are severely depressed, Dr. Sackeim says. Another 5 percent are intensely manic, some so disoriented they have stopped eating. Another 5 percent are schizophrenic. The rest suffer from a variety of ailments.

In all cases, electroshock therapy is typically used with patients for whom medications have proven ineffective or when they are in immediate danger of harming themselves.

"We use ECT when we have a patient so severely depressed and suicidal or emaciated that we can't wait the four to six weeks it might take a drug to work," says Edward Coffey, a psychiatrist at Duke University who wrote an article on electroshock therapy in the May 1991 issue of *Hospital and Community Psychiatry*.

More typical, though, is the patient described by Dr. Coffey: "She was an elderly widow, about 70, who had been deeply depressed for the last half-year. She had crying spells and would pace for hours, wringing her hands. She lost all interest in things; she wouldn't go out to play bridge or even to visit her grandchildren. She could sleep only an hour each night. And for months, she wore the same clothes and wouldn't wash, brush her teeth, or comb her hair."

Marked Improvement Seen

The woman's psychiatrist tried treating her with antidepressants, to no avail. Her family brought her to the Duke University Medical Center, where Dr. Coffey treated her.

"Shock" Treatments, Then and Now

Widely used in the 1950s and '60s, electroshock therapy fell into disfavor because of the severe side effects it could cause. Newer techniques have made it safer, some experts say. A panel of the American Psychiatric Association has recommended changes in the way the treatment is offered.

Length of Electrical Charge:
Reduced from one second to one twenty-fifth of a second.

Intensity of Electrical Charge:
Reduced from up to 200 joules, the power equivalent of burning a 40-watt bulb for five seconds, to as little as will induce a seizure in a patient: not above 103 joules, and as little as five.

Timing of Sessions:
Reduced from as often as every day or even two or three a day to no more than three times a week.

Length of Treatment:
Formerly, up to 20 sessions or more; now, typically 10 or fewer.

Preparation:
Formerly, no anesthetics or muscle relaxants; now, brief, total anesthesia, muscle relaxant, and oxygen.

Monitoring:
Formerly, no monitoring; now, EEG (brain wave) and EKG (heart) monitoring.

Placement of Electrodes:
Formerly, on both sides of the head; now, often on right side only, depending on the patient.

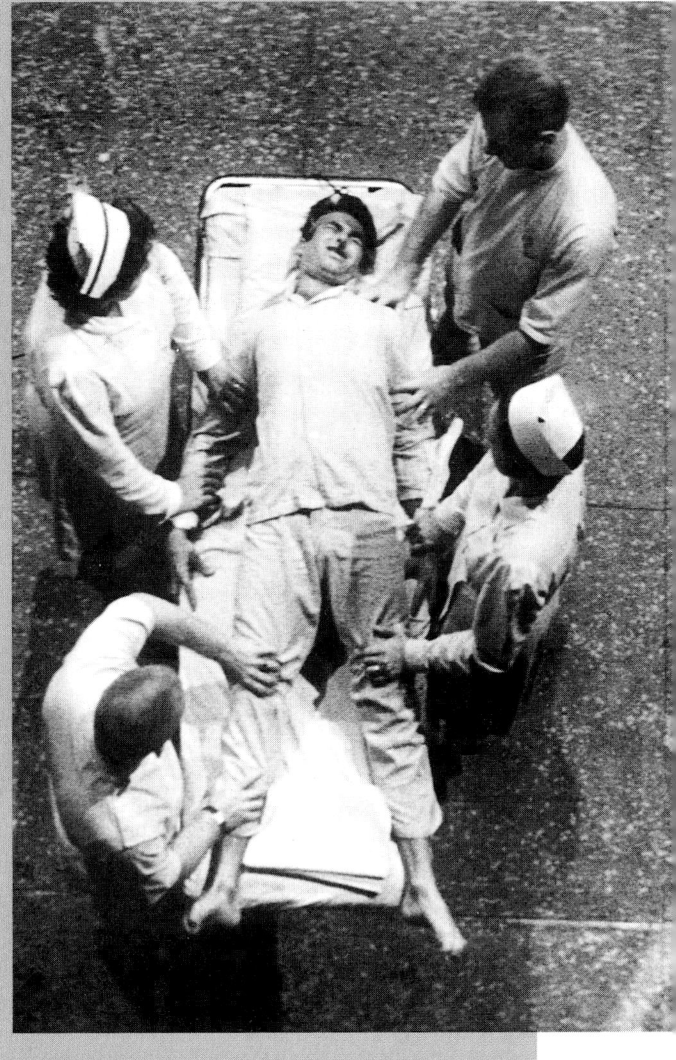

Such movies as Shock Treatment *(above) helped perpetuate ECT's reputation as a barbarous form of abuse indiscriminately used to control psychiatric patients. Recent studies confirm the effectiveness of ECT in treating depression.*

"Within the first week, she began to sleep better, she stopped pacing, and her appetite came back, though she still felt down," he reports. "By the end of the second week, her mood lifted, she started taking care of her personal hygiene, and she started looking forward to going home and visiting with her friends and grandchildren."

Dr. Coffey adds, "It's remarkable, because you see ECT working with a group of patients that nothing else has helped."

For advocates of electroshock therapy, the tragedy is that the treatment is underused, particularly when its effectiveness has been so well documented.

"Most patients who get ECT are better-off patients in private hospitals," Dr. Sackeim says. "It's very underused in public hospitals, partly because the anesthesiology equipment is too expensive. That means there is much needless suffering among poorer patients who could benefit."

TV or Not TV

by Patrick Cooke

If you hate American television, if you believe that it withers all who tumble into its dead blue light, you would have loved Family Pizza Night at Andover Elementary School. Out by the big lake, dozens of children ran through the dusk playing tag; inside the school cafeteria, grown-ups were swapping gossip. Anneliese Reilly was smiling and strangling a wad of pizza dough. "So far, this week without TV has been great," she said. "My family has really been forced to examine how television affects our lives. My son Ian, for instance, watches all those violent cartoons. They get him so worked up that he sometimes bonks me on the head with a book or something. I hate that! But this week he hasn't missed TV at all. I don't think anybody has."

NTV Week they called it in Andover, Connecticut—No TV. "All we're asking is that people turn off the tube and tune into life for seven days," said Dianne Grenier. Some weeks before, Grenier, the town firecracker, had been forced to prove a point to all those friends who accused her of having too much time on her hands. They wondered how she could work a 40-hour week at the Pratt & Whitney aircraft plant in nearby East Hartford, serve on a number of village committees, and still have time to teach at a kids' fishing school and volunteer at two soup kitchens in adjacent towns.

"People kept saying, 'Wow! How do you do all that?' I said: 'Simple, *I don't watch television!*' Look at it this way: Let's say you watch TV 3.4 hours a day. That equals 24 hours a week. If you didn't watch at all, you'd have an eight-day week. Anybody can do what I do!" People knew that Grenier didn't have any kids to slow her down, and that she ran on higher octane than anybody else in town, but when she said, "Try it, just for one week," they started to think, why not?

Now, four days into a fabulously sunny week in April, Andover had begun to understand what Grenier meant. Hundreds of the

town's roughly 2,500 residents had volunteered to take part in the revolt. An organizing committee printed booklets, the *NTV Guide*, listing dozens of alternative activities for the week: wildflower walk, horseshoe competition, puppet show, stargazing, model-rocket launch, fishing derby, historical tour, wine tasting. Dozens of civic groups pitched in, and local businesses offered discounts to NTV participants. Even the owner of the local video store was a good sport: one 99-cent movie rental, *after* NTV Week.

Marcie Miner hung her TV remote control in effigy in her front picture window. The first-grade class at the elementary school posted a list of substitutes for TV: Read a book! Help your mother! Claire Ursin planted a big sign out at the end of her driveway: No Couch Potatoes! There were TV logos—a television with a red slash through it—sprouting out of the daffodil beds all along Route 6, the highway that runs into town. Andover was thinking about television, all right: TV as a waste of time and a hazard to their mental and physical health.

TV addiction, or at the least, obsession—that's what troubled lots of adults in Andover. The town was vegging out. "There are a lot of people here who say they watch all this highbrow stuff. Well, I watch trash," said Danielle Sharlow. "I watch grade-C movies, and the soaps I've followed for years. I admit it." The townspeople also worried that their kids were growing lazy, dumb, and disorderly plopped in front of the tube.

You couldn't blame the people of Andover for believing that TV rots the mind and body, or that it erodes family life, or that it threatens civilization itself. Since the first studies on the effects of television began in the late 1940s, the majority of nearly 3,000 reports have reached these apocalyptic conclusions.

Shadow of a Doubt

But how solid is the evidence? The fact is, after nearly 40 years, research on television's impact remains largely inconclusive or contradictory.

Here, for example, are a few findings from the cumulative wealth of data: Television leads to hyperactivity in children; television makes children passive. Television causes viewer isolation; television comforts the lonely. Television brings families together; television drives families apart.

Ninety-nine percent of U.S. homes have at least one television set, and 98 percent of Americans watch some TV every day. What exactly that means for us as a culture

remains anybody's guess. For decades, however, the accent has overwhelmingly fallen upon the most pessimistic theories.

Purely as a matter of taste, the Connecticut rebellion made sense. In 1961 Newton Minow, then chairman of the Federal Communications Commission (FCC), observed that television was a "vast wasteland." In the past 30 years, TV hasn't improved much. There's just a lot more of it.

Television researchers still regard crass commercial TV as the dark and dangerous heart of the wasteland. The Public Broadcasting Service (PBS) has shed a little light in the darkness, but its share of American viewers remains comparatively small.

America's Most Endangered

In 1978 advertising and public-relations executive Jerry Mander wrote a book that became something of the TV haters' manifesto, *Four Arguments for the Elimination of Television*. Part essay, part exposé, Mander's book gave expression to what many Americans still feel today: "Television produces such a diverse collection of dangerous effects . . . mental, psychological, ecological, economic, political; effects that are dangerous to the person and also to society and to the planet—that it seems to me only logical that it never should have been introduced . . . [Television] qualifies more as an instrument of brainwashing, sleep induction, and/or hypnosis than as anything that stimulates the conscious learning processes."

Other popular books, citing mostly anecdotal material, followed, works like educator and media critic Kate Moody's *Growing Up on Television: The TV Effect*, published in 1980, which described a nation of children in frightening terms: "glazed eyes, drugged, spaced out, zombies. . . ."

That Moody and others turned their critical gaze toward the effects of TV on children is not surprising. There has been no greater source of anguish about the wasteland in the past decade than its supposed effect on kids. For example, Yale University studies have estimated that preschoolers watch between three and a half to four hours of TV a day; school-age children, five to six hours. According to researchers, these children have demonstrated low reading scores because they find it too difficult to stick with the laborious process of learning to read when the temptation of TV is so great.

A University of Oklahoma investigator recently reported that a child who watches 20 to 25 hours a week—the current average for American youngsters, according to the A. C. Nielsen Company—is reduced to a vegetative state, stunting creativity. "Some studies suggest that the more TV a child watches," the Oklahoma researchers cautioned, "the more that child may be at risk for both academic and emotional problems."

The American Academy of Pediatrics, taking aim at commercials aired on kids' programs, warned this year that "obesity and elevated cholesterol levels are two of the most prevalent nutritional diseases among children in the United States, and television viewing has been associated with both."

What IS the Message of the Medium?

Poor reading ability . . . impaired comprehension and creativity . . . elevated cholesterol . . . hyperactive zombies! Just the sort of damage that parents like those in Andover could easily believe emanated from the TV.

Three years ago, however, a monograph funded by the U.S. Department of Education reported a somewhat different side to the study of children and television. The paper concluded that while watching TV may not be the best use of kids' time, a good deal of the documentation claiming it's harmful to them is not very reliable.

The authors of the report, who reviewed scores of studies on television and children's cognitive development before drawing their conclusions, came away convinced that much of the past research is slipshod or biased, and influenced by a prevailing attitude that television is dangerous. Much of what trickles down from academia to the public comes from unrefereed books and unpublished conference papers, or journals with minimal scientific review. Often, the report stated, television researchers are quick to announce a trend, when in fact the research does not justify the claim.

"Beliefs about the negative influences of TV on kids seem to satisfy some kind of need among educated people," says Daniel Anderson, a psychologist at the University of Massachusetts, and coauthor of the Department of Education report. "Those beliefs are easy to reinforce by simply repeating the same things over and over. What we tried to do in this paper was to challenge people with what the actual evidence shows and doesn't show."

So does the evidence prove the nation's children to be zombies? Not really. The report found that school-age children who are observed watching TV look at only about seven and a half minutes of an hour-long program before they begin wandering in and out of the room, and that most of them engage in some other activity—playing with a toy, looking at a book—while the TV is on.

Even while kids are viewing a program, they look away from the television at a rate of about 150 times an hour, the same rate at which adults do. "That's significant for a few reasons," says Anderson. "First, it belies a lot of the language used to describe kids as mesmerized; sure they're engaged with TV temporarily, but we found that they can become equally engaged with a book or a toy. That's the way healthy, inquisitive kids are. Second, some researchers claim that kids get overstimulated because they can't look away from the TV. Well, obviously they do look away. Third, for kids, looking away makes sense. For instance, when an adult male voice is heard, they associate the voice with abstract adult conversation that they don't have any interest in."

The Department of Education report found, too, that there was no good evidence to suggest that a child's creative skills are inhibited by TV. But the widespread claim that reading skills suffer because of the tube is an area where Anderson and his colleagues devoted much of their attention. The report concluded: "Because television is a salient aspect of the home environment, and because children spend so much time with television, it would appear that TV *must* have a major impact on schooling. If it does, the nature of that impact has yet to be determined."

"We bent over backward trying to find some negative effect," Anderson says, "because there is a common belief in a direct trade-off between reading ability and time spent watching TV. But the evidence just isn't impressive."

As for the studies that suggested some beginning readers may be distracted by television, Anderson says, the effect seems only

temporary. "Long-term studies looking at children later on in life found no effect over time."

Parents like those in Andover would perhaps like to believe that if only the television were switched off, kids would become engrossed in literary classics. But, as Anderson says, "That doesn't happen. No way are kids going to run to the library and start devouring Shakespeare."

Television on Trial

In the days that followed Family Pizza Night in Andover, it seemed that more and more townspeople were clambering aboard the NTV bandwagon, and that all 35 TV channels being beamed out over the low hills and quiet lakes of central Connecticut were vanishing unintercepted into the ether. Seventy people turned out for Terri Crimmins's evening of children's stories. The chamber-music concert was unfortunately canceled (not enough rehearsal), but there was a good showing at the safe-boating course, and a full house for bingo.

Less well attended was a seminar called "Television on Trial," held in the town library's tiny basement. Dianne Grenier was there, along with a few other NTV stalwarts who admitted they didn't ever watch much television. A film crew from Channel 8 in New Haven drove up to shoot a spot for the 11 o'clock news.

Ruth O'Neil handed around a list of grim statistics about television before the discussion got under way—number of hours a day that the TV is on in the average household (6.7), the number of American households owning VCRs (75 percent), things like that. At the bottom of the list was this: "By age 16, the average American child has seen 18,000 murders on TV."

That this particular item was included would have pleased the National Coalition on Television Violence. For more than a decade, the alliance, whose leadership includes physicians and educators, has been trying to drive its message home: TV violence is at the heart of the nation's crime woes. "Many criminals admit to having gotten their ideas from TV," reads a pamphlet the group mails out to interested citizens. Murders and rapes are more likely to occur because of TV violence, it says, adding: "It increases the chances that you will be mugged in the street or have your belongings stolen."

It's not just the cop shows that are responsible, either. The coalition notes that " 'Bugs Bunny,' 'The Roadrunner,' 'Popeye' . . . have been proven to have a clear and harmful effect on children." Don't laugh—according to the group, the comic violence on PBS's "Monty Python" is harmful to adults.

Whether it's children, adolescents, or adults being scrutinized, television's presumed propensity for inciting violence in Americans is perhaps its critics' direst concern.

One of the most influential pieces of research on TV and children's aggression is the renowned Bobo-doll study of 1963. A group of children was shown a film of a man commanding a doll—one of those big plastic clowns that pop back up when you bop it in the nose—to get out of the way. When the doll didn't comply, the man sat on it, hit it with a mallet, and punched it while yelling threatening remarks like, "Pow, right in the nose, boom, boom."

When the children were allowed into the room with the doll, they did just what they had seen in the film (as opposed to a control group that had not seen the film and did not display any aggressive behavior). Researchers concluded that children could indeed learn violent behavior from watching a filmed experience.

But not everyone agrees that this and similar experiments prove anything. "My sense of studies like the Bobo doll is that they don't occur in any natural setting," says Paul Hirsch, a sociologist at Northwestern University. Hirsch, something of a nonconformist in the TV-research community, has studied television's effects for more than a decade. "You could argue that the kids were invited to hit the doll. They saw a grown-up do it, and then they were left alone in the room with the doll to do whatever they felt like. I don't know that I'd base any sweeping social changes about TV on research like that."

Since the Bobo-doll era, numerous other studies have suggested that television can increase violence. Most investigators now agree that a link may exist between heavy television viewing and aggression, and the American public has been quick to agree with that conclusion. But except for those who earn their profits from TV, few question the research's logic or bias.

"What's always bothered me about these studies," says Hirsch, "is this willingness to assume the worst about people. Everybody seems to think that we all model our behavior on the guy in the TV show who murders somebody. Nobody ever takes into account that the murderer *always* gets caught, and the message is that crime doesn't pay.

"Someday I'd like to see a content analysis of TV shows that notes how many times people are nice to one another, instead of only measures of the darker side of the human personality," says Hirsch. "Look at a popular show like 'The Wonder Years,' and count how many times somebody does something kind for someone else. Statistically, you could come to an equally misleading conclusion about TV—that over a lifetime a kid will see 180,000 instances of altruism; therefore, TV causes niceness."

George Gerbner, dean emeritus of the Annenberg School of Communications at the University of Pennsylvania in Philadelphia, sees the picture differently, however. Gerbner claims that heavy viewers—generally thought to be those who watch television more than three hours a day—suffer from what he calls "the mean-world syndrome."

"The television in some cases is on seven hours a day from the moment a child is born," says Gerbner, who has studied TV and its effects since the mid-1960s. "Over time, it cultivates a sense of violence as the solution to many problems. It also generates a sense of insecurity and danger in viewers, who see themselves as potential victims."

Seems plausible, but such theories, according to Hirsch, ignore other contributory factors. "Let's say I live in a high-crime area, and I also watch a lot of TV," he says. "There is no way to make a statistical connection between having the set on all the time and my beliefs about crime in America. What is

Fears You Can Turn Off

While controversy still surrounds some of TV's effects, several health concerns have been laid to rest. To help you separate the truth from the static, here are answers to a few of the most common questions concerning television and your health:

Is Radiation a Danger?

Not anymore. In 1967 health regulators became concerned that some color televisions were leaking unacceptably high levels of X rays. The problem (never an issue for black-and-white sets) was remedied in all color models built after 1970, but not before a generation of children was admonished to scoot back several feet from the television screen, "just in case."

Many, in turn, passed the precaution along to their own kids, but such concern today is entirely needless, says William Beckner, senior staff scientist with the National Council on Radiation Protection and Measurements, one of the first organizations to sound the alarm 25 years ago. "Modern receivers are built differently, using lower voltages and better shielding," he says. "No matter how close you sit to the set, X rays just aren't a problem."

Will TV Damage My Vision?

No. The contrast between a bright set and a dark room temporarily tires some people's eyes, as does the reflective glare off the screen from a poorly placed lamp, but neither situation will lead to long-term damage, says Theodore Lawwill of the American Academy of Ophthalmology. Some people with mild cataracts, he says, may even see the screen better in dim light.

Nor is there any need to fear that sitting nose to nose with favorite screen characters will make children nearsighted. Young children are able to focus sharply on objects as close as an inch or two away from their eyes; that distance lengthens as they get older. "Children like to be as close to the action as possible, and would climb into the TV if they could," Lawwill says. "That may block the screen, but it won't hurt anybody's vision."

Will TV Make Me Fat?

Not exactly. The light and sound waves from a television don't carry calories, of course, but many researchers do think there's a subtler link between a voracious appetite for television and developing obesity.

Deborah Franklin

the measure of my reality? I live in a dangerous neighborhood. I'm poor. The TV is always on—even though it's not clear that I'm watching it at all. It's misleading to conclude that because the TV runs all day long, I've learned from television to be fearful of crime."

TV is the conspicuous fall guy for such social ills as violence. But, says psychologist Bernard Friedlander of Connecticut's University of Hartford, in the rush to fault television, psychologists and social scientists have lost sight of the best predictor for violence: American society. "Television is an obvious target, but it's almost irrelevant," says Friedlander. "America was a violent society long before TV came along. All television does is reflect and amplify the violence."

Plato Feared Storytelling

There were so many people eager to attend the NTV closing ceremonies that the final evening's potluck supper was moved from the First Congregational Church to the town school. No one was about to be rooked out of a hard-earned NTV Certificate of Accomplishment for having survived seven days and seven nights without TV.

Dianne Grenier stood at the podium under the blinding lights of TV news cameras and told the 200 or so gathered that, as far as she was concerned, NTV Week had been a "hoot." If the more than 30 media interviews she had given were any indication, Andover—Bumpkinville, as she called the pioneering town—had struck a national nerve with its boycott.

Most young parents at the closing ceremonies were too young to recall that golden era of radio—or the studies of the 1930s and 1940s that warned of the harmful effects radio was having on children's school performance and ability to distinguish fantasy from reality. "This new invader of the privacy of the home has brought many a disturbing influence in its wake," one researcher wrote about radio in 1936. "Parents have become aware of a puzzling change in the behavior of their children. . . ."

Perhaps it's only natural that each generation laments its lost, carefree past and foresees danger in the future. Plato's *Republic* warned about storytellers: "Children cannot distinguish between what is allegory and what isn't, and opinions formed at that age are usually difficult to change." Socrates believed that writing would destroy memory. The early days of film prompted books like *Movies, Delinquency, and Crime*. And comic books, blamed for poor reading skills in children, were thought to cause linear dyslexia, a condition brought on by constantly shifting the eyes vertically from picture to text. (A current medical worry is that children who watch TV excessively have difficulty reading because they do not learn to shift their eyes *enough*.)

Wide World of Entertainment Options

Andover was merely continuing a grand human tradition. Television may not prove any more harmful to the town than radio did more than half a century ago. But that doesn't mean NTV Week was a waste of time. There are plenty of reasons to forgo American television: moronic sitcoms, tabloid news, beach volleyball, talk-show celebrities with their tiresome confessions. . . .

Three hours of television a day works out to more than a thousand hours a year. By the time American teenagers today reach the age of 70, they will have watched more than seven years' worth of TV.

What Andover found is that there are better, or at least other, ways to have a good time. And for a week at least, the fun was in rediscovering the faces of people in town, rather than silently staring at sterile electronic images.

The NTV banner may well in fact be leading a growing parade, if the results of a recent Gallup poll are any measure. In 1974 half of those surveyed named television viewing as a favorite evening activity; in 1991, only about two in 10 did. What's more, those who felt most disillusioned with TV and expected to watch less in the future were younger Americans between ages 18 and 29, precisely the age group television advertisers must retain if the medium is to remain the message.

"You know what my biggest discovery was this week?" asked Corrine Ackerman as people filed out of the cafeteria and headed home on a warm spring night in Andover. "I'm the one who turns on the TV, not my son."

Twelve-year-old Heather Freeto missed her favorite show—"Beverly Hills 90210"—during the week, but not *that* much, she said. "I played a whole lot of volleyball and messed around a lot with my friends. It was cool."

John Whitman said: "We had loads of fun this week with our kids. I don't remember when I talked so much with my wife."

NTV Week had come to a close. Outside in the parking lot, a few people lingered on under a full moon that rose out over Andover Lake. Parents cradled sleepy children in their arms.

Bill Dakin, sitting at the wheel of his camper, was asked what he was going to do when he got home. "Watch TV," he said. "I promised the girls they could." But when he turned and looked at his small daughters in the backseat, they were slumped against one another, nodding fast.

"Besides," he said with a wicked smile, "the basketball play-offs are on."

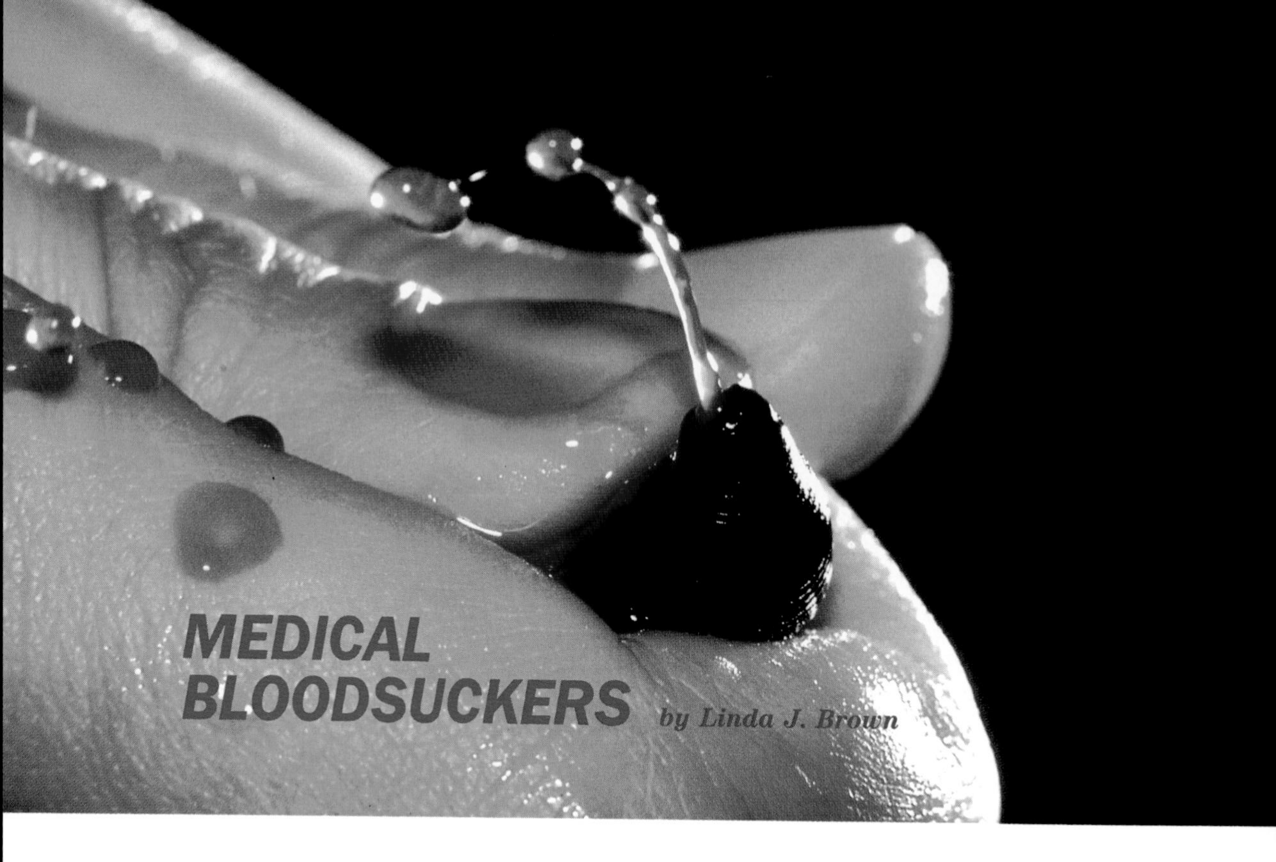

MEDICAL BLOODSUCKERS

by Linda J. Brown

The status of leech therapy has risen in recent years, thanks largely to the greater frequency of "reattachment" operations. Leech saliva (above) contains a variety of medically useful chemicals.

Leeches—just the thought of these slimy, bloodsucking creatures can send shivers up and down the spine. They attach to their prey with either an extendable, tubelike proboscis or with three knifelike jaws. And they are sneaky devils—their saliva contains an anesthetic that prevents pain when they fasten to their victim, so they often remain undetected while they happily engorge themselves with blood. But their bad reputation may be unjustified—leeches actually have some very important uses in medicine and medical research.

For instance, an increasing number of plastic and reconstructive surgeons around the globe use leeches to maintain circulation in small blood vessels during delicate operations, such as for finger, ear, toe, and skin-flap reattachments. The saliva produced by leeches is also the focus of some exciting research that may lead to the development of drugs that can dissolve dangerous blood clots or prevent reformation of clots.

Leeching through the Ages

These medicinal applications have greatly elevated the status of the lowly leech in recent years. But surprisingly, its claim to fame can be traced back to the early days of bloodletting, or bleeding. More than 2,500 years ago, bloodletting was employed by healers who believed that they were ridding the body of evil spirits and poisonous fluids that cause disease. Sharp objects such as animal teeth, thorns, pointed sticks, and bones were used. More often than not, life-threatening infection or serious blood loss resulted.

Using leeches to draw blood was developed as a safer alternative. Members of the genus *Hirudo,* called medicinal leeches, were used in the Far East for hundreds of years before traders passed the information on to Europeans. The first known Westerner to use leeches for this purpose, according to Cheryl Halton, author of *Those Amazing Leeches,* was Nicander of Colophon, a Greek healer who lived 2,000 years ago.

The idea of evil spirits inhabiting the body eventually fell from favor around 400 B.C. But the tradition of leeching continued, ostensibly to balance the body's "humors," the four life-giving bodily fluids: blood, phlegm, yellow bile, and black bile. This practice continued through the years and became quite popular in Europe in the 18th and 19th centuries. Halton writes that during this time as many as 100 million leeches per year were used in France alone. Medicinal leeches were in such great demand that their population dropped drastically, to the point that England and Russia passed laws to protect the creatures.

Americans never practiced bloodletting or leeching with quite the same zeal of the Europeans, although Halton writes that "1.5 million leeches were used each year in the United States during the 50 years following the Civil War."

In 1885 microorganisms became recognized as the true origin of disease. With this shift in thinking, bloodletting fell to the wayside, although the use of medicinal leeches never disappeared entirely. It wasn't until very recently in this century, however, that leeches began to gain renewed respect in the medical community.

Maintaining Circulation

Leech use is still far from widespread, but a growing number of plastic surgeons are using these bloodsuckers with great success. Kuwant Bhangoo, M.D., chief of plastic surgery at Mercy Hospital in Buffalo, New York, had his first leech experience in June 1986 when treating a 46-year-old man who had accidentally amputated his right ear in a fall. Dr. Bhangoo reattached the ear, using delicate microsurgery that involved reconnecting an artery to restore proper blood circulation to the ear. Soon after the operation, Bhangoo noticed that blood was flowing into the ear, a positive sign that the artery was functioning as it should.

However, because of the nature of the injury, the tiny veins of the ear had not been reconnected during the surgery. As a result, the blood flowing into the ear from the reconstructed artery could not flow out. The ear began to swell and turn blue, indicating that the stagnant blood was not circulating and thus not being reoxygenated. If nothing was done, the tissue would ultimately die, and the ear would be lost.

Dr. Bhangoo had read about the recent use of leeches, and in Great Britain while training for general surgery, he had heard reports of leeches being used in European hospitals. He thought the creatures might provide a good solution for promoting blood circulation to the ear while giving the veins from the skull and the ear time to hook up on their own, a process that usually takes about five days. So he presented the idea to the patient, who agreed, although Dr. Bhangoo sheepishly admits that "He was agreeable

Roy Sawyer, a scientist and the head of Biopharm, a leech farm in Swansea, Wales, forages for leeches in their freshwater habitat. Hirudo medicinalis *is the only leech species commercially marketed.*

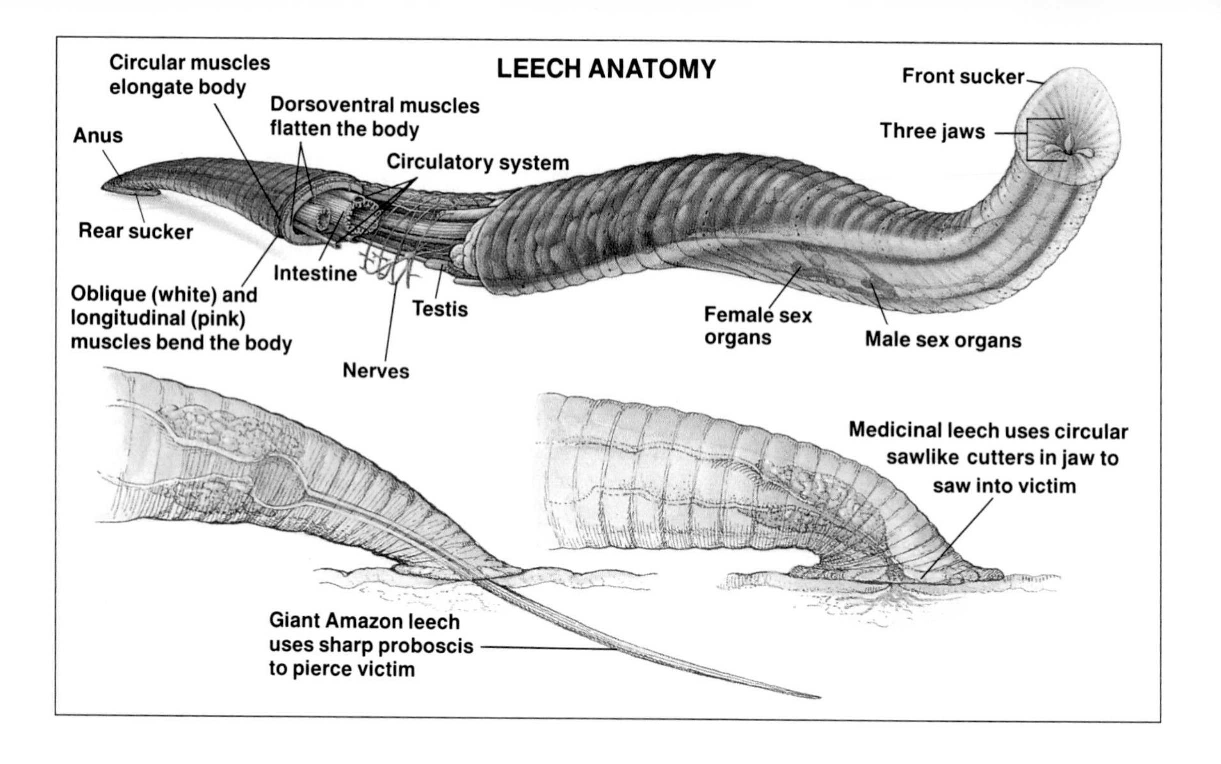

considering the fact that if we didn't do anything, he would lose the ear."

The leech was placed over the attached ear, where it hungrily began sucking blood. Within minutes "the improvement was dramatic," says Dr. Bhangoo. He watched the 1-inch (2.5 centimeter) flat leech grow in front of his eyes for 15 to 20 minutes. Then, engorged with blood and now resembling a cigar, the leech fell off. Even after leeches detach, however, the blood continues to ooze from the tiny bite for up to 24 hours, thanks to the presence of *hirudin,* an anticoagulant (blood thinner) present in the leech's saliva. Another substance in the saliva dilates the tiny blood vessels, permitting more blood to flow from the incision to aid the blood-circulation process.

Over five days, Dr. Bhangoo applied 15 leeches, and the results were wonderful. "The patient still stops in every now and then to say hello and show off his ear."

Besides ear reattachments, medicinal leeches have improved the success of operations to reattach fingers, fingertips, and parts of noses, lips, and scalps. Dr. James Apesos, chairman of plastic surgery at Wright State University School of Medicine in Dayton, Ohio, says, "I personally do not like leeches. I find them creepy and crawly, but I put them on, and they do the trick." He's used them in such varied circumstances as nose reconstruction, breast augmentation where the

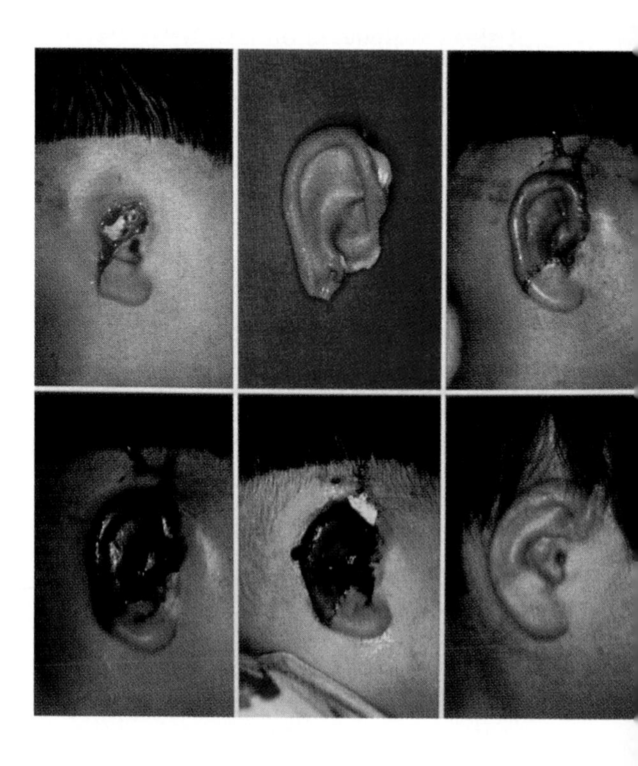

Leeches can make all the difference in the success of an ear-reattachment operation. Used post-operatively, leeches help maintain blood circulation to the area until the crucial veins that nourish the ear have the opportunity to reconnect.

The anticoagulant properties of the saliva of the Amazon leech (left) may be significantly more powerful than those of any drug yet available.

DOWN HOME ON THE LEECH "PHARM"

Driving up to the grand manor house in Swansea, Wales, most first-time visitors have little idea of what's in store for them. The manor house's charming Victorian style belies the fact that here live 60,000 to 80,000 leeches! This is the home of Biopharm, the world's only commercial leech farm.

Roy Sawyer, scientist and head of Biopharm, reports that his company has experienced approximately 30 percent growth each year for the eight years they've been in business. "It's really quite remarkable," Sawyer says. "I keep thinking it's going to plateau one of these years, but so far, it hasn't."

At any one time, Biopharm has on hand some 15 of the 650 species of leeches. But the only leech that Biopharm markets commercially is the European medicinal leech, *Hirudo medicinalis*. The others are kept purely for research purposes. This includes the Amazon leech that Sawyer found in French Guiana, South America. The Amazon, which can grow to 18 inches (45 centimeters) in length, is the world's largest leech. This leech contains a type of anticoagulant called *hementin,* which may be more powerful and effective than any type of anticoagulant available today.

Leech farming is no breezy task. Biopharm's leech building is comprised of three main rooms, each differentiated by temperature. The warm room holds the tropical-leech species in various fresh-water-filled vats, tubs, and tanks. The middle-temperature room is reserved for leech-breeding purposes. And the room with cooler water temperatures stores dormant leeches, which makes them easier to ship.

Biopharm's leech buildings hold vats of fresh water kept at varying temperatures. The warmer water encourages breeding; cooler water promotes dormancy.

The tops of all the leech containers are covered with light cloth secured by Velcro strips. "Velcro is the best thing that ever happened to leech farming," explains Sawyer. "Leeches tend to crawl out; they're more like frogs than fish. So Velcro is a wonderful barrier." Not the picture that pops into your head when you think of a farm, but then, leeches aren't your ordinary farm animals, either.

flow of blood to the nipples was impaired, and in surgery to relieve a pressure sore on a paraplegic child.

Clot Buster

The leech's role in reattachment surgery is quite notable, but leeches may make their biggest contribution in the field of drug therapy for thrombosis, or blood clotting. Blood clotting is a normal response that prevents bleeding to death when the body is injured. But blood clotting becomes life-threatening when a clot forms within a blood vessel and grows to a size that prevents blood from passing through unimpeded. A blood clot that forms within an artery supplying the heart muscle or the brain can lead to a heart attack or stroke.

Scientists have discovered that hirudin found in leech saliva inhibits a key blood-clotting factor in the body, called *thrombin.* For humans, it may have exciting potential as a blood-clot-preventive drug and as a clot-dissolving treatment.

Scientists have now developed a synthetic version of hirudin using DNA recombinant technology. This genetically engineered substance is being tested in laboratories around the world to unlock its potential.

Hirudin may have lifesaving applications after balloon angioplasty, a common procedure performed on people whose clogged coronary arteries put them at risk for heart attack. In this procedure a thin tube with a tiny balloon at its tip is snaked through arteries to the part of the coronary artery clogged with accumulated fatty deposits, known as plaque. The balloon is then inflated, squashing the plaque against the vessel wall and restoring blood flow. Angioplasty saves many lives by preventing heart attacks, but it may cause damage to the vessel wall by triggering a blood clot—exactly what doctors are trying to avoid. To prevent this potential problem, doctors typically give patients an anticlotting drug, such as heparin, which, unfortunately, is not always successful.

In a comparison study of the clot-preventing prowess of hirudin and heparin, Valentin Fuster, M.D., Ph.D., chief of the cardiac unit at Massachusetts General Hospital in Boston, and James Chesebro, M.D., professor of medicine at the Mayo Medical School in Rochester, Minnesota, studied 50 pigs that had undergone balloon angioplasty. The study found hirudin to be 10 times more effective than heparin at preventing the clots that can develop following balloon angioplasty.

A similar study by Ian Sarembock, M.D., and colleagues from the University of Virginia Health Sciences Center in Charlottesville also showed significantly less renarrowing of the vessels after angioplasty in rabbits receiving hirudin versus those that received heparin.

In addition to blood-clot prevention, researchers are also examining the clot-dissolving abilities of hirudin. Fuster and Chesebro tested hirudin's power to dissolve clots that often form following atherectomy, the surgical scraping of plaque from vessel walls. In another study using pigs, the animals were divided into three groups. One group received a placebo as a control; a second group received tissue plasminogen activator (t-PA), a drug used during heart attacks to break up clots impeding blood flow to the heart; and the third group received a combination of hirudin and t-PA. The t-PA/hirudin therapy provided far superior results than the t-PA alone.

The next big question facing researchers: Will hirudin work as well on people as it has on animals? "I think hirudin's potential is enormous, but we must proceed with caution and carefully prove its safety and efficacy on humans," says Dr. Fuster. Two multicenter human studies are now under way in this country. One study is evaluating the clot-busting ability of t-PA combined with either heparin or hirudin on people suffering heart attacks. The other is comparing hirudin to heparin in people with unstable angina, a type of chest pain caused by a blood clot.

In the meantime, Roy Sawyer, scientist and head of Biopharm, the world's only commercial leech farm, is focusing research on the Amazon leech, which produces a different anticoagulant called *hementin.* He and his researchers are in the process of developing a recombinant version of hementin. Sawyer feels that hementin may be able to dissolve particularly tough clots that he says other drugs can't break up.

As if all of these advances weren't enough, researchers are also using genetic engineering to develop antibiotic and anticancer agents found in leech saliva. So stay tuned for more medical advances coming from what certainly must be nature's most unexpected source—the leech.

REVIEW

Behavioral Sciences

A study of twins shows that genetics apparently plays a stronger role in fostering alcoholism among males than it does among females.

THE NURTURE OF ALCOHOLISM

A study of twins suggests that genes play only a minimal role in fostering alcoholism among women of all ages and among men whose drinking problems begin during adulthood. A parent's alcoholic behavior and other family influences may help lead to alcoholism in these groups, said psychologist Matt McGue of the University of Minnesota in Minneapolis, who directed the project.

Genes importantly influence alcoholism that develops among some men during adolescence, often accompanied by illicit drug abuse and delinquent behavior, according to the study.

McGue's team studied 85 pairs of male identical twins, 96 pairs of male fraternal twins, 44 pairs of female identical twins, 43 pairs of female fraternal twins, and 88 pairs of opposite-sex fraternal twins. The researchers chose one twin who had undergone hospital treatment for alcoholism, and then located the sibling.

Alcoholism and alcohol abuse, along with abuse of illegal drugs and delinquency, appeared much more often among both male identical twins than among both male fraternal twins, but only when the treated twin had developed problems during his teens. Female identical and fraternal twins exhibited no differences in the rate of alcohol problems. For opposite-sex twins, only the brothers of treated women displayed an excess of alcoholism.

STRESS AND THE COMMON COLD

Psychologists found that emotional stress increases the likelihood of catching a cold, apparently by interfering with immune responses that fight cold viruses.

Sheldon Cohen of Carnegie-Mellon University in Pittsburgh, Pennsylvania and his colleagues administered psychological-stress questionnaires to 420 healthy adults. Each volunteer reported the number of major stressful events experienced in the past year, feelings of control over daily demands, and the frequency of negative emotions such as depression, anger, and irritation.

Next, 394 participants received nasal drops containing one of five cold viruses; the remaining 26 received drops without a virus. For two days before and seven days after getting nasal drops, volunteers stayed in large apartments (either alone or with a few others) where physicians examined them daily.

A total of 148 virus-exposed volunteers developed colds; none of those who got virus-free drops developed a cold. Volunteers who reported high levels of psy-

Emotional stress may inhibit the immune system, making the body more prone to colds and other minor infections.

chological stress ran twice the risk of getting a cold as those reporting low stress levels. This pattern remained steady when experimenters accounted for other factors that can reduce immune responses, such as advanced age, allergies, cigarette and alcohol use, sleep problems, and housemates with a viral infection.

EMOTIONAL LEGACY OF DIVORCE

Family conflict in two-parent households may exert as much emotional harm on children as parental divorce, according to researchers who analyzed the results of two large surveys conducted in England and the United States. For boys, family conflict before a divorce influenced later school and behavior problems as much as the divorce itself, the scientists said. For girls, predivorce family conflict made a smaller impact on later school and behavior problems. However, the surveys may have missed symptoms of anxiety and depression among girls that did not get translated into blatant misbehavior.

The British project collected parent and teacher ratings of behavior and school-achievement scores for more than 11,600 unrelated children at ages 7 and 11. All children lived in two-parent families at age 7; 239 divorces or separations occurred over the next four years. The U.S. survey collected information on behavior problems for 822 unrelated children at ages 7 and 11 living in two-parent families; four years later, 65 divorces or separations had occurred.

Other researchers have argued that youngsters—especially girls—suffer the worst emotional consequences of divorce during young adulthood. The ongoing British and U.S. surveys will test this assertion, since children in the projects have now reached 16 to 23 years of age.

Married couples who share experiences tend to make well-corresponding adjustments of values and attitudes over the course of their years together.

SHARED EXPERIENCE AND ADULT ATTITUDES

Psychologists who reexamined data from a study that ran from the 1930s to the 1950s concluded that the experiences shared by married couples largely determine their attitudes and values during adulthood. Successfully married couples apparently strive to create shared experiences, such as common involvement in recreational and religious activities, that promote a sense of belonging and connection, asserted study director Avshalom Caspi of the University of Wisconsin, Madison.

In this way, adult personality development differs sharply from that of children and teenagers, Caspi said. Each youngster in a family experiences sibling and parent relationships differently, and these differences—rather than shared experiences—guide personality development through adolescence, he contended.

Caspi's team looked at measures of general values—such as religious and political views—and attitudes toward marriage—such as beliefs about marital fidelity and child raising—completed by 165 married couples just after their engagement and again 20 years later.

Couples did not gradually grow more alike, as many people assume. Instead, their values and attitudes remained moderately similar over 20 years. Husbands and wives also made corresponding adjustments in values and attitudes over the course of their marriage, suggesting that shared experiences of some kind maintained their similarities, Caspi asserted.

Bruce Bower

Food and Population

Numerous local conflicts in Africa have produced millions of refugees, a situation that has compounded the already severe strain on food supplies.

The year 1991 was not a good one for the international food system. Food production, rather than matching the customary increase of about 93 million people (roughly the population of Mexico), declined by nearly 5 percent (86 million tons). Global stocks of food will be drawn down to their lowest level in 15 years—well short of the cushion needed to guarantee consumption requirements.

Because the distribution of even 1990's bumper crop left the world's hungry regions worse off than in 1989, it is clear that the extent of hunger and malnutrition in 1992 will depend heavily on the as-yet-unpredictable 1992 harvest. The 1991 production shortfall took place almost entirely in the developed countries, which generally produce the surpluses that feed the less-productive regions where most of the hungry people live.

Food prices will rise (because the United States and the European Community subsidize them above market levels), and appropriated funds will therefore buy less food to donate to those who lack purchasing power. There will also be less food to trade. This means that last year's decline in food trade will continue, along with the inability of many countries to purchase food because of the heavy burden of external debt.

Southern Africa is facing its worst drought of the century; unusually serious food shortages will occur there and in eastern Africa. In the latter region, the misery is deepened by the presence of nearly 15 million refugees from various local conflicts—mainly in Somalia and Sudan. As is always the case, it is chiefly women and children who suffer. There are also supply difficulties in Iraq, where Gulf War damage and inadequate imports have left even minimum needs unfulfilled.

The dramatic events in Eastern Europe and in the former Soviet Union tended to remove from the public dialogue systemic issues like the extent of hunger and what to do about it. And the upcoming presidential election campaign in the U.S. will probably not focus very significantly on that subject.

Nevertheless, food and agriculture continue to be central to the relations between the industrialized North and the underdeveloped South. Until the hungry countries increase their own food production and improve their food-distribution networks, we are unlikely to witness any significant decline in world hunger.

An additional major set of policy questions bearing on the food system concerns the impact of agriculture and the distribution of its products on the physical environment. Because land, water, energy, and technology are the major supply-side inputs into the food system, and all of them strongly affect the environment, it is expected that agricultural considerations will loom large in the 1992 U.N. Conference on Environment and Development, which will occur in Rio de Janeiro, Brazil, during the first 10 days of June. This conference, the largest international meeting ever held, will focus upon, among other things, the extent to which the use of toxic pesticides, chemical fertilizers, and irrigation water on increasingly fragile soils not only threatens the quality of food, but also raises the possibility of irreparable environmental damage.

Martin M. McLaughlin

REVIEW

Genetics

GENETICS OF HOMOSEXUALITY

In December 1991, a study was published in the *Archives of General Psychiatry* on the role of genes in determining male homosexual behavior. The investigation, conducted by psychologist J. Michael Bailey of Northwestern University in Evanston, Illinois, and psychiatrist Richard C. Pillard of Boston University School of Medicine, studied adult male homosexuals who had either a twin or an adopted brother. Among the twin pairs, identical- and nonidentical-twin pairs were distinguished. Identical twins are derived from a single fertilized egg, resulting in the birth of twins that are of the same genetic constitution, and have also been exposed to the same prenatal and postnatal environments. By contrast, each nonidentical twin is derived from a separate fertilized egg. As a result, they have as many genetic differences between them as any pair of brothers, but they have also been exposed to very much the same prenatal and postnatal environments. The adopted-brother group represents genetically unrelated males who have had as close a common postnatal environment as possible.

The investigators studied 56 pairs of identical twins, 54 pairs of nonidentical twins, and 57 pairs of adoptive brothers. They found that in 52 percent of the identical-twin pairs, both brothers were homosexual, while only 22 percent of the brothers in the nonidentical-twin group were both homosexual. Eleven percent of both brothers in the adoptive-brother group were homosexual. This pattern, in which the frequency of a shared sexual orientation between brothers lessened as the genetic relatedness of the brothers decreased, strongly suggests that homosexuality is genetically determined. But since there was not a complete correspondence of homosexuality between identical-twin brothers, it would appear that there is also an environmental factor involved.

PARENTAL IMPRINTING

Recently W. P. Robinson and colleagues from the Institute of Medical Genetics in Zurich, Switzerland, reported on their studies of two genetic diseases—Prader-Willi syndrome (PWS) and Angelman syndrome (AS). Although mental retardation characterizes both syndromes, they are easily distinguishable based on other abnormal traits.

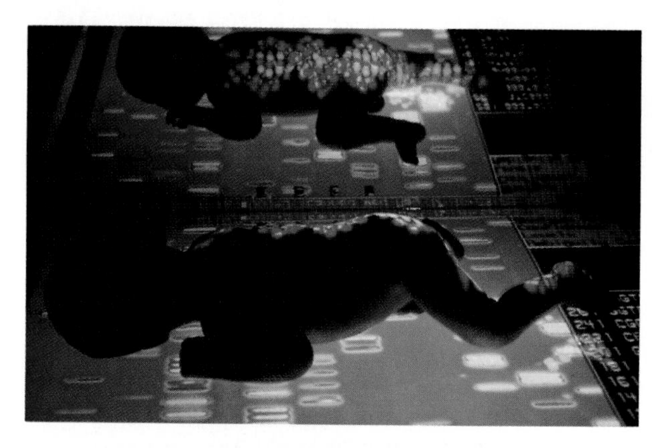

As geneticists gradually unravel the human genome, a wide variety of behaviors, as well as a number of diseases, may be found to have a genetic link.

In both syndromes a section of the chromosome-15 pair is missing. Curiously, in PWS, it is always the chromosome 15 inherited from the father that has the deleted section, whereas in AS, it is always the maternal number 15 chromosome that has the deletion. In these syndromes, different human genetic diseases result from chromosomal deletions that are identical in extent but different in parental origin. The parental determination of a gene's effect is called *parental imprinting,* or *genetic imprinting.*

Further studies show that an intact chromosome number 15 must be inherited from each parent for normal human development to take place.

Louis Levine

Health and Disease

CRISES IN HEALTH CARE

The United States leads the world in health-care spending: over 12 cents of every dollar of national income is used to buy health-care services, and the amount continues to grow. Rapidly growing health spending aggravates budget difficulties of individuals, corporations, and state and federal governments. And while many people receive high-quality health care, millions of Americans lack adequate care. Over 35 million Americans lack health insurance, which means they receive lower-quality care than that received by people with insurance. For example, a study of almost 30,000 infants directed by Paula A. Braveman at the University of California at San Francisco found that sick newborns who lack insurance coverage receive much less medical care in hospitals than insured newborns.

In contrast to the United States, Canada, France, Germany, and Japan achieve near-universal health-insurance coverage while being more successful in controlling costs. U.S. officials have expressed increased interest in health-care reform, and various plans for revamping the nation's health-care system were put forward during 1991 and 1992. The problem also promised to become part of the election-year debate. However, many consumers were skeptical that any significant reforms would soon occur, viewing efforts to date as "all talk and no action."

THE RIGHT TO DIE

On December 1, 1991, the U.S. Patient Self-Determination Act took effect. The law requires that patients admitted to hospitals be asked if they wish to fill out a living will detailing the types of treatment they do or do not wish should they be dying. Patients must also be asked if they wish to designate a power of attorney for health care should they become too ill to speak for themselves.

Final Exit, *Derek Humphry's suicide handbook, sparked heated debate about the ethics of suicide and the right to choose death over pain and debilitation.*

Several recent incidents have forced consideration of the right of people to choose suicide rather than long, painful deaths. The controversial book *Final Exit* by Derek Humphry, which describes various methods of suicide, made the best-seller lists when it was published in 1991. Jack Kevorkian, a Michigan pathologist, helped several people kill themselves by allowing them to push a button to receive a lethal dose of drugs or to suffocate on carbon monoxide breathed through a mask. And in Washington state, voters narrowly rejected (54 to 46 percent) a referendum that would have legalized suicides assisted by a physician.

COMBATING HEART DISEASE

The most expensive drug is not necessarily the most effective. A study by scientists at Oxford University in England evaluated three drugs used to dissolve blood clots in the coronary arteries: streptokinase, eminase, and t-PA (tissue plasminogen activator). The drugs were found to be equally effective in saving lives, though their cost per treatment ranged from $200 for streptokinase to $1,700 for eminase and about $2,500 for t-PA.

At menopause a woman's production of the hormone estrogen declines—and her risk of heart disease increases. Many physicians prescribe estrogen-replacement therapy to reduce the incidence of heart disease, osteoporosis, and such postmenopausal problems as hot flashes. However, the therapy has been controversial because it is believed to increase the risk for breast and endometrial cancer. The pendulum swung in favor of estrogen-replacement therapy following a 1991 report of a 10-year study involving nearly 49,000 nurses. The study found that women who take estrogen after menopause cut their risk of heart disease almost in half. The investigators noted that a white woman age 50 to 94 has a 31 percent chance of dying from heart disease, a 2.8 percent chance of dying from breast cancer, and only a 0.7 percent chance of dying from endometrial cancer.

Two new studies found that Enalapril, a drug widely used to alleviate shortness of breath and other symptoms of congestive heart failure, also cuts the risk of death from the disease. A third study found that Enalapril can actually reduce the risk of developing congestive heart failure by 37 percent among people with impaired heart function.

CANCER

One of the most promising new cancer drugs is taxol, produced from the bark of the Pacific yew tree *(Taxus brevifolia)*. Studies indicate that taxol halts tumor-cell growth by preventing cell division and stimulating tumor-cell death. It appears to be particularly effective in treating ovarian, breast, and lung cancers. Unfortunately, it takes about 20,000 pounds (9,090 kilograms) of bark to produce 2.2 pounds (1 kilogram) of taxol. Pacific yews are relatively rare evergreens found in Washington and Oregon. Loggers have long considered them to be weeds, burning them after clear-cutting forests for other types of wood. In 1991 the U.S. Forest Service ordered an end to the burning of downed yews. Meanwhile, scientists are trying to create synthetic taxol. Other researchers are extracting taxol from species of yews that are more plentiful and more easily grown than Pacific yews.

New tests are helping physicians detect cancers at an earlier stage. A simple blood test combined with the standard rectal examination increases the chances of detecting prostate tumors before they spread into other parts of the body. A newly created monoclonal antibody helps diagnose specific types of leukemia. And scientists are coming closer to developing a simple urine test that would detect bladder cancer, which is difficult to diagnose in its early stage.

HEART AND BREAST IMPLANTS

In 1991 surgeons at St. Luke's Episcopal Hospital in Houston, Texas, implanted the first fully portable heart pump into a 52-year-old man. It worked "successfully" for two weeks before the patient died.

Over a period of nearly 30 years, millions of women received silicone breast implants, mostly for cosmetic reasons. The implants had been marketed without proof of safety; when manufacturers were finally required to submit safety data in 1991, the data proved to be inadequate, and the varied risks of the implants—including scarring and hardening of breast tissue, leakage of silicone, autoimmune reactions, and even cancer—became widely publicized.

Manufacturers argued that 30 years of experience showed that implants were safe. However, five of seven U.S. implant manufacturers announced that they would stop production. In 1992, the FDA sharply restricted use of silicone implants until long-term safety studies are conducted. "These are not approved devices, and any woman who wants one will have to be in clinical studies," said FDA Commissioner David A. Kessler.

NEW WEAPON FOR SMOKERS

The single most preventable cause of death in the United States is tobacco smoking. Each year, some 435,000 Americans die from heart disease, lung cancer, and other illnesses caused by their smoking; thousands more die as a result of inhaling secondhand smoke. In 1991 smokers received new help in fighting

The announcements by sports stars Earvin "Magic" Johnson and Arthur Ashe that they have tested HIV-positive helped many Americans to realize that anyone can become infected with the AIDS virus.

their addiction. Several companies introduced adhesive skin patches that release a continuous stream of nicotine into the blood, thereby reducing the desire to smoke. Users can concentrate on fighting behavioral aspects of their addiction to tobacco while the patch relieves symptoms of physical withdrawal. Then the ex-smokers can gradually end their nicotine addiction by applying patches that release successively smaller doses.

AIDS VIRUS CONTINUES RAPID SPREAD

By early 1992 more than 200,000 cases of AIDS, including over 126,000 deaths, had been reported in the United States. Although sex between men accounted for the majority (58 percent) of U.S. cases, the experiences of two famous athletes helped stress that the HIV virus that causes AIDS is also spread by other means.

Basketball's Magic Johnson announced that he had become infected with the HIV virus through heterosexual contact. Heterosexual transmission accounts for 90 percent of all new infections worldwide, according to the World Health Organization (WHO), which estimates that there are currently 2 million people with AIDS, with an additional 10 million people infected with the HIV virus.

Johnson's announcement led many worried people to undergo tests for the HIV virus. The standard test identifies antibodies to the virus, which become detectable in a person's blood about six months after infection. The virus can remain latent for a long time: it may take 10 years or more before an infected person begins to show symptoms of AIDS.

Another increase in requests for tests occurred in mid-1992, after tennis great Arthur Ashe revealed that he had contracted the disease from a blood transfusion, either after a 1979 operation or a 1983 operation. Today tainted blood is a minor risk factor. Since 1985 all donated blood must be tested for the HIV virus before it is transfused.

Until late 1991 the only drug approved by the U.S. Food and Drug Administration (FDA) to fight the HIV virus was azidothymidine (AZT). Then the FDA gave its approval for dideoxyinosine (DDI) as a therapy for patients who cannot tolerate or are not helped by AZT. The step was controversial because tests to fully establish DDI's safety and effectiveness had not yet been completed.

Preliminary trials of vaccines intended to prevent infection by the HIV virus have been promising; 11 experimental compounds have been injected into about 600 human volunteers to determine their safety and ability to stimulate immune responses. Some of the vaccines may be ready within the next two years for testing on people who are at high risk of becoming infected.

Jenny Tesar

Physiology or Medicine

In 1991 Dr. Erwin Neher and Dr. Bert Sakmann, two German scientists, shared the Nobel Prize for Physiology or Medicine. Working together over the course of many years, the scientists confirmed the way in which cells take in and release ions through the cellular membrane. The laureates' development of the so-called *patch-clamp technique* has made it possible for cellular biologists to probe the causes of a number of such diseases as cystic fibrosis, epilepsy, and diabetes.

Scientists around the world have hailed the decision of the Nobel committee, agreeing that in 1991 the prize was for a discovery that truly revolutionized biological research and made possible the tailoring of new drugs to achieve optimal effect on certain diseases.

OPENING A DOOR TO THE CELL

Scientists had long theorized that ions somehow pass through a cell membrane to regulate activity within the cell. It was hypothesized that cells controlled some sort of channel between the internal and external environments. There was no way to demonstrate or record this movement, however, and no way to measure how many ions flowed through how many channels, and how much electrical charge each ion carried. The best researchers could do was to puncture the membrane with an electrode and measure total internal ionic change as it happened.

Drs. Sakmann and Neher worked for seven years to develop a technique that precisely analyzes the nature of ion channels. These narrow, water-filled passages, they discovered, can alter their shape to control the type and quantity of ions allowed into or out of a cell.

The technique uses a micropipette, a slender glass tube filled with a saline solution that tapers to a point just wide enough to cover—and isolate—one of a cell's 20 to 40 ion channels. The experimenters used the tip of the pipette as an electrode connected to a powerful array of sensitive electronic equipment. After setting up the initial experiment over the course of a full year, they were able to detect ionic currents of mere picoamperes passing through a channel in a matter of millionths of a second.

Dr. Neher and Dr. Sakmann went on to discover that they could use the micropipette to gain access to the interior of the cell. By attaching the tip to a section of the membrane and applying suction, they were able to tear up a tiny patch of the membrane, opening a narrow door (1 micron—0.00004 inch—wide) through which they could inject or extract substances.

The scientists published their findings in 1976. When Dr. Neher presented the new technology at a meeting of the

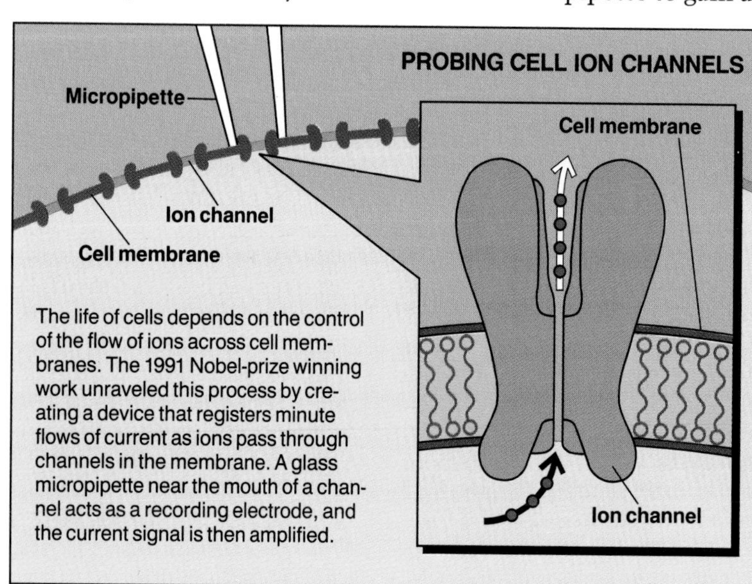

The life of cells depends on the control of the flow of ions across cell membranes. The 1991 Nobel-prize winning work unraveled this process by creating a device that registers minute flows of current as ions pass through channels in the membrane. A glass micropipette near the mouth of a channel acts as a recording electrode, and the current signal is then amplified.

German scientists Erwin Neher (left) and Bert Sakmann (right) shared the 1991 Nobel Prize for Physiology or Medicine for their patch-clamp technique, which has permitted the detection and analysis of ionic channel activity in the cell membrane. This technique has been instrumental in studying such diseases as cystic fibrosis and diabetes.

Biophysical Society, the audience cheered in recognition of how the discovery would revolutionize the study of cells and cell behavior.

The two young researchers—Dr. Sakmann was 34 at the time; Dr. Neher was 32—still had work to do, however. Their patch-clamping technique did not create a perfect seal between the cell membrane and the pipette. Leakage around the edge generated electronic "noise" that obscured important, low-level amperage.

It wasn't until 1980 that Dr. Neher stumbled onto a solution. At first not understanding how it happened, he noticed the noise readings on his oscilloscope had dropped to near zero. In his words, it was "chance favoring a prepared mind." He investigated the unexpected but much-desired phenomenon, and found he'd created an electrical seal 100 times stronger than that of the original technique. He was able to reproduce the seal by using a fresh, fire-polished pipette. The seal is now known as the *gigaseal* because of its high electrical resistance.

Dr. Sakmann says that once they had these new tools adequately refined, the real fun began. Working with specialized research teams, they were able to alter the genes for ion channels to identify the specific biomolecular structures that open and close the channels. Gene manipulation also identified the various structures that control which ions are admitted into the channels.

The technique has also allowed researchers to analyze the defective ion channels that seem to cause cystic fibrosis. The malfunctioning channel, they discovered, is unable to control the passage of chloride ions, which explains why the perspiration of people with cystic fibrosis contains abnormally high levels of that compound.

Patch clamping has also been used to explore the body's regulation of hormone levels. Of special significance was the discovery that the release of insulin from the pancreas is, in part, a function of ion channels that are sensitive to ATP (adenosine triphosphate). Inadequate insulin production being the cause of diabetes, researchers now have a better understanding of how certain drugs control diabetes by influencing ATP ion channels.

Dr. Erwin Neher, 47 when he won the Nobel prize, formerly worked at the University of Washington and at Yale University, and is currently the director of the membrane biophysics department at the Max Planck Institute of Biophysical Chemistry in Göttingen, Germany. His wife, Eva-Marie, is a microbiologist; they have five children.

Dr. Bert Sakmann, 49 when he became a Nobel laureate, is a physician and research scientist at the Max Planck Institute for Medical Research in Heidelberg. His wife, Christianne, is an ophthalmologist; they have three children.

Glenn Alan Cheney

Nutrition

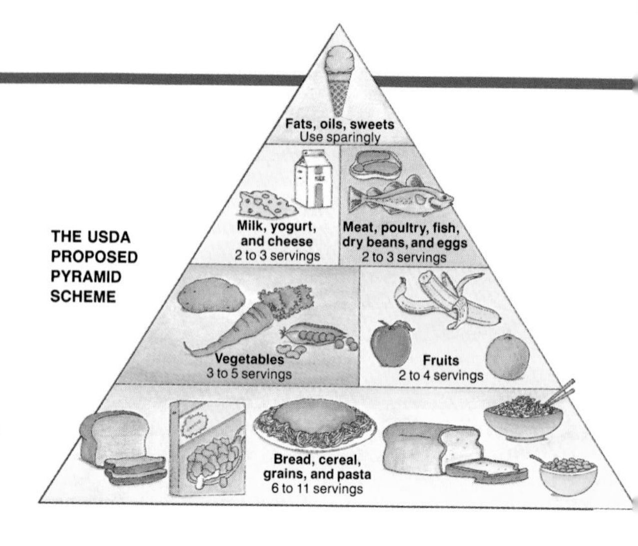

DIET AND DISEASE

A growing number of studies suggest that certain food substances may have a protective effect against cancer. Several teams of scientists have found that green tea, a popular drink in Japan, significantly inhibits stomach, liver, lung, and other types of cancer in mice. Epidemiological evidence supports these findings: Japanese people who drink large amounts of green tea have lower death rates from cancer than other Japanese.

Green tea contains large amounts of polyphenols, which are believed to act as antioxidants, combating molecules created in the body called free radicals. Free radicals are believed to damage cells and to play a role in the development of cancer. Beta carotene and vitamins C and E are other known antioxidants.

Researchers at Johns Hopkins University have isolated a chemical in broccoli that also appears to be a strong anticancer compound. Called sulforaphane, it appears to remain intact even when broccoli is steamed or microwaved.

A British study confirmed earlier work showing that pregnant women who take the B vitamin folic acid greatly reduce the risk of having babies with neural-tube defects such as spina bifida. The study followed 1,195 pregnant women who had previously given birth to a child with a neural-tube defect. In the second pregnancies, 72 percent of such defects were prevented by daily folic-acid supplements. The researchers believe folic acid would also prevent neural-tube defects in first pregnancies, since the causes of the defects are likely to be the same.

FOOD PYRAMIDS AND LABELS

Since 1958 the major nutritional message from the U.S. Government has been to eat from the "basic four" food groups diagrammed on a circular chart: milk, meat, vegetables and fruits, and bread and cereals. Nutritionists stress that this chart is misleading because it overemphasizes milk and meat. In 1991 an attempt to replace the chart with a pyramid ran into strong opposition from the meat and dairy industries. Cereals, grains, fruits, and vegetables—which nutritionists say should comprise the bulk of a person's diet—occupy the broad lower portion of the pyramid. Meat and dairy products occupy the much smaller upper portion. At the tip of the pyramid, to be eaten "sparingly," are fats, oils, and sweets. The proposed pyramid scheme appears above.

Responding to pressure from health and consumer groups, the U.S. Government began to crack down on food products with misleading labels. Labels with terms such as "fresh," "lean," and "lite" are among those receiving particular scrutiny. For instance, the Food and Drug Administration (FDA) seized 2,000 cases of Procter & Gamble's Citrus Hill Fresh Choice orange juice, which was made from concentrate rather than from fresh juice. New regulations requiring labels to contain detailed nutritional information are scheduled to take effect in 1993. Among the facts the new labels must carry: the number of calories, calories derived from fat and saturated fat, cholesterol, total carbohydrates, complex carbohydrates, sugar, fiber, protein, vitamins A and C, sodium, calcium, and iron. It is also proposed that the labels advise consumers how much of each listed substance they need each day.

Jenny Tesar

Public Health

TUBERCULOSIS SPREADS AGAIN

A recent jump in death and illness from tuberculosis (TB), caused by drug-resistant strains of TB bacteria, is forcing health agencies to scramble to respond to the spread of this contagious illness.

Drug-resistant strains of TB bacteria arise partly from TB patients failing to complete drug-treatment regimens, and thus culling out the most resilient of these microorganisms. TB, which slowly destroys a victim's lungs, spreads through the air via tiny particulates generated by coughing, or through more-intimate respiratory contact. One can contract the disease after a few days' exposure to someone with an active infection, although a month's exposure is usually required.

Tuberculosis is an opportunistic infectious agent, tending to thrive in individuals with weakened immune systems. Many people with healthy immune systems may never know they've been exposed unless they take a TB test, because the bacteria may remain dormant for decades. Only about 5 to 10 percent of these people will ever develop active TB.

Most occurrences of the resistant form of TB in the United States recently have been in AIDS patients, in whom the infection can occur at many different sites and is difficult to diagnose. Poverty-stricken communities with a large homeless population, where people live in crowded conditions or with little medical care, also provide ideal conditions for a resurgence of TB. Nonetheless, TB has also now begun to slowly spread beyond these high-risk groups.

Many questions remain unanswered about how TB bacteria function and how the disease can be opposed. Some experts are suggesting reintroduction of the tuberculosis "sanatorium," a once-common type of convalescent center that helped separate TB victims from the general population while they were contagious.

Wearing a helmet greatly reduces the likelihood of head injury in the event of a bicycle accident.

BICYCLE HELMETS

Bicycling accidents send more than 600,000 people to emergency rooms every year, 1,300 of whom die as a result of their injuries. Most of the deaths and hundreds of thousands of the injuries are due to trauma to the head.

Despite these frightening statistics, studies show that, overall, only one out of every 10 bicyclists wears a helmet, and only one out of every 50 youngsters does.

The large percentage of bicycling injuries that involve the head or face has led emergency-room personnel to coin the term "street face" to describe the result of the impact of a moving face on unmoving asphalt or cement. Abrasions to the face, though, are one of the less serious outcomes of bicycling head injuries. The majority of deaths are from collisions with automobiles, most occurring in young males ages 10 to 14.

It appears that most fatal crashes involving head injury would be survivable with a helmet. Studies have found that with a helmet a bicyclist is 4 to 20 times less likely to suffer a head injury in an accident.

A recent study conducted by the Centers for Disease Control (CDC) showed that between 1984 and 1988, if all bicyclists had worn helmets, the helmets would have prevented one death a day and one head injury every four minutes.

Russ Allen

PAST, PRESENT, and FUTURE

CONTENTS

20
30
40
50
60
70

The legendary lost city of Ubar, fabled in poetry and song, was identified on maps dating from the second century A.D. (below). Modern archaeologists may have uncovered this key center of the ancient frankincense trade using sophisticated satellite technology.

The Lost City of UBAR

by Lee Galway

Lost cities are the stuff of archaeologists' dreams. For archaeologists who do their spadework in the Middle East, no lost city has embraced their imaginations more than the city of Ubar.

The Koran mentions a "city of towers" called Iram in the Shabbah region of the Arabian peninsula (or Sheba—as in the English translation of the Bible), which was destroyed by a flood as punishment for its people's decadence. Many Islamic scholars believe Ubar was another name for Iram.

Buried over the centuries under the shifting sands of the Omani desert, Ubar's legend grew, its towers fabled to have been leaved in gold and encrusted with gems. The city kept cropping up in the works of poets and historians. Lamentably, as a tale in the *Arabian Nights* puts it, "Allah blotted out the road that led to the city." The 12th-century

Rubáiyát of Omar Khayyám affirms, "Iram indeed is gone with all his rose. . . ."

The 20th century witnessed a renewed interest in Ubar's whereabouts. T. E. Lawrence, better known as Lawrence of Arabia, mused about finding this "Atlantis of the sands." Bertram Thomas, explorer and first British governor of Oman, began a more serious effort. His memoirs, published in 1932, tell of finding traces of a road to Ubar. Others mounted expeditions in the ensuing decades, but to no avail.

Then, in 1981, Nicholas Clapp, an Emmy-winning documentary filmmaker, chanced upon Thomas's memoirs. His curiosity about Ubar piqued, he tracked down every reference he could find to the place, including a map by Claudius Ptolemy, the 2nd-century Alexandrian geographer. Ptolemy located the "Omani Marketplace," as he called Ubar, in the Empty Quarter, an aptly named expanse of desert along the Arabian Peninsula.

Though Ptolemy pointed Clapp in the right direction, the filmmaker needed much more sophisticated maps as a guide. This need would be fulfilled by "remote sensing"—the technique of taking (and sometimes computer-enhancing) photographs of landscapes from high-flying aircraft, satellites, or spacecraft to uncover hidden geological or archaeological features.

Aircraft have been employed by archaeologists for decades in preliminary site surveys. From an airplane—and given the right conditions—hidden or buried features stand out as if in bas-relief. Readily apparent features, such as ancient but still-visible roads in the U.S. Southwest, have been discovered in this fashion.

Seen from the air, sunken or raised ground, or "shadow lines," can reveal ancient wells or walls. If the land has been tilled, the dirt over buried building foundations may have slight color variations. If there are crops, the growing plants themselves can be markers, as plants in soil above a buried stone wall will not be able to tap as much water as their neighbors, and will appear stunted and discolored. Even in winter, frost and light snow will lie differently over land concealing buried archaeological features. However, conventional photography fails to reveal archaeological treasures hidden under jungle canopy or shifting sands.

Fortunately, advances in photographic techniques, combined with American and other nations' nascent space technologies, are providing a wealth of new archaeological-survey information for field-workers.

Air Search

When Nicholas Clapp had exhausted traditional approaches to his search for Ubar, he, too, resorted to the new remote-sensing technologies. He had read about radar images of Egypt taken by instruments on the space shuttle *Challenger.* These images revealed many ancient roads and riverbeds never before seen.

Clapp convinced the scientists at Caltech's Jet Propulsion Laboratory (JPL) to include radar photography of southern Arabia in the July 1984 *Challenger* mission. The outcome looked promising. The faint white

Once satellite images pinpointed the area thought to be Ubar, archaeologists began the painstaking work of excavating the site. Among the finds: a walled fortress that probably supported 150 people.

Archaeological Space Photography

Archaeologists Nicholas Clapp (left) and George Hedges use a satellite image to show the convergence of old caravan roads at a point thought to be the site of Ubar.

The first man to think of using space photography as an aid to archaeological research was Tom Sever, who almost single-handedly created this advanced technique for archaeologists.

The idea came to Sever in 1977 while he was surveying a complex series of 41 Incan astrological lines radiating out from the Temple of Gold in Cuzco, Peru. Along each line were strung eight shrines, or "Wakas." Multiplying the number of lines times the number of Wakas along each line yields 328—the total length in days of the Incan lunar cycle.

It took Sever and a colleague three months to survey two and a half lines. He reasoned that to survey the rest of that site and 28 similar sites throughout South America would take 100 years.

"That's when I thought about NASA [National Aeronautics and Space Administration] satellite technology," Sever said in a 1987 article in *Space World* magazine. "Couldn't this help archaeologists out of their dilemma?"

That question landed Sever a job as the agency's only practicing archaeologist—a job he still holds—at NASA's Earth Resources Laboratory in Mississippi. Using satellite- and spacecraft-mounted microwave radar, infrared and near-infrared cameras, thermal scanners, and "false-color" enhancements, geologists and oceanographers had gotten the jump on archaeologists in building remote-sensing databases. Sever applied the same data to his field.

The first available images, from 1972 Landsat satellite shots, could distinguish objects no smaller than 262 feet (80 meters) across. Commercially available satellite images today can reveal objects in the 66- to 98-foot (20- to 30-meter) range. Aircraft equipped with the proper sensing devices can resolve objects as small as 3 feet (1 meter) across.

The results of the new technology thus far are impressive:

—Infrared photography of the Costa Rican highlands shows 1,500- to 3,000-year-old footpaths leading between ancient villages and graveyards.

—Radar images of the Andean highlands are revealing a sophisticated mountaintop society that predates the Incas.

—Satellite data have enabled Israeli archaeologists to survey the 3,000-year-old City of David outside Jerusalem's walls without resorting to "invasive" excavation that would disturb Palestinian settlements.

—Multispectral scanners (instruments that use visible and infrared light) have uncovered dozens of Mayan sites, including a causeway in the Guatemalan jungle.

Archaeologists have been slow to adopt this new technology. This is due, in part, to the novelty of the process, but more so to the cost of prints, which can run $3,000 to $4,000 apiece. The price, however, will likely fall as more countries, such as Japan and France, launch remote-imaging satellites.

lines of old caravan roads crisscrossed photographs like spider tracings. Many of the roads converged on the same spot marked by Ptolemy on his map.

Cross-checking the *Challenger* images with those from earlier Landsat missions and from newer French satellite images, Clapp and his team were able to assemble a striking map of the Omani desert trade routes and watering holes. Finding those features at ground level would prove a much more daunting task, however.

Land Search

To that end, Clapp enlisted the aid of famed British explorer Sir Ranulph Fiennes and Arabian archaeological expert Juris Zarins of Southwest Missouri State University. Fiennes, who once served in the Omani army, had no trouble convincing the sultan of Oman to let the trio sift the sands of the Empty Quarter. With financial aid from both Omani and British firms, they set out on a three-month expedition in 1991.

They almost came home empty-handed. Fortunately, nearing the end of their expedition, they explored an oasis crossed by a road that Bertram Thomas had described in his memoirs.

Soon after they started digging in November, they made a stunning discovery. Buried below the oasis was an octagonal, walled fortress consisting of eight 30-foot (9-meter)-high towers and a complex of storerooms, living areas, and assembly halls. Archaeologist Zarins estimates that the fortress probably supported 150 people year-round. Surrounding the fortress are some 40 campsites, a finding that matches the literature about Ubar's role as a marketplace for frankincense and other goods.

But perhaps most interesting of all, the fortress appears to have fallen due to a natural catastrophe, which matches the Koran's account of Ubar's demise. Caverns underlie the oasis, and Zarins speculates that the fortress collapsed into the caverns—perhaps when the water table was low—sometime between A.D. 300 and 400.

Despite the numerous coincidences, and despite Clapp's enthusiasm, Zarins is not sure the expedition has found the lost city of Ubar. "The references to the city in the Koran and in Islamic histories are vague," Zarins says, "and there's no real proof that Iram and Ubar were the same city. The local tradition [of travelers who frequent the oasis] is that Ubar is somewhere near the coast."

Rewriting History

Zarins thinks, however, that the history of the area may have to be rewritten. The Arabian Peninsula economy revolved around the frankincense trade in the centuries around the first millennium—indeed, up to the early Middle Ages. What petroleum is now to the Arabian peoples, frankincense was then.

Frankincense, an aromatic gum resin, is harvested from the small trees of the genus *Boswellia*, the finest specimens of which cling to the southern side of the mountains along the Arabian Peninsula. Deep incisions made in the trees yield small, tear-shaped, translucent globules that are scraped off and gathered by laborers.

Considered precious by most and a holy substance by some, frankincense was burned during religious and civil ceremonies by nearly every religious and ethnic group in the area. Adding to this its utilitarian purposes as a medicine and as an insect fumigant, frankincense became the most important commodity of the region.

Because historians of Arabia figured the Empty Quarter to be too forbidding an area for caravan travel, they presumed that the frankincense trade was carried north along the Arabian Peninsula, as much of it was.

But if the planned fortress excavation over the next three to five years pans out according to Zarins's presumptions, more of the trade might prove to have been funneled through the Empty Quarter. The newly rediscovered fortress might have been, in Zarins's words, "the last major consortium place in the surrounding countryside."

If the fortress proves to be a major trade center, then artifacts from all corners of the ancient world should turn up there. The preliminary dig has already uncovered stone, metal, and ceramic artifacts from a wide geographic area. The site runs chronologically deep as well, containing artifacts that may date back as far as 5000 B.C.

Whatever the fortress was, the story of its discovery is appealing on a variety of levels—as an archaeological romance, as a fable, and as a tale of technological sleuthing. What kind of history lesson it becomes remains to be seen.

TAKING BACK THE PAST

by Grace Lichtenstein

Claudia and Ralph Haynie live outside the town of Mancos, in the juniper-covered hills of southwestern Colorado. Their spread, at the edge of the Mesa Verde plateau, is like many here in the Four Corners, where Colorado, Utah, New Mexico, and Arizona meet. Their mobile home sits at the base of a scrub-covered rise, the front yard cluttered with old cannibalized cars, fodder, or Ralph's auto-body shop. Out back behind the trailer, the land gently rises, broken only by several small, deliberate mounds of rocks. Beneath the mounds, the Haynies have discovered, lies a 700-year-old Anasazi dwelling.

Surprise Excavations

Years ago, nearly everyone in the Four Corners who had mounds on their land would invite friends over for hot dogs and digging. In fact, the ruins are so numerous that folks here like to say that a farmer can hardly plow a straight furrow without turning up an Indian site. Centuries before Europeans settled or even explored this region, the Anasazi were building architecturally astonishing cities at places like Chaco Canyon and Mesa Verde, now both federal parklands. The tribe abandoned its settlements by A.D. 1300, but its monuments survived: pueblos, the circular kivas where the Anasazi prayed, rock piles beneath which they buried their dead, the ceremonial pots and tools that were placed in the burial chambers, and even skeletal remains of the Anasazi themselves.

Ralph Haynie stumbled onto his first "excavation" eight years ago, when he plowed a backhoe into one of the mounds in his backyard and unearthed a kiva. Claudia Haynie, a practical woman who had lived in Alaska during the oil boom and once drove a truck for a living, figured she'd better learn something about archaeology before she and Ralph tried any more digging. She read up on prehistoric art and learned enough to use a trowel instead of a backhoe and to scoop the dirt away more cautiously. When she uncovered the first pieces of a few elaborately painted jars and bowls, she was thrilled. She fitted them together with Elmer's glue.

"The Anasazi were interesting people," Claudia says as she, Ralph, their pack of yipping dogs, and I walk up to a site behind the trailer. A backhoe sits next to the path. Pottery shards crunch underfoot as we climb a pile of dirt that Ralph has cleared to reveal the stone outline of a pueblo. Claudia tells me that one memorable afternoon, she found an almost complete skeleton here. "It was a weird feeling, like opening a coffin," she says. Her uneasiness, however, did not prevent her from selling the skull for a few hundred dollars.

"Contaminated Sites"

Ten years ago, Claudia probably could have sold that skull for a good deal more. Anasazi pots were going for as much as $60,000; surely, the Haynies thought, they could get close to that for some of their backyard artifacts. They began spending all their weekends out back with trowels. Hoping to attract investors, they paid $10,000 to have a video made of the ruins on their land. "We were gold-digging," Claudia says, dragging on a Kool. "We thought we'd get rich."

The Haynies never came close to getting rich. The most they ever received for a pot was $1,400 from a tourist from California. They couldn't even interest archaeologists, all but two of whom scorned Claudia's requests for help in excavating, telling her that she had "contaminated" the sites with her random digging. Now, because of exposure to air and the elements, the kivas are collapsing. The Haynies have been told that it would cost over $1 million to stabilize them.

Spending that kind of money, of course, is impossible for the couple, but the point became moot last year when Colorado's legislature passed a law restricting removal of any Native American artifacts from public and private land. The Haynies quit digging; it wasn't worth the red tape they'd have to cut through if they found anything. Now their favorite pots sit on display in their mobile home, along with a copy of Louis L'Amour's *Haunted Mesa* and a diorama that Claudia crafted of an ancient pueblo. The law has the Haynies confused and angry. The artifacts are buried on their land, they figure; what right does the government have to say they can't dig around on their own property?

"It's like saying it's against the law to have a Rembrandt in your living room," says Claudia.

For many years, the Haynies (below) and other landowners dug up Indian gravesites on their property and sold any artifacts they found to collectors and tourists. Indian activists call this practice grave robbing.

Pot Patrol

Years ago, on my first visit to Chaco Canyon, I pocketed a black-and-white shard similar in pattern to Claudia's pots. I'd had an almost spiritual experience walking through the ruins of the Anasazi's elaborate metropolis, and I wanted the shard as a memento. I was covering the region for a national newspaper, yet I did not know that by taking the shard, I was committing a crime. Neither did I know that by the tenets of some Indian tribes, what I had done was almost sacrilegious.

Pot hunting, as illegal excavation and removal of artifacts is called, has been going on in the Southwest since a Colorado cowboy named Richard Wetherill stumbled upon the fantastic stores of the Mesa Verde cliff dwellings in 1888. While recent laws have made the penalties for digging harsher, pot hunting has never completely stopped. The Bureau of Land Management's (BLM's) tiny "pot patrol," which investigates both kinds (ceramic and organic), generally has to give priority to the latter. And there has always been private land, that inviolable quantity in the West, where pot hunters could legally dig. Sometimes landowners would lease a few acres to commercial diggers, either for a fixed price or for a percentage of the take when the "grave goods"—skeletons, pottery, and ceremonial objects—were sold, often to galleries. Still, Native American art experts estimate that at least half of the artifacts are likely to be falsely pedigreed, having come from federal land before being legitimized with phony papers.

"For years, archaeologists and museums have conveniently ignored where the artifacts came from. Now the shoe is on the other foot," says James Reid, a New Mexico art dealer and lobbyist for the Antique Tribal Art Dealers Association.

Reid is referring to the federal "repatriation" bill that President Bush signed into law in 1990. The law, considered a huge victory by Indian-rights groups, makes trafficking in human remains from Native American, Alaskan, or Hawaiian graves on national and tribal lands a federal crime punishable by a fine, up to five years in jail, or both. Perhaps more significant, the law allows tribes to retrieve and rebury their ancestors' remains and grave goods from federally funded museums, universities, and historical societies. One implication of this is that the return of these bones and artifacts may result in a serious loss to the historical record. Another, more troubling one is that the artifacts could fall into careless or corrupt hands and end up neglected or in private collections rather than back in the ground.

Depleting Collections

Some institutions were returning artifacts and remains even before the federal repatriation law passed. Notable among them are the Smithsonian and Stanford University, both of which had extensive Native American collections. But some academics at Stanford protested the decision.

"What they gave away was my life's work," says Bert Gerow, professor emeritus of anthropology at Stanford. Between 1948, when he joined the faculty, and 1988, when he retired, Gerow collected 550 skeletons thought to be those of Ohlone Indians, some of which were 3,000 years old. The bones

The Anasazi conducted religious ceremonies at circular kivas, some of which are under excavation (left) by government-approved organizations. The kivas and other Anasazi monuments are rich sources of pottery shards (above), skeletons, and a variety of other Anasazi relics.

were stored at Stanford until a university staff member sympathetic to repatriation efforts contacted Ohlone tribe members and suggested they ask for the bones for reburial. Stanford decided to return the bones, but Gerow objected, saying that, because the bones were thousands of years old, there was no way to prove that they belonged to the Ohlone's ancestors.

"The idea that there were descendants of these bones was ridiculous. I wasn't digging in anything that was a known cemetery," he says. "Many of the bones I unearthed were salvaged from housing-development construction. They would have been destroyed if I hadn't removed them.

"Eventually the Native Americans are going to come to their senses and say that they want these bones to be studied because they're the only record of the past. Won't they want their children to know their history? To have a database as far back as 10,000 years is an enormous benefit."

Anthropologists like Gerow fear that the repatriation law will encumber or even halt digs on federal land, if not dry up federal grants altogether. Art dealers like Reid are worried that the tribes will take the mandate a step further and push for legislation that will allow them to go after private collections.

Walter Echo-Hawk, senior attorney for the Native American Rights Fund, and a member of the Pawnee tribe, has little patience for such concerns.

"This is massive looting of Indian graves in the name of entertainment as well as science," he says flatly. "These are not archaeological resources. They're people." The grave goods of the dead, he says, are "philosophically, religiously, culturally, logically, and legally one and the same."

Echo-Hawk is a decidedly modern Native American, straddling Indian and Anglo worlds. His conservative suits and ties conform to the uniform of his profession, yet his graying hair, combed straight back, falls to his shoulders. Born on a reservation, Echo-Hawk is also a product of the establishment and works well within it. He is a leading spokesman for the repatriation movement, which has lobbied successfully in 31 states for laws that restrict digging on Indian burial sites, and he was a key witness in congressional testimony on the federal law.

"If you desecrate a white grave, you wind up in prison. But desecrate an Indian grave, and you get a Ph.D.," says Echo-Hawk. People like the Haynies are pot hunters, he says, but so are archaeologists and art dealers. And pot hunters are grave robbers.

Analyzing Artifacts

Bill Lipe is an affable, lanky anthropology professor from Washington State University, and he is taken aback by this hostility toward his field, the science of humankind.

Lipe spends his summers in southwestern Colorado, studying the cliff dwellings and kivas of the Anasazi. Between A.D. 900 and 1200, the northern part of the Four Corners was in many respects an American Athens. As many as 100,000 Anasazi occupied its cliffs and riverbanks, farming the plateau and praying in their underground kivas. Lipe has brought me to Goodman Point, once an Anasazi community in the midst of this civilization. He works with the nonprofit Crow Canyon Archaeological Center, and on this hot July day, several Crow Canyon participants are hard at work, carefully scooping soil with trowels from the ruins of an Anasazi pueblo. Having already exposed one wall, which has been buried under centuries of accumulated dirt, they delicately sift through the loose soil, seeking bits of pottery, bone, and clothing fiber. Lipe tells me that they are looking for clues as to how these Anasazi lived in the 13th century, roughly 100 years before they mysteriously began abandoning their dwellings.

Lipe points out a pile of rocks that is similar to the mounds in Claudia and Ralph Haynie's backyard. Called a midden, it marks where the Anasazi buried their dead and grave goods, usually adjacent to a kiva. The Crow Canyon participants will undertake the tedious tasks of labeling and measuring the artifacts and mapping where they found them. Bones will then be taken to a laboratory for analysis. Such analysis by physical anthropologists could yield information on the Anasazi's health, what they ate, their life expectancy, and which diseases they contracted.

"What we do is important," Lipe says. "It's good for us as a species to study history by any means possible."

Vine Deloria, Jr., disagrees. The author of the 1969 book *Custer Died For Your Sins,* which touched off the Native American militancy of the 1970s, he is now a professor of American Indian studies at the University of Colorado. Archaeologists, he says, "dig in these ruins and classify Indian societies according to the kind of pottery they had, and then make up all these incredible stories about what they've learned from it. Now you can go to the dump and look at layers of Pepsi and then layers of Coke bottles. What would you say? The Coke people invaded and took over the Pepsi people? It's absurd."

Bill Lipe chuckles, without mirth, when I tell him Deloria's opinion of archaeology. "People who depend on traditional religious beliefs tend to feel that science is a pain," he says. "The key difference between pot hunters and archaeologists is that pot hunters sacrifice the context to get to a few valuable artifacts. You have to chew through a lot of archaeology to find the pots. It's like burning the woods to toast marshmallows."

Revisionist archaeologist Doug Bowman admires Anasazi artifacts in their natural setting; he sees no need to excavate them for museum use.

Dire Statistics

Southwestern Colorado, unfortunately, is full of chewed-up sites, and it's Jon Wesley Sering's impossible job to find the culprits. Matinee-idol-handsome with his Marlboro-man mustache and polished badge, Sering is a BLM ranger and a member of the agency's overextended pot patrol. He and Max Witkind, a BLM anthropologist, are taking me to a recently vandalized Anasazi burial ground not far from Bill Lipe's digs.

Sering stops his truck atop a remote mesa, and the three of us bushwhack through sagebrush to a spot sprinkled with shards, bones, and small, deep holes that obviously have been poked by a ski pole or a probe. A plastic Pepsi cup and a cigarette butt, along with the holes, are enough for Sering to dub the place a crime scene.

"You can see that they've just tossed everything all over the place," Sering says as he surveys the litter.

Witkind, though, is mesmerized by the ceramic bits. "I've never seen so many shards," he says. Or bones. "Left clavicle," he notes, pointing at one.

In accordance with the new Colorado law, Witkind notifies the state archaeologist about any newly uncovered skeletons. If the remains prove to be Anasazi, the Hopi, who are considered to be the Anasazi's descendants, can recommend that a spiritual leader conduct reburial services.

Odds are Sering will never find the guilty party: after nine months on the job, the only perpetrators he's caught are arrowhead collectors. It would be a miracle if he could do otherwise. Sering is the only ranger for a territory covering 2.25 million acres (900,000 hectares), and he's far outnumbered by commercial pot hunters. At spots like Goodman Point, anyone with a trained eye and binoculars can sweep the horizon and quickly identify an outcropping that marks a burial site. In 1988 a House subcommittee warned that the number of sites vandalized in the Four Corners could be as high as 90 percent, and that the looting was getting worse.

There are perhaps 2 million prehistoric sites in the area, with only 7 percent surveyed; just a handful of BLM, Forest Service, and National Park Service officers to watch over them; and almost no cases, let alone convictions. (Gallery owners contend that the Indians themselves have been among the

A law enacted in 1990 allows tribes to retrieve and rebury Native American grave goods now held by museums, universities, and other institutions that receive federal funds. Some scholars object to the statute, likening the loss of Indian archaeological artifacts to a scientific catastrophe.

biggest commercial diggers.) Anyone with a little archaeological know-how and a big backhoe, it seems, can loot Anasazi pueblos with impunity.

Despite such dire statistics, neither Sering nor any other law-enforcement officer can show me a site recently gouged with earth-moving equipment. Education campaigns, plus a few highly publicized raids on private collections, may have made looters more wary than before. "People are starting to call me at home if they see someone moving a rock," Sering says.

It may also be that the financial rewards of pot hunting no longer outweigh the risks, especially for commercial diggers. Typically, fewer than 1 percent of the ceramics dug up are intact, with only a fraction of them salable. The pot hunter receives an average of $200 for each pot. The jobber he sells to might earn $1,000 per pot from a gallery owner or dealer, who then marks the price up to as much as $10,000.

Even that scenario is fairly optimistic, according to one antiquities specialist who spoke only on condition of anonymity. "The average piece of prehistoric pottery today sells for under $1,000 retail," he said. "If a guy with a truckload sold 10 pieces after a month's worth of excavation, he might do well to get a few thousand dollars." Items sold at major auction houses like Sotheby's, in New York City, get the most ink. Still, such pieces are the exception, not the rule.

Auctioning Artifacts

The Armani-suited *Bonfire of the Vanities* crowd that normally dominates Sotheby's auctions has given way to scores of pony-tailed, fringe-jacketed art dealers and Indian-art enthusiasts who have flown in from all over the world. Sotheby's is conducting its biannual Indian-artifacts auction, and the art dealers flit around the exhibits of pottery, kachina dolls, baskets, blankets, and other Native American objects, waiting for the auction to begin.

The hot topic of conversation, of course, is repatriation. Will the new legislation have an effect on sales or on the availability of newly "discovered" masterpieces, the assembled wonder? One southwestern art expert puts it a little less euphemistically: "After they're through stripping the museums, they'll attempt to do the same thing to private collections."

The room grows quiet as Robert Woolley, Sotheby's suave auctioneer, begins the bidding. A black-on-white effigy vessel is announced at $3,025. It looks like a first cousin to one Ralph Haynie cradled like a football outside his kiva a few weeks ago. It's a beautiful pot, but Woolley can coax the bidding only up to $2,750—too low for Sotheby's—and the auction house declines to sell.

When Woolley intones "Number 39," however, a hush comes over the room; Number 39 is the star of the show. The piece, a rare Sikyaki polychrome olla, circa 1400–1625, is an intact vessel of pale yellow decorated with a painted band of "elaborate highly stylized avian motifs." It was crafted by an unknown artist who is reverently described to me by one dealer as the Van Gogh of North American pottery. Its preauction estimated price is between $80,000 and $120,000.

The olla was discovered 20 years ago—accidentally and totally illegally, a dealer assures me—by an Indian on Hopi land. The Indian sold it to someone, who then resold it. It made its way through various hands, eventually somehow acquiring the necessary papers for it to be permitted at Sotheby's. An Arizona businessman is reluctantly selling this pot and others from his extensive collection because he and his wife want to finance a new llama-raising venture. The word in the room, however, is that the businessman wants to unload the pots—fast. "People are nervous about the changing tide here of political and emotional feelings toward excavated material," one Santa Fean puts it. Maybe the day will come, he says, when repatriation activists will challenge such an auction on the grounds that it is trafficking in stolen grave goods.

Prodded by Woolley, the bidding on the olla jumps from $50,000 to $55,000, $60,000, $65,000, and the gavel falls. The final price of $71,500 (the auction house automatically adds a 10 percent commission), is to be paid by an Aspen, Colorado, dealer. The room buzzes. A pot that rare ordinarily wouldn't have gone for so much less than the low estimate. Is it the shaky economy? Or are the "bone laws" already having an impact? It strikes me that most of the people in this room are probably genuinely interested in Native American culture. What they've never considered, what I hadn't considered when I pocketed my pottery shard, is the question of who, if anyone, owns the past.

Revisionist Approach

Doug Bowman is a late-blooming archaeologist for the Ute Mountain tribe, whose land abuts Mesa Verde to the south. For years, Bowman worked for mining companies in Colorado and Utah, but he eventually came to feel that he was, as he puts it, "on the wrong side of the table." So he went back to college at age 56, got a degree in archaeology, and became the superintendent at the Ute Mountain Tribal Park, a huge swath of preserved land rich in Anasazi ruins. He is giving me an on-site demonstration of how to appreciate Indian artifacts without stepping on scientific, religious, or aesthetic toes.

We leave his four-wheel-drive van at the head of a canyon, not far from Mesa Verde, but light-years from the motorized hordes in the park. We are hiking in to a site far from the trail, a place Bowman says he would never have found unless the Ute had decided to tell him about it. Bowman has an appreciation for the dramatic. He wears an Indiana Jones hat and a knife (Swiss, not bowie), and he clearly relishes the building suspense as we climb up a wooden ladder and into a hidden cliff dwelling. Then, with the finesse of a waiter removing the lid from a dazzling tureen, he lifts a rock slab from the dirt floor.

Beneath it sits an entire Anasazi pot, half-buried, the wind of eight centuries having literally blown its cover. The pot, about 10 inches (25 centimeters) in diameter and 10 inches tall, is gray with a complex corrugated finish. Bowman tells me that the Anasazi pressed their thumbnails into the wet clay to create a textured pattern. Inside the pot is a doeskin bag filled with petrified, 800-year-old corn, squash, and beans. The Ute have kept the pot secret, and it is the most memorable Anasazi artifact I have ever seen.

It is so quiet in this arid sandstone dwelling that it feels a bit like being in church. Indeed, there is something about the half-buried pot that gives it a context and a power that almost certainly would be lost in the pre-Columbian wing of some museum.

"People have been coming through here for 10,000 years," Bowman says, "but the Ute have 125,000 acres [50,600 hectares] of pristine wilderness. No development. No concrete. We want to keep it that way."

The Search for Amelia Earhart

by Elizabeth McGowan

"We must be on you, but cannot see you. Gas is running low. Have been unable to reach you by radio. . . ." Amelia Earhart's voice, usually calm and deliberate, sounded strained and tired to the crew of the Coast Guard cutter *Itasca* as her words crackled over the radio the morning of July 2, 1937. Small wonder. Earhart and her navigator, Frederic J. Noonan, had been flying for 19 hours straight in a twin-engine Lockheed Electra plane that was as cramped as it was noisy. Neither had been able to stretch their legs since takeoff; with Earhart in the cockpit and Noonan in a rear compartment, the two could only communicate by way of notes attached to a fishing pole.

The route Earhart and Noonan were running was a pilot's nightmare—2,500-plus miles, half over unrelenting ocean with no landmarks to aid navigation. The Electra had taken off the previous morning from Lae, New Guinea, en route to Howland Island, the longest and most difficult leg of a highly publicized equatorial flight around the world.

Theories abound as to why aviatrix Amelia Earhart vanished over the Pacific in 1937. Richard Gillespie (inset) claims that a piece of metal he discovered on a remote island may help unravel the mystery.

Traveling west to east, Earhart and Noonan had already flown some 22,000 miles (35,400 kilometers) from Oakland, California to Miami, Florida, then on to the Caribbean, South America, Africa, India, Burma, Singapore, the Dutch East Indies, and Australia. They had 7,000 miles (11,265 kilometers) to go. Howland Island, their destination on this overcast day, and the second-to-last stop on their journey, was a barren speck measuring a mere 0.5 mile (0.8 kilometer) at its widest, barely a raindrop in the vast waters of the Pacific. The *Itasca*, anchored offshore Howland, had orders to guide Earhart by radio to a safe landing on the easy-to-miss atoll. With the technological equipment available in 1937, most experts agree that even the best of pilots and navigators would have had trouble hitting the small target without external aid.

And few would argue that Earhart—or A.E., as she preferred to be called—was the most accomplished of aviators. Though A.E. was, and is to this day, the most famous female pilot in the world, many of her peers believed she flew on guts rather than natural talent. Courage, they conceded, Earhart had in abundance, and her love of flying and determination to fight for equality for women drove her to push herself to her limits, both on land and in the air.

Early Years

Earhart relished breaking the gender rules even when she was still in pigtails. Born in Atchison, Kansas, in 1897, Earhart as a child dismayed her prim grandparents by climbing fences, sleigh-riding belly-down like the boys, and wearing functional bloomers rather than more-ladylike skirts and crinolines. In her 20s, she cut her fair, waist-long hair an inch at a time, pinning it up each morning in hopes that her mother wouldn't notice that her daughter's mane was shrinking. At 23, mesmerized by an air circus she attended with her father, she decided to learn to fly, "knowing full well I'd die if I didn't," she later recalled. Earhart sought out Neta Snooks, a female pilot, to give her lessons; a year later, she bought a brown leather flyer's jacket (which she confessed she slept in to make it look like the jacket had already logged many hours of airtime) and her first airplane, a Kinner biplane. Soaring through the skies in that bright yellow flying machine, *The Canary*, Earhart set an altitude record for women, climbing to an extraordinary 14,000 feet (4,269 meters).

In 1928 she captured America's heart and imagination when she flew from Newfoundland to Wales, the first woman to cross the Atlantic in a plane. Though she traveled as a passenger, she returned home a heroine to a ticker-tape parade and a blizzard of newspaper headlines and speaking-engagement requests, a riotous welcome engineered in part by media whiz George Palmer Putnam, who was to become her husband in 1931.

The tall, slender Earhart was frankly embarrassed by the attention she didn't believe she deserved, well aware that she had served only as "luggage" on the flight, a fact several newspaper columnists delighted in pointing out. If the unearned public adulation disturbed Earhart, it also fueled her resolve to live up to her media-manufactured moniker "Lady Lindy." Though she detested the nickname, she admired the man from whom it derived, famed pilot Charles Lindbergh. If Lady Lindy was what they wanted, Lady Lindy was what she was going to give them. Within a year, Earhart did the name proud, setting a women's closed-course speed record of 181 miles (290 kilometers) per hour and winning third place in the first-ever women's cross-country flying derby, an event she helped to organize, competing against a field including the top female talent in aviation.

In 1932 A.E. made front pages again, flying solo across the Atlantic, limiting her personal luggage to a thermos of soup and a toothbrush. When she landed in a cow pasture in Ireland, she broke another barrier for her sex, proving to naysayers—and to herself—that she was more than just decorative feminine baggage.

That matter settled, Earhart vowed to go where no woman—or man—had gone before. Three years after her Atlantic success, Earhart became the first person, male or female, to fly solo from Hawaii to California, a distance of 2,400 miles (3,840 kilometers). Next she soloed from Los Angeles to Mexico City, stopping south of the border just long enough for the weather to clear before heading east on a nonstop flight to Newark, New Jersey.

A.E. kept to a rigorous schedule on the ground as well, rushing from city to city, east

Amelia Earhart had become the most prominent woman in aviation years before her attempt to fly around the world. Earhart earned the distinction of being the first woman to fly solo across the Atlantic Ocean on May 21, 1932, when she successfully landed her plane in Ireland (left). For this accomplishment, President Hoover presented Earhart with a special gold medal from the National Geographic Society (below).

side to west side, church meeting to Hollywood bash, lecturing on the two subjects close to her heart: equality for women and the burgeoning field of aviation. To ensure that women remained contenders in that field, Earhart started the Ninety-Nines, an organization of female pilots. She also served as vice president for the Ludington Line, the first airline to offer regular passenger service between New York and Washington, D.C.

Flight Around the World

Earhart's flight around the world was to be both a personal pièce de résistance and a scientific contribution to the field she loved so much. Her Lockheed Electra, donated by Purdue University, was designed to be a "flying laboratory" in which Earhart could document the effects of prolonged flight and various altitude and weather conditions on the human body and the machinery. "Records such as these," Earhart said, "be they of success or failure, can do much to safeguard subsequent flights." Before she left from Oakland, California, A.E. confided that after traversing the equator, she planned to limit her whirlwind flying and lecture schedule. She wanted to spend more time enjoying her California home, her books, and her friends. "I feel," she explained, "like I have one last long flight in my system."

On the morning of July 2, 1932, the crew of the *Itasca* frantically tried to ensure that the Lae-to-Howland Island leg of that last long flight ended without incident. Attempt after attempt was made to communicate with the Electra. No response.

Earhart's next transmission—*"We are on a line of position of 157-137. . . . We are running north and south."*—confirmed the *Itasca*'s worst fears. Using celestial navigation, Noonan had apparently fixed the Electra's position at a sun line of 157-137, but the message made it clear that he had no idea of where along that line they were flying. The question: Were they north or south of Howland Island? The Electra obviously wasn't picking up the *Itasca*'s return messages, or remaining on the radio long enough for the Coast Guard cutter to take a bearing on the plane's position. (It was later learned that neither Earhart nor Noonan knew Morse code, and that Earhart had removed a crucial low-frequency antenna, which likely would

Only after careful planning did Earhart and navigator Fred Noonan set out on their round-the-world flight. When they disappeared, the largest search ever undertaken for civilians was launched.

have solved communication problems, because she found it a bother to reel in and out.) The *Itasca* radio operators stood by, anxiously waiting for word from Earhart; crewmen on deck trained binoculars skyward, hoping to spot a glint of the Electra. On both counts the *Itasca* was out of luck. And so, it seemed, were A.E. and Fred Noonan.

The *Itasca* reported the Electra lost. Over the next two weeks, the U.S. Navy, following direct orders from President Franklin Delano Roosevelt, launched the largest search ever undertaken for civilians before or since. A flotilla of ships, as well as planes launched from the aircraft carrier *Lexington*, combed 250,000 square miles (647,500 square kilometers) of the Pacific. Countless sorties and $4.5 million later, the search was called off. The most famous female pilot in the world and her navigator had vanished without a trace.

Theories Aplenty

What happened to Amelia Earhart? That depends on whom you talk to. Her disappearance has been the source of countless magazine articles, several books, a Rosalind Russell film, a host of made-for-TV documentaries, and plenty of cocktail-party conversation. The simplest answer to the question—and many, including A.E's husband, George Putnam, who personally researched other theories in the quest for his wife, believe the only logical answer—is that the Electra ran out of fuel, ditched in the sea somewhere in the vicinity of Howland Island, and that Earhart and Noonan sank in the deep waters with the wreckage. After the Electra was reported lost, Jacqueline Cochran, Earhart's friend and fellow aviatrix, boasting extrasensory perception, told Putnam that she had had a vision of A.E. floating in an area already being "well combed" by naval personnel. Cochran maintained that Earhart died after three days of drifting.

Relying on less-extraordinary channels of information—and, more often than not, gossip and rumor—a number of theorists have proposed more-elaborate, often bizarre—and, in some cases, sinister—scenarios to explain the mystery. One account had Earhart suffering from amnesia and working as a prostitute in a fishing village in Japan (the man who claimed to have seen this tall, slender Caucasian lady of the evening found no trace of her upon his return to Japan). Another story held that Earhart and Noonan were secret sweethearts and flew to a desert island to escape the anticipated vengeance of George Putnam. Presumably they lived—or live—happily ever after (love letters written by Noonan to his bride, whom he married shortly before the doomed flight, throw a bit of cold water on this passion-filled drama). Still another theory claimed that the infamous World War II disc jockey Tokyo Rose was actually Earhart. (In his desperation for an answer to his wife's disappearance, George Putnam traveled to the front lines of the Pacific theater to monitor a Tokyo Rose broadcast. He determined that the voice of the femme fatale of the airways and Earhart's voice were not one and the same.)

In the book *Amelia Earhart Lives*, author Joe Klaas went so far as to recount the claim by ex-U.S. Air Force officer Joe Gervais that Amelia Earhart was alive and well and living in New Jersey (Gervais also managed to find a stateside look-alike for Frederic Noonan). In this version, Earhart was recruited by the U.S. military to spy on Japanese-controlled islands. Shot down by the Japanese on Hull Island, part of the Pacific's Phoenix group, Earhart was held prisoner until the end of World War II, when she was released in exchange for a U.S. promise not to prosecute Emperor Hirohito for war crimes. Earhart returned to America and assumed the alias Irene Bolam, hoping to stay out of the public eye. If that's true, Gervais threw a serious wrench into her plans—Bolam, who denied allegations that she was Amelia Earhart, was forced to obtain a court order to keep the intrusive investigator at bay, and later sued the author for disrupting her life.

The most persistent theory to date hypothesizes that Earhart and Noonan went down west of Howland in the Marshall Islands, controlled at the time by the Japanese. The two were then taken prisoner by the Japanese and shipped to Saipan, where they later died.

Military Spy?

Fred Goerner, a former radio broadcaster for KCBS in San Francisco who spent six years investigating the case with the support of his radio network, was a strong proponent of this theory. In his book *Search for Amelia Earhart*, Goerner suggested that Earhart was on an unofficial, Pentagon-authorized mission to fly over the Japanese-controlled islands. Her job: to record signs of Japanese militarism on the islands, activity strictly forbidden when the League of Nations handed control of this territory over to Japan. (The spy theory was bolstered by Earhart's mother, Amy, who in 1949 told the press she believed that her daughter was on a secret mission. Mary S. Lovell, author of *The Sound of Wings*, a comprehensive biography of Earhart, believes that to Mrs. Earhart, "the possibility that her daughter had died in the service of her country was more palatable" than the thought that she had simply run out of gas.)

According to Goerner's version, U.S officials were suspicious because the Japanese would not allow foreigners to visit the mandated islands, and they were concerned that Japan's increasing aggression in Asia would lead to an attempt at military dominance of the Pacific. When high-level military and governmental officials learned of Earhart's planned flight, they requested that she make a detour to the Japanese base of Truk in the Caroline Islands on her way to Howland. Before she left Oakland, California, to begin her round-the-world flight, Goerner believed her plane was secretly equipped with stronger engines and larger fuel tanks than publicized. This way, she could fly over Truk and still make it back to Howland Island on schedule, nobody the wiser.

According to Goerner's scenario, uncooperative weather conditions confused navigation on the return trip to Howland, however, and the plane ran out of fuel. Earhart was forced to make an emergency landing on the Marshalls' Mili Atoll. For several days, Earhart radioed for help before the Electra's battery died (thus explaining SOS messages many ham operators claim to have heard after Earhart went down). The Japanese then appeared on the scene and took Earhart and Noonan captive. Fearful of what the two might have seen, the Japanese neglected to tell the rest of the world that the missing flyers were prisoners. They moved Earhart and Noonan to military headquarters on Saipan, where they both were executed or died of disease, depending on which witness you believe, and were buried on the island.

Thomas E. Devine, author of *Eyewitness: The Amelia Earhart Incident*, believes that he was shown the common grave of Earhart and Noonan by a native woman who identified the resting site as that of two white people "who came from the sky."

Devine, who was stationed on Saipan as a sergeant in the U.S. Army Postal Corps during the war, also claims to have seen Ear-

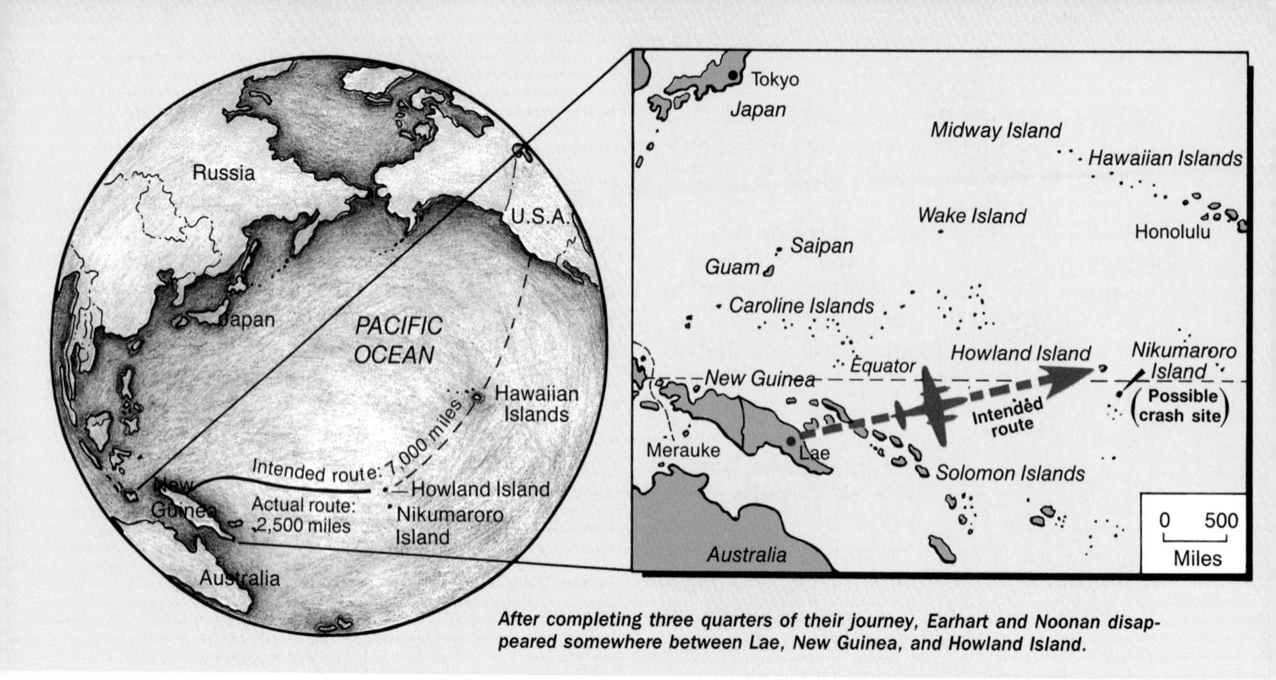

After completing three quarters of their journey, Earhart and Noonan disappeared somewhere between Lae, New Guinea, and Howland Island.

hart's plane torched by the U.S. military, part of what he, Goerner, and others believe to have been a cover-up of her presumed spy activities. Devine participated in Goerner's investigation, which began after Goerner heard testimony from Josephine Blanco Akiyama, a California woman born in Saipan whose account is also covered in Paul L. Briand, Jr.'s, *Daughter of the Sky*. Akiyama remembered seeing a plane land in the water in Saipan's Tanapag Harbor when she was a child. Japanese soldiers, she recalled, took from the plane into custody a white man and a white woman with short hair. Soon afterward, Akiyama heard gunfire, and assumed the captives had been executed.

Intrigued by Akiyama's tale, Goerner traveled to Saipan four times and spoke to a couple dozen natives who reported hearing about or seeing on the island in 1937 a tall Caucasian man and a woman who fit Earhart's description. American G.I.'s who served on Saipan during World War II also recalled learning tales that corroborated Akiyama's story. Some witnesses said the woman died of dysentery, others that the man was beheaded. Some said the woman was held in a hotel, others that she was confined to a prison cell. But most agreed that the two were white and held as spies.

Goerner perseveres to this day, though he's hit a number of dead ends in his investigation. For example, human remains that he discovered on Saipan with the help of Devine, which he hoped were Earhart's, were determined to be those of natives; a twin-engine plane, resembling an Electra, excavated from Saipan's harbor turned out to be Japanese. Goerner's theory, predictably, has evolved with time. He no longer believes that Earhart and Noonan were spies, though the Japanese might have mistaken them as such. The release of thousands of pages of classified U.S. information has convinced him that there was no covert mission.

Joseph Loomis, an ex-Air Force officer who wrote *Amelia Earhart: The Final Story*, concurs. "Amelia was a pacifist," he explains. "There's no way she would have agreed to have been a spy." Loomis also agrees that Earhart and Noonan went down in the Marshalls by Mili Atoll, although he differs with Goerner about the route they took to get there. Loomis began his research in the Marshalls rather than Saipan, interviewing natives who claimed to have seen or heard about a "lady pilot" who crashed in the waters off their homeland. A Japanese medical corpsman told Loomis he treated Noonan aboard a Japanese ship for a leg wound. Loomis also obtained Japanese ship logs, eyewitness accounts, and official documents that he says prove that the Japanese never conducted a search of their territory for Earhart and Noonan, though they told the world they had assigned several ships to the job.

The reason? Why look for two people if you already have them in custody?

Loomis, who is hard at work on another book, says that since publication of his first account, he has returned to the Marshalls and uncovered more material to back his conclusion. Exactly what does that material consist of? Earhart buffs will have to wait for the sequel.

New Clues Uncovered

Meanwhile, others are hoping to snatch final story honors from Loomis. Richard Gillespie, cofounder of The International Group for Historic Aircraft Recovery (TIGHAR), maintains he already has. Gillespie says he has proved that Earhart landed during low tide on a coral reef surrounding the Phoenix island of Nikumaroro (formerly known as Gardner Island), which lies southeast of Howland.

Gillespie decided to visit the island after analyzing the calculations of two retired military aviators. In a *Life* magazine article, Gillespie says that the aviators "re-created the exact navigational situation faced by Noonan as fuel ran low, radio navigation proved useless, and no island appeared ahead." Their evidence pointed in the direction of Nikumaroro. Though the military had flown over Nikumaroro during the 1937 search and reported "signs of recent habitation," no Navy personnel actually landed on the island to investigate further. When the circling planes failed to spot inhabitants, they flew off, never to return. In 1937 Nikumaroro had not hosted any known residents since 1892. Gillespie therefore believes those "signs of recent habitation," which may have included an unexplained water-collection device noted by a Coast Guardsman during World War II, were left by Earhart and Noonan.

During two trips to Nikumaroro, Gillespie's team found, among other things, the tattered remains of a woman's size-nine oxford shoe (Earhart's size and preferred style), a cap from a type of container commonly used to hold a stomach-ailment medication popular in the 1930s (during her round-the-world flight, Earhart complained several times of unusual queasiness, leading some to speculate that she was pregnant when she started the trip), and a piece of metal Gillespie believes came from the underbelly of the Electra. He hopes to find the rest of the plane on a return trip to the island in 1993.

Gillespie asked Herman Stevens, a retired Lockheed shop foreman, and Joseph Epperson, senior metallurgist for the National Transportation Safety Board, to examine the piece of metal. According to Gillespie's *Life* magazine article, Stevens, who used a photograph and Lockheed repair orders to conduct his analysis, believes that repairs that were made to the belly of Earhart's plane would display a rivet pattern identical to that found on the metal sheet recovered on Nikumaroro. Epperson concluded that "the materials in that specimen were consistent with those used in Electra planes from that time [the year 1937]."

Irrelevant, rebuts Frank Schelling, who heads the P-3 Aircraft Structures Branch at the U.S. Navy Aviation Depot in Alameda, California. "They are still using this material in aircraft today. It could have come from any aircraft." Schelling, who examined an almost life-size photograph of the fragment, as well as an engineering report containing the overall dimensions of the fragment, disagrees with Stevens and says that "Gillespie's case doesn't stand up. . . . That fragment did not come from the Electra."

At this writing, the controversy continues to rage, with Gillespie maintaining that the cases built by his competitors are based on hearsay and contradictory eyewitness accounts, while his is based on "historical documents, contemporaneous information, and physical evidence."

To that, Fred Goerner retorts, "We investigated the possibility that Earhart landed on Nikumaroro a decade ago. Mr. Gillespie has taken a button and sewn a suit on it." Joseph Loomis simply storms, "Gillespie doesn't know what he's doing."

Whether the theory of the embattled Gillespie proves right or wrong, he's probably correct about one thing: even if the case is eventually solved (and he, rather smugly, maintains he *has* solved it), there are some who simply won't believe it. As Gillespie puts it, "Amelia is like Elvis and JFK. She was a martyr, the Joan of Arc of aviation, lost at the prime of her life and her career. There are people who won't ever relegate her to the history books."

Until irrefutable evidence—bones, dental work, a piece of wreckage with an imprint of the Electra's serial number—has been recovered, maybe that's the way it should be.

What Happened to the Passenger Pigeon?

by Jerry Dennis

Not so long ago, the grassy hill overlooking our little city was a good place to gain perspective on the world. On spring days, you could drive up a gravel road, park in a meadow, and climb a short distance to the peak for a nicely symmetrical view of the sky above and the town on the lakeshore below. If you went there often enough, you could see winter retreating in tiny steps up Lake Michigan, toward Lake Superior and the Canadian frontier.

From that hilltop, you could watch the approach of thunderheads and misty streamers of rain, see squalls pierced by beams of sunlight long before they blundered across the lowlands. You could also see the corridors of industry, the new roads and highways, the geometric patterns of subdivisions creeping up the slopes.

Without these hilltops, we can't see valleys. Without long views, we're blind. Sometimes, from a hilltop, it's possible to see across centuries.

Eclipse by Pigeons

A few hundred years ago, people climbed that very hill above my hometown to watch vast flocks of passenger pigeons on their spring migration to northern Michigan. To those of us who have never seen a passenger pigeon, the stories about the numbers that once inhabited the central and eastern United States seem like bald hyperbole. Written accounts dating back to the 1600s describe flocks of passenger pigeons so large that they blackened the sky for hours on end, flocks that contained so many individuals that when the birds settled in forests to roost, their combined weight stripped limbs from trees. Observers claimed to have seen flocks containing millions, even billions, of pigeons. How could so many passenger pigeons have existed? More important, how could such an enormous population be extinguished?

John James Audubon was fascinated by the "wild pigeons" so common in his adopted home of Kentucky. In *Birds of America,* he described a species larger but similar in appearance to the mourning dove, with a bluish head and back, and a throat and breast the color of red wine, with wings and body designed for long migrations at speeds up to 60 miles (95 kilometers) per hour. But it was their sheer numbers that made the most powerful impression on Audubon. "The multitudes of Wild Pigeons in our woods are astonishing," he wrote. "Indeed, after having viewed them so often, and under so many circumstances, I even now feel inclined to pause and assure myself that what I am going to relate is fact. Yet I have seen it all, and that too in the company of persons who, like myself, were struck with amazement."

In the autumn of 1813, Audubon witnessed a 1-mile (1.6-kilometer)-wide flock of pigeons near Louisville, Kentucky, that passed overhead, without interruption, from noon until sunset, and so filled the sky that daylight "was obscured as by an eclipse." By his estimate, that single flock, one of dozens that Audubon saw that season, contained 1,015,036,000 birds.

Another ornithologist, Alexander Wilson, in 1806 visited a pigeon breeding area in Kentucky that measured 40 miles (64 kilometers) long and from 1 to 3 miles (1.6 to 5 kilometers) wide. The trees within the breeding ground each held up to 100 nests, and the surrounding ground was covered with broken branches, smashed eggs, and dead squab. He estimated the area contained 2,230,272,000 birds.

Such descriptions were common as early as the beginning of the 17th century, when Samuel de Champlain and other explorers commented on the pigeons they encountered in "thickened clowdes" of "countless" or "infinite" numbers. As late as 1870, when much of the pigeon population had been reduced by market hunting and habitat destruction, witnesses near Cincinnati, Ohio, saw a single flock estimated to be 1 mile (1.6 kilometers) wide and 320 miles (515 kilometers) long. Various observers have calculated that at one time, there were probably well over 3 billion passenger pigeons. That amazing number means that they would have represented 25 to 40 percent of the total bird population in the United States.

In the mid-19th century, passenger pigeons existed in such enormous numbers that a single flock might obscure the Sun for hours. By 1914, relentless hunting had wiped out the species.

OTHER VANISHING ACTS

- Nobody knows the precise rate at which animal and plant species are becoming extinct, but the estimates range from as few as one to as many as several hundred per day. Most are disappearing as tropical rain forests are cleared, and they are vanishing before biologists even have a chance to identify and name them, let alone study their roles in ecosystems. One thing biologists do agree on: we are witnessing one of the greatest mass extinctions in the planet's history.
- It's been estimated that the tropical rain forests contain as many as half of all Earth's plants and animals—about 2 million species—in an area covering only 7 percent of the total landmass. According to biologists Paul Ehrlich and E. O. Wilson, the rate of rain-forest loss doubled between 1979 and 1989, with approximately 1.8 percent of the remaining forests disappearing each year. As those forests vanish, so do plants and animals—at least 4,000 species each year, estimate Ehrlich and Wilson.
- Birds and mammals are the glamour species on the lists of endangered and extinct organisms, but there are hundreds of insects, reptiles, amphibians, fish, crustaceans, and plants that disappear each year with little or no fanfare. And not just in the rain forests. In the past 10 years, the United States alone has lost dozens of species of vertebrates and invertebrates, from the San Felipe leopard frog of California, to the Carolina elktoe mussel of North Carolina, to the chestnut ermine moth of New Hampshire and Vermont. The Hawaiian islands have lost at least 20 species of butterflies and moths, 18 bees, and 24 land snails.
- Why should we care? The loss of biological diversity has three basic consequences for humans. A wide variety of plants provide us with many of our foods and medicines, yet only a tiny percent of the world's flora have been tested for their medicinal and nutritional values. Although some skeptics may scoff at the idea that we have an ethical responsibility to protect Earth's inhabitants, nobody can deny that our lives would be immeasurably poorer without otters and warblers and cheetahs. Perhaps most important, no one knows the precise role each species plays in the scheme of things, but it is certain that all life depends on association with other life. Individual species have been compared to the rivets that hold an airplane together. How long can the rivets keep popping before the entire structure comes crashing down?

Relentless Killing

The unbelievable abundance of the passenger pigeon was the species' undoing. Market hunting was a widely accepted means of supplementing income in areas where the birds were abundant, and few people worried that the population could be seriously diminished by shotguns or nets. Even Audubon, witnessing the wholesale killing of the pigeons, concluded that "no apparent diminution ensues."

Yet the killing was too widespread and relentless to continue without consequences. Thousands were shot, netted, and clubbed wherever migrating or nesting pigeons congregated during spring and fall. Commercial netters took huge numbers by baiting small clearings with grain or salt, waiting until pigeons had settled in the space to feed, then launching large nets over them. Gunners frequently killed dozens of massed birds with a single shot. At a nesting site near Shelby, Michigan, one trapper admitted shipping out 175,000 pigeons himself during the spring of 1874, and estimated that in one 30-day period, an average of 100 barrels per day (with 500 birds per barrel) were shipped to restaurants in Chicago and Detroit.

By the late 1800s, the birds were threatened by both commercial hunting and the cutting of the hardwood forests where they nested and fed. The massive flocks were forced to move north to find places where they could nest unmolested, and, in the harsher climate, may have fallen victim to cold spells, or may have perished when they were caught in storms over open water. Suddenly, almost before anyone could imagine it happening, great flocks of passenger pigeons no longer obscured the daytime sky.

Attempts were made to save them, but the efforts were too few and too late. The Michigan legislature in 1897 passed a law that prohibited the killing of passenger pigeons, but that was 10 years after anyone had seen a sizable flock in the state, and in the years since 1890, even sightings of single pigeons had been notable enough to be mentioned in ornithological journals. In 1894 a flock of 500 was reported in Aitkin County, Minnesota; two or three flocks of four to six birds each were seen during the summer of 1893 at Elk River, Minnesota; and a few single birds were spotted in Michigan's Upper Peninsula in 1895 and in southern Wisconsin in 1896. Just three years after the well-meaning legislation was passed, the last confirmed wild passenger pigeon in the United States was shot by a young boy in Pike County, Ohio.

Efforts to establish nesting populations in captivity were never successful. Audubon in 1830 sent live birds to a British nobleman, who managed a small flock of passenger pigeons for a number of years before it finally died out. A male and female passenger pigeon captured in Wisconsin in 1888 produced 15 birds, but, weakened by inbreeding, they all died by 1910.

Last Survivor

In 1813, when Audubon counted 163 enormous flocks of passenger pigeons pass overhead in 21 minutes, he was moved to predict that nothing could threaten the survival of this incredibly abundant species. Yet, in fewer than 100 years, the great flocks were gone forever, and in their place was only a single female, named for Martha Washington, at the Cincinnati Zoological Garden. Although it has been widely reported that Martha had hatched in captivity and lived to the age of 29, those facts are not at all certain. She was also said to have been caught in the wild in Wisconsin, and to have been 17, 18, 19, 20, 27, and 28 years old. It is relatively certain, nonetheless, that Martha was the last passenger pigeon.

None have been seen since she died on the afternoon of September 1, 1914. There is little to be gained by brooding over past mistakes. It is better by far to remember the lessons of the passenger pigeon, the great auk, the dodo, the sea mink, and the rufous gazelle, and to direct our energies toward slowing the further loss of species throughout the world. Indeed, the rapid demise of the passenger pigeons shows just how tenuous wildlife survival is. And it is still difficult for Americans not to feel cheated of the opportunity to view the vast flocks of passenger pigeons in flight.

As with many birds, the male passenger pigeon had more colorful plumage than did the female. A typical passenger pigeon measured about 16 inches in length, including its long, tapering tail.

The hill above our city was bought by developers a decade or so ago, and is built over now with sprawling, expensive houses, each with a view of the valley, the city, and the bay. I still go there sometimes, looking for perspective, but it is not a place I can visit for long. The view is so cluttered. The sky is so empty.

THE RISE OF THE INTERSTATES

by T.A. Heppenheimer

To see the dream, you could visit General Motors' Futurama exhibit at the 1939 World's Fair in New York City. As visitors glided through a scale-model world, a recorded voice murmured compelling promises in their ears. By 1960, it said, 14-lane expressways would carry traffic "at designated speeds of 50, 75, and 100 miles [80, 120, and 160 kilometers] an hour." The cars would enter and leave at high speed via sleek interchanges. "One marvels at the complete accord of this man-made highway with the breathtaking scenic beauty of its route," the voice proclaimed. Futurama quickly became the fair's most popular attraction.

Such highways hardly seem farfetched today—even if the speed limit was set too high—but the reality in 1939, and for a number of years thereafter, was worlds different. It could be seen along a stretch of U.S. 1 between Baltimore, Maryland, and Washington, D.C. There the road was a four-lane highway, built as recently as 1930. Even though widespread use of turn signals was still years in the future, not only was the road undivided, but cars were allowed to make left turns across traffic along the route's entire length. This deficiency reflected the demands of local merchants, who had insisted that a center barrier and a ban on left turns would cut their business in half. The result, through the years, was a large number of collisions.

Nor was this all. Approximately 1,000 driveways ran directly onto the main road, so drivers had to watch out for cars slowing unexpectedly for a turnoff or entering the highway without warning. The lanes were only 10 feet (3 meters) wide, and the lack of a median barrier caused many head-on collisions on curves. Motels, hamburger joints, nightclubs, used-car lots, and the occasional house stood virtually at the roadway's edge, often with signs that ran to the edge itself, tending to squeeze motorists even farther inward. Twenty-ton trucks added to the dangers. Yet this stretch of highway was part of the best America had to offer.

A stretch of interstate highway cutting through the gentle hills of upstate New York represents just one scenic leg of the largest network of engineered structures on Earth.

Between the reality and the dream lay new opportunities in road building that were already demonstrating their worth. The basic elements of Futurama's freeways had been in use for more than a decade. These were the cloverleaf interchange, first built in 1928 at Woodbridge, New Jersey, and the parkway with a separating median between opposing directions of traffic, which dated to New York City's Bronx River Parkway in 1922. In 1939 America's first true freeways, the Pennsylvania Turnpike and Connecticut's Merritt Parkway, were already under construction (part of the Merritt had already opened). An even more far-reaching venture abroad, the 1,260-mile (2,027-kilometer) network of German autobahns, offered roads of similar quality and had been under way since 1929. There was little doubt by 1939 that a similar array of highways could cover the United States. What was lacking was the money.

Blueprint of Today

Through the 1920s and 1930s, there was still no hint of a national network of roads that could accommodate large volumes of traffic at high speeds. Long-distance travelers had to contend with whatever the various states and localities along their routes had managed to come up with, and conditions in some places were virtually unchanged since the dawn of the automobile.

The first stirrings of interest in a major system of multilane divided highways arose during the Depression, when their construction was touted as a public-works effort that would provide jobs. In 1938 the Senate considered but rejected a plan for an $8 billion network of toll highways. Then, during the following year, Thomas MacDonald, chief of the Bureau of Public Roads, began promoting a plan for a 30,000-mile (48,000-kilometer) system of expressways. Two years later President Roosevelt appointed the seven-member Interregional Highway Committee to work with MacDonald and to plan for road construction after the war.

MacDonald and his committee drew up a proposed interstate-highway map, whose roads ran to 32,000 miles (51,500 kilometers). The proposal bore an extremely close resemblance to today's network of interstates. All the major north-south routes were on the 1941 map, very nearly where we see them now. Their designations then lay decades in the future, but we would recognize them as I-15, 25, 35, 55, 75, and 95. The same is true of the principal east-west routes, today's I-10, 40, 70, and 80. The path of the future New York Thruway showed clearly, as did a triangle joining Dallas, San Antonio, and Houston, along with another triangle linking Atlanta, Birmingham, and Montgomery. There is even close correspondence in some of the fine detail: the future I-4 as a spur route to Tampa, the eventual I-25 jogging eastward near Albuquerque.

Some sections were missing, to be sure. There was no Massachusetts Turnpike. Today's I-25 was to run only as far north as Cheyenne, instead of extending across Wyoming to meet I-90; I-70 was to go west only to Denver, instead of joining I-15 in Utah. But in a nation with barely half its current population, and concentrated far more than today in the Northeast and Midwest, the 1941 map would prove a remarkably accurate guide to the highway routes of the next half-century.

Congress made no immediate move to fund the ambitious plan. Then, in 1943, the

At the 1939 World's Fair in New York City, General Motors' Futurama exhibit envisioned an interstate highway system consisting of 14-lane expressways and speed limits of 100 miles per hour.

American Association of State Highway Officials (AASHO) began lobbying for a bill that would provide a billion dollars a year for road construction. The AASHO called for a network of expressways similar to MacDonald's, totaling 40,000 miles (64,000 kilometers), with Uncle Sam to provide three-fourths of the construction cost, not only for this new system, but for a host of other roads as well. The final bill, passed late in 1944, indeed took note of this network, which it christened the National System of Interstate Highways. But the bill provided only half the money the AASHO had sought, and divided it entirely among conventional road projects. If any of these new interstate highways were to be built, then, the states would have to do the job.

Tollways

The war was not long over before several states began to take the necessary initiatives. The preferred approach was the tollway, to be built with funds from the sale of state-issued bonds, with the bonds' principal and interest to be paid from the tolls. Maine was the first; in 1947 the 47-mile (75-kilometer) Maine Turnpike opened between Portsmouth and Portland. During the next two years, Pennsylvania sold bonds to finance eastward and westward extensions of its own turnpike. That route had originally covered only 160 miles (256 kilometers); when the extensions were completed, in 1950 and 1951, it reached 327 miles (525 kilometers) to span the entire state. Also in 1950 a bond issue launched what would set the pace for postwar tollways: the New Jersey Turnpike.

This $255 million, 118-mile (188-kilometer) route took shape with remarkable speed, even though parts of it ran through cities and swampland. In Elizabeth, some 240 buildings were torn down or moved to clear the needed swath, with little opposition. In crossing the Passaic River, the highway soared to a height of 110 feet (33 meters) above water level. Near Linden the builders had to move tanks and pipelines within an oil refinery without hindering plant operations. One of the biggest challenges came in 100-foot (30-meter)-deep marshland near Secaucus. Engineers drove forests of 20-inch (50-centimeter) pipe into the muck, through several layers of silt, and down to the bottom of the soft clay. They filled the pipes with sand, then covered the right-of-way with dirt and pulled out the pipes. This left freestanding columns of sand, which acted as wicks. The weight of the dirt squeezed water from the bogs, causing the land to settle by as much as 12 feet (3.6 meters). With the water removed, the sand columns and the layer of dirt provided a solid foundation for the roadbeds. The result was a highway with no grade greater than 3 percent, curves designed to be handled at speeds up to 70 miles (110 kilometers) per hour, and, to combat drowsiness among mo-

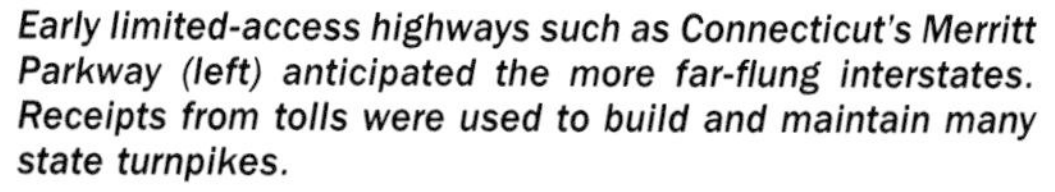

Early limited-access highways such as Connecticut's Merritt Parkway (left) anticipated the more far-flung interstates. Receipts from tolls were used to build and maintain many state turnpikes.

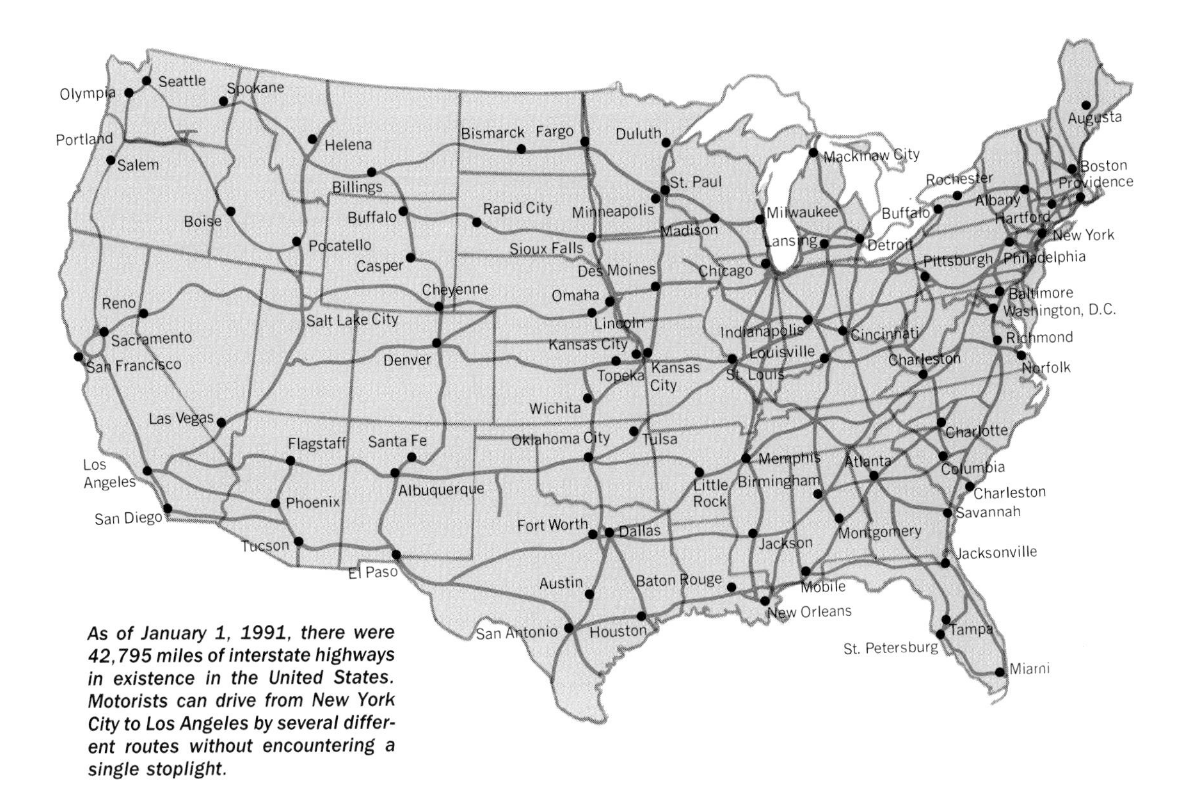

As of January 1, 1991, there were 42,795 miles of interstate highways in existence in the United States. Motorists can drive from New York City to Los Angeles by several different routes without encountering a single stoplight.

torists, no straightaway longer than 3 miles (4.8 kilometers).

The turnpike opened early in 1952 and was an immediate success. It cost $1.75 in tolls for its full length, roughly matching the cost of gasoline for the trip. But it cut the time to drive its route, New York City to Wilmington, Delaware, from five hours to two. It eased the wear and tear on autos and drivers alike while boosting gas mileage. Within months, turnpike officials found to their delight that traffic was 51 percent higher than expected, and revenues were 21 percent higher. Soon these officials would be proceeding with improvements they had not expected to pursue until far in the future, while making plans to pay off the bondholders years ahead of schedule.

This success was highly encouraging to people in other states who were promoting or pursuing similar tollways. In June 1952, with the success of New Jersey's project clear, Ohio floated a $326 million bond issue to finance its own turnpike. The New York Thruway, another toll route, was already seeing its first concrete, and Indiana commissioners soon announced that they would build their own route to link up with Ohio's. In a matter of months, then, motorists could see a dazzling prospect: through a combination of routes already completed, under construction, and planned, they would be able to drive from New England to Chicago without encountering a single stoplight.

Tollways were by far the most striking achievement of the road-building trade during the first postwar decade. But even their proponents knew they could not be more than a small part of a total system of interstate highways. They could be built only where traffic was heavy enough to generate the needed revenues, and in the mid-1950s, traffic engineers estimated that no more than 9,000 miles (14,500 kilometers) of highway could pay for themselves through tolls.

Pay-as-You-Go

In April 1954, President Eisenhower directed two of his top officials—his chief of staff, Sherman Adams, and Arthur Burns, chairman of the Council of Economic Advisers—to find ways to speed up the federal highway program. The Korean War had ended, and the time was right for such an initiative. Ike's intention was to pursue a program that would cost as much as a war; in his words, he was seeking "a 'dramatic' plan to get $50 billion worth of self-liquidating highways under construction." These roads would pay for their construction through motorists' taxes.

The Pavement Problem

Today's interstate highways are built according to a well-established procedure. Typically, a gravel subbase is first covered with concrete. Then an asphalt mixture is applied and compacted; or, for concrete highways, a steel mesh is laid on to help absorb stresses from expansion and contraction, and a layer of concrete is poured on top. These basic materials and methods have been in widespread use for decades.

At the turn of the century, though, much less was known about paving materials. Before the automobile era, most roads, if they were paved at all, had a surface of macadam (compacted stones bound with clay) or gravel. These were fine for horses and iron-wheeled wagons—if it didn't rain too much—but when cars started whizzing along at 20 miles (32 kilometers) per hour on pneumatic tires, the roads were quickly torn to shreds. Local authorities had to improvise and experiment.

Even before the automobile, many different paving methods had been tried. Cobblestones were the principal urban pavement until mid-century, but animal wastes, garbage, and diseases collected on their uneven surface. Wood-block paving seemed an attractive idea in lumber-rich areas, but it absorbed water and less-pleasant liquids, grew fungi, and usually had to be replaced within five years. Granite proved quite durable, but also quite expensive; sandstone was cheaper, but tended to wear smooth and slippery, especially when wet. Iron blocks were tried on New York City's Cortlandt Street, but they were noisy, and hexagonal projections intended to lend traction turned out to be too effective, causing horses to trip and stumble. Even glass blocks were installed, in Lyons, France, but they chipped and split under heavy traffic.

Brick roads were first introduced in the United States in 1870 in Charleston, West Virginia, and became quite popular over the next several decades. They were attractive, offered a good foothold for horses, and were easily repaired and cleaned. They were expensive to build, however, and the rumble they produced when an automobile drove over them was deemed unacceptable. In 1881 London surfaced two roadways with an inch-thick layer of rubber; it was quiet and easy to clean, but again too costly for general use. Also in London, ground cork embedded in asphalt was tried near hospitals, where quiet was necessary; it passed all the tests except durability.

Road builders often turned to local industries in their search for autoworthy pavements. In Newton, Massachusetts, in 1908, the newly created U.S. Office of Public Roads tried a covering made from blackstrap molasses and quicklime in a petroleum binder. It held up well under traffic but proved to be, as one reporter put it, "somewhat soluble in water."

Other cheap and plentiful local materials that were tried and found wanting as pavement

The preautomotive macadam roads (left) rutted under the weight of an automobile. Brick roads (below) produced a loud rumble when cars drove over them. Although wood-plank roads (right) muffled noise well, they deteriorated too quickly to ever gain widespread use.

included seashells, coal, tar, burnt clay, sand-clay, something called sand-gumbo, slag from blast furnaces, and even compressed straw. Towns lucky enough to be close to stone quarries had it made. Gouverneur, New York, paved its roads in 1911 with crushed marble sandwiched around asphalt. The asphalt cost $500 a carload; the marble was only $1. Around 1900, lumber interests came up with a process of treating wood blocks with preservatives to avoid the problems of earlier wood pavements. Minneapolis, controlled by timber magnates like the Walkers, invested heavily in creosote-treated wood-block pavement from 1902 to 1912. Other cities turned to wood in an effort to alleviate noise problems. Unfortunately, even with the preservative process, wood blocks were too expensive for cities remote from major forest areas. In addition, the sticky chemicals soon seeped out and got all over everything. Within a few years, much of the preservative would be gone, and then the pavements buckled, bulged, and rotted just like untreated wood.

In retrospect the eventual popularity of concrete seems inevitable. It was of moderate cost, required little maintenance, was easily cleaned, and held up well. But concrete took quite a while to gain favor. The first American concrete pavements were laid down in Bellefontaine, Ohio, in 1884, and for the next several decades, concrete lagged behind brick, asphalt, and wood. One reason was poor workmanship: a few early contractors, unfamiliar with the material, produced improperly built roads that crumbled along the expansion joints. These problems were soon corrected, but they gave concrete a bad reputation that took years to overcome.

Noise was another problem; horses were only slightly quieter on concrete than on stone or brick. There was also considerable local resistance, because pressure to use the products of nearby firms was ever present. In Chicago, where such things as price and quality were minor factors in the awarding of contracts, officials called for treated wood blocks, with the specified preservative being a type of oil produced by only one company in the United States—located in Chicago, of course.

But eventually the soaring popularity of automobiles made concrete the clear favorite. Motorists enjoyed the smooth ride they got on concrete roads; local residents found them to be quieter than brick roads. Concrete proved to be the most cost-effective durable surface when exploding auto traffic placed demands on roads that were inconceivable at the preautomotive turn of the century. Today's America would find it hard to function if its interstate highways were made of anything less tough than asphalt or concrete. But back in 1905, it was not implausible to imagine someday driving from Boston to St. Louis on molasses and wood.

Steven R. Hoffbeck

The 1950s and 1960s were decades of intensive interstate-highway construction. During those years the cloverleaf exchange became a symbol of American progress and ingenuity.

The entire federal budget in 1954 was $71 billion, and people spoke of $50 billion somewhat the way we would speak of a trillion dollars today. So Ike picked as a key adviser on highways a man who was accustomed to thinking in sweeping terms: General Lucius Clay, one of his wartime commanders in Europe. Clay would head up a committee to develop a specific plan.

Late in 1954 Clay was ready with his proposal. The interstate highways would receive $25 billion, with their construction to take place during the decade 1955–64. Substantial sums would also be allocated for other roads. The $25 billion would be raised largely through a sale of bonds, with interest at 3 percent, to be issued by a new Federal Highway Corporation. Their principal and interest, in turn, would be covered entirely through income from existing gasoline taxes. Indeed, because traffic would increase, this arrangement would prove so lucrative that the federal government could cover 90 percent of the construction costs. States would have to put up less than $3 billion.

The plan drew opposition from Harry Byrd, chairman of the Senate Finance Committee, who saw the proposed bonds as a trick to allow the government to borrow money while keeping the borrowings separate from the national debt. Deficit financing of any type was anathema to Byrd, much more so if its nature was concealed. Faced with such opposition, a bill incorporating Clay's plan went down to congressional defeat in July 1955.

However, Byrd was quite willing to support an alternative: pay-as-you-go financing. This had long been the favored approach at the federal level; the 1944 highway act had been pay-as-you-go, as had a host of other such enactments. The funds collected for allocation under such laws had always gone for legislators' favorite conventional roads and highways, which often turned out to be pork-barrel projects. But the flood of new tax money accompanying the expected increase in traffic would be enough to pay for not only conventional roads, but the new interstates as well.

This idea was the basis for what became the Federal-Aid Highway Act of 1956, the work of Representatives George Fallon and Hale Boggs. In the words of historian Mark Rose, "the key to success was promising something for everyone without imposing high taxes on truckers." A principal feature of the act was the establishment of the Highway Trust Fund, which would receive all federal gasoline and automotive taxes. By law, all money in this fund was to be spent on roads and highways, thus addressing a longstanding complaint: that the federal government received more income from these taxes than it spent on roads, with the balance being diverted into general expenditures.

The taxes would rise under this act: from 2 to 3 cents a gallon on motor fuel, from 5 to 8 cents per pound on tires, from 8 to 10 percent excise tax on new trucks, buses, and trailers. But these, along with other minor imposts, drew little opposition. In return, there would be abundant sums for roads of all types. For the new interstates, Washington would pay 90 percent of the cost. Yet state highway departments would have the right to choose their routes, with no fear of being overruled by a federal agency. And the entire package would indeed be pay-as-you-go. The combination was irresistible; in June 1956, the bill was passed by voice vote in the House and by an 89-to-1 margin in the Senate.

New Type of Construction

The phrase *interstate highways* was already in use. It referred to a set of existing U.S. highways designated in 1947 as a result of recommendations from state-highway departments. These routes served 209 of 237 cities with populations of more than 50,000, and had a total length of 37,600 miles (60,500 kilometers). Outside the cities, they included only a little more than 1 percent of the nation's total mileage, yet they were carrying 20 percent of all rural traffic. In the cities the designated roadways constituted two percent of all streets and avenues while carrying 11 percent of urban traffic. These highways, the nation's major trunk lines, would represent the point of departure for the new interstate system.

Contemporary accounts stated that these existing routes would be "modernized completely" or "largely rebuilt" or would undergo "reconstruction." Such phrases reflected the expectation that much road building would consist of upgrading existing two-lane routes and then building another set of lanes on the other side of a median strip. The Clay report, however, had stated that "on a considerable portion of the interstate network, especially in urban and suburban areas, it will be more economical to relocate than to acquire the additional land necessary to permit control of access." This meant that the new interstates would largely replace the old system, reducing those U.S. highways in many cases to little more than local roads.

Indeed, the new interstates would differ dramatically from any of America's existing roads other than major tollways. The Baltimore-Washington stretch of U.S. 1, for instance, ran for 30 miles (48 kilometers) and had cost $1.76 million to build. The new interstates soon would cost an average of more than a million dollars a mile. This was mostly due, not to inflation, but to the fact that they represented a very different type of construction. Old-style highways had pavement 4 to 5 inches (10 to 13 centimeters) thick. It cracked and degraded under heavy traffic. The new ones would have pavement 9 to 10 inches (23 to 25 centimeters) thick, atop a prepared roadbed that by itself could be as much as 50 inches (127 centimeters) deep. Rights-of-way would be up to 300 feet (92 meters) wide, to provide ample room for future expansion. Earth moving would take place on a heroic scale. Along the route from San Francisco to Sacramento, for instance, designers wanted to link up with an existing bridge across the Carquinez Strait. A 500-foot (150-meter) hill lay in the path, and an existing highway and railroad ran meekly along its base. For the new route, however, engineers boldly ripped out a gash 360 feet (110 meters) deep and nearly half a mile

The interstate highway system indirectly helped California develop its "automobile culture." The daily concentration of exhaust-producing cars on freeways has contributed to that state's pollution problems.

long. This was described in 1957 as the biggest human-made cut since the Panama Canal was built.

The construction equipment itself was on the same scale. Some two dozen types of heavy machinery were in standard use: power shovels, excavators, large-capacity dump trucks, bulldozers, graders. There also was a paver that straddled an entire roadbed and laid down concrete at a rate of 8 to 10 feet (2.5 to 3 meters) a minute. It would scoop up paving material dumped in its path and spread it in a layer up to 8 inches (20 centimeters) thick over a grid of reinforcing steel rods that itself sat on several inches of concrete. One witness described the paver as a "giant, ungainly spider . . . spinning its strand of highway."

The work sped along. No Golden Spike ceremony marked the date, but slightly more than a century after completion of the first transcontinental railroad, it became possible to drive from New York to San Francisco without encountering a stoplight. The new freeways, however, were not entirely free. Existing toll roads such as the New York Thruway had been incorporated into the system, giving it an instant 2,100 miles (3,380 kilometers) of completed interstate. And freeways were not uniformly popular in rural areas. Farmers sometimes found themselves driving miles out of their way to find an interchange merely to get to their acres on the other side of the right-of-way. If their inconvenience was great enough, as determined in lawsuits or other proceedings, highway commissioners would often build a small underpass.

People who lived near interstates were also mollified by rapidly escalating prices for their land, which they could sell to developers. This represented a turnabout from earlier days, when a railroad had a "wrong side of the tracks" downwind of its soot, and when buildings adjacent to the line deteriorated. Instead, the new freeways emerged as major catalysts for growth and development. Property values soared: by a factor of 87 during 12 years along a route near San Francisco, and by a factor of 20 during a 28-year period along New York City's Grand Central Parkway.

Industrialists also were quick to appreciate the freeways' advantages. Near Boston, one stretch of what would become I-95 had 39 firms in June 1955; by September 1958, there were 209. By 1959 a syndicate was planning to build 200,000 houses along New Jersey's Garden State Parkway. At about the same time, a state report declared that at least $650 million in new development had resulted from the construction of the New York Thruway, including a $28 million electronics center built by General Electric in Syracuse. Company officials said the thruway would act as a pipeline, bringing in raw materials and taking out finished products to the markets of the Northeast.

Environmental Concerns

Were freeways an affront to the environment? In New Hampshire, I-93 provided a dramatic example of how engineers could reach a compromise even in a highly sensitive area. The route was to pass through scenic Franconia Notch, which features a natural granite profile called the Old Man of the Mountains, so much a symbol of the state that it appears on its license plates. A standard four-lane design would have meant cutting into the side of the mountain. The compromise involved upgrading an existing 3 miles (5 kilometers) of two-lane road that already ran through the Notch, while adding parking lots for sightseers. The state got its freeway, and the environmentalists won full protection for one of their most cherished treasures.

Did freeways encourage individual drivers to use their cars wastefully while discouraging buses and car pools? This issue has been addressed, near Washington, D.C., in California, and elsewhere, by building high-occupancy-vehicle lanes. These are restricted during rush hours to multipassenger vehicles, and those who use them often find themselves proceeding at full speed while single-passenger traffic in adjacent lanes remains badly clogged. Indeed, I-66 near Washington is set aside entirely for high-occupancy traffic during hours of heavy use. For many commuters, it is the most convenient route into northern Virginia, but they cannot use it unless they join a car pool.

More than 35 years after passage of the Interstate Highway Act, then, the nation's freeways can be counted among America's successes. These expressways had recapitulated the achievements of the railroad builders of the previous century. They were what the nation needed, what we paid for and built. And they worked.

Salem's Darkest Hour by Bruce Watson

In February of 1692, something was gripping the young girls of Salem Village, something neither doctors nor ministers could divine. Betty Parris, the local minister's own daughter, was in a trance, hands frozen in place, uttering the most hideous gargles and growls. Prayers did no good. "Our Father which art . . ." set her screaming. Soon her cousin Abigail Williams began crawling around the house, under chairs, barking like a dog, stomping her feet.

None dared call it witchcraft, but Salem's girls were up to some mischief. Thrusting a key into a Bible, they read the verse it touched as prophecy. And they read palms until Abigail Williams spread the word of a better oracle. Come to the Parris kitchen and meet Tituba. She floats an egg white in a glass. It tells your fortune. All that winter, girls met with the Reverend Samuel Parris's West Indian slave to learn "what trade their sweethearts should be of."

Among the futures in Tituba's glass, the milky shape of a coffin spread panic. Only a few weeks later, Dr. William Greggs, unable to find a medical cause for Betty and Abigail's fits, declared: "The evil hand is upon them!" Suddenly events quickened, and the girls were no longer mere girls. Ministers surrounded them, demanding, "Who torments you?" When the girls gave no names, names were suggested. On February 29, three women were charged with witchcraft. Then, as spring gave way to summer, Salem itself fell into a series of fits that has made its name synonymous with sorcery ever since.

Three hundred years after the most infamous witch-hunt in North America, covens of tourists are descending on Salem to hunt for history. With ongoing symposiums and exhibits and productions of Arthur Miller's *The Crucible* and other dramas, Salem and nearby Danvers are commemorating the tercentenary of the trials. All summer, explanations and theories will fly like broomsticks.

In a 19th-century artist's fanciful engraving of a Salem witch trial, the coincidental occurrence of a lightning storm during the proceedings provides sufficient evidence to damn the accused as a witch.

Assertion of innocence was considered tantamount to witchcraft in the late 1600s. During the Salem witch trials, any woman who dared defend herself against the accusations of her hysterical, swooning accusers was considered a probable victim of demonic possession.

Witch-Hunt Beginnings

For centuries, historians have branded Salem's witch-hunters as hysterics, fanatics, and liars. But does anyone understand why, in one hellish summer, the town filled its jails, turned a bucolic knoll into "Gallows Hill," hanged 19 people, tortured another to death, and then, only four years later, hung its head in shame? If this is witch-hunting, where should the hunt begin?

Begin here: "Thou shalt not suffer a witch to live"—Exodus 22:18. Despite God's commandment to Moses, witches and wizards practiced their craft from biblical times until the Middle Ages. Only when the medieval world began to crumble in the 1400s, with scapegoats needed and power up for grabs, did the hunt begin in earnest. In 1486 witch-hunting received its own bible, *The Hammer of Witches*. Written by two Dominican scholars with papal approval, the book became a best-seller of sorts by explaining in painstaking detail how witches torture, possess, and kill the innocent, especially children, and how authorities should judge and execute the wicked.

In one of humanity's darkest chapters, fires blazed beneath tens of thousands of witches throughout the Protestant Reformation in the 16th and 17th centuries. In England, where witchcraft was a crime against the state, punishment was more merciful—hanging. In Presbyterian Scotland, where the hunts were especially rabid, witches were often hanged, then burned. By 1660 the worst was over in Europe. But New England's Puritans, still few in number and isolated on the edge of forests primeval, had their own demons to hunt.

On March 1, 1692, most of Salem Village took a holiday. Dressed in its Sunday black and white, an overflow crowd pressed into an icy, wooden meetinghouse. Two stern judges had come from nearby Salem Town to examine the accused. Amid commotion and gossip, they set the bewitched girls on a bench and at last brought the first witch to the bar.

Judge John Hathorne instructed the girls to look at Sarah Good. Casting their eyes on the twisted woman, Betty, Abigail, and their friends screamed, writhed, swooned.

"Sarah Good," he asked, "do you not see now what you have done? Why do you not tell us the truth? Why do you thus torment these poor children?"

Clearly, something was torturing the girls. Sarah Good claimed that "it was Osborne" who did it. Sarah Osborne, ill and dragged out of bed into custody, claimed to be more bewitched than bewitching. In a dream a "thing like an Indian, all black," had pricked her neck and pulled her out of bed. Everyone knew Osborne's "thing." He was a celebrity in Salem, cursed by every tongue, the archnemesis behind every form of malice. But never did goodmen and goodwives expect to see Satan torture the innocent in the Salem Village meetinghouse. The crowd pressed closer as Tituba herself stepped to the bar.

Beaten by the Reverend Parris, the slave had learned to tell Puritans what they wanted to hear. "The Devil came to me and bid me serve him," she confessed. She told tales of blood-red cats, red rats, witches' Sabbaths, and demonic creatures kept by Sarah Good and Sarah Osborne. Satan had offered Tituba a book to sign. Sarah Good's blood-red mark was in it; Sarah Osborne's, too. And when Tituba signed, she saw six more marks.

Victims of Social Strain

The Puritans' daily lives were marked by hardship. In such a world, what today would seem trivial were deadly sins. Malice, envy, anger, pride, opened the door for Satan and his advance guards. Witches entered bedrooms late at night. A devil had sat at Christ's table, and now Tituba's magic had loosed the Devil among them. "We are either saints or devils," the Reverend Parris preached. "The scripture gives us no medium."

Were the girls truly possessed, or merely possessed by fear of the spells and curses floating around Salem? Were they "great actors," mischievously deceiving their elders, as a Salem mayor charged in his 19th-century history? Did they suffer from "mass hysteria" caused by Puritan repression, as Freudian analysts have suggested? Or did a hallucinogenic fungus in bad rye twist their arms and prick their skins, as a psychologist hypothesized in 1976? Each theory has its converts and its foes.

Witchcraft, writes Yale historian John Demos, was a gauge of "social strain." Historians have long noted that the Salem outbreak occurred during a time of trouble. A power struggle between the Puritan Colony and the king had left Massachusetts adrift, without a governor, a royal charter, or any legal system. The faithful were shaken by epidemics, Indian wars, and ideological clashes between thriving merchants and simple Puritan farmers. Driven by such tensions, Puritans brought witchcraft out of the closet and into the courtroom.

Most New England "witches," Demos writes in *Entertaining Satan*, were poor women over 40 years old. Some were healers or midwives who had lost a patient or miraculously saved one. They were abrasive people, quarreling with families and neighbors, often finding themselves involved in lawsuits. They were, in a word, misfits. The label fits Salem's first accused like a pointed hat. Sarah Good, a beggar; Sarah Osborne, a hermit and heretic; Tituba, a slave—these three women were dragged into court, where neighbors came forth to testify about their every evil deed.

Spectral Evidence

Salem had been warned about specters. Because the Devil might come clothed in anyone's shape, even Cotton Mather discounted the conclusiveness of purely "spectral evi-

Nineteenth-century illustrator Howard Pyle depicts an accused witch praying for mercy while her accusers point and howl from the front row. The horseshoe-pronged staffs are meant to represent antiwitch devices.

dence" in witch trials. A witch is proved a witch, Mather said, only by confession, possession of voodoo dolls, pins and "poppets," or witch's teats—flaps of flesh used to suckle Satan himself, searched for and found in the merest birthmark. But Salem Village chose to believe its children. Through most of the trials, the judges took evidence as thin as a ghost as concrete proof of the worst crime imaginable.

Once it became clear that anyone's specter could be blamed, anyone was fair game. Anyone. Rebecca Nurse was a pious, gray-haired churchgoer. "I never afflicted no child," she said meekly, "no, never in my life." But before the icy stares of her neighbors, she cried out, "Oh, Lord help me!", spread her hands, and the girls swooned.

"Rebellion is as the sin of witchcraft," Mather preached, and fearing the Devil more than each other, Salem made rebellion tantamount to sorcery. A constable who refused to arrest more witches was accused and jailed. John Proctor said if the girls were not stopped, "we should all be devils and witches." Proctor and his wife were next. "Dear child, it is not so," Elizabeth Proctor pleaded to the accusing Abigail Williams. "There is another judgment, dear child." But the fear was contagious. Seeing specters, the girls called out in court, "There is Goodman Proctor going to Mrs. Pope," and Mrs. Pope fell into fits. "There is Goodman Proctor going to hurt Goody Bibber." And Goody Bibber fainted. The witch-hunt was on.

From out of their dens came neighbors renewing timeworn vengeance, settling old vendettas. In April and May, more than 60 people were accused. In June, formal trials began. For those who could explain their specters, confess, or escape, life might go on. For the rest, a sturdy gallows on a nearby hill was fitted with a noose.

Before Salem, New England witches had been the subject of rumor, gossip, and an occasional trial. Why did Salem alone turn witch-hunting into a craze?

At the heart of Salem's discord was a Hatfield-McCoy-type feud. Throughout the 1680s, the Porter family gained land and influence in the church, while the Putnam family lost both and vowed vengeance. Come 1692 three "afflicted" girls in the Putnam household, together with parents, in-laws, and cousins, testified against at least 46 accused witches, many from the Porter camp. The feud, coupled with factions split by the contentious Reverend Parris, made Salem a house divided against itself.

During that zealous summer, there were moments when reason might have triumphed. After Bridget Bishop was hanged on June 10, one judge resigned in protest. Boston ministers cautioned against spectral evidence, and neighbors signed petitions supporting the accused. At the next trial, despite testimony by several Putnams that Rebecca Nurse's specter had murdered six children, the jury returned with its verdict. Not guilty. Bedlam tore through the courtroom. The girls howled, their convulsed limbs snapped, until the startled judges ordered the jury to reconsider. After further questioning, the jury gave a different verdict. The governor, petitioned by Nurse's family, granted the old woman a reprieve, but when the girls heard about it, they fell into fits and were said to be near death. The reprieve was rescinded.

On July 19, Rebecca Nurse and four others rode a creaky oxcart past jeering crowds to Gallows Hill. With the noose around her neck, the muttering witch Sarah Good no longer muttered. "I am no more a witch than you are a wizard!" she shouted at a minister. "If you take my life away, God will give you blood to drink!"

Following July's hangings, the Devil knocked on other doors. After the Salem girls were called to Andover to detect witches, more than 50 were jailed there, and a dog was executed for witchcraft. Crammed into jails, the accused dared not hope for justice. On the eve of their trials, a few escaped with the help of friends. Others, noting that Tituba still hadn't been tried, confessed and were spared. But the most devout believed that lies brought a more lasting judgment, and rode their innocence toward Heaven.

On August 19, five more were hanged, another eight on September 22. "What a sad thing to see eight firebrands of hell hanging there," said a minister.

Systematic Violence Against Women

In a society that believed in its own shining eminence, 19 had been hanged and one tortured to death. Three more had died in jail. Something beyond factionalism and tension must have possessed Salem, some other devil, perhaps in feminine form.

Three-fourths of Salem's accused were women, including 14 of the 20 executed. Witchcraft, writes Carol Karlsen, a University of Michigan historian, "confronts us with ideas about women, with fears about women, with the place of women in society, and with . . . systematic violence against women." In *The Devil in the Shape of a Woman,* Karlsen explains how Puritans both constrained and exalted women. Woman was wife, mother, "helpmeet," handmaiden of the Lord, and those who did not fit this holy mold were surely handmaidens of the Devil. Among the accused throughout New England, Karlsen finds women who dared to fight for an inheritance, childless women, adulteresses, single women, and women involved in lawsuits. Their common sin, Karlsen writes, was refusal to knuckle under to the Puritan patriarchy.

During the summer of 1692, 19 people in Salem were hanged as witches. Four years later, a judge and 12 jurors publicly repented.

Accused men were most often witches' husbands or sons, but men themselves leveled three-fourths of the accusations. They told of dreams in which female figures sat on their chests, making it difficult to breathe. And girls, the most frequently bewitched, used witchcraft to vent their jealousies. Many of Salem's "afflicted" girls had lost their fathers, Karlsen notes. Fatherless girls might wonder "what trade their sweethearts should be of," but having no father and therefore little or no dowry, they were mostly bound for spinsterhood. Claiming vengeance as theirs, they lashed out. It was an approved form of rebellion, Karlsen writes. Blasphemy led to whipping, but if the Devil made them do it, the young girls became the most important people in town.

"The little crazy children are jangling the keys of the kingdom," Arthur Miller has John Proctor proclaim in *The Crucible*, "and common vengeance writes the law!" More accused might have gone the way of all devils had not Truth at last put on its boots.

Increase Mather, Cotton's father and president of Harvard College, believed in witchcraft, but was skeptical about Salem. When a parishioner took his sick child to Salem for a diagnosis, the elder Mather had seen enough. "Is there not a God in Boston that you should go to the Devil in Salem?" he demanded. On October 3, his sermon condemned the trials. "We ought not to practice witchcraft to discover witches," he said. "It is better that ten suspected witches should escape than one innocent person should be condemned." In sermons, other ministers doubted that so many could "leap into the Devil's lap at once." During the trials as well, almost 300 citizens had signed petitions or testified on behalf of the accused.

The girls struggled to maintain their power, "crying out" on ministers' wives. But soon after they accused his wife, Governor William Phips stopped the trials, on October 29. Of the remaining witches, 52 were tried in 1693 without the use of spectral evidence. Three were convicted but pardoned. The hunt was over.

When the bloodletting stopped and time soothed Salem's fears, the guilt poured forth. In 1696 first a judge, then 12 of the jurors, publicly repented. Even Ann Putnam, Jr., one of the most afflicted girls, apologized. In 1711 the state paid survivors of the executed sums of £7 to £150, and reversed the convictions of all but seven, who, having no family come forward to demand a pardon, remain guilty on the records.

REVIEW

Archaeology

AMAZON EXCAVATIONS

Recent excavations at Santarém in the Amazon Basin of Brazil have yielded shards of pottery that seem to be the earliest in the Americas, dating to some 7,000 or 8,000 years ago. Intensive archaeological investigation of the Amazon Basin has begun relatively recently, and the discovery of very early pottery there is something of a surprise. Results of earlier research suggested that pottery and other signs of cultural complexity developed first in other places in the Western Hemisphere, such as the highlands of Mexico and in the Andes. Dr. Anna C. Roosevelt of Chicago's Field Museum of Natural History has found that the soils of the Amazon region were as rich as those of major river valleys of the Old World, where the first civilizations apparently developed. The new evidence suggests that the rich Amazonian environment may have supported substantial agricultural populations much earlier than had been previously thought.

PYRAMID BUILDERS

In Egypt, new research is teaching us about the lives of the common people who lived at the time of the construction of the great pyramids. Archaeologists working at Giza discovered the oldest known bakery in Egypt, dating from the time of the pyramids, around 2500 B.C. It probably supplied bread for the pyramid laborers. The molds and pots found in the archaeological excavations match bread-baking equipment depicted in paintings on the walls of tombs.

Nearby, archaeologists have also discovered the cemeteries of these workers. In the study of ancient Egyptian civilization, scholars have concentrated on the monumental architecture, such as the pyramids, and on the elite segment of society. With the discovery of these cemeteries, archaeologists will be able to study the lives of the majority of the ancient Egyptian people. Some of the newly discovered tombs are similar to those of the elite, although smaller and simpler. Inscriptions identify some of the graves as those of persons who directed different aspects of the building of the pyramids.

NEOLITHIC MAN DISCOVERED

In September 1991, climbers in the Alps found the body of a man who died some 5,000 years ago. The frozen body was discovered on the Similaun Glacier at an elevation of 10,500 feet (3,200 meters) above sea level, on the border between Austria and Italy. The body had been extraordinarily well preserved under the snow: skin, muscles, and internal organs were still intact, as was the skeleton. The man, who scientists believe was between 20 and 40 years old, wore a leather coat lined with hay, and lined leather boots. He had tattoos on his back. Equipment he carried

Archaeologists have gained new insight into ancient Egyptian society thanks to the discovery of cemeteries in which pyramid laborers were buried.

included a copper ax, a wooden pack, a stone knife, a leather quiver with 14 arrows, a bow, a stone bead, and, attached to his belt, a tool for striking a spark to make fire. Radiocarbon dating suggests that the body is about 5,000 years old, thus belonging to the Late Stone Age, or Neolithic period. It is not clear what this man was doing so high in the mountains when he died. Investigators have suggested that he may have been a shepherd, a hunter, a mineral prospector, or a traveler hiking through the mountain pass. While we know much about the Late Stone Age from excavations of settlement sites and cemeteries, this find of a man, struck down in life by an as-yet-undetermined cause, with all of his personal equipment intact, is unique. Ongoing study will produce new information about this man and perhaps about why he was so high in the mountains when he died.

The remarkably well-preserved remains of a prehistoric man have provided clues as to the dress style and level of tool sophistication of 5,000 years ago.

MAYAN ARCHAEOLOGY

Scientists continue to make exciting new discoveries in the lands of the ancient Maya in Central America. Focusing on the site of Dos Pilas in northern Guatemala, an expedition led by Professor Arthur Demarest of Vanderbilt University discovered an apparently undisturbed grave of a Mayan king. The tomb was found deep beneath a pyramid, and contained a man's skeleton outfitted with ornaments of jade, pearl, and conch shell, along with pottery bearing hieroglyphic writing and blades made of obsidian. A monument in front of the pyramid identifies the buried person as a Mayan king, designated by archaeologists as "Ruler 2" (his name is not yet known), who reigned between the years A.D. 698 and 725. Demarest suggests that the unknown king may have played an important role in the growing warfare between Mayan communities in that period.

New evidence from inscriptions indicates that much more violence occurred between Mayan communities than archaeologists had known about earlier, and much of the art depicts scenes of conflict. The recent excavations at Dos Pilas revealed major fortifications, suggesting a need for defense on a large scale. Demarest and other archaeologists have interpreted the accumulating evidence derived from inscriptions, art, and fortifications. Their studies indicate a rapid increase in large-scale warfare during the 7th and 8th centuries A.D., which may have contributed to a breakdown in the ecological balance between the agricultural practices and the natural environment. Together, warfare and ecological deterioration may have been major causes in the long-debated "collapse" of Mayan civilization in lowland Central America.

ROANOKE METALLURGY

In research supported by the National Geographic Society, archaeologists working at the site of the "Lost Colony" on Roanoke Island, North Carolina, discovered remains of a laboratory that was used in 1585 to test New World metals. The small building was situated inside a fort constructed by the colonists, the first English settlers in North America. Fragments of metal, slag, ceramic crucibles, glass, and other equipment indicating metallurgical testing confirm written sources about early attempts to find commercially valuable metals in North America. Ongoing analyses of the newly discovered materials are being conducted in the facilities at Williamsburg, Virginia; they are likely to yield new information about this early laboratory in the New World.

Peter S. Wells

REVIEW

Anthropology

EARTH MOTHER

In an intensive analysis of DNA patterns from 189 living individuals belonging to many different races, scientists have suggested that all modern humans have a common ancestor—a woman who lived in Africa around 200,000 years ago. The study was based on comparing DNA structures among peoples who developed in different parts of the world. The conclusion of the analyses was that the oldest group originated in Africa, around 200,000 years ago. The recent results derive from reworking studies first made public in 1987, with revisions based on criticisms made by other scientists. This model depends upon the hypothesis that some of the early modern humans (the group descended from this woman and her community) moved out of Africa between 50,000 and 100,000 years ago. They did not mix genetically with the other types of early humans that then lived in other parts of the world, but instead replaced them. Some scientists do not believe that the modern humans from Africa could have eliminated the earlier humans so completely with little or no genetic mixing. Many investigators think that substantial interbreeding between different groups of early humans took place. They emphasize the importance of local developments in early human populations, rather than relying exclusively on migration and replacement to explain the observed changes in the fossil record.

NEANDERTAL

Debate continues regarding the place of Neandertal (formerly spelled Neanderthal) in the evolution of modern humans in Europe and southwestern Asia. Neandertal populations lived between 35,000 and 200,000 years ago. After that time the only form of hominid is the modern human, *Homo sapiens.* Current discussion concerning the place of Neandertal focuses on two main models. One suggests that Neandertal became extinct when that form lost out to competition from modern humans, who had developed in Africa sometime between 100,000 and 200,000 years ago, and spread out from Africa around 100,000 years ago. Proponents of this model envision little interbreeding between Neandertal and physically modern groups.

The competing model sees Neandertal as part of the evolutionary development of modern humans in Europe and southwestern Asia. Neandertals in different parts of western Eurasia may have interbred with other groups of humans. In the skeletons of early modern humans in parts of Europe, some anthropologists cite physical features that apparently indicate genetic links with the local Neandertal populations. In Israel, recently discovered evidence suggests that Neandertals and physically modern humans may have been living in the same region for as long as 50,000 years. The artifacts made and used by the two physically distinct groups suggest that their cultures were remarkably similar.

Peter S. Wells

Are Neandertals our ancestors? Some theorize that Neandertal brain size, locomotor ability, and other traits indicate genetic links with modern humans.

Paleontology

INJURY-PRONE DINOSAUR

Paleontologists discovered the largest known specimen of the ferocious dinosaur *Tyrannosaurus rex* in rocks from South Dakota. Preserved in excellent condition, this fossil reveals that even the king of the dinosaur world suffered debilitating injuries. This particular *T. rex*, named Sue in honor of its finder, had healed wounds on its rib, its skull, and its lower jaw—all apparently incurred during a fight with another beast. Growths on the fibulae indicate that the dinosaur may have broken both its legs at different points in its life. Paleontologists know that Sue survived these traumas, because the injured bones show signs of healing. The dinosaur's luck eventually ran out, though, when it suffered a bone-crushing blow to the skull, possibly from the bite of another *T. rex*. These bones did not heal, suggesting Sue never recovered from that final battle, according to the paleontologists at the Black Hills Institute of Geological Research.

Considered the fiercest of the carnivorous dinosaurs, Tyrannosaurus rex *was nonetheless subject to crippling injuries, most likely incurred through fights with its enemies.*

EARLIEST FEET

While looking through some drawers in the British Museum, researchers discovered the earliest animals with feet—a find that helps clarify how our early ancestors left the sea to take up life on land. The bones, which date back 370 million years, had formerly been identified as fish remains. But the new discovery of tiny leg bones indicates that these creatures had primitive limbs, making them look like a fish with legs. Some scientists believe the leglike appendages helped fish grab on to plants while waiting to attack some prey. That theory challenges the traditional view that fish evolved legs for the purpose of locomotion.

GREAT EXTINCTIONS

A study of dinosaur fossils support the theory that the dinosaurs died out over a very short geologic span, possibly because a meteorite or comet struck the Earth. Evidence discovered in the past year now suggests that the impact occurred at a large circular formation on Mexico's Yucatán Peninsula. Rocks drilled from inside the formation have yielded pieces of shocked quartz—mineral grains with tiny fractures that are known to form during meteorite impacts.

Aside from the dinosaurs, many land and marine organisms became extinct at the end of the Cretaceous period. A new study in France and Spain revealed that a major group of marine creatures, the ammonites, also disappeared quickly 65 million years ago. The same study also showed that a prominent type of bivalve died out gradually some 2 million years before the ammonites and dinosaurs. That discovery indicates that a meteorite impact was not the only trauma causing extinctions at the end of the Cretaceous period. Climate change and other factors may have contributed to the die-offs.

Richard Monastersky

PHYSICAL SCIENCES

CONTENTS

MG
R
YL
G
CY
–I

THE GREAT BRIDGE CONTROVERSY

by David Berreby

A picture of wounded innocence, Joe McKenna wonders why the civil engineers of America are so angry at him. All he did was suggest that the Golden Gate Bridge might fly into pieces during an earthquake. And that other suspension bridges the world over might be vulnerable, too. And that, really, the people who build suspension bridges don't understand what they're doing. Where's the harm? "Mechanical engineers find it interesting," he says of his arguments. "Civil engineers, it's a different story."

McKenna, a 43-year-old mathematics professor at the University of Connecticut at Storrs, says his equations and computer models suggest that the bridges don't always behave in the way their designers assume they will. Applying new mathematical techniques, he and his Florida colleague A. C. Lazer of the University of Miami now have what they claim is a more accurate model of the way a suspension bridge vibrates. Like Einstein's physics coming after Newton's, McKenna says, their model confirms the old approach in most everyday instances, but is superior in accounting for certain special cases. And the cases involved are spectacular: instances when the vibrations get out of control, and the bridge flies apart and crashes to Earth.

Bad Vibrations

Any structure that must endure the buffeting of wind, ground tremors, or passing traffic will vibrate in response to the forces assaulting it. Bridge-building engineers have always sought an exact knowledge of how much vibration their works could tolerate. But when in doubt, they could overwhelm most assaults with sheer weight of building material, making flimsy parts heavier and stiffer.

Suspension bridges, however, require a defter touch. In a suspension bridge, the roadway hangs from cables—typically, ca-

bles draped in graceful parabolas between widely spaced towers. Giving the weight-bearing work to cables rather than to massive trusses or arches saves material, thus making the bridge lighter. When the great 19th-century bridge builder John Augustus Roebling completed his railroad bridge over the rapids above Niagara Falls in 1855, it spanned 821 feet (250 meters), yet used only one-fourth as much material as the 460-foot (140-meter) span of the Britannia rail bridge over the Menai Strait in North Wales.

Unfortunately, a lighter bridge is also a more flexible bridge. Many a 19th-century suspension bridge shook itself to pieces, making the whole breed's reputation unsteady. Even Roebling's work was tarnished by the general unease. "You drive over to Suspension Bridge," Mark Twain wrote of the Niagara River span, "and divide your misery between the chances of smashing down 200 feet [61 meters] into the river below, and the chances of having a railway train overhead smashing down onto you. Either possibility is discomfiting taken by itself, but, mixed together, they amount in the aggregate to positive unhappiness."

Although the Niagara bridge held, to a later generation, Roebling's bridges seemed stiffer and heavier than some thought necessary. He had been especially wary of high winds, which, he once explained, could make a suspended, flexible roadway undulate so wildly that it would snap the cables and fly apart.

Trying to guarantee safety and yet push their craft beyond Roebling's technical conservatism, 20th-century bridge builders have used wind tunnels, computer simulations, and mathematics to create sophisticated models of a bridge's vibrations. (Many of the lessons have been applied to aircraft and skyscrapers as well.) But now, McKenna says, the engineers' math is insufficiently sophisticated, their models inadequate. It doesn't make for good feelings.

"The word I would use is 'unfortunate,' " says Yusuf Billah, who teaches engineering at Princeton. "Off his trolley," says Robert H. Scanlan of Johns Hopkins, engineering's dean of experts on wind effects on structures.

Scientific Turf

Scanlan and other engineers first became aware of McKenna's theory in the summer of 1990, when news about a paper McKenna and Lazer had written for a math journal was circulating. McKenna had spoken with a reporter at the *San Francisco Examiner* about the implications of his work for the Golden Gate Bridge. The result of the interview was a front-page story on June 7 about how a severe earthquake could destroy the bridge. Scanlan, who won't meet a reporter because "these things really have to be discussed in professional journals," was horrified. Advancing an alarming theory that's wrong was bad enough. Going to the general public was unforgivable. "I think what McKenna's done is disgusting," he says.

Over the past year and a half, McKenna and the engineers have been engaged in a classic struggle over scientific turf, with each side claiming that its discipline holds the key. Asked what physical force causes suspension bridges to collapse, McKenna replies, "I don't care." He and Lazer have yet to build a physical model, and aren't likely to soon. Instead, they explore equations that describe simple oscillating systems like a pendulum or a bar suspended from springs. When they run those equations on a computer, they find that a single large disturbance—as might come from a high wind or an earthquake—can lead to unpredictable, wild oscillations that sometimes get bigger and bigger rather than slowly dying out, as smaller vibrations generally do. "Basically, we discovered a mathematical phenomenon," McKenna says. "And then we said, 'If this is happening in the math, then it's probably happening in the real world.' "

Bridge builders also use mathematics, of course, but traditionally they have worked the other way around. They observe the behavior of real objects, then go looking for math that will fit. To them, McKenna's breezy indifference to hands-on experience immediately brands his work as suspect. A few bridge experts concede that McKenna's ideas may have a purely theoretical interest. "I think in 10 years, McKenna will be better than us at predicting exactly what part of the bridge hits the water first—if a bridge falls apart," says Mark Ketchum, a senior engineer at the San Francisco consulting firm of T. Y. Lin International. "So what? My job is to make sure the bridge never flies apart and hits the water in the first place."

The clash is also stylistic. As McKenna points out, mathematical convention dictates that one list the weaknesses and holes in one's argument in the very paper that presents it. So mathematicians go where their thoughts take them. (It was a mathematician, after all, who recently proposed that one way to improve the world's weather might be to blow up the Moon.) After he'd heard that Scanlan thought his calculations were all wet, McKenna says, "I thought he would ask me to give a talk. That's what a mathematician would do." He's more than a little offended that he got the cold shoulder from the engineering community.

Engineers say it's McKenna who won't engage in a dialogue, refusing to slog through the nitty-gritty math with Scanlan. McKenna admits ignoring a long letter from Scanlan on the subject of why McKenna and Lazer are wrong. "I wasn't going to write 20 pages explaining the error of his ways," he says dryly, his vowels and r's gently hinting of his Dublin birthplace.

Professor McKenna (below) views a film of the 1940 collapse of the Tacoma Narrows Bridge. More than a half-century later, engineers still cannot adequately explain how it happened.

The Tacoma Narrows Collapse

There are a few things, though, that both sides agree on. One is that Exhibit A in any argument has to be the infamous collapse of the Tacoma Narrows Bridge in Washington

state. When it was built in 1940, engineers had strayed far from stiff, heavy structures like Roebling's famous Brooklyn Bridge—the sort of bridge, McKenna exaggerates, that "if you took away the cables, would still stand just as well"—and were building bouncier, jouncier suspension bridges.

"A highly flexible, very pretty, thin ribbon of steel," as aerodynamicist Theodore Von Kármán described it, 1-mile (1.6-kilometer)-long Tacoma Narrows Bridge hung 190 feet (58 meters) above Puget Sound, linking the Olympic Peninsula with the rest of the state. It opened on July 1, 1940, and was soon famous for its springiness. Even in winds of only 3 to 4 miles (5 to 7 kilometers) per hour, the center span would rise and fall as much as 4 feet (1.2 meters). The bridge was soon nicknamed Galloping Gertie, and drivers would go out of their way either to avoid it or to cross it for the roller-coaster thrill of the trip. "People said you saw the lights of cars ahead disappearing and reappearing as they bounced up and down," McKenna says.

Engineers monitored the bridge closely, but concluded that the motions were predictable and tolerable. In fact, a story got around later that the Tacoma authorities had been planning on reducing their insurance coverage. (Their insurance agent was confident, too—he pocketed the premiums instead of buying a policy, and ended up in jail.)

On November 7, 1940, four months after the bridge opened, a storm whipped up wind speeds to 42 miles (67 kilometers) per hour by 10:00 in the morning. The powerful winds were striking the bridge broadside. The oscillation of the center span increased, but the up-and-down motion was familiar and considered nothing to worry about. At a little past 10:00, though, some braces connecting the center span to the towers gave way, and the span suddenly stopped oscillating up and down and shifted into a violent twisting motion. One observer said it "appeared to be about to roll completely over." Everyone got off, including the thrill seekers.

The "writhing corkscrew motion" got worse, Von Kármán wrote later. "At one moment, one edge of the roadway appeared to an observer to be as much as 28 feet [8.5 meters] higher than the other edge. The next it was 28 feet lower. The cables in the main span, instead of rising and falling together as in the usual bouncing motion, pulled and wrenched in opposite directions, tilting the deck from side to side as much as 45 degrees from the horizontal." Finally, at 11:00, the center span buckled, and a 600-foot (183-meter)-long piece of the bridge plummeted into the sound.

In the aftermath of the collapse, engineers went into wind tunnels with models of the bridge, and concluded that the flexible deck, with its solid-steel paneling on each

On November 7, 1940, a storm whipped the newly opened Tacoma Narrows Bridge with 42-mile-per-hour winds. With a pronounced corkscrew motion, the bridge began to undergo wild side-to-side oscillations. Within an hour, a 600-foot piece of the bridge collapsed.

side of the roadway, was inherently unstable in flutter—the twisting and jerking that results when an object is gaining and losing lift. It was such wind-tunnel testing that produced the design for the replacement Tacoma bridge and for many others. The new bridge is undergirded with Eiffel Tower-like struts that allow the wind to pass through, rendering the whole structure more aerodynamically stable, yet making the roadway much stiffer. "I stood on that bridge last summer, and a huge truck went by full of logs," McKenna says. "The bridge didn't move at all. I was disappointed; I thought it would cause some interesting motions."

Meanwhile, a remarkable film of the bridge's collapse—made by an engineer who regularly filmed the bridge's dramatic motions—has become the *Texas Chainsaw Massacre* of classroom movies, a physics-teacher cult classic. Generations of high-school and college students have cheered and hooted as they watched the bridge fly apart.

Was Resonance at Fault?

As for the cause, they've been told that the culprit was simple resonance. Like a weight hanging on a spring, goes the explanation, the bridge had a natural up-and-down oscillation, and was thus subject to the same principles that underlie such familiar back-and-forth oscillations as those of pendulums and swings. If a child on a swing gets a push at exactly the right moment in each back-and-forth trip, the swing will go higher and higher rather than slowly stop. The key is to add energy as the swing completes each cycle. This precise match between pushes and oscillations is called resonance.

According to this explanation, the Tacoma bridge was buffeted by a force that kept hitting at exactly the right time to pump up its normal oscillations. Of course, there's a limit to how much energy any structure can absorb. Just as a wineglass can be shattered by a soprano's high C, the bridge eventually passed its breaking point. As for what caused such a precisely timed series of pushes, the standard explanation is aerodynamic: wind striking the unstreamlined deck, instead of passing smoothly over it, broke into hurricane-shaped spirals of quickly rotating air called vortices. These vortices created a rhythmic alternation of pressures above and below the deck as they spun past it. The frequency at which the deck was pushed first in one direction and then the other was exactly right for a fatal resonance.

It's about as dramatic a real-world example of resonance as physics and math teachers could ever hope for, but McKenna has long been skeptical. "I always hoped that some smartass would ask how you could get such perfect resonance in the middle of a howling gale," he says. On this point, at least, the engineers agree. "We now know," Scanlan and Billah wrote in 1990, "that this is not what occurred at Tacoma Narrows." The standard classroom explanation (which Scanlan and Billah found in 38 physics and eight math texts) is wrong.

The Self-excitation Theory

According to the engineers, what really happened was a different, more complex process involving a feedback loop. Usually when wind shoves a bridge, every slight change of position alters the wind's effects. Parts of the deck that were sheltered from the wind's full force become more exposed as it twists or swings, and the wind pressure on the structure changes. Ideally, the wind's effects will oppose the deck's normal oscillations, helping to damp them out just as they do on a weather vane. If a freak gust pushes a weather vane broadside into an otherwise steady

The Brooklyn Bridge (shown here under construction in 1881) was the first suspension bridge to use steel-wire cables. Some engineers think that the bridge is strong enough to stand even without cables.

wind, the vane's own motion changes the balance of forces operating on it: the farther it swings out of line from the wind, the more of itself it exposes to wind pressure that forces it back into line.

A key step in the design of any bridge is the creation of a scale model to test the proposed structure's resistance to excessive oscillation. The model is subjected to conditions more severe than the bridge is likely to ever face.

Wind-tunnel work on a one-fiftieth scale model of the original Tacoma bridge has shown, Scanlan and Billah wrote, that it didn't always self-correct like a weather vane. Instead, there was a particular frequency of twisting motion that, once started, got stronger in the wind rather than dying out. The farther the bridge twisted, the harder the wind pushed it away from its original position. When oscillating up and down, and when pushed in addition by a strong crosswind, the Tacoma Narrows was more like an umbrella in the wind than a weather vane. Caught just right, perhaps during a single twisting maneuver, the bridge, by its own motion, created the aerodynamic forces that tore it apart. It was such "self-excitation" that led to catastrophe.

"I know self-excitation sounds like some sort of erotic condition," Ketchum says. The lesson engineers have learned, he says, boils down to this: "In a well-designed bridge, an upward motion creates aerodynamic effects that force the bridge down, and a downward motion creates effects that force the bridge up. At Tacoma, the opposite happened."

The key difference between this correct explanation and the textbook one, says Billah, is that self-excitation can get started from only a single unlucky wiggle. After a small oscillation begins at just the right frequency, the bridge's motion itself amplifies the wind's effects. Hence, the modern theory doesn't require engineers to claim that a perfect resonance between wind and bridge could exist for more than 45 minutes in the middle of a storm.

McKenna, though, doesn't believe that the new explanation is much better than the old explanation. Both, he says, are based on naive mathematics.

Slackening Cables

The problem, according to McKenna, is that engineers rely too much on linear equations to describe bridge behavior. A linear process is one in which effects are neatly proportional to forces. For example, he says, "if you're pulling on an object hanging at the end of a spring, when you pull twice as hard, it'll pull back twice as hard." But other processes—involving aerodynamics, say—aren't so neatly proportional. In such nonlinear situations, "you can't assume that twice as much force will cause twice as much vibration."

"What distinguishes suspension bridges, we claim, is their fundamental nonlinearity," McKenna and Lazer wrote in *SIAM Review,* the journal of the Society for Industrial and Applied Mathematics. Once bridge behavior has entered a nonlinear mode, the effects of the forces that hit it are not proportional, or even predictable. In fact, exactly the same amount of force can produce very little vibration or a huge amount—depending on slight variations in other variables, such as the angle at which the wind strikes the bridge.

Thus, engineers trying painstakingly to figure out what frequency of vibration caused a particular bridge to collapse are barking up the wrong tree, McKenna says; they're trying to make linear equations fit nonlinear circumstances. Self-excitation doesn't grow from a tiny wobble at a single fatal frequency. Instead, any kind of push can get it started—even a single gust—provided that the push is a big one.

What kicks a bridge into a nonlinear condition, McKenna claims, is simply an upward lurch. "Classical engineering says cables must behave like incompressible, inextensible rods," he says. But if the roadway is accelerated upward, cables begin to go slack, and all bets are off. McKenna's idea that cables go slack is the central disputed fact in his running feud with bridge engineers. "It doesn't happen," snaps Scanlan.

Ketchum is willing to elaborate: "If I'm holding you up by a rope, and you're pushed upward for a second by a giant hand, McKenna's assumptions would say that the rope immediately goes slack. That's not what would happen." It takes time, and the rope doesn't always go completely slack. "If the giant hand pushes you up with a force of 100 pounds, and you weigh 200, the rope still has 100 pounds of tension in it." Likewise, for a bridge cable to go truly slack, the bridge roadway would have to be knocked upward with tremendous force, long enough for the steel cables to lose all tension.

In simulations of earthquake effects on the Golden Gate Bridge, Ketchum says, only when the earthquake is assumed to be a 90-second jolt as powerful as the record 1906 quake, and a number of structural failures elsewhere on the bridge are also plugged in, does a single cable go slack. And, he stresses, even then, it's only one cable out of many.

But McKenna says eyewitnesses at Tacoma Narrows claimed they saw slackening. He doesn't have solid reports on other bridges yet, but he's looking. "I would say that eyewitnesses who report slackening are laypeople who mistake the very fast vibrations of the cables for a slackening," says Ketchum. A vibrating guitar string might look loose, too, he adds, even though it still has plenty of tension.

No matter who is right, the point neatly illustrates McKenna's Catch-22 situation. Disallowing the accounts of frightened, excited nonprofessionals leaves only one other possible source: the accounts of professionals. McKenna must ask the very people he's made so angry to look for something they haven't yet seen in order to undermine their own position. "I have to get more data on when it is that cables loosen and tighten," McKenna says. "People telling me it doesn't happen makes that rather hard."

Engineer Mark Ketchum is reengineering the Golden Gate Bridge to make it more resistant to oscillations caused by earthquakes.

Anchoring Above and Below

Although McKenna and Lazer don't have the data yet to simulate particular cases, they have already suggested a general-purpose cure for nonlinear oscillations. Because cable slackening is the key, they say, nonlinear states can be easily avoided by anchoring a span with cables below as well as above. Their ideal suspension bridge would look the same upside down or right-side up—it would be held up by cables above, but also held down by cables below. The Tacoma Narrows case supports this idea, McKenna claims: although the middle section, between the

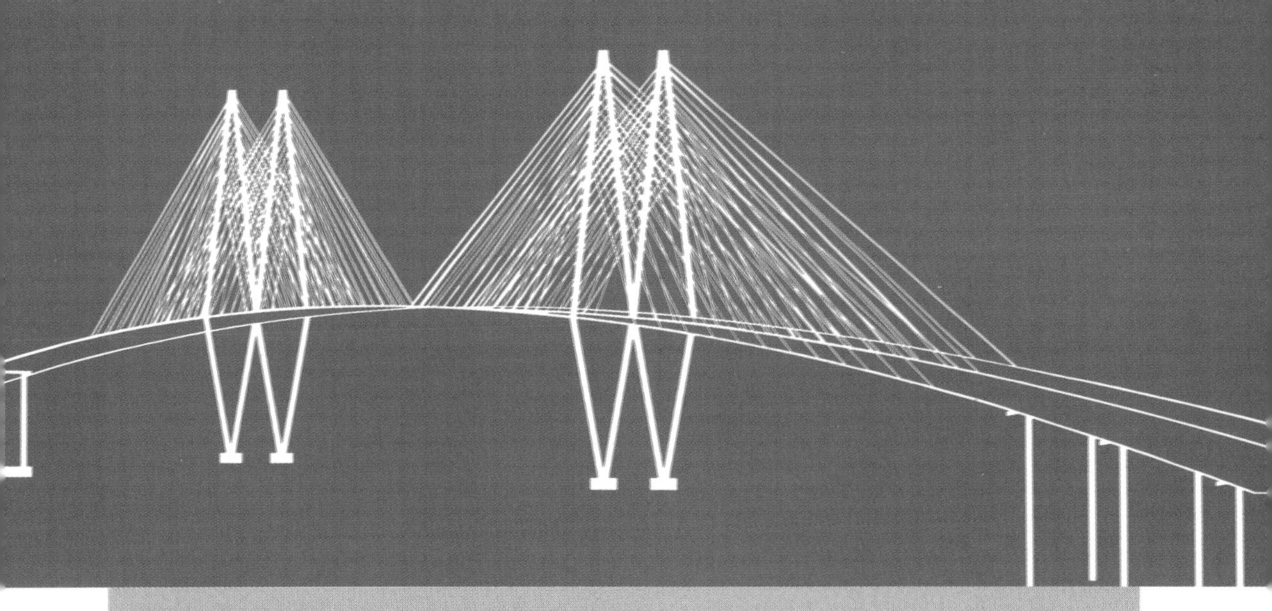

Stay By Me

Always concerned that his bridge decks not jump and twist, John Augustus Roebling would install extra cables to hold them in place. These cables, called stays, stretched directly from tower to deck, like the hypotenuse of a right triangle. Such "inclined stays," when combined with the usual suspension cables, created a weblike look, as on the Brooklyn Bridge. But in later bridges—whose decks were stiffened with steel trusswork—engineers discarded inclined stays, leaving the simpler harp-string pattern of suspension cables alone, as on the Golden Gate Bridge.

Nowadays, though, stays are back. In fact, it's the other cables that are being dispensed with, so that the stays actually hold up the deck. A prime example (above) is under construction over the ship channel just east of Houston: a 2,475-foot (754-meter) span with two four-lane decks that's expected to open sometime next year. The wide decks will present a hefty 8-acre (3.25-hectare) surface for a hurricane to batter if one of the giant Gulf storms ever gets them twisting, but the stays are supposed to prevent that. Wind-tunnel mock-ups and computer simulations indicate that the bridge should stand up to even 150-mile (240-kilometer)-per-hour winds, says engineer Thomas Lovett, one of the Houston span's designers.

Greater stiffness is one reason engineers are going for the cable-stayed approach, says Lovett. Every stay is anchored directly to a rigid tower (in the Houston bridge, they are superrigid, "double-diamond" towers). But the main appeal is cost. "For a given span, it would require about twice as much cable to use a suspension approach as it would to use inclined stays," he says. "It's so much more economical that you probably won't see any suspension bridges for spans under 2,000 feet [610 meters] anymore." (Longer bridges pose such technical challenges that most proposals involve some combination of stays and traditional suspension cables.)

Modern materials tipped the balance in favor of all-cable-stayed bridges, which were first built in Germany after World War II. "Regular-grade steel can't bear the tension," Lovett explains. But with high-strength steel available, the Houston bridge will be one of some 200 "suspension" bridges in the world that don't actually work by the principle of suspending a roadway from free-hanging cables.

two towers, collapsed, the spans between each tower and the shore did not fail or even launch into wild oscillations, although they had the same aerodynamic characteristics as the center span. And both of those spans had been anchored with cables from below.

To Ketchum, such a suggestion just illustrates McKenna's impractical, unworldly outlook concerning bridge design. "If you anchor the bridge below as well as above," he asks, "how are you going to get ships under it? Why not just use landfill?"

McKenna hasn't come up with any other suggestions. But with a more accurate picture of what makes a bridge fall apart, he argues, engineers would not have to build them with so much reinforcement. "My calculations suggest that it is possible to build light, flexible, undergirded suspension bridges," he says. If the engineers were using his more sophisticated equations, he claims, they could add material where it's needed, rather than make the whole bridge heavier and stiffer. On the other hand, his calculations also suggest that the sharp blow of a major earthquake may be exactly the sort of force that could send a naively designed bridge into a nonlinear state. "I ask what's going to happen in an earthquake," he says, "to bridges built without a knowledge of the nature of large oscillations."

Safe to Simplify

Why, he wonders, can't civil engineers at least consider his arguments?

The short answer, says Ketchum, is that they have, and that McKenna is hacking away at a straw man. "McKenna's point is that engineers need to take nonlinear effects into account," he says. "Well, guess what? We do. We consider every one of the nonlinearities that McKenna says we should."

Just like McKenna, Ketchum says, he uses computers solving racks of nonlinear equations to calculate the chaotic effects of ground displacement, cables snapping, and towers buckling. The conditions change every second, which means thousands of equations must be solved over and over to represent a complete simulation. If modern engineers sometimes use simplifying assumptions, Ketchum says, it's because they've made the nonlinear calculations needed to find out when it's safe to simplify.

McKenna disagrees. Although he believes that Ketchum did do a nonlinear analysis, he argues that most engineers' assumptions are actually made beforehand, to keep the math linear. For example, calculations in a recent paper by Scanlan and Billah for the angle of deflection in a twisting bridge assume that the angle is small. Linear equations fit small angles pretty well. But when a bridge kicks into a nonlinear mode and the twisting angle becomes large, McKenna says, engineers will still apply the standard linear equations because that's all they'll find in most papers and textbooks. "The problem with this branch of the engineering literature," he says, "is that they don't identify their assumptions."

But the engineers point out that McKenna's done some simplifying himself. His equations may be more sophisticated, but he applies them only to simple structures. It's a long jump from a bar suspended by two springs, they say, to a 64,000-ton steel-and-asphalt roadway hanging by hundreds of vertical cables that are suspended themselves from heavier cables draped between towers. "He assumes, for example, that every vertical cable will behave independently of what the others are doing," Ketchum says. Practical experience, both with real bridges and with scale models, has convinced Ketchum that "that's not how it works."

McKenna's equally long experience with the *math* of complex systems leads him to the opposite conclusion. "My claim is that if the simple model shows nonlinear behavior, the more complex one will be even more nonlinear."

New Projects

Personal pride and what Ketchum calls professional chauvinism are at stake in all this, but that's not all. In San Francisco, approval has been given for a $128 million project planned by Ketchum's firm to make the Golden Gate Bridge earthquake-resistant, using a combination of stiffening structures to make the bridge sturdier and giant shock absorbers to take up seismic energy. What's more, plans are on the boards for suspension spans of more than a mile in both Denmark and Japan, and Ketchum is looking into the prospect of building a 6-mile (9.6-kilometer)-long bridge from Spain to Morocco across the Strait of Gibraltar. Indeed, bridge builders are so confident of their post-Tacoma corrections that they're taking on hurricanes: nearing completion is a bridge in Houston that's designed to withstand 150-mile (240-kilometer)-per-hour winds.

So the stakes are high. If the engineers are wrong, millions will have been spent on unneeded extra stiffening that could nonetheless fail to protect a bridge from mathematical and literal chaos. On the other hand, if McKenna is wrong, he'll have spread a lot of confusion about a subject on which everybody ought to be absolutely clear.

The discovery of buckyballs will likely revolutionize chemistry. The soccer-ball geometry of the giant carbon molecule was deciphered by Richard Smalley (above).

BUCKYBALLS: The Magic Molecules

by Edward Edelson

A revolution in chemistry is taking place in a small room in a converted mining building in Tucson, Arizona, where a woman wearing a soiled smock and a face mask is painstakingly scraping soot off a metal container.

Although it's not too exciting to look at, this is the world's first production facility for a newly discovered, exotic material, dubbed "buckyball," that has such extraordinary potential that chemists and physicists around the country are lining up to pay $1,200 a gram for the stuff, roughly 100 times the price of gold.

"This is the biggest news in chemistry I could have imagined," exclaims Robert Whetten of the University of California at Los Angeles (UCLA).

THE THREE KNOWN FORMS OF CARBON

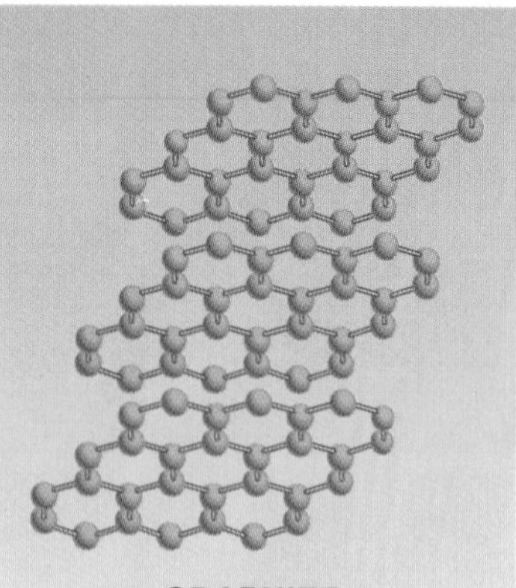

GRAPHITE

Graphite has molecules that form flat sheets of atoms arranged in a hexagonal pattern. It is resistant to heat and is a good conductor of electricity.

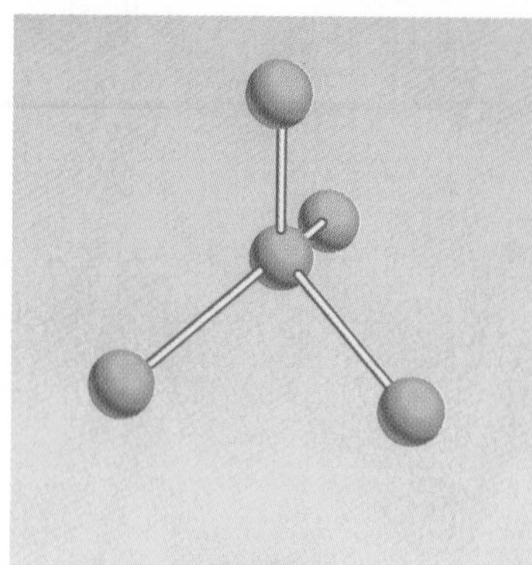

DIAMOND

A diamond has a pyramidal atomic structure called a regular tetrahedron. This rigid structure makes diamonds the hardest substance known to humans.

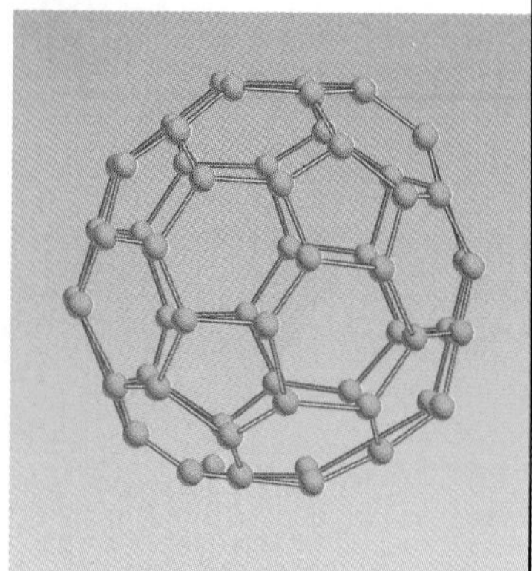

BUCKYBALL

Buckminsterfullerene has 60 atoms arranged in regular pentagons and hexagons. Its chemical properties include superconductivity.

A New Kind of Carbon

The discovery of a new kind of carbon came as a stunning surprise to most scientists. Carbon is the most intensely studied of all the elements because it is the basis for most of the molecules of life—the organic molecules. Look in any high-school or college chemistry textbook, and you'll read that for centuries, research showed carbon came in just two basic structures: hard, sparkling diamond, whose carbon atoms are arranged in little pyramids; and dull, soft, slippery graphite, which consists of sheets of carbon-atom hexagons.

Those chemistry textbooks are now obsolete. There's a new basic form of carbon with an almost unbelievable structure: its 60 carbon atoms form something that looks like a hollow soccer ball. It is the only molecule of a single element to form a spherical cage.

The molecule's official name is buckminsterfullerene, because it is shaped like the geodesic dome invented by that American original, Buckminster Fuller. Informally, chemists call it buckyball, or C-60. Its atoms are arrayed in a collection of regular pentagons and hexagons—12 pentagons and 20 hexagons, to be precise. It's one of a newly discovered family of similar molecules that has a related geometry, but different multiples of carbon atoms. Scientists have called this whole family the fullerenes; scores of chemists and physicists are working full blast to unravel their properties.

It's not just the intellectual kick of a major advance that is energizing the scientific community. It's the prospect that buckyball's properties will make possible a cornucopia of valuable applications.

"To a chemist, it's like Christmas," exults Richard Smalley of Rice University in Houston, Texas, one of the key players in the buckyball game. To explain, he harkens back to the discovery of benzene in 1825. The benzene molecule is a relatively simple six-carbon ring, yet it's the parent of countless compounds, from aspirin to nasal decongestants to paints, dyes, and plastics—all made by working with that unique six-carbon-ring structure. Now chemists hope to perform the same magic with this family of new carbon molecules that is at least 10 times bigger than benzene, with, therefore, even greater possibilities.

"This isn't 1825," Smalley says. "It's like discovering benzene, only, now you have all the techniques and the scientific instruments of the 1990s available."

Infinite Applications

It is now clear to researchers that the C-60 molecule is exceptionally stable and resistant to radioactivity and chemical corrosion. It also greedily accepts electrons, but is not reluctant to release them. These and other attributes have scientists and engineers already speculating about microscopic ball bearings, new cancer treatments, lightweight batteries, powerful rocket fuels, and the infinite possibilities in plastics and other organic compounds that have carbon atoms as their backbones.

One proposal for antitumor therapy in cancer patients is to enclose radioactive atoms inside buckyballs. The carbon barrier would help maintain the integrity of the radioisotopes after injection. Smalley has already replaced some carbon atoms in the ball with other elements to create semiconducting "dopeyballs." Doping silicon with foreign atoms is what turns silicon into the semiconductors found in transistors.

Another idea Smalley talks about is creating a superpowerful battery by wrapping lithium and fluorine atoms, which create energy when they combine, inside a buckyball cage to protect them from being attacked by oxygen in the air.

Scientists speculate about stringing buckyballs together to form the basis of new types of plastics. They dream of altering the molecule in a million ways by hanging different atoms or chemical groups from the 60 carbons. "It's the starting material for making a whole new family of organic compounds," says chemist Fred Wudl of the University of California at Santa Barbara.

A supercomputer simulation shows the electron distribution in a single buckyball molecule. The name buckyball is derived from Buckminster Fuller, the inventor of the geodesic dome.

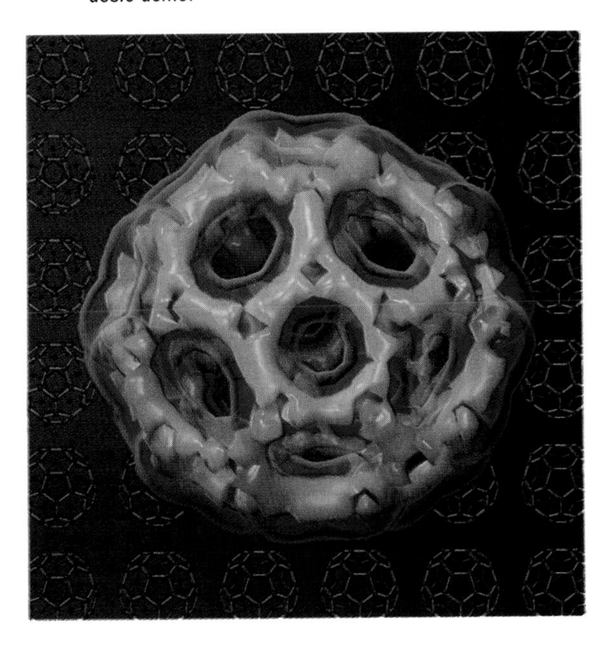

Buckyball's Discovery

Go back to 1984 to Rice University, where a team headed by Smalley was investigating the properties of atomic clusters, groups of atoms larger than molecules but smaller than visible solids. The Smalley team was using an unusual device it had invented, called the laser supersonic cluster beam apparatus. It's a steel vacuum chamber that holds a hollowed-out steel block. A sample placed inside the block is zapped by a very intense, short pulse of laser energy that vaporizes it. At the moment of zapping, a whiff of inert helium gas carries the vaporized material toward another laser, which ionizes the clusters by stripping away electrons. The clusters are then pushed into an analytical instrument called a mass spectrometer, which gives a reading of their size. Smalley was using the machine on a variety of elements, including silicon.

At that time, Harry Kroto of the University of Sussex in England was visiting Rice, and suggested that carbon be added to the list of elements Smalley's team had been zapping. When Smalley's group, joined by Kroto, zapped carbon in the apparatus, the results were astonishing. They had expected a similarly random, and uninteresting, assortment of carbon clusters like that found by a team doing similar work at the Exxon Research and Engineering Corporation. Most of those contained from 2 to 30 carbon atoms, with some much larger clusters of even-numbered atoms. There were also increased amounts at 10-carbon intervals: 50-, 60-, 70-carbon clusters.

But there was something strange about the 60-carbon cluster that drew their attention. Much more of it appeared in their samples than could be explained by random formation—three times more than any other even-numbered cluster. Intrigued by that finding, Jim Heath, one of Smalley's graduate students, worked over a weekend to develop a way to increase the yield of C-60 clusters: he found that he could tinker with the experiment so that the amount of C-60 yielded was 40 times as great as any other even-numbered cluster.

60 Vertices

As the Rice chemists kicked these results around, they asked two questions: Why even-numbered clusters, and why so much carbon 60? One explanation was that they were making carbon "sandwiches," flat sheets of material that contained large numbers of atoms, made up of graphitelike hexagonal groups. But, Smalley recalls, such a flat molecule would have unattached dangling chemical bonds at its ends with no apparent way to tie them up. Besides, why should such an open-ended cluster have exactly 60 carbon atoms, no more and no less?

One of the Rice-group members—no one remembers who—suggested that the carbon-60 cluster wasn't actually a cluster, but a molecule in the shape of a hollow ball. Maybe those flat sheets actually curled around to form a sphere and would turn out looking like a geodesic dome. That would take care of the dangling-bond problem. Smalley had seen a photo of one of Buckminster Fuller's geodesic domes, and thought the geometry worth trying.

Meanwhile, Smalley sat down at his computer and tried to generate a model structure for a 60-atom ball of carbon. After hours of work, he got nowhere. Frustrated, he began cutting regular hexagons out of legal paper, 1 inch (2.5 centimeters) on a side, and tried to make a sphere out of them. No dice. As he reached for an after-midnight beer, he remembered Kroto saying that he had at one time built a geodesic dome for his children, and that it contained regular pentagons as well as hexagons. So Smalley cut out a pentagon and began arranging hexagons around it, adding more pentagons and hexagons, taping the flimsy paper shapes together as he worked, and finally, halfway through, saw he had something.

"My heart leaped," Smalley recalls. In fact, the paper model formed a ball; it even bounced when dropped on the floor. It had 20 hexagons and 12 pentagons. Each of the 60 vertices, or corners, representing one carbon atom, was identical to the others; each occurred at the joining point of one pentagon and two hexagons.

Smalley called the head of Rice's mathematics department, William Veech, and described what he had built. Eventually Veech responded: "What you've got there, boys, is a soccer ball."

The structure is technically called a truncated icosahedron, one of an infinite number of spheroidal cages that can be formed with hexagons and pentagons. Buckminster Fuller realized that many of these structures are endowed with unusual rigidity for their mass because of their geometry. Thus, the strong, lightweight geodesic dome was born.

The day after their epochal discovery, the Rice chemists thought of names like "soccerene" and "ballene" for the C-60 molecule, but finally decided on "buckminsterfullerene." Today it is also known as buckyball. The other even-numbered, geodesic-dome-shaped carbon clusters are collectively known as "fullerenes." Smalley and his colleagues announced the discovery of C-60, the theory of its structure, and the structure of other fullerenes in a scientific paper published in 1985.

Buckyballs by the Bucketful

After announcing their exciting discovery, the Rice people were in a bind. They had only fractions of a milligram of C-60, not enough to confirm its existence. How could they convince the doubters and substantiate their theory of C-60's structure? Obviously, they had to produce a whole lot of buckminsterfullerene, enough of the stuff so it could be thoroughly analyzed. Smalley assigned the job to Heath. Smalley called it "the search for the yellow vial," because theory indicated that the C-60 molecule would be yellowish. It seemed a simple job, but it turned into a nightmare. For two years, Heath mixed the material coming out of the nozzle of the cluster-beam apparatus with benzene, hoping the solvent would concentrate appreciable amounts of C-60. The effort was a bust.

Ultimately, the answer came from Tucson and Heidelberg, Germany, in a way that demonstrates the inexplicable nature of scientific breakthroughs: the two men who found the way to make buckyballs by the bucketful were studying something else.

Donald Huffman of the University of Arizona and Wolfgang Krätschmer of the Max Planck Institute for Nuclear Physics were working with carbon clusters, but with a totally different perspective and with different goals than Smalley's.

Huffman and Krätschmer were studying how all kinds of small particles absorbed

A FULLERENE FAMILY PORTRAIT

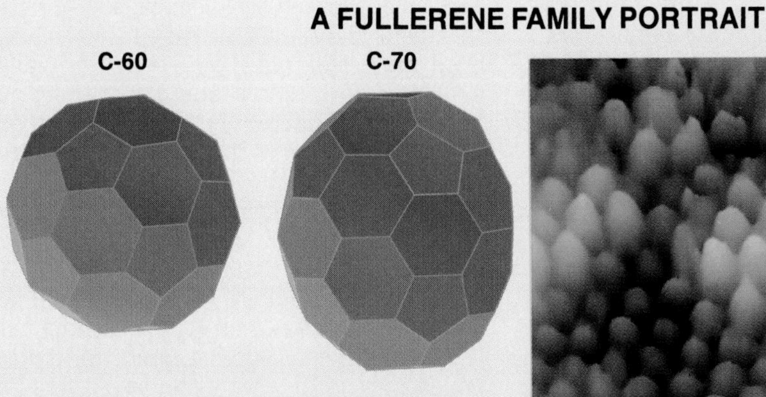

The first visual evidence for the existance of fullerenes cames from experiments with the scanning tunneling microscope, which works by dragging a diamond tip, just one atom in width, across a surface and detecting the current that is generated when electrons "tunnel" between the tip and the surface. The resulting image of C-60 and C-70 (right) provided the first direct evidence of the proposed shape of fullerenes. The C-60 molecule is spherical, like a soccer ball, while the C-70 has a more oblong, rugby-ball shape (above) that juts higher, appearing lighter in the image.

light: biological particles, soot particles, any very small particle of carbon.

After trying a number of methods, Huffman and Krätschmer had developed an ingeniously simple device for making lots of small carbon particles. Their machine consisted of two graphite rods connected to a high-electric-current circuit surrounded by a helium atmosphere. A hacksaw blade acted as a spring to push the rods together. Where they touched, carbon vaporized, forming lots of carbon clusters—soot to you.

It's a dirty business, working with soot, but this time it paid off. The reward came from methodical work that measured how carbon clusters absorb visible light.

"We were the first to measure directly the optical-absorption spectrum of very small carbon particles," Huffman says. "And when we did, we saw this feature."

The feature was a peak indicating that light at the wavelength of 2,200 angstroms was being absorbed by the carbon—almost, but not quite, like the peak astronomers were seeing in interstellar dust.

Huffman and Krätschmer didn't completely understand the finding. "So we went back to the lab and started making more carbon clusters," Huffman says. "It was then that we started seeing new and funny things in this peak. In fact, we saw three little wiggles in it." Krätschmer immediately called it the *kamel* sample (the German word for "camel").

That was in March of 1983, and Krätschmer and Huffman began arguing about what it might be: "Maybe it's a new form of carbon. That's ridiculous. Maybe it's some sort of cluster of carbon atoms. Maybe it's just junk. Mostly we thought it was some kind of junk," Huffman recalls.

When Huffman read the 1985 Kroto-Smalley paper that discussed a new 60-carbon molecule, a light flashed on. This strange new stuff could explain all the funny things he and Krätschmer had been seeing. Quickly the focus of their research on carbon changed radically. Huffman and Krätschmer weren't at all convinced they had made buckminsterfullerene, but they began to point their work toward that direction. To be on the safe side, in 1987 Huffman put in a patent-disclosure memo through his university for "a proposed way of making macroscopic amounts of C-60." When the patent attorney called back, in February 1988, Huffman found he could no longer make samples with the camel feature. To increase the yield of C-60, his graduate student, Lowell Lamb, began tinkering with the experiment, changing combinations of conditions, mostly the helium pressure. The result was large amounts of C-60—milligrams of it, more than anyone else had ever made.

Predicted Properties

They couldn't yet take a picture to prove they had carbon-60, but they could work on the basis of its predicted properties. Organic chemists had become interested enough in Smalley's proposal to figure out how buckminsterfullerene would absorb infrared light. They conjectured that most of the infrared light would go right through the carbon molecule, except for four wavelengths that would be absorbed. Plotted on a graph, the absorption spectrum was a mostly smooth curve, with only four strong peaks. When Huffman and Krätschmer beamed infrared energy through their sample, they saw the predicted four peaks. Bingo!

Well, almost. Vacuum-pump oil, used to lubricate their experimental apparatus, has two peaks of its own—almost dead-on the ones predicted for buckyball. Krätschmer performed an experiment that eliminated the possibility that two of the peaks had come from the oil. He made buckyballs out of carbon 13, which is slightly heavier than the dominant isotope, carbon 12. The heavier atom is predicted to shift the infrared peaks by a predictable amount; it won't shift any peaks attributable to contamination. The predicted shift appeared. Buckyball lived.

For a meeting, Huffman and Krätschmer wrote up a small paper modestly titled "The Possibility of Carbon-60 in Laboratory-produced Interstellar Dust Analogues." It was published in a fairly obscure journal in September 1989. By early 1990 Krätschmer and Huffman had relatively pure samples, not only of C-60, but also of another fullerene, C-70. Now at last they could reveal to the scientific world what they had been doing.

They did it in full-fledged style in the journal *Nature* in September 1990. Huffman and Krätschmer described their method for making buckminsterfullerenes, and showed photos of the actual crystals.

Word that something big was happening had already leaked out. The real surprise was that buckyballs were so easy to make. But they were still not being made in large enough quantities to enable scientists to pin down their structure. That task fell to others among the by-now droves of investigators who were playing buckyball.

"We always regarded its shape as the most likely, and it was so attractive that everybody talked about it as though it was proven," says Whetten, who by then had his own lab at UCLA. When Whetten and a colleague, François Dederich, read the *Nature* paper, they shifted gears and began working on the Huffman-Krätschmer method.

Something similar was going on with Don Bethune at the IBM Almaden Research Center in San Jose, California. Inspired by the Kroto-Smalley paper, he had begun work on carbon clusters using a machine developed by another IBM scientist, Heinrich Hunziker, to study contamination of disk-drive heads. That machine used laser pulses to lift organic molecules off a clean surface and transfer them into an analytical instrument called a spectrometer to study their masses.

Bethune was having the same trouble as Smalley: he couldn't get enough of the carbon-60 clusters to do a useful experiment. So he cast about for another method.

One evening, Bethune and a colleague were talking about this problem in a telephone conversation with someone who was using a Smalley apparatus at Lawrence Livermore Laboratory in California. Maybe, Bethune suggested, if you held some small object in front of the laser and tried pulsed beams, that might work. The response was, "That can't really be done. You might as well just light a match and put some soot on a metal plate. That's as stupid as what you're asking me to do here."

Mass Production

The IBM scientists hung up the phone, exchanged glances of recognition, and looked around for something to burn. The first thing they tried was methanol—wood alcohol—which burns with a nice, clean, soot-free flame. Then they tried a piece of paper. No soot again. Then Bethune spotted a polyethylene lid from an empty can of peanuts. That gave him the soot he wanted. The mass spectrometer showed the desired peaks in the region of the 60-carbon atom.

Bethune and his colleagues cleaned up the experiment, burning pure carbon, and saw a major peak of carbon-60 clusters. Just about that time, they saw the Huffman-Krätschmer paper and realized what they had.

They then began an intensive set of studies on their carbon-60 samples: nuclear magnetic resonance, Raman spectroscopy, infrared spectroscopy. They cooled the samples to liquid nitrogen temperatures to slow down the buckyballs, which spin madly at room temperature, and made scanning-tunneling-microscope pictures showing the overall shapes of both C-60 and C-70 molecules, but not the arrangements of their atoms. The IBM group quickly published a paper confirming the Huffman-Krätschmer finding.

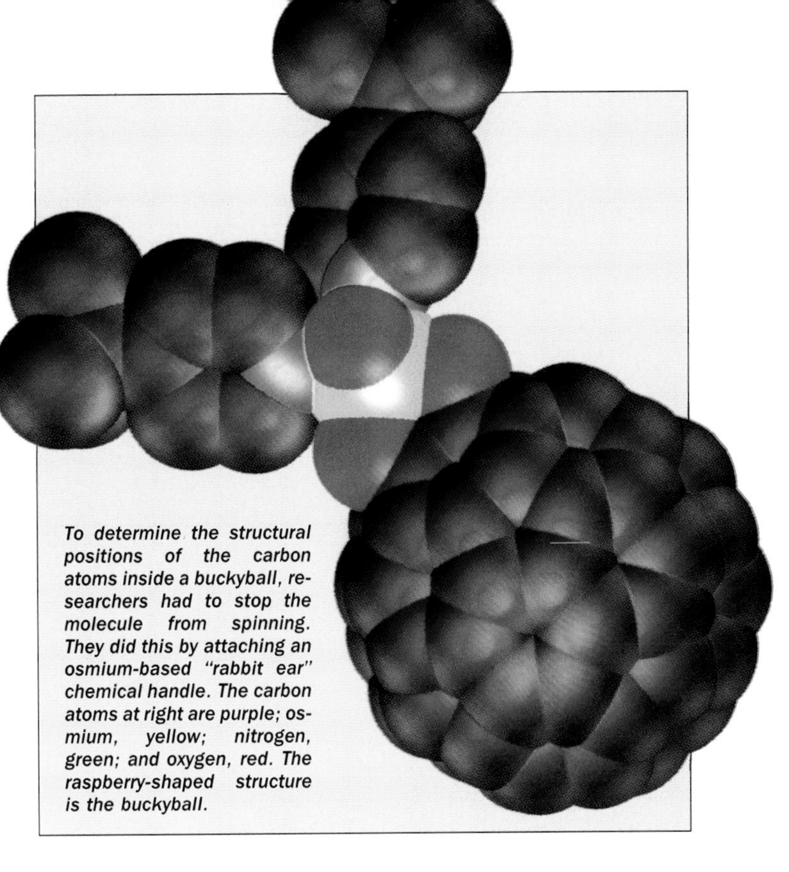

To determine the structural positions of the carbon atoms inside a buckyball, researchers had to stop the molecule from spinning. They did this by attaching an osmium-based "rabbit ear" chemical handle. The carbon atoms at right are purple; osmium, yellow; nitrogen, green; and oxygen, red. The raspberry-shaped structure is the buckyball.

The world's first buckyball production facility came on line early in 1991 at the Materials and Electrochemical Research Corporation in Tucson, which was assigned the patent to produce research-quantity amounts.

"At the moment the problem is that they can't keep up with the demand," Huffman says. "They're making more than a gram a day, but it's time-consuming." Down the hall, though, is the equipment for a tenfold scaleup, with bigger plans on the horizon. "If there's a really big demand," Huffman adds, "C-60 ultimately could be produced for pennies a gram. I really think that 10 or 20 years down the road, there will be large factories producing this material."

The absolute, complete confirmation of the soccer-ball geometry of C-60 came in April 1991, when chemist Joel Hawkins and colleagues at the University of California at Berkeley published the first X-ray pictures of the molecule's crystal structure.

Meanwhile, researchers have found even more curious and potentially valuable properties of buckyballs. In April, scientists at Bell Laboratories in New Jersey planted potassium in buckyballs and found that they became superconductors at a temperature of −427° F (−255° C). That's the highest superconducting temperature of any organic compound, and it opens a whole new field of buckyball research.

In California, Whetten fired buckyball molecules into a stainless-steel wall at 15,000 miles (24,000 kilometers) an hour. They bounced back unharmed. "It's resilient beyond any particle that's been known," Whetten says—resilient enough, maybe, to be used as rocket fuel, which must withstand enormous pressures.

Arthur Ruoff, who works in high-pressure-materials science at Cornell University, has made theoretical calculations that show buckyballs to be far stiffer than diamonds at moderate pressures, although they are "mushy" at atmospheric pressure. He believes this property could be a way to extend the range of high-pressure research. So-called "diamond anvils" are now used to create pressures of 4 million atmospheres. Ruoff is thinking about putting the material to be tested inside buckyballs to achieve even higher pressures.

VOLATILE VACUUMS

by Owen Davies

Imagine a world in which endless, nonpolluting, and virtually free energy powers our cities, our automobiles, and our homes. Try to envision laptop computers more powerful than today's largest, most sophisticated mainframes, and tiny X-ray machines that can enter the body and kill tumors without harming surrounding cells.

Interstellar space may not be as empty as it seems. Some physicists theorize that abundant power exists in the vacuum of space—power that if harnessed could provide an endless supply of energy.

All this and more may be possible within the next 10 years, according to physicist Hal Puthoff, currently with the Institute for Advanced Studies at Austin, Texas. The source of these marvels? Something Puthoff calls *zero-point energy*—the abundant power that he says can be found in the vacuum of space. Puthoff's articles on the subject have been published in the prestigious *Physical Review.* And he has attracted heavy-hitting business associates, including Ken Shoulders, the man credited with developing much of the technology for microcircuits, as well as superrich Texas entrepreneur Bill Church. Rumor has it that their new company, Jupiter Technologies, may soon try to manufacture zero-point-energy machines. There's more: zero-point energy could be the Rosetta stone of physics, explaining everything from gravity to atoms to the origin of the cosmos itself.

Missing Links

In a sense, Puthoff's search for order in the universe started 20 years ago, when he was a freshly minted Ph.D. from Stanford University. One day, he now explains, he was thinking about tachyons, hypothetical particles that appear to travel backward in time. If they existed, he reasoned, the particles—if they were not frauds—might be the "missing link" that allowed psychics to intuit events at distant locations or future times. Puthoff sought funding to study the problem, and wound up heading a new parapsychology-research program at the Stanford Research Institute, now known as SRI International. Studying telekinesis and ESP was intriguing, Puthoff says. Yet in 1985, after 13 years at SRI, Puthoff was ready to make a change.

Enter Bill Church. An ex-math major from the University of Texas, Church dropped out of college when his father died. By the mid-1980s, the trim, personable entrepreneur had made millions with a regional chain of fried-chicken restaurants. Eager for new challenges, the energetic Church vowed to spend his wealth promoting the kind of high-risk, potentially high-payoff research that government and corporate bureaucrats were too unimaginative to fund.

To that end, he founded the Institute for Advanced Studies, housed in a two-room office in a new building along the Capital of Texas Highway in Austin. Then he lured Puthoff, also a respected laser scientist, away from SRI.

Soon after Puthoff arrived in Austin, he and Church recruited a third member to their team: star inventor and electronics genius Ken Shoulders. A born tinkerer, Shoulders wanted a new research project, something that would probe the unknown regions at the borders of physics and electronics, where strange and wondrous discoveries might yet be made. He also needed the appropriate funding for such a product. Puthoff and Church, on the other hand, wanted someone who could turn the theoretical work of the institute into nuts-and-bolts technology. When the three sat down to ponder their first project, they came up with an impressive goal: exploring the vacuum, referred to by some early physicists as "the tranquil void."

A Hotbed of Forces

The institute trio knew that vacuums were not really empty and certainly never tranquil. In fact, most physicists casting their eyes toward the cosmos believe that the vacuum is a hotbed of forces. Phantom particles flicker into existence and then disappear. "Empty" space itself seethes with what physicists call vacuum fluctuations: vast amounts of energy that suddenly burst forth, jiggling particles to and fro. One fluctuation is not very powerful, but cumulatively they can be intense. In fact, physicists John Wheeler and Richard Feynman calculated that there is enough energy in the vacuum of a single light bulb to boil all the seas.

It was City College physicist Timothy Boyer of New York, however, whose work convinced Puthoff that the vacuum was a good place for the institute to begin. Most physicists, Boyer pointed out, tried to explain the somewhat random movements of atomic particles through the theories of quantum physics. Quantum physics states that even under precise conditions, atomic particles may assume any one of a variety of positions. To determine with greater certainty where a particle could be found, however, physicists developed "probability equations." The equations predicted the likelihood of any given particle landing in any given place.

Electronics whiz Ken Shoulders is studying the Casimir effect in a vacuum, a process during which objects that come together produce enormous heat and energy.

Boyer held a different point of view. Perhaps, he suggested, the uncertain nature of the subatomic realm was due, not to the nebulous mathematics of probability equations, but rather, to vacuum fluctuations. We could not pin down the location of subatomic particles, Boyer suggested, because vacuum fluctuations jiggled them around.

Puthoff felt Boyer's notion could be used to explain other vexing problems as well. Writing in *Physical Review D,* Puthoff suggested that the zero-point energy of the vacuum might prevent atoms from collapsing, allowing the world as we know it to be to exist. Puthoff says, "Electrons should radiate their energy as they circle in their orbits. Eventually they should drop into the nucleus like a satellite falling back to Earth. Quantum mechanics never really explains why this does not happen."

Zero-point energy does. According to Puthoff's theory, electrons do radiate their energy away as they circle in their orbits. But they also absorb enough energy from vacuum fluctuations to make up for the loss. Calculations presented in *Physical Review* appear to back him up. Says Puthoff: "It seems that the stability of matter itself depends on the zero-point-energy sea."

Rethinking Einstein

Puthoff's next *Physical Review* paper was even more daring: it attempted to rewrite the theory of gravity proposed by Einstein himself. "Einstein described gravity as a warping of space-time caused by the mass of objects within it," Puthoff says. To understand Einstein's version, imagine the fabric of space-time as a taut rubber diaphragm. Place any given weight in the diaphragm, and it makes

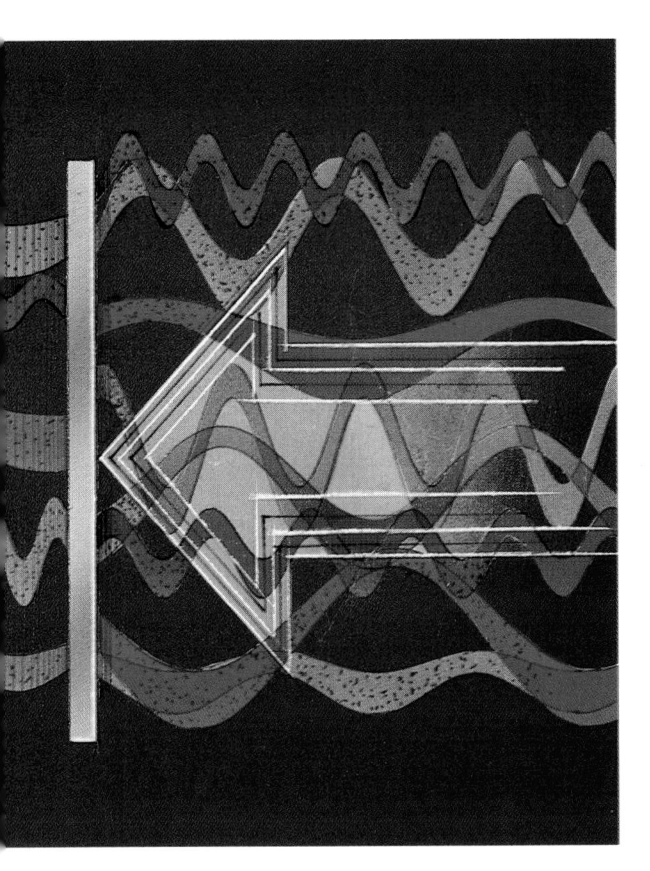

an indentation. Roll a marble onto the diaphragm. No matter how the marble is rolled, it ultimately winds up at the weight. This, according to Einstein, is how gravity works. Objects bend space-time just as the weight bends the rubber diaphragm, so two objects "roll together" with a force that depends on the objects' mass and distance.

"This shows how gravity acts," Puthoff says, "but doesn't really explain the mechanism behind it." That's where zero-point energy comes in. If two physical bodies are relatively close, he theorizes, the first shields the second from zero-point energy coming from its direction; in a similar fashion, the second object will shield the first. The objects will nonetheless continue to be pressured by zero-point energy coming from all other directions. The two bodies thus move toward each other in what scientists have dubbed the Casimir effect, named after Hendrik B. G. Casimir, the Dutch physicist who first described the phenomenon. What classical physics defines as gravity is the result, according to Puthoff.

Extracting Zero-Point Energy

It is the Casimir effect, Puthoff believes, that may help us extract zero-point energy from the void. Puthoff gives an example: Bring two smooth metal plates extremely close together, he explains, and they seem to attract each other so strongly that they are virtually welded to each other. Move them still closer, and they collide with a metaphorical boom, generating enormous heat. Use that heat energy, and the conversion of vacuum energy to usable energy has occurred.

This scheme, first proposed by veteran California physicist Robert Forward in *Physical Review,* has a problem: once the plates collide, they can no longer be used to generate energy, becoming a sort of one-shot device. "To recycle the generator," Puthoff explains, "one would have to return the plates to their original positions; that would require as much energy as the machine produced in the first place. As a result, not even break-even operation could be achieved."

His solution: "an inexhaustible supply of such devices, each to be discarded after the Casimir collapse." Puthoff concedes this would not be possible with metal plates, but suggests that engineers try designing zero-point-energy machines with a cold, charged plasma, or gas. "The Casimir effect would pinch the plasma together," Puthoff says, "and energy in the form of heat and condensed, charged particles would result."

At least one such device, Puthoff says, may be in the works. Moscow physicist Aleksandr Chernetsky has built a plasma generator that reportedly takes 700 watts of electricity from a wall socket and gives back 3,500 watts, creating a little more than 3 horsepower out of nothing. The Soviet government was impressed enough to back his research with several hundred thousand dollars' worth of equipment.

"I went to the Soviet Union to look at Chernetsky's work," Puthoff says. "I couldn't tell in a couple of days whether his equipment really works, or whether there is some fallacy in his experimental design. But it is plausible that it might be extracting zero-point energy."

Condensed-Charge Technology

Whether or not Chernetsky's power system works, other equipment apparently based on zero-point energy and the Casimir effect is under development. The inventor: Ken Shoulders, who hopes to create the next generation of circuits for laptop computers, telephones, and large-screen TVs.

Shoulders hopes to create these new appliances through a phenomenon he has discovered and put to use. Called condensed-charge technology, or CCT, the phenomenon occurs when electrons crowd together much as in Chernetsky's plasma or Puthoff's metal plates. "When electrons are packed densely enough, they no longer repel each other," says Shoulders. "Instead, they form charge clusters that hold together even without a wire to carry them. That lets us build circuits from grooves in a sheet of ceramic or plastic. Condensed charges can move through these grooves 1,000 times faster than electrons travel through a semiconductor chip." What is more, says Shoulders, it's easy to generate condensed charges: just make a spark.

His first major trick, Shoulders hopes, will be replacing today's silicon computer chips. If anyone else were to make so unlikely a claim, few would listen. But the 62-year-old Shoulders, formerly of the Massachusetts Institute of Technology (MIT) and Stanford Research Institute, possesses extraordinary credentials: in the early 1960s, he made the world's first vacuum microelectronic circuits and the very first prototypes of the equipment now used to manufacture silicon chips.

According to Shoulders, his new circuits will render silicon-based technology obsolete. "It looks like there is nothing in electronics that you cannot do a whole lot better with clustered charge," he says.

For an amiable Texan, Shoulders is remarkably closemouthed about the product he is said to be developing. But he is open about the advantages of condensed charge. "Using beads of condensed charge, we have already made transistor-type switches with speeds of less than one-trillionth of a second. That's 10,000 times faster than you can buy, and I think we're going to get a lot faster than that," Shoulders says. In fact, engineers working with conventional chips a couple of inches long are having trouble figuring out how to speed the passage of electrons from one side to the next. With condensed-charge technology, electrons move so rapidly that a single circuit could be a foot across.

Long, compact circuits working at high speed would enable us to build machines

with far less bulk than today's technology. For instance, Shoulders says, we could build "a 100-horsepower motor no bigger than the shaft it takes to deliver the torque [power], or a flat-screen TV with all the electronics built right into the display. You could use the screen for anything from high-definition TV to computing. Simpler yet: an X-ray machine that fits inside a hypodermic needle. You could put it into the patient's body to irradiate a tumor, say, without exposing the other organs to X rays."

Perhaps most incredible, CCT may be available soon. Condensed-charge devices are astonishingly easy to make, Shoulders says. "We can get rid of the complicated photographic techniques I had to invent to make microchips, and use simple etching and stamping. This is really low tech."

Though Shoulders works closely with Puthoff, he is reluctant to admit that CCT derives from zero-point energy for sure. "There are at least four competing theories that might explain condensed charges," he says, "and though zero-point energy is a likely candidate, I can't say which theory will turn out to be right."

If physicist Hal Puthoff is proved correct, we may someday obtain energy by tapping the force of random fluctuations that jostle atomic particles in a vacuum.

Mixed Reviews

Other scientists give Puthoff's work on zero-point energy mixed reviews. Timothy Boyer, whose papers inspired Puthoff in the first place, for instance, disagrees with Puthoff's explanation of gravity. "As far as I am concerned, the idea is fuzzy, and the calculations ambiguous," Boyer says. "To think in terms of the curvature of space-time is a much more useful, extensive idea."

Physicist Alfonso Rueda of California State University at Long Beach, on the other hand, is sympathetic to Puthoff. Rueda studied vacuum fluctuations, using them to explain both the enormous power of cosmic rays and the dense concentration of stars at certain intersections of the universe. Rueda feels Puthoff has presented some powerful evidence for his idea that zero-point energy holds atoms together. And he is "impressed with Puthoff's treatment of gravity. I think he is on the right path."

New York University physicist Benjamin Bederson, editor of the respected *Physical Review A,* where most of Puthoff's work has been published, has an opinion as well. "Many articles that appear in *Physical Review* turn out to be wrong," Bederson says. "Like any journal, we rely on the judgment of our referees. Some expressed doubts about Puthoff's conclusions, but they all agreed that it was stimulating work and deserved a wider audience."

As for Puthoff, he is confident indeed. A new series of experiments, he says, should deal with Boyer's criticisms and move his own research along. He looks forward to the day we tap the power in the void, using it to energize our cities and propel starships beyond the solar system without an ounce of onboard fuel. "Only the future," Puthoff says, "can reveal the ultimate use to which humans will put the remaining fire of the gods, the quantum fluctuations of empty space."

THE MYSTERY OF MAKING SCENTS

by Timothy S. Green

Dawn has just broken, but it is a gray, drizzly morning, with mist hanging heavy on the hills. Perfect weather, actually, for the thousands of men, women, and children already busy on gentle, sloping terraces in the valley below. Wrapped in parkas, shod in boots, they are moving slowly along waist-high hedges of pale pink roses. Working dexterously with both hands, they nip off the blossoms and put them in large plastic sacks. The sweet scent of roses pervades the damp air.

The preservation of that scent is the purpose of this harvest in the Valley of the Roses, near the town of Kazanluk in Bulgaria. For more than two centuries, the damask roses grown here have been the world's prime source of attar of roses, or rose oil, one of the most treasured ingredients of the perfumer's craft. It is blended into such fragrances as Joy, Femme, and Yves Saint Laurent's Paris. Already perfume makers from France, the United States, and Japan have telexed their bids for this year's rose oil, offering up to $2,500 a pound for the top blend.

A moist, cool dawn is essential to the unique aroma, or bouquet, of Kazanluk's roses; it ensures a profusion of rose oil in the petals, which would be dried out too quickly by a hot morning sun. "Climate is the dominant factor," explains Spas Kurtev, director of Kazanluk's Research Institute for Roses, Aromatic and Medicinal Plants. Stephen Manheimer, a major importer of raw materials for the perfume industry, considers Bulgarian oil "the standard by which others are measured."

The desire to capture that scent in a bottle has made perfumers seek out Bulgarian rose oil since the early 1700s. Around that time the Bulgarians apparently found the secret of steam distillation of rose oil from petals. The oil from Bulgaria's rose gardens was soon exported to France, where the perfume industry was just getting into its stride. The marriage has endured. The international perfume business, whose sales notched up over $10 billion in 1989, still depends on the nimble fingers of those early-rising Bulgarians for a unique floral oil. As a French buyer remarks, "You cannot substitute another rose oil for Bulgarian; it is irreplaceable."

The creation of a successful perfume is determined by the quality of oil harvested from flowers (such as the rose petals opposite), the intricate blending of scents, and eye-catching packaging and marketing programs.

The Valley of the Roses, therefore, is a good starting point on the perfume trail. Not that the history of perfume begins in Bulgaria. The art has flourished from the earliest civilizations, as primitive man often believed that scent from flowers contained the presence of a nature deity. The first perfumes personified gods or tried to please them with sweet fragrances. The very word "perfume" comes from the Latin *per fumun,* "through smoke," a reminder that it began in burnt offerings, like incense. Perfume's origins are apparent in India and the Middle East, where fragrant plants—such as cinnamon, frankincense, myrrh, jasmine, and violets—or fragrant trees—such as orange, lemon, and sandalwood—grew in abundance. Fragrance often had a symbolic function in important events in life and death: guests at ceremonies and banquets were sprinkled with fragrant waters, and mummies were impregnated with aromatic substances. Unguent pots in King Tutankhamen's tomb in Egypt, sealed around 1352 B.C., were still fragrant when it was opened in 1922.

Society's Frenzy for Fragrance

In this millennium, the development of perfume in Renaissance Italy and then in France may have had much to do with the need to cover up bad smells, but over time the motive became increasingly focused on sexual attraction. The modern history of perfume, based upon the use of traditional skills to blend those ingredients into a tantalizing aroma, begins in 17th-century France, in Grasse, the hill town west of Nice. By the mid-18th century, the court of King Louis XV was dubbed *la cour parfumée.* "Society went into a frenzy over perfume," notes the guide to Grasse's International Museum of Perfumery. "Custom dictated everyone leave a distinct aura in their wake."

The skills acquired have never been lost. Several French perfume houses date back to those heady days. Jean-François Houbigant (1772–1807) set up shop at 19 Rue du Faubourg St.-Honoré in Paris to sell perfume. The company is now located in New Jersey, but its Quelques Fleurs captures all the floral smells of Provence. Roure, a multinational firm that buys raw materials to create and

Certain areas are blessed with just the right soil and climate to grow uniquely aromatic plants. Above, jasmine blossoms are collected on the Côte d'Azur. In the Comoros Islands off the African mainland, ylang-ylang blossoms are loaded into a homemade still (right).

manufacture perfumes for fashion houses or movie stars, was founded in Grasse in 1820 by Claude Roure. The present head, Jean Amic, is the sixth generation of the family to be a creator of perfumes. Guerlain boasts five generations since Pierre François Pascal Guerlain became a perfumer in 1828. The firm still sells a perfume called Jicky, a blend of lavender, sandalwood, lemon, rosemary, and rosewood, created in 1889. Its latest, Samsara, was perfected precisely a century later by Jean-Paul Guerlain, drawing on skills taught him at age 17 by his grandfather.

The creators of perfume, or "noses," as the trade calls them, are a small clan. Only a dozen or so might merit the reputation of *un grand nez,* "a great nose." A nose must have not only an exceptional olfactory sense, but also the ability to retain a memory bank of more than 2,000 different scents from which he or she can draw between 100 and 500 to blend into a new perfume. The tricks of the trade are often handed down in the family. "It's easier to teach them early to a son or nephew," confides Guy Robert, the nose behind such fragrances as Calèche and Madame Rochas. He learned them from his uncle, father, and grandfather. "It's the little secrets that count," he says. "I've trained several perfumers myself, but I gave the most information to my son because I was able to teach him every day."

But it does not have to be all in the family. Many noses lacking a family background graduate from a training school in Grasse run by Roure. Among them is Akiko Kamei, a sparkling Japanese woman with impeccable French, who started her career as a dietitian but found her talents more suited to perfume than food; she helped create a fragrance for Hermès. James Bell, who majored in economics at Fordham, in the Bronx, New York, signed on as a clerk at a nearby Roure plant; he found himself captivated by the smell of his job and headed for Grasse.

Noses are likened to artists, composers, even chefs or winemakers, as they blend subtle potions to turn you on. The range of ingredients at the noses' disposal is broad; to obtain it, they scour the world. Jean-Paul Guerlain searched India for the choicest sandalwood while concocting Samsara. Guy Robert confides to me that he is experimenting with a lemon tree from the coast of Japan; its aroma is akin to a cocktail of lemon, lime, and mandarin.

Essential Oils

The basic ingredients produced from natural sources are the essential oils, resinoids (gums or resins purified with a solvent), and absolutes (aromas, in the form of viscous liquids, achieved by extraction with solvents) that are formed within the cells of plant tissues. Barely 2,000 of the 250,000 known species of flowering plants contain the oils. They may be in the entire plant, as with lavender, geranium, or rosemary; in the flower, as with jasmine or rose; in the roots of irises; in the fruit of vanilla or in an herb such as anise; in the rind of lemons and oranges; in cedar or sandalwood; in the bark of cinnamon; or in lichens such as oakmoss.

The aroma of each essential oil or absolute will vary considerably, not just from one country to another, but within a small region. For instance, the finest oil of bergamot, a pear-shaped citrus, comes from a few square miles along the Calabrian coast in southern Italy, and is produced by a unique environment of chalky soil, intense sunlight, and strong sea air.

Most countries have some particular combination of climate and soil that encourages a certain plant to flourish and win the attention of the perfume trade's raw-material buyers. Go to Guatemala for fine lemongrass, to Comoros for the spicy flowers of the graceful ylang-ylang tree, to Malaysia or Sumatra for the exotic herb patchouli; go to Finland for pine resins, to Yugoslavia for oakmoss, to China for eucalyptus, to India for ginger and sandalwood, to Madagascar for vanilla. The United States contributes lemons and oranges from California and Florida, peppermint from the Pacific Northwest, and clary sage from North Carolina. The bouquet of one year's crop may, as with wine, differ from the next. A trained nose can often distinguish, from a single whiff, the exact place of origin of the oil, perhaps even the "vintage."

The ultimate aromatic, usually attained by pressure, extraction, or steam distillation, is a remarkable reduction from the original. The mathematics for jasmine alone are prodigious. It takes 45 minutes to pick 5,000 jasmine flowers, which weigh about 1 pound; 800 pounds of flowers provide only 1 pound of jasmine absolute.

Harvesting may be only the beginning. Iris absolute, whose rich and elegant aroma is an essential ingredient of such perfumes as Chanel 19 and Guerlain's Shalimar, is obtained only after the roots, grown mainly in Italy and Morocco, are peeled by hand and dried for 18 months. Iris absolute commands the top price of any natural perfume ingredient: about $7,000 a pound.

The cost of picking natural ingredients can become prohibitive, especially such tiny flowers as violets, now rarely used. A crisis may arise as Bulgaria makes the transition from communism, under which the roses of Kazanluk were produced on cooperative farms where pickers received only a nominal wage or were paid in foodstuffs for their animals.

Synthetic Aromas

Does that mean a greater swing to synthetic aromas produced in the laboratory? Synthetics are not new. Even in the 18th century, early scientific analysis of natural oils led to attempts to isolate an oil's crucial "odorous principle," which produced the vital aroma. The first chemical aromatic was nitrobenzene, prepared from nitric acid and benzene, which was used to give an odor of almonds to scented soaps. The key breakthrough came in 1868, when an Englishman, William Perkin, synthesized coumarin, which captured the new-mown-hay scent of the South American tonka bean. Soon after, Ferdinand Tiemann of the University of Berlin produced synthetics of vanilla and violet. And in the United States in 1889, Francis Despard Dodge pioneered citronellol (an alcohol) with a rose or geranium odor, and a variation, citronellal, which could be variously adapted to give the scents of lily of the valley, hyacinth, narcissus, and sweet pea.

Synthetics offered two advantages. The perfumer was no longer at the mercy of a bad year when the harvest of many flowers might be poor. More important, synthetics widened the range of the perfumer's palette. With real jasmine oil, for instance, there used to be very few constituents to work with; synthetics increased the range greatly. But they cannot be a complete substitute in high-quality perfumes. "It's like silk against polyester," explains Françoise Marin, Roure perfumer and director of its perfumery school. "You cannot replace the touch of nature."

The touch of nature from animal products, however, has been largely replaced by synthetics as the perfume industry has bowed to pressure from conservationists in recent years. Extracts from the pungent scent glands of the musk deer, a small deer found in the northern Himalayan regions of Tibet and China; of the civet cat, which is bred mainly in captivity in Ethiopia; and of the beaver, from Canada and Siberia, were used for centuries. So was ambergris, from the intestines of the sperm whale, an endangered animal protected, with all its products, since 1970 in the United States, and by the Convention on International Trade in Endangered Species since 1977. All these substances helped fix the various essential oils to ensure that fragrance remained long after the bottle was opened.

The Nose at Work

The challenge for the nose is to conjure a subtle combination of perhaps 100 or more scents from the galaxy at his or her disposal into a new perfume. "We are dream merchants," says one. Setting to work, a nose is often smartly turned out in a white coat and seated before a U-shaped desk in a laboratory, surrounded by rack upon rack of little vials of *le jus*—"the juice." This setup is known as the "organ" because the composition of a perfume is described in terms of "notes." Notes are the perfumer's code. Each essential oil and each perfume has three notes. *Notes de tête* ("top notes"), composed of volatile aromas, such as lemon or orange, give the initial tang of a perfume when it is opened or put on the skin, but vanish swiftly. Next come *notes de coeur* ("central notes," or "heart notes") from rose, jasmine, or iris, which provide the richness and body. Finally, there are *notes de fond* ("base notes") from sandalwood or cedar, which provide a long-lasting bouquet. In blending, the perfumer may also add "spicy notes," "woody notes," "Oriental notes," "green notes," or "fruity notes." The ultimate blend gives a perfume its distinctive "signature."

The nose begins by dipping narrow strips of blotting paper known as *mouillettes* into the vials and passing them lightly before the nostrils for an initial impression, then attaching them to a rack on the desk to see how the aroma evolves as they dry. Making careful written notes, the perfumer adds and mixes and smells again, drawing on experience of countless previous concoctions. At big manufacturers such as Roure, with scores of clients, the team of noses can draw, too, on "maybe 20,000 fragrances in the [computer] bank," explains Geoffrey Webster, president of Roure's U.S. operations in Teaneck, New Jersey. "So we change a top note here, a woody note there." The final perfume concentrate is not, of course, what you get in the bottle. A high-quality perfume will usually contain from 18 to 25 percent of the concentrate dissolved in alcohol, and a trace of distilled water; an *eau de parfum* may contain 10 to 15 percent; an *eau de cologne,* 5 to 8 percent; and an *eau de toilette,* 2 to 4 percent.

Competition is intense. Perfume houses like Guerlain, Jean Patou, or Rochas have their own in-house noses. Fashion houses or movie stars rely on free-lance noses or multinationals—such as International Flavours and Fragrances (IFF), Florasynth, or Firmenich—that create fragrances on a grand scale, for everything from perfume to shampoo to detergent. Many of the best-sellers are created in these houses. IFF made Beautiful for Estée Lauder, and Paris for Yves Saint Laurent; Florasynth dreamed up Giorgio for Fred Hayman; and Firmenich did Anaïs, Anaïs for Cacharel.

Passion and Obsession

In 1921 Coco Chanel's nose, Ernest Beaux, presented her with a few numbered bottles. Coco sniffed them; she liked bottle No. 5, so Chanel No. 5 was born. Such classics enjoy a long life. Worth's Je Reviens dates from 1932, Guerlain's Shalimar from 1925, Patou's Joy from 1931.

Yet new perfumes proliferate. Originally perfume houses created and sold perfume.

"Nose" Jean-Paul Guerlain relies on exceptional olfactory sense, indelible scent memory, and skills passed from generation to generation to conjure an appealing combination of scents for a new perfume.

The Paris fashion houses commissioned them, initially as little gifts to clients, but soon as major lines in their own right. So Chanel, Dior, and Yves Saint Laurent became as much identified with fragrance as with fashion.

Men's fashion has joined in: witness Calvin Klein's Obsession and Giorgio Armani's Armani. From high fashion, it was a short step to enlarge the perfume parade to young fashion and life-style, with Colors from Benetton, and Guess by Georges Marciano, while Hermès and Gucci added perfumes like Calèche and Gucci No. 1 to their accessories. Jewelers Boucheron, Cartier, and Van Cleef & Arpels commissioned their own brands, as did cigar makers Davidoff and Dunhill.

Movie stars and personalities, of course, find the notion of a perfume in their image irresistible, although customers are often less persuaded; many "personality" perfumes sink without a trace.

But there is one star who has really pulled it off. Since 1987 Elizabeth Taylor's Passion has been among the five top sellers in the United States. It was created for Taylor by Roure, but she followed the waft of virtually every *mouillette.* "She was involved with our perfumers for months to get it right," recalls Roure's Geoffrey Webster. They finally agreed on a heady cocktail of ylang-ylang, black currant, gardenia, and marigold. Then Taylor went on a nationwide tour, quoting the Shakespeare-like "Bee to the blossom, moth to the flame, each to his passion." People loved it, despite an initial price tag of $165 an ounce. Annual profits of $70 million rolled in. So did the lawsuits. The initial problem was with Paris perfumer Annick Goutal, who just happened to have created a perfume called Passion. No sooner was that settled than Taylor's ex-boyfriend Henry Wynberg claimed a slice of the action, charging that Passion was a copycat version of a perfume he had developed and presented to her in a heart-shaped bottle when they were stepping out together between her two marriages to Richard Burton. The case was all set for trial in Los Angeles in 1991, with each side having mustered some distinctly unsavory evidence to discredit the other. But a last-minute, out-of-court settlement left the issue moot.

Still, no one has revolutionized the American perfume scene as much as Estée Lauder, a young Hungarian-American who started marketing four pots of skin products in New York City in 1946. She created her first perfume, Youth Dew, in 1953, and then seven other best-sellers, such as Estée, Cinnabar, and Beautiful, in the ensuing 35 years. She convinced the American woman that perfume could be worn all day, not just on special occasions, starting a trend that has made the United States the world's leading consumer. Out of $10 billion spent on perfumes worldwide in 1989 (two-thirds on women's fragrances), $3.7 billion is spent in the United States, compared with $4 billion for the whole of Europe, a modest $1.2 billion in Japan, and $800 million in duty-free shops around the world.

Hitting just the right formula, as Taylor did with Passion, is not easy. The first question from a nose is always, "What is the concept?" It may be a vague brief, or profile of the prospective consumer, for a "kill me" perfume, for a fresh floral fragrance, or for a zesty sea-breeze appeal. Some briefs are disarmingly explicit. One specified that the fragrance should "smell like my grandmother's

Through the ages, perfume containers have added élan to their contents. Above from left to right are a svelte opaque glass container on which a lizard chases a fly, circa 1912; Admiral Nelson's engraved crystal scent flasks, circa 1870; a Roman blown-glass bottle from the first or second century A.D.

closet." Another, more boldly, demanded "the metallic, leathery smell of a couple who had made love in a Ferrari car."

Saint Laurent's Opium was a trendsetter for the perfumes of the past 20 years that sought to capture the spirit of the sexual revolution. As a Paris perfumer put it, "There was a rush to meet the demand of the American woman who wanted a strong, aggressive perfume that said, 'Look guys, I'm here, see me and smell me.' " The trade labeled them "knock 'em dead" fragrances.

But times change. The olfactory trend for the 1990s is to less-overpowering perfumes, hinting at floral freshness or the healthy tang of the sea. "Most of the briefs come in now for discreet, feminine, sensual rather than sexy fragrances," says Roure's Anne Cantacuzene. "Scents reflect the cocooning spirit of family. There are lots of natural, healthy 'green notes,' as if we've just mowed the lawn."

The Total Package

Behind the launching of a new perfume is an immense marketing effort—and cost. If perfume is to sell for $150 or more an ounce, it must be packaged around a complete concept. While the nose sniffs *mouillettes,* others huddle on the name, the design of the bottle, a launch party, and advertisements to outdo all predecessors. The perfume itself may account for only $5 to $10 of a $150 package. In planning such a coup, secrecy is intense: even the nose probably does not know the name destined for the perfume he or she is creating.

The rules of the name are that it should be easily remembered, create its own image, and, ideally, be universally pronounceable: the simplicity of Paris or Chanel No. 5 is perfect. Tweed conjures an English country aura for the older woman; Safari promises African adventures. The name should also reflect an era. Today perfumes like Balmain's Vent Vert (Green Breeze) are being relaunched to join the green revolution.

The search for a name often takes more-pretentious directions. Guerlain, for instance, sifted through 50,000 names before settling on Samsara for its newest fragrance, formulated around the concepts of serenity and harmony. "In Sanskrit the word *samsara* means 'everlasting return,' " explains Bernard Fornas, who masterminded the launch. "And it's an international name, easily pronounceable by Anglo-Saxons, Latins, or Japanese." But is a name "free"? Perfumers habitually register thousands that they might conceivably use one day, often just to block

Above left is a 1907 René Lalique frosted flacon, commissioned by François Coty. Above right is a unique 1825 opaline crystal bird-shaped flask highlighted in gold. Today, perfume bottles are first modeled from Lucite (right) before the final delicate flask is created from glass. The elaborate trappings surrounding a new scent may attract a consumer initially, but it is the unique essence that will make consumers return for more.

competitors. Even Samsara had been registered in several countries, and it took Guerlain three years to negotiate the rights to use the name.

The perfume bottle, too, must match the overall image of its contents. For Samsara's Oriental concept, sculptor Robert Granai came up with a red bottle based on a Tibetan figure in the Guimet Museum in Paris; the yellow bottle cap represents the deep, closed eyes on many Buddhist statues. Bijan's bottle is egg-shaped, echoing nature's most perfect form. Poison, naturally, comes in a sinister little flask, and Panther has black cats leaping out on either side.

A little extra help from marketing people with large checkbooks comes in handy. Flip through almost any magazine, pass through any airport duty-free shop—the images are there. Such hype is costly; budgets have soared. Megamarketing really kicked off with Yves Saint Laurent's Opium. Christian Dior's Poison got about $20 million for its launch; the stake money doubled for C'est la vie and Elizabeth Taylor's Passion. Not to be outdone, Guerlain in 1989 raised the stakes again for Samsara. Says Bernard Fornas, sipping black coffee in his office above the Champs Élysées in Paris: "We all used to launch perfumes discreetly, with very little promotion. For Samsara, I've set the level quite high with $50 million."

To get in the mood, Guerlain redecorated the Guimet Museum of Oriental Art in Paris for the initial launch party. In Asia, they flew 300 journalists to Bangkok's Oriental Hotel for a bash. Bernard Fornas and his marketing team made 39 presentations everywhere from New York and Chicago to Tokyo and Yokohama. Saleswomen worldwide were dressed in Tibetan-red uniforms to parcel out the red fans, wrapping paper, and shopping bags flaunting the Samsara label. Did it pay? Admits Fornas: "We hope to break even rapidly."

Still, a caution: the customer may be wooed once by an exotic name, a fancy bottle, and millions spent on media hype, but will not come back for a second bottle if the perfume itself does not please. As French nose Guy Robert puts it, "In the end, the perfume decides."

THE PHYSICS OF CAR ACCIDENTS

by Tim Folger

Arthur Damask's garage is dark, cool, and uncluttered. The wooden walls are unadorned, the concrete floor bare save for the hulking form of a car under a blue plastic tarpaulin, which Damask is busy pulling off. As the plastic slides to the floor, what emerges is neither a mint-condition antique roadster nor the latest Italian sports coupe. Rather, the unveiled vehicle looks as if it had been in a monster-truck rodeo: its roof is crushed, its windows smashed, its doors punched in on one side.

Crash Connoisseur

Damask walks around the wreck, reading its scratches and dents with a practiced eye. "The car was struck from the side while stopped at a light on a two-lane highway," he says by way of history. He then points to the crushed metal above the front wheel—the exact spot where the impact occurred. "It rolled over twice, coming to rest on its roof. This car should have slid sideways. But it rolled. The poor driver is a vegetable—the roof penetrated his skull."

Damask is a connoisseur of car crashes. When he's not teaching at the City University of New York, the soft-spoken 67-year-old physicist investigates accidents as an expert court witness, using the bloodless laws of matter and energy to reconstruct an often tragic sequence of events. The crippled vehicle in his Long Island garage is the central piece of evidence in a lawsuit brought by the relatives of the driver against the car's manufacturer. Damask, who's working for the plaintiffs, can't yet disclose the make or model of the car—but he's more than forthcoming about its physical properties.

"Go to the right rear wheel," he says, crouching in front of it. "Now, you see, the wheel is bent inward about 18 degrees." This inward bend, he explains, decreases the "track"—the distance between the rear

wheels—and makes the car unstable. The laws of physics say the car still wouldn't be unstable enough to flip over. But as with many laws, there's a loophole. "Metal has an elastic component—it snaps back," says Damask. To get a permanent deformation of 18 degrees, he says, the deformation during the collision had to have been greater.

How much greater? To find out, Damask did some bending of his own. From an identical car, he took a wheel and the support it was mounted on. The support was a T-shaped, hollow tube of metal with walls about 0.1 inch (0.25 centimeter) thick; it is used in lieu of an axle in some car models, and it costs much less to manufacture. "I bent the T-support with a metal press, push-

The cause of a car crash in broad daylight (above) can be as difficult to determine as that of an accident with no witnesses (right). Police departments and insurance companies are increasingly turning to the laws of physics to explain what went wrong.

ing it in and letting it go until I got this 18-degree bend in the wheel," Damask says. "I had to push it in 27 degrees before it would snap back to 18." If a 27-degree deformation occurred during the collision, it would have narrowed the track by as much as 25 percent—enough, Damask calculates, to make the car roll on impact.

Damask believes that the car's stability would not have been compromised if the rear wheels had been mounted on an axle, and he points to the car itself to buttress his argu-

ment. "During the collision the T-support buckled, and the tire slipped from its rim, which dug into the concrete." The scars on the rim, Damask says, testify to the strength of the impact. "The front rim is in even worse shape"—in this case from the direct blow of the collision—"but the front tires are on a normal axle: big, stiff metal that didn't bend at all." The track of the front wheels wasn't affected by the impact, even though the force of it was greater there than in the back.

Damask thinks it's pretty clear that an axle on the rear wheels as well as the front would have kept the car from rolling, and would have saved its passenger from a debilitating head injury. He plans to present that argument in court this fall. "I have a bunch of lawyers coming here—I don't know if they'll want to get down on the floor and look at that wheel, though." To save his visitors that inconvenience, Damask often puts a mirror on the floor beneath his subjects.

Conflicting Versions

Damask himself has spent a lot of time on the floor of his garage. In the past 25 years, he's investigated hundreds of car accidents—and then there's a plane crash, an elevator mishap, and some muggings thrown in for good measure. The physicist has been a crash enthusiast ever since a friend who had helped an attorney with an auto-collision case introduced Damask to forensics. "Physicists like to work out puzzles," Damask says. "It's our shtick. And here my friend had discovered a field where you could get a daily high by solving a problem."

Damask now averages a case a week—and $150 an hour—for his particular brand of problem solving. His most memorable investigation, however, is one he got paid "practically nothing" for—nothing, that is, except the satisfaction of saving an innocent person from 60 years in prison. The incident in question happened on a wet March night in 1983, when five young men set out from a bar in the Bronx. On their way home, they drove down a long, sloping straightaway that veered to the left at the bottom; the curve was cast in darkness, however, because the streetlights above it weren't working.

The boys never made it past that curve. When police arrived at the scene, they found the youths' car, a 1973 Mustang, rent almost entirely in two in the middle of the road. The five occupants were strewn about the wreck. Only one was still alive, and he couldn't remember anything about the accident.

A New York City district attorney assumed that the young survivor had been driving, because he was the car's owner. The district attorney charged him with four counts of manslaughter, each of which carried a maximum 15-year sentence. By the time the defense attorney brought Damask into the case early in 1984, the police had already developed their own explanation of the accident.

Physicist Arthur Damask studies wrecked cars (below) and visits crash scenes for clues as to why an accident happened.

Basing their version on a skid mark at the accident site, the police assumed that the car had been traveling too fast to make the turn, and had slid and jumped up on to the grassy shoulder of the curve. The driver lost control when he tried to steer back on to the road, swinging the car sideways into a perpendicular collision with the end of a 3-foot (1-meter)-high concrete guardrail lining the highway. The crash broke the Mustang in half, the police said; its two pieces bounced into the road 24 feet (7.3 meters) beyond the point of impact, and the driver was thrown from the wreck.

An officer claimed to have determined the Mustang's speed be-

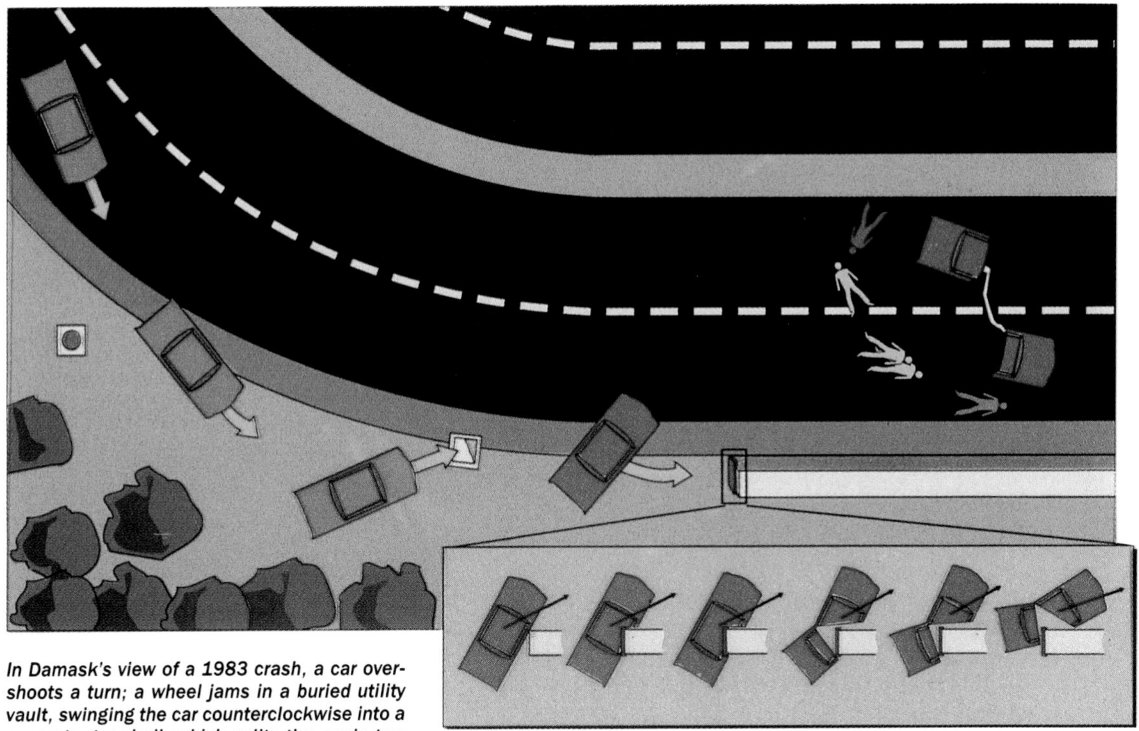

In Damask's view of a 1983 crash, a car overshoots a turn; a wheel jams in a buried utility vault, swinging the car counterclockwise into a concrete guardrail, which splits the car in two and throws the occupants onto the roadway. His testimony proved that the only survivor (in orange) was not driving the car.

fore the crash by measuring the skid mark; he came up with a figure of 82.24 miles (132 kilometers) per hour. The accident looked like a clear case of reckless driving. But was it? After visiting the site and examining police photos, Damask realized that the official interpretation was seriously flawed. "If the accident happened that way," he says, "it would be as clear a violation of the laws of physics as if a mouse overbalanced an elephant on a seesaw."

Had the car hit the guardrail with the speed and trajectory suggested by the police, for example, it would not have rebounded at all, but would have wrapped around the rail. At 82 miles per hour, Damask says, cars that hit concrete walls straight on don't bounce off like a rubber ball; instead they stick, like a ball of putty. The force of the collision completely overwhelms the elasticity of the sheet metal.

But Damask also knew that the Mustang couldn't have been going that fast in the first place. He'd charted the alleged path of the vehicle, and saw that at such a high speed, the car could never have turned sharply enough to hit the guardrail. Instead, the Mustang's momentum would have carried it straight into a stand of trees that runs parallel to the highway about 12 feet (3.6 meters) behind the rail. For the driver to have made the turn he did, Damask estimated, the car could not have been going more than 35 miles (56 kilometers) per hour.

The officer who had calculated the Mustang's speed had done so by assuming that the skid mark traced out the arc of a circle. A simple formula will indeed give the velocity of a car traveling in a circular path if the weight of the car and the radius of the circle are known. But Damask says the radius couldn't be determined from the skid mark. "The mark was practically straight," he says. And it wasn't even clear that the skid mark had been left by the Mustang.

Damask had his own theories about what had happened. In the ground next to the guardrail, he'd found an uncovered utility vault—a sort of buried concrete electric box measuring 3 feet by 2 feet (91 centimeters by 61 centimeters) across at the top—and realized that the car had to have gone over it to hit the end of the rail. Furthermore, one rim of the vault was chipped, and the right front tire rim of the Mustang was dented. Damask deduced that it was the vault, rather than the car's speed, that had caused the driver to lose control. The youths had overshot the darkened curve and strayed onto the shoulder of the road; when the front tire got caught in the uncovered vault, the back end of the car swung around counterclockwise, and the Mustang hit the guardrail at an angle. The blow fractured the car and ripped off the front passenger door, but did not arrest the forward motion of the vehicle's halves, which slid out onto the road.

Who Was Driving?

Damask's reconstruction was persuasive enough to invalidate the prosecution's version of the accident. But he still had done nothing to challenge the assertion that the defendant was driving the car—the most important part of the case. To settle this issue, Damask turned to the coroner's records and began studying the mechanics of the victims' and the survivor's injuries. In his grand philosophy of automobile accidents, Damask would say, he had solved but one collision; he had two more to go.

"There are really three collisions in any one accident," says Damask. "When a car strikes something that brings it drastically to a stop, that's the first collision. But the occupants are still going along at the same speed as the car, until they are brought to a stop by striking something in the interior of the car or by being ejected and striking the road." That's the second collision.

"The heart, the brain, the internal organs—the viscera—have also been going at the same speed as the car. They are stopped by the internal walls of the body. The brain is stopped by striking the skull, the heart by striking the chest, and the spleen and the liver by striking the abdominal muscles. That's the third collision."

Damask knew that in an automobile accident, the driver usually suffers a characteristic set of injuries from the second collision: the head and torso strike the steering wheel; the right ankle is wrenched when the leg gets stuck between the dashboard and the floor; the knee is injured when it hits the bottom of the dashboard. And the third collision—the result of sudden, high deceleration—will sometimes tear the aorta, the great artery on the left side of the heart. Damask calculated that when the car slammed into the guardrail, the young men trapped inside experienced a 240-G deceleration, more than 150 times what space-shuttle astronauts face on reentry, and more than enough to tear an aorta. "That would cause death within a heartbeat or two," he says.

When Damask examined the medical records, he found that the four young men who had died in the accident did in fact suffer ruptured aortas. But only one occupant of the car had the spectrum of injuries expected of the driver, and that person was not the accused. Damask argued that the defendant survived because he had escaped the lethal effects of the second and third collisions—and the only way he could have done that was by sitting in the front passenger seat. He must have been thrown from the car when the right front door was sheared off by the concrete guardrail, Damask concluded; he probably flew through the air for 9.5 feet (2.8 meters) and slid on the road for another 10.5 feet (3.2 meters), avoiding the brutal deceleration that killed his four friends.

When Damask testified, he held up a chart he had made based on the autopsy reports that listed the injuries. "I put up this chart and concluded, 'The accused wasn't even driving the car.' All hell broke loose in the court, like I dropped a bomb." After four hours of deliberation, the jury returned a verdict of not guilty on all four counts of manslaughter.

As Damask's tale draws to a close, the contrast between the mayhem he analyzes and the life he leads in a tranquil seaside community seems more striking than before. His career has given him a dozen gruesome anecdotes, each an object lesson in its own right on the dangers of risky driving and hasty assumptions. But Damask's personal experience of accidents has been limited to fender benders. What of the lessons he's learned? Has his work, at the very least, made him a more cautious driver?

"Oh yes," he says, "very much so. I stopped being in a hurry. I don't even start the engine until everyone in the car has their seat belt fastened." And Damask, who used to drive smaller cars, now pilots a heavy Lincoln Continental for safety's sake. Too many times, he says, he's discovered in the course of an investigation that a tragic accident could have been easily prevented. "The unfortunate thing about accident reconstruction," he reflects, his gaze on the peaceful ocean outside his study window, "is that you end up with what might have been."

THE MAGIC OF CARD SHUFFLING

by Tim Folger

No matter how good a cardplayer you are, you probably don't shuffle your cards enough, at least according to mathematicians Persi Diaconis of Harvard University and David Bayer of Columbia University. They made headlines recently with their announcement of a mathematical proof that it takes at least seven riffle shuffles to thoroughly mix a deck of 52 cards. More than seven won't make much difference.

The problem had long stumped other mathematicians. A 52-card deck can be arranged in an enormous number of ways. Any of the 52 cards can be first in the deck, any of 51 can be second, and so on. The number of possible arrangements is $52 \times 51 \times 50 \ldots$ all the way down to 1. The resultant number, roughly 10^{68}, is far larger than the number of seconds that have elapsed since the universe began. Trying to figure out how many shuffles it takes to make each one of these arrangements equally likely, and to erase all influence of any previous arrangement, had proved an intractable task.

Bayer and Diaconis took an unusual approach: they did lots of fieldwork. They played cards. They haunted casinos. They even tape-recorded decks being shuffled, and analyzed the sound to see how the cards were interleaved. And in the end, they found a simpler way of measuring the randomness of a deck.

Sequences

As even an amateur cardplayer knows, you can tell a deck hasn't been thoroughly shuffled when a particular sequence of cards from one hand crops up in the next. To transform this intuition into a general measure of randomness, Diaconis and Bayer defined something that they call a rising sequence. If you number the cards in a deck from 1 to 52, and forget about the faces, a rising sequence is a sequence of consecutively numbered cards, such as 1, 2, 3 or 17, 18, 19, 20. The cards don't have to be next to each other to be considered part of the same rising sequence. To find the rising sequences in a deck, you start with the top card and then proceed toward the bottom, skipping over cards until you find the card numbered one higher. When you reach the bottom of the deck, the first rising sequence is over. The next sequence begins with the first card you skipped.

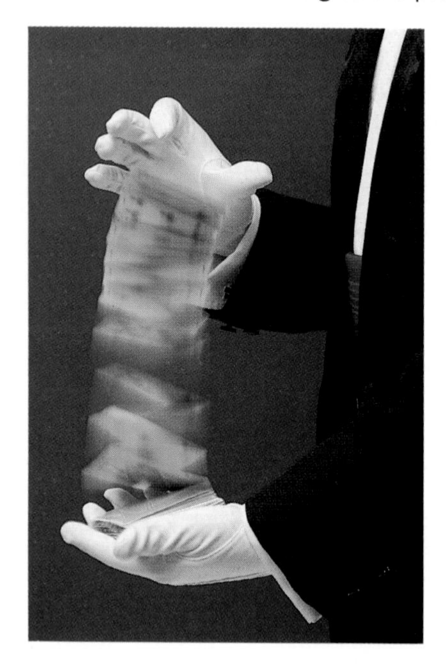

Now imagine that the deck starts out in perfectly consecutive order, with card number 1 at the top and card number 52 at the bottom. That deck has one rising sequence. If it is cut evenly into two stacks, one stack will contain cards numbered 1 through 26; the other, cards 27 through 52. After the shuffle the deck will look something like this: 1, 27, 2, 28, 3, 29. . . . The shuffled deck has two intermixed rising sequences: 1, 2, 3 . . . and 27, 28, 29. . . . With each shuffle, in fact, the number of rising sequences increases, and the sequences get shorter. A thoroughly shuffled deck has 26 rising sequences, and any trace of the original order has been eliminated entirely.

How many shuffles does it take to get there? Diaconis and Bayer were convinced that seven was the magic number. Through educated guesswork, they came up with a formula for calculating the probability that a given number of shuffles would produce a given number of rising sequences. The formula predicts that seven shuffles will generate close to 26 rising sequences—in other words, that seven shuffles will randomize the deck. "We were guided in our calculations by knowing what the right answer was," says Diaconis. "That is, we hoped we knew what the right answer was. Mathematics has a way of surprising you."

Hypercubes

The researchers still had the formidable task of proving that their formula was correct. That's when they made their second innovation: a geometric description of shuffling. "We had a real breakthrough," says Diaconis. "It wasn't just a matter of using bigger computers. We had a completely new way of looking at the problem."

The new method is abstract but visual: it uses a 52-dimensional hypercube to represent all possible arrangements of a 52-card deck. A point inside the hypercube represents the current state of the deck. Each axis corresponds to one card, and the position of the point along that axis corresponds to the card's position in the deck—the farther you go along the axis, the closer that card is to the top of the deck. Rank the point's 52 coordinates according to their magnitude, and you will have written down the sequence of the cards.

Each of the 10^{68} possible sequences is represented in the hypercube by an individual "volume element," within which the ranking of the coordinates does not change. If you can visualize a 52-dimensional volume ele-

According to a new theory proposed by mathematicians, a deck of cards must be shuffled seven times to be well mixed. A person able to figure out a probable sequence of cards from a well-shuffled deck would likely make a bundle at the blackjack table.

ment, you're certainly not from planet Earth, but it's not hard to visualize in three dimensions—that is, for the case of a three-card deck. In that case the hypercube reduces to an ordinary cube with three axes—*x*, *y*, and *z*—that originate at one corner. The volume elements are six tetrahedrons that have in common the diagonal that cuts across the cube from the point of origin. Within each tetrahedron the ranking of the three coordinates—say, *x* is greater than *y*, and *y* is greater than *z*—doesn't change. Each tetrahedron corresponds to one of the six possible sequences of three cards.

Now go back to 52 cards, and to shuffling. In the hypercube model, shuffling corresponds to moving the point from one volume element to a different one. Diaconis and Bayer represent the process this way: first they stretch all 52 axes of the hypercube to twice their length, then they cut the doubled axes in half and mash each new half back into the old one. (The operation is similar to stretching and folding dough, which is why the researchers have dubbed it the "baker's transformation.")

When the axes are doubled, the volume element containing the point is stretched to 2^{52} times its size. It can then be divided into 2^{52} new volume elements. All the new volume elements are carbon copies of elements in the original hypercube; they represent card arrangements that the deck can be put into by that shuffle. Some elements are copied more than once, reflecting the fact that a single arrangement of the deck can be arrived at by several different ways of cutting and interleaving the cards. And here's the key: when the axes are doubled, the point's coordinates are doubled, too. In the process the point can travel into any of the new volume elements, but then it inevitably gets mashed into the corresponding element of the original hypercube—and the deck winds up in that arrangement of the cards—when the axes are cut back to their original length.

Probability

To figure out the probability of the cards ending up in a particular arrangement, all you have to do is add up all the volume elements corresponding to that arrangement after the original volume element has been stretched. After one stretch, there are more copies of some elements than of others. What Diaconis and Bayer found is that you have to double the axes seven times (without cutting them back) before all 10^{68} of the original volume elements are represented almost equally—which is the same as saying you have to shuffle cards seven times before all 10^{68} theoretically possible arrangements of the deck become equally likely. "Once we had a geometric representation of shuffling," says Diaconis, "the problem boiled down to counting up volume elements."

Diaconis is tickled at the reaction his work has gotten. "The most you can usually hope for in mathematics," he says, "is the grudging acknowledgment of a few people who know what you're talking about. But shuffling cards seems to have captured people's attention."

NIKOLA TESLA:
The Forgotten Physicist

by Curt Wohleber

Nikola Tesla was born, it is said, during a thunderstorm at the stroke of midnight between July 9 and 10, 1856, in the village of Smiljan, Croatia. He was the son of Milutin, a minister, and Djouka, who was illiterate, but had a prodigious memory and could recite vast amounts of poetry. It was from her, Tesla maintained, that he inherited his inventive abilities; she had contrived many household devices.

As Tesla remembered his childhood, it was filled with many narrow escapes from death: on several occasions, he nearly drowned; once he was almost boiled alive in a vat of hot milk, and he survived perilous encounters with dogs, hogs, and angry crows.

The phenomenon of static electricity entranced him. He once stroked the family cat and saw that its "back was a sheet of light, and my hand produced a shower of crackling sparks loud enough to be heard all over the place." His mother told him to stop before he started a fire. His father explained that it was electricity, just like lightning.

Nikola was fascinated. "Is nature a giant cat?" he asked himself. "If so, who strokes its back?"

Mechanical things interested him early on. He took apart his grandfather's clocks. At the age of five, he built a bladeless waterwheel. (Interestingly, one of his last inventions of note was a bladeless turbine.) He also fashioned popguns, and arrows that he said could bore through a 1-inch (2.5-centimeter)-thick pine plank.

Once he built a tiny motor powered by June bugs glued to a cross-shaped assembly

Through his various inventions, Nikola Tesla (above, as a young man) made possible today's electricity-driven society. He was at the height of his creative powers when he posed for the picture at top.

mounted on a spindle. When the bugs tried to fly away, they turned a pulley, which drove the engine. The experiment ended when a local boy came by and began to eat his power supply.

In the classroom, Tesla was a star pupil, especially—almost alarmingly so—in mathematics. He would blurt out the answers to involved problems without putting pen to paper. In his autobiography, *My Inventions*, Tesla boasts of an amazing mental ability: if a particular object or figure came to his mind, he would see a vivid image of it before his eyes. This gift disturbed him, because sometimes particularly distressing scenes, such as a funeral, would suddenly intrude on his vision. But this odd capacity also enabled him to carry out complex arithmetic operations without a blackboard, and later to design whole machines solely in his mind. From time to time, he would also see strange flashes of light. "In some instances," he wrote, "I have seen all the air around me filled with tongues of living flame."

Audacious Concepts

In 1870 the 14-year-old Tesla went off to school in the city of Karlovac, where he stayed with an aunt and uncle. He finished the four-year term at the Higher Real Gymnasium in three years. He also caught malaria, and he was still weak from it when he returned to his family in Cospic, Croatia, where they had moved when he was six. He came down with cholera. Things looked bleak on all fronts. Even if he recovered from the cholera, Tesla faced a compulsory three years of military service. Worse still, his father insisted that he enter the ministry instead of studying engineering. The predicament sapped his will to live. "I was confined to bed for nine months with scarcely any ability to move," he recalled.

His father tried desperately to rouse him from his malaise. Then, when it looked as if he was going to die, Nikola whispered, "Perhaps I may get well if you will let me study engineering." Overcome, Milutin Tesla told Nikola that he would go to the best technical institution in the world. That promise, along with "a bitter decoction of a peculiar bean," restored Tesla's health: "I came to life like another Lazarus."

At the advice of his father, Tesla spent the next year roaming the mountains, regaining his strength and, not incidentally, evading the draft. During this time, he conceived a number of audacious inventions. One was an ocean-spanning hydraulic tube for transporting mail at high speeds. Another was an elevated ring circling the Earth at the equator. The ring would remain stationary as the Earth turned within it, thus providing a simple means of rapid travel.

When his *Wanderjahr* was over, he entered the polytechnic institute in Graz, Austria, in 1875. Fiercely determined to penetrate the mysteries of electricity, he studied with an intensity that alarmed the dean of the engineering faculty, who wrote to Milutin that his son was working too hard.

During his second year in Graz, the school acquired a Gramme machine, a device consisting of a rotating conductive coil mounted between the ends of a horseshoe magnet. It could function as either a motor or a generator. If one turned the coil, its movement through the magnetic field would induce a current. Conversely, supplying an electric current to the coil would magnetize it, making it rotate within the field, thus providing mechanical power.

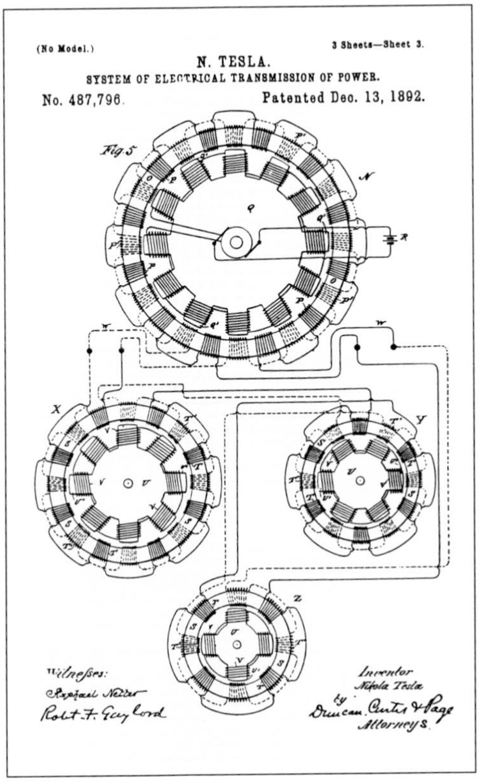

(No Model.) 3 Sheets—Sheet 3.
N. TESLA.
SYSTEM OF ELECTRICAL TRANSMISSION OF POWER.
No. 487,796. Patented Dec. 13, 1892.

After much work, Nikola Tesla invented the system that made the transmission and distribution of electricity possible. The patent diagram at left describes several phases of the system.

When it was functioning as a generator, the strength and direction of the current induced in the coil changed according to the coil's orientation to the magnetic field. As the coil turned, the current increased to a maximum, decreased to zero, then began to increase again, but in the opposite direction.

What was created, then, was an alternating current (AC). A commutator—a hollow cylinder split lengthwise, with each segment connected to one end of the coil—enabled a set of brushes to exchange connections with the coil terminals twice per rotation, resulting in an intermittent current of constant direction.

The device intrigued Tesla, but the violent sparking where the commutator and brushes met struck him as intolerably inefficient.

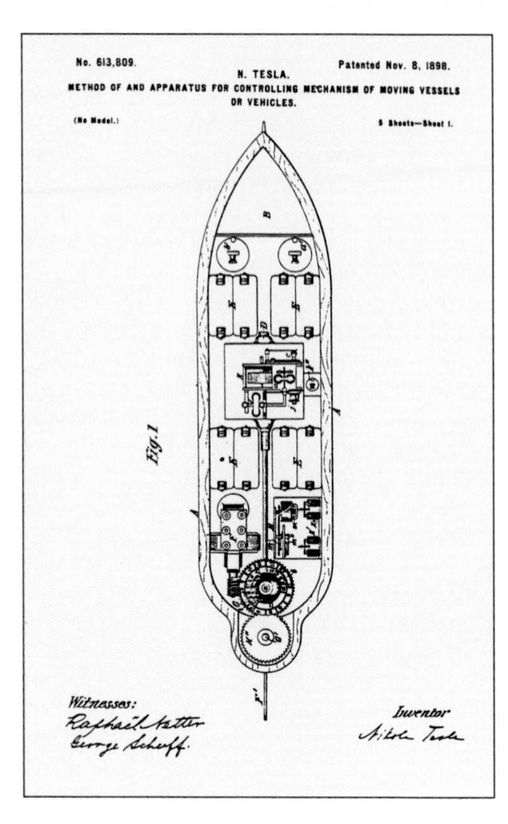

Tesla patented several methods that dealt with the propulsion of boats. A system he developed for a remote-control boat was turned down by the U.S. Navy during the Spanish-American War.

Mental Laboratory

Not until 1881 did Tesla finally figure out how to build an electric motor that dispensed with the commutator and brushes. His solution was to run two or more out-of-phase alternating currents through the stator, or stationary outer portion of the motor. These currents would induce currents in the rotor, or armature—the inner, rotating portion. With a properly controlled phase relationship, the currents in the stator would create a rotating magnetic field. The induced currents in the rotor would create another magnetic field, at an angle to the one created by the stator currents. This would cause the rotor to turn, as the opposite poles of the two magnetic fields attracted each other. This breakthrough became known as the polyphase system.

Retreating again to his mental laboratory, Tesla built a series of imaginary motors, dynamos, and transformers. He later said that he was able to test his prototypes by running them for weeks on end and periodically checking them for signs of wear.

Soon afterward, he got a job with the Continental Edison Company in Paris. He tried to interest his employers in the polyphase system, but Edison's antipathy to alternating current was well known. Tesla nevertheless distinguished himself as a crack troubleshooter, directing the repair of Edison power stations in France and Germany. While on assignment in Alsace, he built his first actual prototype, a two-phase AC induction motor. It worked beautifully, but the only person who expressed any commercial interest in it was the mayor of Strasbourg, who tried without success to recruit local investors.

In Paris, Charles Batchelor, a friend and close business associate of Edison, told Tesla that he ought to seek his fortune in America. Tesla arrived in the United States in 1884. He had lost his luggage on his way to the steamship, so he arrived in New York City with just the clothes on his back, a few coins, and a bundle of papers, mostly technical articles and poems he had written. He earned $20 on his first day by fixing a machine for a grateful Manhattan shopkeeper. The next morning, armed with a letter of introduction from Batchelor, he called on Edison at his laboratory on Pearl Street. The inventor, impressed with Tesla's credentials, hired him to repair a malfunctioning electrical system on a steamship.

The cultured, fastidious Tesla and the slovenly, folksy Edison were uniquely mismatched. Unable to find Croatia on a map, Edison once asked Tesla if he had ever eaten human flesh. They differed in inventive philosophy as well as personality. Edison took an almost perverse pride in his plodding trial-and-error approach to problem solving. Tesla, whose inventions existed fully formed in his head before he began to build them, found Edison's "empirical dragnets" distasteful. "If Edison had a needle to find in a haystack," Tesla said later, "he would proceed at once with the diligence of the bee to examine straw after straw until he found the object of his search. I was a sorry witness of such doings, knowing that a little theory and calculation would have saved him ninety percent of his labor."

Edison and Westinghouse

Despite their differences, Tesla impressed Edison with his skill and hard work. For months the young immigrant worked from 10:30 A.M. to 5:00 A.M. the next day, seven days a week. He presented a plan to improve the efficiency of Edison's dynamos. Edison said he would give him $50,000 if he could make the scheme work. Tesla labored feverishly on the project, and the improved dynamos did everything he said they would.

Then he asked Edison for the $50,000 he had been promised. Edison was taken aback. "Tesla, you don't understand our American humor," he said. Tesla was unamused. Edison made a counteroffer: a $10-per-week raise. Tesla quit on the spot.

Soon afterward a group of investors approached Tesla with a proposal to form a new company to develop and market arc lights. As head of the Tesla Electric Light Company, he devised a safer, more reliable arc lamp than the ones generally in use at the time. For his effort the investors paid him in stock certificates, which were virtually worthless because the United States was in the grip of the economic crisis that followed the Panic of 1884. He was forced out of the company, and he spent the next year working as a street laborer.

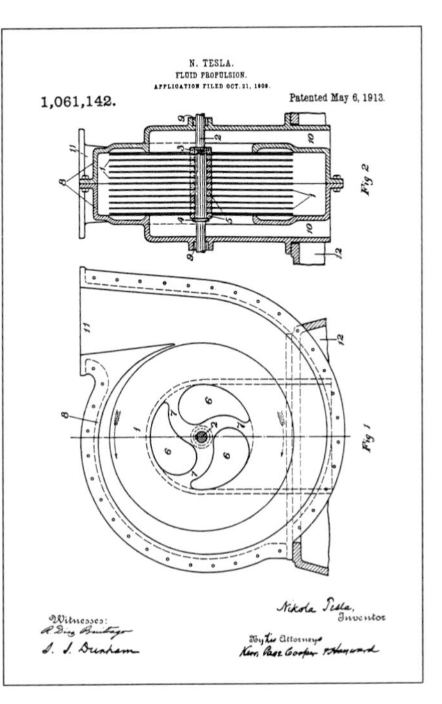

The foreman of his ditchdigging gang was moved and impressed by Tesla's story, and introduced him to a Mr. A. K. Brown of Western Union, who helped Tesla form a new firm, the Tesla Electric Company. With financial backing and his own lab just a few blocks from Edison's, Tesla set to work building motors, dynamos, and transformers. In addition to his polyphase system, he developed single- and split-phase AC motors, which were less efficient, but useful for special applications, or where a polyphase current was unavailable or impractical. Over the next several years, Tesla received 40 patents related to his AC system.

In May 1888, he lectured to the American Institute of Electrical Engineers. His address, "A New System of Alternate Current Motors and Transformers," created a sensation within the profession. Among those impressed by Tesla's vision was George Westinghouse, the Pittsburgh businessman and inventor of the Westinghouse air brake for trains. Westinghouse had secured American rights to a transformer, patented by Lucien Gaulard and John Gibbs, which was used to supply high-voltage alternating current for arc lighting.

William Stanley had greatly improved the Gaulard and Gibbs transformer, and in 1886 Stanley electrified 18 businesses in Great Barrington, Massachusetts. Westinghouse declared that AC had arrived. Alternating current had a great advantage over the direct-current system marketed by Edison. Low-voltage direct currents dissipated rapidly, limiting the transmission radius of every generating station to a few miles. Stanley's transformer made it feasible to step up an alternating current to a high voltage, at which it could be transmitted efficiently across long distances.

Among Tesla's later patents was an important method of fluid propulsion (left). The patent was granted by the United States government in 1913. Altogether, Tesla received 112 patents over the course of his lifetime.

But the system suffered from the lack of a practical motor it could power. Tesla's key patents filled that void. Westinghouse visited Tesla at his laboratory and nodded appreciatively as the inventor demonstrated his quietly humming machines. The two men made a deal: Tesla would get approximately $60,000 in cash and stock, plus a $2.50 royalty for every horsepower of motor or generating capacity sold.

"War of the Systems"

As word of Westinghouse's plans spread, Edison, sensing a threat, prepared to strike back. The "War of the Systems" was one of the most down-and-dirty public battles in the annals of American business. Edison set out to convince the American people that alternating current, with its high voltages, represented a menace to anyone who dared let it into his home, a tactic Edison had used with considerably more justification when persuading people to give up gas-lighting for DC power and incandescent bulbs.

In West Orange, New Jersey, the home of his sprawling new invention factory, Edison offered local children a 25-cent bounty for stray dogs and cats. Pets suddenly disappeared. Edison had hired a New York engineer named Harold Brown to document the menace of alternating current and explore its use as a means of execution.

One corner of Edison's renowned center of beneficial innovations became a chamber of horrors as Brown "Westinghoused" dogs and cats and later calves and horses.

In 1888 New York State adopted electrocution as its official mode of capital punishment. Brown surreptitiously secured a license on several Tesla patents, and two years later William Kemmler, a convicted murderer, was the first man to be executed in Sing Sing Prison's new electric chair. It was a grisly, protracted affair. Kemmler survived the first jolt, and the current had to be administered again. "They could have done better with an axe," commented Westinghouse.

World's Fair

The American public accepted alternating current anyway. And eventually so did Edison's company. A ruthless price war forced Edison to merge with Thomson-Houston, which used AC for arc lighting, and a new company—General Electric—was born.

In 1893 Westinghouse underbid General Electric and won the contract to light the Chicago World's Fair. Working on short notice, Tesla cobbled together a dozen 1,000-horsepower AC generators. Celebrating (a year late) the 400th anniversary of the discovery of America, 25 million visitors mobbed Chicago and beheld such technological wonders as Thomas Edison's early motion-picture system, music transmitted live from New York via telephone, and an early prototype of the zipper.

Tesla, nattily dressed as always, amazed audiences with an electric clock, glowing phosphorescent tubes, and spectacular discharges from mammoth electric coils. He sent 200,000 volts of high-frequency current through his body. A newspaper reporter wrote that Tesla was surrounded by "dazzling streams of light."

That same year, Westinghouse won the contract to build three AC generators at Niagara Falls. Years earlier, Tesla had daydreamed about harnessing the enormous power of that thundering curtain of water. Now it was to become a reality. Huge turbines were built, and in 1895 the first three 5,000-horsepower, two-phase generators went on-line. One of the first customers was the

Chicago's 1893 Columbian Exposition gave Tesla a showcase for his electrical wizardry. By the turn of the century, Tesla had become a household word.

Using his "Tesla coil" (above), Tesla proved that alternating current at very high voltages would be harmless if the frequency was high enough.

Pittsburgh Reduction Company, which later became the Aluminum Company of America (now Alcoa). Cheap, plentiful electric power from the Falls made possible the large-scale commercial production of aluminum, and in the years that followed, Niagara Falls became the center of the electrochemical industry. By the end of 1896, a 26-mile (42-kilometer) transmission line carried current to power the lights and streetcars of Buffalo.

Zenith of Creativity

The last decade of the century saw Tesla at the zenith of his creative powers. In 1891 he had demonstrated the Tesla coil, an air-core step-up transformer and capacitor capable of converting high currents at low voltage to low currents at high voltage, all at high AC frequencies. He also began to experiment with ways of transmitting electricity wirelessly. Tesla had noticed that sending a current through a coil of a specific frequency would elicit sparks from other coils tuned to either the same frequency or one of its harmonics. He foresaw a day when telegraph signals and electric power would be transmitted all over the world without wires, and he filed several key patents describing wireless transmitters and receivers.

Tesla's experiments were interrupted in 1895, when fire reduced his laboratory to a smoking, half-melted ruin. He had no insurance. Fortunately, Edward Dean Adams, whose Cataract Construction Company had awarded Westinghouse the Niagara Falls generator contract, came through with $40,000, and suggested forming a new company with Tesla, capitalized at $500,000. Tesla, prizing his independence, declined the offer. He set up shop on East Houston Street and returned to work, agonizing as the new equipment trickled in by rail.

Three years later, at the Electrical Exhibition in New York City's Madison Square Garden, 15,000 people watched Tesla demonstrate a remote-control model boat, an astounding feat considering that radio technology was still embryonic. War had broken out with Spain earlier that year, and Tesla hoped to sell his "tele-automatic" system to the government, but the military men found the idea of robot boats too exotic.

Mounting Hardship

Though Tesla would survive into the 1940s, his important scientific work was largely completed by the turn of the century. In the last decades of his life, he faced mounting financial hardship. He was evicted from his beloved Waldorf-Astoria and took up residence in a series of progressively less fashionable hotels. Growing old and desperately in need of money, he turned his hand to some less visionary projects, and enjoyed modest success with an improved locomotive headlight and a speedometer. His bladeless turbine showed great promise, but it never caught on. In the past two decades, however, interest in his turbine has grown, and that invention may soon find application in generators and jet aircraft.

Tesla died in his hotel room in January 1943. Less than a year later, the U.S. Supreme Court voided Marconi's primary wireless patent in favor of a 1900 Tesla patent.

Today Tesla's name, while hardly forgotten, is unknown to many. His accomplishments, significant as they were, are not concretely present in everyday life the way Thomas Edison's light bulb is. Yet Tesla's voice can still be heard, on quiet nights when the sounds of traffic recede and when one can hear the steady hum of power lines, carrying on the work of the world.

REVIEW

Chemistry

In 1985 scientists predicted the existence of a new form of carbon—a giant molecule made of 60 carbon atoms arranged in the shape of a soccer ball. This fullerene molecule was dubbed buckminsterfullerene, or buckyball for short, after the late R. Buckminster Fuller, the inventor of the geodesic dome. Today buckyballs are synthesized in the laboratory. (See also *Buckyballs: The Magic Molecules*, page 235.)

Researchers at Harvard University in Cambridge, Massachusetts, Allied-Signal, Inc., in Morristown, New Jersey, and NEC Corporation's Fundamental Research Laboratories in Tsukuba, Japan, reported successes in using metals to make superconducting buckyballs. Paul W. C. Chu of the University of Houston also reported that after eliminating the superconducting property of niobium by adding buckyballs to the metal, the buckyball-niobium compound regained its superconductivity when put into a magnetic field. Heating the material abolished superconductivity, and, unlike most other superconducting materials, it did not regain this property after being cooled.

Paul J. Drusic at Du Pont Central Research and Development Experimental Station in Wilmington, Delaware, found that buckyballs appear to soak up charged molecules very readily, which suggests that fullerenes might have great potential as catalysts. And researchers at Northwestern University in Evanston, Illinois, used buckyballs to lay down diamond films on silicon, thus providing insight into how diamonds are formed.

But on the darker side, Robert Whetten and his colleagues at the University of California, Los Angeles (UCLA), found that in the presence of light, the 60-carbon buckyball transforms molecular oxygen into atomic oxygen, an especially reactive state that is known to disrupt biochemical functions of cells and tissues.

In other areas of chemistry, scientists at the Howard Hughes Medical Institute at the Whitehead Institute for Biomedical Research in Cambridge, Massachusetts, showed how stretches of amino acids—previously called the leucine zipper—can bind two coiled protein chains together to form a variety of substances, such as keratin, a basic ingredient of skin, hair, bones, and muscle fibers. Such proteins also help regulate gene expression by interacting with sections of DNA. The work shows that chemical knobs from one chain fit into holes in the other chain.

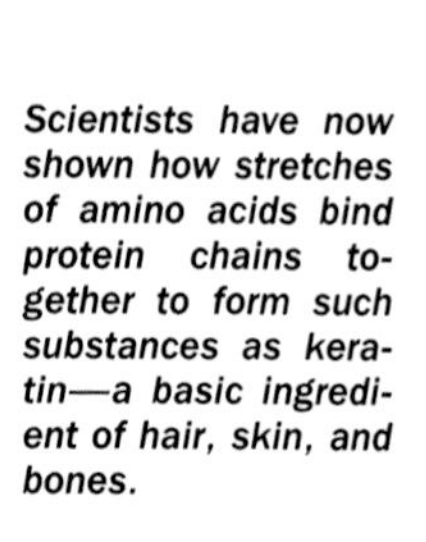

Scientists have now shown how stretches of amino acids bind protein chains together to form such substances as keratin—a basic ingredient of hair, skin, and bones.

Critics and proponents of cold fusion, one of the hottest topics in the field just a couple of years ago, continue to debate whether tabletop experiments can produce more energy than is injected into the experimental apparatus. For example, Bruce E. Liebert and Bor Yann Liaw at the University of Hawaii obtained equivocal results in their own work, and failed in most attempts to repeat their findings. Meanwhile, the National Cold Fusion Institute in Salt Lake City, Utah, has shut down for lack of funds.

A series of experimental and theoretical studies showed that on certain surfaces, deposited atoms can trade places with surface atoms, and that, through a sequence of such exchanges, an atom can move across the surface, alternating its position as it goes. The work, performed by Peter J. Feibelman and Gary L. Kellogg at the Sandia National Laboratories in New Mexico, adds a new dimension to the studies of surface diffusion for materials scientists who are attempting to grow crystalline materials layer by layer.

The movement of water also came under scrutiny as researchers at Boston University and the University of Dortmund, Germany, discovered how water, which exists in networks of four molecules held onto a central water molecule by stiff hydrogen bonds, can flow as easily as substances without such tight connections. The scientists found that the normal pattern in which one water molecule is linked to four others is sometimes weakened when a fifth water molecule crowds into the matrix, increasing the overall mobility of the water. The observation helps to explain why water under pressure—during which its molecules are forced together—flows faster.

Marc Kusinitz

Water under pressure flows faster because the linkages between the water molecules are weaker than they are when no pressure is applied.

Mathematics

Surprisingly, many mathematical advances with everyday applicability have their foundation in the mathematical riddles that amused people before the days of TV, radio, and other wonders of modern entertainment. These puzzles were very difficult (as might be expected!), and they required fairly sophisticated reasoning to find answers to them.

One such example is called the Kirkman schoolgirl problem. Simply stated, it asks if 15 girls can be arranged in 5 rows of 3 apiece on 7 consecutive days so that no 2 girls are in the same row more than once. Try to do it! (It is possible.) Questions like this arise naturally in other contexts, such as arranging round-robin sporting tournaments and designing statistical experiments. Mathematicians have developed general techniques to construct distinct subsets of a set, all of the same size, in such a way that any pair of points will belong to exactly one of these subsets. These techniques fall under the broad classification of *combinatorics.*

These combinatoric techniques have far-ranging applications, from compact-disc players to deep-space communication. These connections have been explored very actively over the past 40 years, corresponding to the explosion of information technology related to computers.

We can use combinatorics to generate what are called error-correcting codes. The word "code" often brings to mind cloak-and-dagger imagery. But mathematical coding theory has nothing to do with cat-and-mouse games. Instead, the object is to send a message with enough redundancy so that anyone who sees the message will be able to understand what was being sent. The person receiving the message should be able to know what was sent even if some mistakes are made in the transmission. The simplest example of this involves a situation where a yes-or-no response is required. A mathematical way to respond would be to use 0 for no and 1 for yes. To avoid the possibility of a mistake, 00000 might be used for no and 11111 for yes. That way, even if a mistake is made and 00001 is received, it is likely that the message "no" was what was intended, since there are more 0's than 1's.

COMPACT-DISC TECHNOLOGY

To understand how arranging schoolgirls can help make Mozart sound better on compact-disc players, consider the problems involved in trying to make music digital. Compact-disc players take musical sounds and convert them into strings of 0's and 1's. For example, the musical note middle C might be represented by 0011101. This string is then sent to a processor, and 0011101 is converted back to middle C. Suppose an electronic error is made, and 0011001 is sent instead of 0011101; if 0011001 represents high E, an odd note will come out of the speaker. There is no time to go back and retransmit the correct sound because the symphony marches on. Thus, to have the capability to correct errors on the fly requires an efficient way to determine what was intended as the most likely sound. If the patterns of 0's and 1's are different enough, then we should be able to tell what the original pattern was, even if there are mistakes made during transmission.

The mathematical technique of combinatorics is used in a number of applications, including the digitalization of music for compact discs.

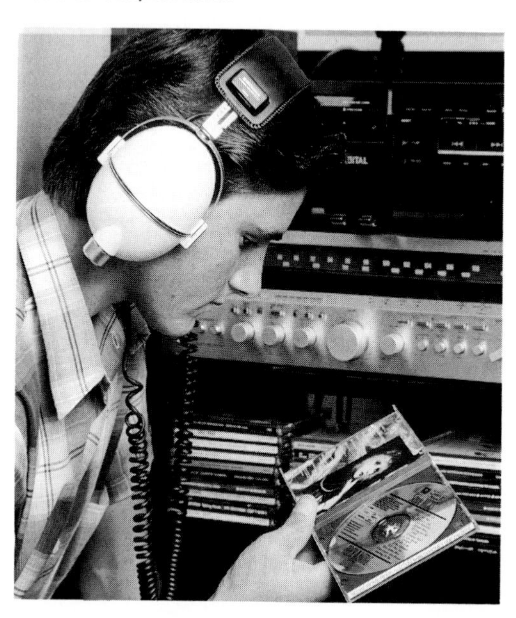

DEEP-SPACE COMMUNICATION

Similar problems are faced in deep-space communication. It takes many hours to send a message back to Earth from Neptune, and the distance means that the signal is very weak when it gets here, making it likely that the proper message may not be received. Consequently, error-correcting capabilities need to be placed into the code. However, if we put too much redundancy in, the rate at which information can be transmitted will be significantly slowed. Thus, compact ways to encode messages are needed that can still be recognized if quite a few mistakes are made. This again requires that the pattern of 0's and 1's be different enough from each other, and we can use combinatoric arrangements like the schoolgirl problem to accomplish this.

SEARCHING FOR GOOD CODES

Many tools of modern mathematics are used in the search for good codes. One particularly effective tool is a high-powered computer, such as the Cray supercomputer. But a computer's success in figuring out math problems has been a topic of much discussion and controversy in the mathematical community—how can we know a proof is true if no human can actually check that the proof is correct? Aside from this controversy, though, it is clear that computers will be used in many different ways in the future.

Thus, from mathematical principles based on riddles from the past century, we have succeeded in studying our solar system and digitally recorded music. The next time that you listen to a new compact disc or see new photographs sent back from deep space, remember that much of the technology that is being used behind the scenes in those situations is based on the same sorts of ideas used to solve the Kirkman schoolgirl problem!

James A. Davis

Physics and Chemistry

Discoveries at the molecular level, achieved through complex applied mathematics, led to Nobel Prizes in both Physics and Chemistry for two European scientists in 1991.

Dr. Pierre-Gilles de Gennes of France was awarded the physics prize for his discoveries in the directional ordering of molecules in liquid crystals and other substances with similar orientational characteristics. Dr. de Gennes' theories demonstrate remarkable and potentially very useful similarities between liquid crystals and superconductors.

Dr. Richard R. Ernst of Zurich, Switzerland, won the prize in chemistry for his exacting research in a method of chemical analysis known as nuclear magnetic resonance (NMR) spectroscopy.

By coincidence, both prizes were for research involving the alignment of atoms and molecules under electromagnetic influence. The Nobel committee also noted that the use of mathematics in both laureates' research and explanations helped bridge the increasingly indistinct areas of chemistry and physics, further blurring the boundaries between the two.

Pierre-Gilles de Gennes won the Nobel Prize for Physics for his discoveries involving liquid crystals.

THE PRIZE IN PHYSICS

Dr. Pierre-Gilles de Gennes looked deeply and carefully into molecular behavior to explain many different phenomena. He focused primarily on liquid crystals, which include not only the materials commonly found in "LCD" (liquid-crystal display) watches, but also organic materials, including components of blood.

The molecules of liquid crystals are oriented either randomly in a "disordered state" or in "ordered-state" patterns determined by force fields, such as those created by electromagnetism. Some molecules behave like magnets, leaping to align with the forces of a magnetic field.

Liquid crystals are not the only substances that can shift between states of order and disorder. Dr. de Gennes and his team at the University of Paris, Orsay, also investigated the structural behavior of complex polymers, including the "superglue" used in aircraft.

One of Dr. de Gennes' most tantalizing discoveries was his explanation of the structural behavior of helium 3. As a liquid, helium 3 flows without frictional resistance, a property it shares with superconductors.

Awarding Dr. de Gennes the $1 million prize, the Nobel committee called the professor "the Isaac Newton of our time," and praised his subtle mathematical wizardry for the ability to unify various physical systems in a single general description.

In his response to the award, Dr. de Gennes attributed his success to the team with whom he worked at Orsay Liquid Crystals Group for 10 years during the 1960s. During that time, France became a leader in liquid-crystal technology.

Richard Ernst was awarded the Nobel Prize for Chemistry for his extensive research into nuclear magnetic resonance spectroscopy.

THE PRIZE IN CHEMISTRY

Dr. Richard R. Ernst, a professor at Zurich's Eidgenössische Technische Hochschule, was awarded the 1991 Nobel Prize in Chemistry for research somewhat similar to that which brought the prize in physics to Dr. Pierre-Gilles de Gennes. Both scientists looked at the phenomenon of atoms aligning themselves in the direction of a magnetic field.

Dr. Ernst's research made possible the development of all the many applications of nuclear magnetic resonance, a spectroscopic technique usually abbreviated NMR. The technology is used in medical examination spectroscopy to image the human body; chemists and physicists also use NMR to perform extremely precise chemical analyses of various compounds.

In the NMR process, the atoms of a target subject are put into an ordered state by an intense magnetic field. Short, intense radio impulses are then shot at the atoms, nudging them out of position. The emitted radio waves are tuned to the natural frequencies of the target atoms. When the radio waves stop, the atoms snap back to their ordered positions, an action that emits a pulse of energy. By measuring these signals, scientists can determine the type and structure of the atoms that sent the signals.

Dr. Ernst worked with an American scientist, Dr. Weston A. Anderson, to refine the sensitivity of the technology. Together, the researchers learned that timed sequences of radio signals gave technicians different perspectives on the sample material. Dr. Ernst brought the technology so far forward that today's NMR equipment can detect three-dimensional molecular structures made up of hundreds of atoms.

Mathematics made it all possible. A mathematical operation called a Fourier transformation made it possible to identify and extract faint bits of numerical data from the masses of data that the NMR process produces. With this mathematical tool, Dr. Ernst made NMR spectroscopy 100 times more sensitive than it had been previously.

Dr. Ernst was flying 30,000 feet (9,000 meters) above the Atlantic Ocean when he first learned he had won the prestigious Nobel Prize. The message came via radio and was passed along to the laureate by a flight attendant. The somewhat surprised scientist, who was en route to receive the Louisa Gross Horwitz prize for research in biochemistry, said he was "very pleased" with the news of his Nobel Prize.

Dr. Pierre-Gilles de Gennes was born in Paris in 1932. He earned his Ph.D. in France, then went on to postdoctoral work at the University of California at Berkeley. He is a member of the Atomic Energy Council (France), and recipient of the Prix Ancel and the Médaille D'Argent. He is currently a professor at the Collège de France in Paris.

Dr. Richard R. Ernst, 58 years old, is a citizen of Switzerland. He worked as a research scientist in Palo Alto, California, from 1963 until 1968. He was the winner of the Wolf Prize in 1991.

Glenn Alan Cheney

Physics

The neutrino continued to baffle researchers during 1991. While studying the energy spectrum of beta particles emitted from decaying carbon-14 atoms, Eric B. Norman of the Lawrence Berkeley Laboratory in California observed an anomaly that suggested that the neutrino is electrically neutral, and has a mass of about 17 keV—much larger than expected; the neutrino also interacted weakly with ordinary matter. Although many researchers obtained similar results, physicists at California Institute of Technology (Caltech) failed to detect this anomaly.

The Soviet-American gallium experiment, which was designed to capture neutrinos hurtling to Earth from the Sun, detected significantly fewer of these particles than had been predicted by theory. This caused some theorists to suggest that either the Sun's core temperature is lower than was thought, or that neutrinos undergo a transformation on their way to Earth that makes them hard to detect.

Stanford University physicists made strontium vapor transparent to light at wavelengths that the gas ordinarily would not absorb. Using a green laser light at 570 nanometers (nm), the researchers were able to excite electrons in the strontium vapor from one energy level to a higher level. This interfered with the ability of a 337-nm beam of ultraviolet (UV) light to excite electrons from a third level to the same higher level. This rendered strontium gas incapable of absorbing the UV light, and made the vapor transparent to the UV light. This technique may eventually be modified to make even solids transparent to certain wavelengths.

Particle accelerators enable physicists to study the composition of subatomic particles by smashing protons apart in high-speed collisions.

Physicists at Bell Communications Research in Red Bank, New Jersey, made a crystal that excluded photons of microwaves; in so doing, the physicists created a primitive semiconductor that might become the basis of highly efficient lasers and solar cells. This so-called photonic crystal prevents atoms embedded within it from spontaneously reemitting light at wavelengths that fall within a range determined by the structure of the crystal.

In a significant break with previous strategies for producing energy from nuclear fusion, scientists at the Joint European Torus (JET) laboratory in Culham, United Kingdom, generated a one-second burst of 1.7 million watts of energy by fusing deuterium and tritium nuclei in a magnetically confined plasma (a gas containing about equal amounts of positive ions and electrons). Fusion scientists usually use two deuterium nuclei. But because fusion of tritium with deuterium occurs at a lower temperature and more rapidly than the reaction between two deuterium nuclei, their success represents a major step forward in the development of fusion as a viable source of energy.

Researchers from the Massachusetts Institute of Technology (MIT) in Cambridge, Massachusetts, the University of Konstanz in Germany, Stanford University in California, and the Federal Agency for Technical and Scientific Research in Braunschweig, Germany, used a new technique called atomic interferometry to divide and recombine single atoms, taking advantage of their dual particle/wave nature. This technique lays the groundwork for atom-based gyroscopes that are much more sensitive than the laser varieties.

Marc Kusinitz

PLANTS and ANIMALS

CONTENTS

Ancient Spell of the SEA UNICORN

by Fred Bruemmer

For Europe's medieval princes and potentates, poison was a major professional hazard. Aconitine, from the deep-blue monkshood plant, paralyzed a victim's heart. Hemlock brought on a gradual demise through slow suffocation. A pinch of henbane produced mad exhilaration, followed by violent cramps and death.

Back in those superstitious times, only one substance—a miraculous horn—was said to detect and neutralize these lethal elixirs, and for 1,000 years the rich and noble paid fortunes for it. Dipped into poisoned food or drink, the horn turned the suspect substance dark, made it froth and bubble, and robbed the poison of its potency.

For centuries, buyers attributed the source of this miracle antidote to the unicorn, a horselike animal with a single horn jutting from its head. None, however, had ever laid eyes on the beast itself, and for people of the time, it remained a perplexing puzzle. "The Animal must be on Earth," wrote Konrad von Gesner, a 16th-century zoologist, "or else its horn would not exist."

Unicorn of the Sea

As it turned out, the magical horn was not the armament of a hoofed land creature, but the fantastically enlarged tooth of a strange, torpedo-shaped whale that haunts the circumpolar Arctic seas. Through the centuries, sailors and explorers brought home tales of a creature that pierced the waves with an awesome spiraling tusk. Eventually, they dubbed it "unicorn of the sea" for its eerie resem-

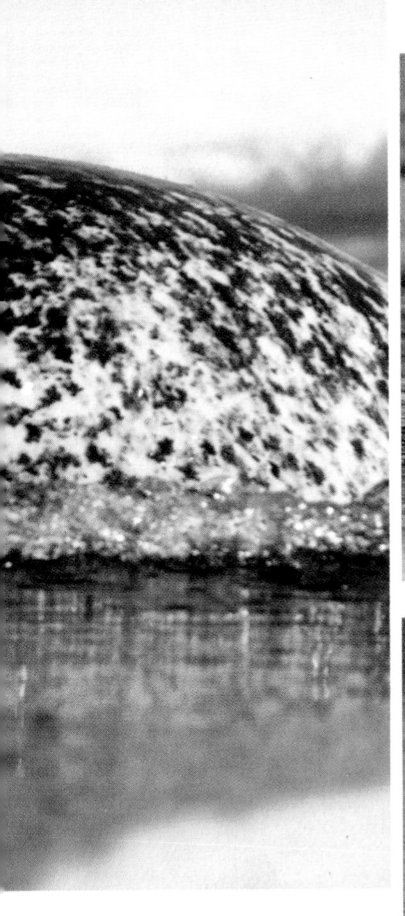

Unlike other whales, the narwhal grows a magnificent, tightly spiraled tusk. The narwhal may use its tusk in courtship displays, as a weapon, or for some other purpose not yet determined.

blance to the legendary but mythical equine that, ironically, it had inspired.

The whale was the narwhal, but so little could be learned of the creature, and so few people had seen it, that the narwhal itself soon became the stuff of fantasy. Over time the human imagination had transformed the timid, retiring sea mammal into a bloodthirsty killer. Even today, in spite of efforts to learn more about the animal—including a recent attempt to radio-tag it—the one true unicorn still remains largely shrouded in mystery.

The narwhal, often called "whale of the ice," follows the ice fields that advance and retreat with the season. Of the world's estimated total population of between 27,000 and 30,000, some live in the Chukchi Sea between Alaska and Siberia. Others are found in the Arctic Ocean and its seas north of Russia, although Russian scientists say the creature's numbers there are small. One population of several thousand lives off the ice-girded east coast of Greenland. By far the largest narwhal concentration is in Davis Strait and Baffin Bay, between Greenland and Arctic Canada.

The narwhals spend winter in polynyas, ice-free regions of the far north kept open year-round by ocean currents and vast masses of upwelling water. As the ice recedes in spring, the narwhals follow it northward. About 19,000 of them turn west into Lancaster Sound and a maze of adjacent straits, channels, inlets, and bays of Canada's Arctic Islands.

Shy and Retiring

Not long ago, researchers led by sea-mammal specialist Wybrand Hoek of the Arctic Biological Station flew over the still-frozen Peel Sound there. They came upon more than 1,000 narwhals in a broad, ink-black gash of open water in the gleaming ice (unlike other whales, narwhals may lie at or near the surface, sometimes for hours). Females and calves gathered in clusters. Nearby, groups of males had assembled, some in rosettes, all tusks pointed toward the center. Far below, other males rolled and jousted in a strange, fluid sea ballet.

A terrible fear of killer whales may push narwhals into the ice. Killer whales, with their high dorsal fins, avoid the frozen floes, but narwhals are designed for ice. They do not have a dorsal fin, only a 2-inch (5-centimeter)-high, 4-foot (1.2-meter)-long dorsal ridge. Every June, at places like Pond Inlet, narwhals come toward land to flee gangs of killer whales, which hunt with relentless speed and military precision.

Until 1577 the narwhal, living in a remote Arctic habitat, was unknown to most Europeans. The tusks, believed to be unicorn horns, commanded a high price for the healing properties associated with them.

In remote ice-spattered places such as Koluktee Bay, the narwhals find relative safety. And there, in July, after a 15-month gestation period, their young are born. Calves are about 5 feet (1.5 meters) long at birth, weigh 180 pounds (80 kilograms), and are a dark, slaty blue.

As they grow older, the belly region becomes lighter: first gray, then cream, and finally gleaming white. In adults, back and sides are streaked, blobbed, and blotted in dark brown or black. The spots fuse in the center of the back so it appears nearly solid black, but the color pales in old animals.

Adult females grow up to 14 feet (4.2 meters) long, and rarely exceed a weight of 2,000 pounds (900 kilograms). Adult males, more massive and missile-shaped, can reach lengths of up to 17 feet (5 meters), and a few weigh in at 4,000 pounds (1,800 kilograms). Their heads are round and domed, their mouths small and prim, and their eyes deep brown and sloe-shaped.

Narwhals feed all summer in high Arctic bays, eating Greenland halibut, Arctic cod, pelagic shrimp, and squid. Come fall, as the temperature plunges and ice hems bays and inlets, the animals migrate toward the open sea. Sometimes the whales linger too long and are trapped by the encroaching ice. "One can hear the sound the unfortunate animals make for miles," said explorer-writer Peter Freuchen, adding that to find a *savssat*, as these whale traps are called, "is the dream of every Eskimo."

Shy and easily frightened, narwhals are not at all the monsters of early reports and perfervid imagination. "The narwhal," wrote 18th-century French naturalist Georges Louis Leclerc, comte de Buffon, "seeks out carnage, attacks without provocation, has no equal in battle, and kills without need." In Jules Verne's *Twenty Thousand Leagues Under the Sea*, the narwhal had become "the most terrible creature God ever created."

That tusk, which is in fact quite brittle, is the narwhal's emblem. When Swedish taxonomist Carolus Linnaeus came to the narwhal in 1758, he named it—not so accurately, it turned out—*Monodon monoceros*, "one tooth, one horn." Yet narwhals each have two teeth, both in the upper jaw.

In the female, these 8-inch (20-centimeter), finger-thick ivory rods remain hidden. However, when a male calf is about one year old, the left tooth pierces the upper lip. Eventually it sprouts into a hollow, tapered, spiraled ivory tusk that can be up to 10 feet (3 meters) long, 8 inches (20 centimeters) around at the base. It looks like a cross between a corkscrew and a jousting lance, and it is unique in nature.

During the 1800s, many narwhals were killed for their blubber, meat, skin, and, of course, tusks. Today, perhaps due to strict hunting regulations, narwhals do not appear to be endangered. Nonetheless, owing to their remote habitat, few Americans have ever seen a narwhal.

Tusk Function

Attempts to fathom the tusk's purpose have produced a plethora of theories and fanciful conjectures, many ignoring the fact that the female gets along fine without one. It has been suggested that the narwhal uses its elongated tooth to poke up flatfish, a favorite food, from the seabed, or as an ice pick to chip breathing holes through overlying ice. Some scientists think the tusk's polished tip acts as a lure for light-sensitive prey, or that males use their tusks like rapiers to duel for a female's favor.

Perhaps the most intriguing notion has been advanced by Canadian bioacoustician Peter Beamish, who suggests that narwhal males use their tusks as sound sabers in bouts of "acoustic jousting." That is, they focus high-intensity sounds through the hollow tusk at an opponent's sensitive ear. In these duels the whale with the longer tusk usually wins, as he is the first to zap and deafen his rival.

None of these theories found universal favor, as most scientists believed the tusk was merely an ornament, a secondary sexual characteristic like a lion's mane or a cock's comb. Recent observations, however, indicate that males use the tusk as a weapon and in dominance displays. They cross horns with an oddly wooden, clacking sound, and sometimes even spear each other. The heads of many narwhal males are scarred, and in one male's upper jaw, researchers found the embedded tip of another whale's tusk.

To some Inuit groups, narwhals were—and still are—immensely important, and not just for the horn. The blubber was rendered into oil that burned with a clear, hot, smokeless flame in the Inuit's stone lamps. The maroon-black meat was food for the people and their sled dogs. "Muktuk," the thick, crunchy skin, was—and is—a delicacy that tastes like fresh hazelnuts and contains more vitamin C per ounce than lemons. The narwhal's sinews were fashioned into an exceptionally strong thread to sew boots, clothing, and kayaks. Its leather—which remains soft and pliable when frozen—was ideal for making harpoon lines and dog-team traces.

But the ivory tusks were especially useful. In a land without trees, they took the place of wood. When British explorer John Ross encountered the polar Inuit in 1818, their sled runners and harpoon shafts were made of narwhal tusks.

Indeed, it was the tusk that for centuries also captured the imagination of the world. When the Vikings settled on Greenland 1,000 years ago, they became avid hunters of the narwhal, whose tusks were sold in Europe as "unicorn" horns. Of course, they didn't let on that the horns were in fact whale teeth. That would have spoiled the myth and lowered prices.

Everyone, it seemed, from kings and queens on down, was willingly, happily, thoroughly duped. Holy Roman Emperor Charles V is said to have settled an enormous debt with the margrave of Bayreuth by giving him two unicorn horns. When Catherine de Médicis, who later became queen, married the dauphin (heir apparent) of France in the mid-16th century, her uncle, Pope Clement VII, reportedly presented her father-in-law with a unicorn's topknot.

Not only did this wonderful horn detect and destroy poison, they thought, it cured everything from ague to plague. (Poison detection may not be as farfetched as it sounds; Russian scientists have analyzed the horn, and attribute its reputed ability to neutralize poison to the presence of calcium salts.) One 15th-century apothecary's advertisement promised that its new supply of unicorn horn would relieve heartburn, cure corns and sore eyes, vanquish epilepsy, and dispel "evil vapours."

As an aphrodisiac, the material reportedly made barren women abundantly fertile, and impotent men magnificently potent. Unicorn horn was sold in the pharmacies of Britain until 1746, and as the wonder-working drug *ikkaku* in Japan until the 1950s. To this day, many pharmacies in Germany are called *die Einhornapotheke*, the unicorn drugstore.

Naturally the horn did not always live up to the grand claims. When 62-year-old Martin Luther was gravely ill in Eisleben, Germany, he took doses of powdered unicorn horn—but died nonetheless.

Even so, faith in the unicorn's horn was nearly universal, and few doubted there was such an animal. The unicorn is mentioned seven times in the Bible, giving it divine recognition. Aristotle and Pliny vouched for its existence, and Caesar was said to have spotted one of the shy, horselike creatures in the Hercynian forest. Leonardo da Vinci even wrote a treatise on the capture of unicorns by using virgins as "bait."

The Age of Exploration revealed the real origin of the unicorn's horn. In 1577 Elizabethan explorer Martin Frobisher sailed into what is now Frobisher Bay on Baffin Island and found "a dead fish floating, which had in its nose a horne streight and torquet, of length two yards lacking two ynches, being broken in the top." A few of his sailors, noticing that the horn was hollow, dropped spiders inside it. Spiders were thought to be poisonous, and unicorn horns destroyed all poisons. So when the spiders died, "we supposed it to be the sea Unicorne."

Though faith in the horned horse gradually receded, the tusk of the sea unicorn retains its widespread appeal. In the late 1800s, a narwhal tusk sold for $50 in New York City. As late as 1903, China was importing huge quantities of tusks to make medicines and poison-absorbent cups.

Prices rose sharply just within the past few decades. In London in 1978, for example, narwhal tusks sold for £500 ($960 U.S. at the time) per meter. That same year an Inuit at Pond Inlet got $5,000 for a twin-tusked skull. In fact, the price of a single tusk at Pond Inlet jumped from $2 per pound in 1960 to $400 by 1972. (A tusk weighs about 20 pounds—9 kilograms.)

After that, prices dropped. In the United States, the Marine Mammal Protection Act, established in 1972, made it a crime to import narwhal ivory. Eleven years later the European Community (EC) imposed a similar ban, causing the price of narwhal tusks in Pond Inlet to slip to $80 per pound.

Canada, under the guidelines of its Narwhal Protection Regulations, established a quota system in 1976 that permits Inuit hunters to take 542 of the animals per year. At first the quota was frequently exceeded. In 1978, for instance, Pond Inlet hunters with a 100-whale quota hauled in 152 narwhals (they probably killed about 350, but the others sank).

Now, however, they usually adhere to the quota. And sometimes, because of bad weather, bad ice, or lack of whales, villages do not even catch the allowable maximum. During 1987, for example, the Pond Inlet hunters caught only 20 of the whales.

Still a Mystery

More worrisome than the number of narwhals legally harvested is the percentage that are killed and never retrieved. Some scientists estimate that for every narwhal shot and recovered, one is killed and lost. Some inexperienced hunters shoot wildly and at too great distances at the whales, wounding many and getting few. In one study, 42 percent of the narwhals examined were bullet-scarred.

For now, narwhals do not appear to be threatened; extensive surveys made in the past two decades do not show a population decline. Some developments, however, appear to threaten the narwhal's future.

Lancaster Sound, summer home of most narwhals, is the entrance to the Northwest Passage, that holy grail of explorers that may soon become a major sea-lane. Narwhals are easily frightened by motor noises, and ship traffic may change or disrupt vital migrations. Furthermore, the value of narwhal tusks has again begun to rise, which probably will lead to increased harvesting. Demand for narwhal tusk has picked up considerably in Japan, where it is still sold as the miracle-working horn of the unicorn.

But the phantomlike creature that has eluded the relentless probing gaze of science for millennia still clings to its anonymity. In the summer of 1988, two Canadian scientists—Michael Kingsley of the Department of Fisheries and Oceans, and Malcolm Ramsay of the University of Saskatchewan—set out to unravel the mystery in Koluktoo Bay, a summer home of the narwhal on northern Baffin Island.

They stretched a great net far out into the bay, then waited. In time, they heard the explosive, squeaking sound of surfacing narwhals, then a deep, melancholy tuba note, drawn out and infinitely sad. A small group of whales swam near, all passing by except one, their captive.

Careful of the animal's powerful, slamming fluke, Kingsley clamped a small, tubular device onto its 4-foot (1.2-meter) tusk. It was the first time a narwhal had ever been fitted with a radio transmitter.

For two days the scientists tracked the creature's movements from a helicopter. Then, suddenly, inexplicably, the signal was gone. The narwhal had disappeared under the icy water, a mystery still.

Narwhals gather in herds of 15 to 20 individuals, often in areas of patchy, broken ice. Young narwhals are a blue gray color. By adulthood, they turn a mottled gray. Old narwhals may be entirely white.

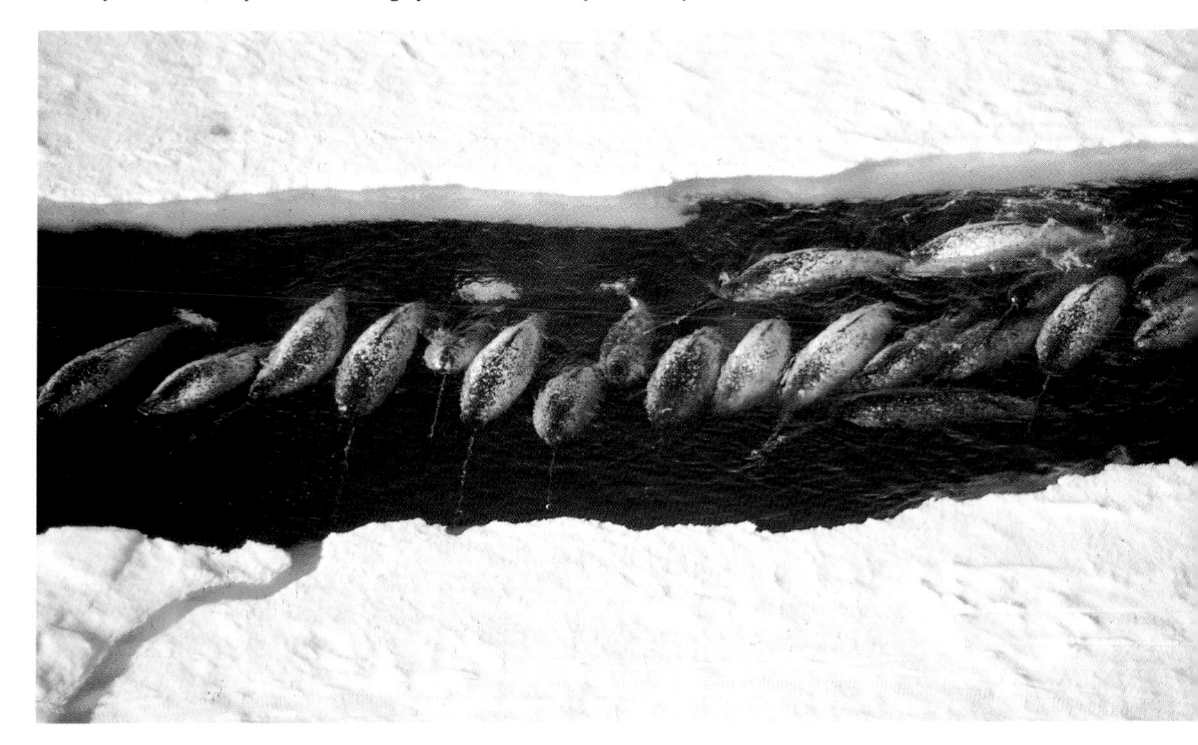

THE WORLD'S BIGGEST FUNGUS

by Natalie Angier

Giant fungi fields have been discovered that experts say may be the largest living organisms on Earth, even dwarfing the blue whale and the towering sequoia.

Scientists have discovered what could be the largest and oldest living organism on Earth, an individual mightier than the blue whale, the giant sequoia tree, or such past pretenders to size supremacy as the dinosaur.

The organism is a giant fungus, an interwoven filigree of mushrooms and rootlike tentacles spawned by a single pair of mushroom spores that mated 1,500 to 10,000 years ago. The fungus giant now extends for more than 37 acres (15 hectares) in the soil of a forest near Crystal Falls, Michigan, along the Wisconsin border.

The fungus, called *Armillaria bulbosa*, has many tiny breaks in it, but it has been found to be genetically uniform from one end of its expanse to the other, which is why scientists say it rightfully deserves to be called a single individual. They suggest it has been growing possibly since the end of the last Ice Age, making it older than any other known organism on Earth. If all its mushrooms and tendrils are considered together, the fungus weighs about 100 tons, about as much as a blue whale.

Dr. Myron L. Smith and Dr. James B. Anderson of the University of Toronto in Mississauga, Ontario, and Dr. Johann N. Bruhn of Michigan Technological University in Houghton reported their discovery of the mammoth *Armillaria* in the journal *Nature.*

"This is a fascinating report," says Dr. Thomas D. Bruns, an assistant professor of plant pathology and a fungal researcher at the University of California at Berkeley. "The catchy part of it is, when you really begin to appreciate how large this thing is, it's mind-boggling. People usually think of a mushroom as a little creature, but most of the action of a fungus is underground."

Armillaria bulbosa is one of about 10 *Armillaria* species in North America. While all these species produce very similar-looking mushrooms, they do not interbreed and they all have very different ecologies. For example, some of these fungi prefer to feed on hardwood while others prefer conifer wood. Some grow faster in warmer temperatures, while others grow better in cooler climates. *Armillaria bulbosa* prefers to feed on hardwood trees and grows better in warmer temperatures.

Uniform Genetics

The organism survives by feeding on deadwood and other detritus, spreading outward right beneath the surface as it senses the presence of nutrients nearby. But scientists believe that the fungus has probably reached its maximum dimensions; at one, and possibly several, of its borders, the *Armillaria* is bumping up against competing fungi, which are blocking the older giant's further colonization of the forest.

Researchers said the finding will force biologists to rethink their assumptions about what constitutes an individual, a fundamental problem in the study of the natural world and its ecosystems. Scientists normally view a single organism as something bound by a type of skin, whether of animal flesh or plant cellulose. But fungi, along with other organisms like coral and some types of grasses, grow as a network of cells and threadlike elements whose boundaries are not always clearly identifiable.

What is more, some of the interconnected elements of the newly discovered *Armillaria* grow independently, thus straining the idea that the entire fungal patch can truly be considered an individual. Nevertheless, biologists said that, given its uniform genetic makeup, the mold merited its ranking as a single giant creature.

"The individual is the basic unit of biology," says Dr. Rytas Vilgalys, an assistant professor of botany at Duke University. "Fungi like *Armillaria* offer us an opportunity for reexamining what the basic unit might be."

Scientists say the new work is particularly significant because it used detailed genetic analysis, similar to the techniques of DNA fingerprinting, to prove that the 37-acre fungus was a discrete being, which had grown over the years by sending out clonal shoots of itself. Other extremely large fungal growths have been identified in the past, but researchers could never be sure that the growths represented individual fungi, rather than populations of smaller molds whose edges had become smeared together.

Larger Growths Are Possible

"We used genetic markers to distinguish between these two possibilities," Dr. Anderson says. "It shocked us to have found such a large fungal entity that is so ancient.

"A lot of people have asked us if this is an April fool's joke," he continues. "I've assured them it is not."

As startled as they were to discover the colossal patch of fuzz, the researchers say their *Armillaria* is probably not the largest fungal clone around.

"It's the most successful one we're aware of, but this is in a mixed forest with many kinds of trees," says Michigan's Dr. Bruhn. "We would think where there was a stand of pure trees like birch or aspen, a sin-

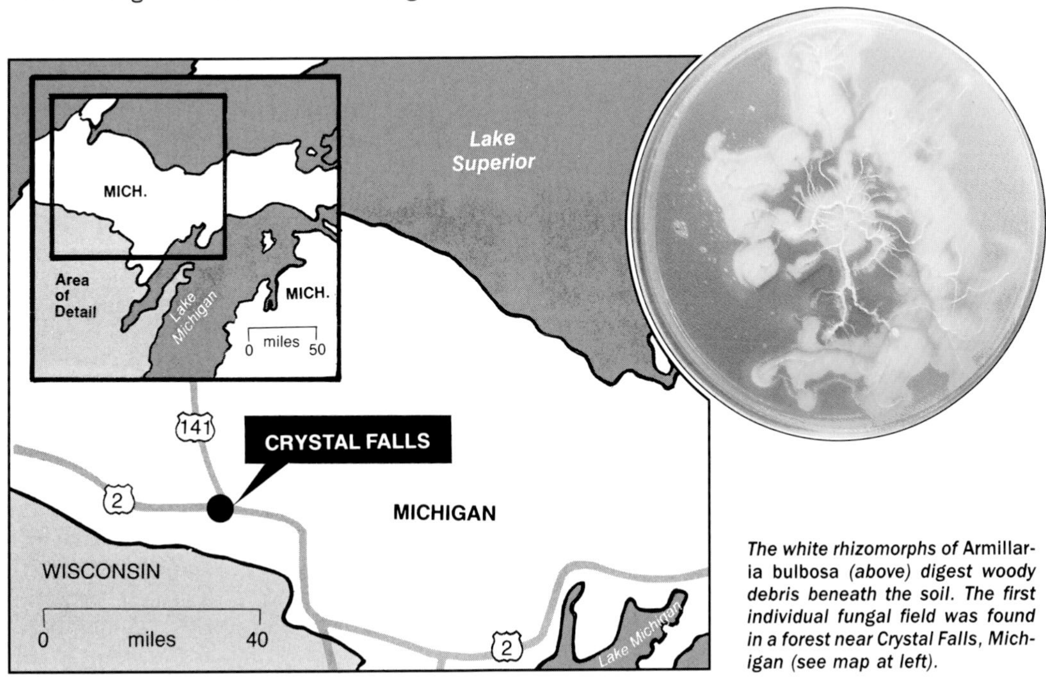

The white rhizomorphs of Armillaria bulbosa *(above) digest woody debris beneath the soil. The first individual fungal field was found in a forest near Crystal Falls, Michigan (see map at left).*

Armillaria bulbosa ***produces edible honey-colored mushrooms above ground or on tree stumps. Each mushroom produces millions of spores, which spread the fungus to new areas.***

gle fungus might be more successful still." In that case a fungus with a taste for a particular type of tree might be able to proliferate especially quickly and over the entire area before encountering any competitors.

Indeed, about one month after the *Nature* article was published describing the Michigan fungus, other scientists revealed an even larger 1,500-acre (607-hectare) fungal covering in Washington State. This fungus, called *Armillaria ostoyae,* grows in a forest of a single species of pine trees, making it particularly easy for the fungus to kill trees and spread quickly.

These new discoveries also underscore the ubiquity and power of the planet's fungi, a kingdom of organisms quite distinct from the plant and animal kingdoms.

Mapping a Discovery

"Fungi are the base of all terrestrial ecosystems," says Berkeley's Dr. Bruns. "No ecosystem on the planet would continue to operate without fungi to decompose and recycle wood and plants." In addition, the decomposition of woody debris reduces the risk of catastrophic forest fires and also adds valuable organic matter and nutrients to the forest soil.

But fungi are not always innocuous; they sometimes attack healthy tissue. A few virulent fungal species, like the Dutch elm pathogen, have managed to devastate entire populations of trees.

The scientists came upon the Michigan megafungus in 1988 while studying tree pathogens for the U.S. Navy. Walking through a forest on the Michigan-Wisconsin border, they collected samples of *Armillaria* mushrooms, familiarly known as button or honey mushrooms, together with the underlying shoestringlike structures called rhizomorphs, which gather nourishment for the fungus. They examined 16 genes from the fungal DNA, and realized that all their samples were the clonal offshoots of a single being.

Over the next several years, they returned to collect ever more *Armillaria* samples. In continuing to push back the borders of that fungal clone, the scientists at last realized that they had something incredibly enormous on their hands.

Through experiments measuring the growth rate of the fungus on wooden stakes, they were able to estimate how long it would have taken the clone to reach its current dimensions.

Fungal Spread

The scientists now believe that at some point in the distant past, a fertilized spore, blown from a parent *Armillaria* mushroom, settled into the soil, germinated, and extended reddish-brown rootlike rhizomorphs downward, seeking wood debris to feed on. Eventually the fungal webbing also began sprouting mushrooms to stretch above ground and disperse new fungal spores to the wind.

THE ENIGMATIC SLIME MOLD

by Sylvia Duran Sharnoff

It had been a wet spring. Now, in the summer of 1973, panic was spreading in the Dallas suburbs, on Long Island, around Boston. Pulsating yellow blobs were crawling across people's lawns and even up onto their porches. The blobs broke apart when they were blasted with water, but the pieces continued to crawl and grow—reinforcing fears that they were indestructible aliens from outer space, or, at the very least, menacing mutant bacteria.

In due time, scientists assured the public that the "unidentified growing objects" were merely a stage in the bizarre life cycle of remarkable but terrestrial creatures called slime molds. Alarming as they can be to those who come across them unexpectedly, slime molds in the crawling stage are so entertaining that some people keep them as pets. In their spore-bearing stage, many slime molds are astonishingly beautiful, overlooked natural treasures.

Slime molds are classified with the most primitive of organisms, largely because their spores germinate into single-celled creatures. The spores develop from often colorful sporophores, which may resemble coral (top) or chocolate balls (left). Threadlike fibers (above) aid spore dispersal.

Slime molds go through a variety of stages during their life cycles. Plasmodia (top left, on leaf) are made up of many cells, all of which divide at the same time. They develop pronounced veins (above) as they creep about, searching for food. Plasmodia transform into sporophores (left) by a mechanism not fully understood by scientists.

Mycological Marvels?

Early botanists first noticed slime molds in the spore-bearing stage and mistakenly believed they were puffballs, those spherical fungi that sometimes release smoky clouds of spores. Hence, slime molds have traditionally been the province of mycologists, people who study mushrooms and other fungi. The experts who identified the mysterious invaders in 1973 were mycologists. While slime molds are studied by mycologists, and while they may resemble fungi during a part of their life, they take on the appearance of other living things during other portions of their life cycle. They sometimes appear as slimy blobs capable of moving as fast as slugs; in an earlier stage, they take the form of microscopic protozoans. Because of this Jekyll-and-Hyde development, slime molds have been a difficult group for scientists to pigeonhole into a specific taxonomic category.

All living things used to be assigned to either the plant or the animal kingdom. Most single-celled organisms that could move and had no chlorophyll were called protozoans and were considered animals. The classic protozoan was the amoeba, which moves—and engulfs its food—by constantly changing its shape. With life on Earth now divided into five kingdoms, controversy lingers about where certain organisms belong. Kingdoms Animalia and Plantae have lost numerous members at the simpler end of the scale. Fungi have their own kingdom. Bacteria, along with what used to be known as blue-green algae, are in the kingdom Monera.

The final kingdom, Protoctista, comprises all the ancient protozoans, true algae, and some simple organisms that used to be thought of as fungi. Slime molds are included: their spores germinate into single-celled creatures. Classifying slime molds under Pro-

The sporophores, or fruiting bodies, are by far the most striking feature of a mature slime mold. Most fruiting bodies, such as the immature ones above and right, develop spores internally. A few (above right) bear the spores externally. After the spores are dispersed, the remaining skeletal structure decays.

toctista is also appropriate for the crawling stage, but the fruiting stage still seems more like a fungus than anything else.

Slime-Mold Life Cycle

The life cycle of slime molds reads like the sort of complex flowchart that computer people use. Under the right conditions of moisture and temperature, a spore germinates and turns into either of two kinds of cells. If the surface on which the spore germinates is simply damp, such as on wet vegetation, in soil or humus, or inside rotting logs, the spore develops into a myxamoeba, which, like its namesake, can creep about its habitat. If the spore germinates in water, it develops into a swarm cell, which uses its whiplike flagella to propel it about. What is particularly intriguing about this stage of a slime mold's life is that the two types of cells are interconvertible: if its environment is inundated with water during a rainstorm, for example, a myxamoeba will develop flagella and transform into a swarm cell within a matter of minutes. On the other hand, when its watery habitat begins to dry up, a swarm cell will absorb its flagella and change into a myxamoeba. Several such transformations may occur repeatedly during a slime mold's development.

Myxamoebas and swarm cells eat by engulfing such things as decaying organic matter, bacteria, and other microorganisms. If conditions become too unfavorable, the myxamoeba can secrete resistant cell walls and survive for long periods as a round microcyst, ready to emerge again when the environment becomes hospitable. When environmental and nutrient conditions are just right, myxamoebas and swarm cells "mate."

Sexual reproduction in slime molds is a unique variation on the sperm-meets-egg theme. Myxamoebas and swarm cells both come in different mating types—genetic variations that can be thought of, loosely, as different genders. One species of slime mold is known to have at least 18 mating types. Two myxamoebas or two swarm cells of different mating types may fuse sexually and begin to develop into the amazing creature called the plasmodium.

Plasmodia, the "blobs from outer space" that so alarmed those homeowners, are typically fan-shaped, with a delicate, open network of veins and a ruffled leading edge. In some species, they stay microscopic and colorless. Others are large and colorful: violet, pink, or red, but most come in shades of yellow. A few species occasionally grow very large, more than 3 feet (1 meter) across. One mentioned in the scientific literature covered an entire decaying log that was 1 yard (0.9 meter) in diameter and 10 yards (9 meters) long.

Even though it appears to be a single cell, like the myxamoebas or swarm cells that give rise to it, a plasmodium has multiple nuclei. All the nuclei, which can number in the millions or even billions, divide simultaneously. This rare trait, as well as the various other unique characteristics of slime molds, make them extraordinarily interesting to cell biologists.

Plasmodia creep about, eating bacteria, fungi, protozoans, and algae, and sometimes cannibalizing sibling myxamoebas. They generally live under leaf litter on the forest floor or inside rotting logs, where they can squeeze through narrow spaces in search of food.

Like myxamoebas, plasmodia can develop into hard, resistant resting states, called sclerotia, when conditions are bad. Some remain viable for at least seven years. In cold climates, many slime molds overwinter as sclerotia and come to life again in warmer weather.

Scientists who study slime molds do not entirely understand how food supply, moisture, and temperature interact to induce plasmodia to transform themselves into the dry fruiting bodies, the sporophores, that are the final stage of their life cycle. It is these tiny spore-bearing structures—the ones best seen with a magnifying glass—that are spectacularly beautiful. They are prizes worth searching for, no matter how silly you might feel inspecting a rotting log, scrutinizing fallen leaves, or carefully examining all the dead twigs around the melting edges of a mountain snowbank. In the desert, look for sporophores on the woody base of cactus plants, on decaying cacti, or on animal dung. Your reward may be a 3-inch (7.5-centimeter) cluster of iridescent purple balls, each smaller than the head of a pin, or a patch of diminutive red cotton candy on tiny stalks. Other fruiting bodies resemble miniature leather buttons or golf balls for lilliputians.

Slimy Pets?

Slime-mold spores from stationary, seemingly lifeless, fruiting bodies, blown aloft on the wind, washed along by rain, or carried about by the insects and mites that eat them, begin the next life cycle when they germinate as mobile cells.

Aside from the occasional frightening encounter with large plasmodia, most people have little knowing contact with slime molds: the exceptions are the scientists who study them, and the aficionados who keep them as pets. In the early 1900s, an eccentric Japanese folklorist and biological collector named Kumagusu Minakata kept slime molds in his garden. At first he reportedly had to get up every two hours during the night to protect his pets from garden slugs, but then he trained a succession of cats to drive away the vermin.

Forty years ago a woman named Ruth Nauss kept a "garden" of slime molds in jars and dishes in the living room of her home. She fed them ground oatmeal flakes, their usual diet in laboratories. When she went on vacations, she took the most delicate along with her, tucking them in with a "warm water bottle" on cold nights. She withheld moisture from the hardier ones for several weeks before a vacation in order to induce them to harden into sclerotia, so she could leave them unattended. At the time she wrote about them, her oldest plasmodium had been crawling around in its dish for more than nine years.

A scientist who studied slime molds wrote in 1892: "Personally, it is not a matter of prime importance whether it be eventually shown that I have been a botanist or a zoologist." A hundred years later, their place in the great chain of life is still imperfectly resolved.

WILD HORSE REFUGE

by Bil Gilbert

Our dealings with the species have been complicated, practically and philosophically, by the question, "Is *wild horse* a zoological oxymoron?" In a generic sense, few things are wilder than a free-ranging mustang. But the prevailing scientific opinion has it that truly native horses have not existed on the North American continent for eons, and the progenitors of what we now call wild horses were domesticated animals that arrived on this continent only 500 or so years ago in Spanish galleons. Therefore, opinions vary about whether they should be treated as good native American beasts or as strayed domestics—trashy interlopers comparable to feral pigs, dogs, cats, and pigeons.

Because of this confusion, wild horses were traditionally managed as the people who lived nearest to them saw fit. On the grounds that they ate food that would otherwise support cattle, sheep, and trophy game, wild horses were frequently killed as varmints or driven into poor forage areas. On the other hand, the wild herds provided free-for-the-taking remounts to anyone who could catch and break them.

But as the livestock industry and military were mechanized, and rangeland became much sought after by ranchers and developers, wild horses became valuable principally as pet food on the hoof. In the period following World War II, the methods of commercial horse hunters who supplied the pet-food market began to outrage people concerned with the humane treatment of animals. The animals became so besieged that

The wild horses that roam the western states are descended from once-tame ancestors who arrived with the Spanish in the New World. Laws protecting these animals have led to a dramatic increase in their population.

in 1971 Congress passed the Wild Free-Roaming Horse and Burro Act to protect these creatures. The legislation prohibited private parties from killing, catching, or otherwise molesting these animals.

Burgeoning Population

Since then the U.S. population of wild horses, which have no significant predators other than humans, has increased from an estimated 17,000 to about 58,000. Most of the horses are found in or immediately to the north and west of the Great Basin country, with the largest number, upwards of 30,000 head, living in Nevada, and another 4,000 or so in Wyoming.

Because most of the wild horses currently inhabit tracts under the administrative jurisdiction of the U.S. Department of the Interior's Bureau of Land Management (BLM), that federal agency has become largely responsible for them. As far as the BLM is concerned, the animals have prospered a bit too much since the passage of the Wild Free-Roaming Horse and Burro Act. The BLM is required by law not only to protect wild horses, but also to maintain an ecological balance in the areas where livestock and wild horses and burros share the land. As the horses—hardy, reproductively vigorous creatures—multiplied, it became increasingly difficult to maintain the balance of various eaters and owners of grasslands.

It was a relatively simple undertaking for BLM wranglers to round up and remove horses from areas where they had become too numerous and were overgrazing the land, but what to do with the animals then was not so simple a proposition. The number of places where wild horses can live or, more important politically, where people want them to live is very limited.

Caught between a biological rock and a legal hard place, the BLM came up with the idea of selling surplus horses (currently for $125 each) to anyone who wanted to keep and cherish a mustang in circumstances that the agency certified as humane and healthful. This became widely known as the Adopt-A-Horse program, and it has been very successful. At the rate of 4,500 to 5,000 horses a year, the BLM has found homes, many in the eastern part of the country, for more than 100,000 formerly wild mustangs.

Unfortunately, there is a glitch in this system. Among the horses rounded up each year, there are always several hundred that nobody wants to adopt because they are too

To remedy the overpopulation problem, the government initiated the Adopt-A-Horse program. Helicopters are used to help herd the horses into holding pens prior to adoption.

People seeking to adopt a horse tend to choose the more striking specimens. Those horses considered ugly or infirm are usually passed up. The horse below was barely healthy enough to survive the trip to its holding pen.

old, too infirm, too ugly, or too pugnacious. Legally obliged to keep these rejects, the BLM contracted with private feedlot operators to hold the animals in cattle pens. This proved to be expensive—an estimated $14 million a year. There were several instances of maltreatment of horses that enraged animal-rights groups, but just the thought of incarcerating born-free animals in small corrals was enough to cause concern.

Emanating Energies

Until two and a half years ago, the BLM was continually whipsawed by the costs of feedlot maintenance on the one hand and by charges of inhumanity on the other. Then, like a deus ex machina—at least from the standpoint of the beset agency—a man named Dayton Hyde appeared with what he said was a cheap, attractive solution for the unwanted horses.

Hyde, 66, stands 6 feet 5 inches (195 centimeters) and is built along the lines of a retired NFL tight end. Still, his interests and ideas are notably larger than his stature. As a boy on the Upper Peninsula of Michigan, he roamed the bush, responding, he says, to the energies emanating from rocks and trees, communicating with animals, and being generally preoccupied with the intricate relationships that connect animate and inanimate natural phenomena. When he was 11, he went to live with and work for an uncle who owned a cattle ranch in central Oregon. In 1958, when he became the owner of the property, Hyde devised a profitable system for the integrated management of the people, livestock, water, soils, and wild flora and fauna that were on his own 5,000-acre (2,025-hectare) property, as well as for the management of another 85,000 acres (34,400 hectares) he leased. Not given to mealymouthed modesty, Hyde says the spread is now "one of the world's great ranches, economically and ecologically."

While setting up the Oregon place, Hyde—who has a degree in English with a minor in zoology from the University of California—traveled extensively, more or less in the role of a free-lance writer and endangered-species expert. In 1974, while surveying whooping-crane breeding grounds in the wilderness south of Great Slave Lake in Canada's Northwest Territories, he was lost for three weeks. Without food supplies or gear, he roamed the taiga and muskeg happily, he recalls, eating snails and wrapping himself in sphagnum moss for protection against biting insects. Over 21 years, he has written a dozen books: novels, nonfiction, and children's books. In his spare time, he was active in various conservation organizations, and he served as a member of the board of the Defenders of Wildlife from 1976 to 1984.

Dayton Hyde arranged the acquisition of huge tracts of land to be used as sanctuaries for the wild horses that were once confined to holding pens.

In Oregon, Hyde had learned to ride and to work cattle with the mustangs he caught and broke. Later, when he sometimes found starving horses on poor public lands, he moved them to better pastures on his ranch. By the mid-1980s, he was speaking before various organizations, trying to prompt action on the wild-horse situation. But his interest in these animals was fairly casual until the spring of 1988. Then, while attending a conservation meeting, he was asked about the BLM horses being held in feedlots.

"I hadn't thought much about it before," says Hyde. "But some of my best ideas have come on the spur of the moment. As we were talking, I could see how this could be handled—that it was something worth doing. I said what was obviously needed was a private sanctuary. The animals and the public would do better if somebody other than the feds managed them in feedlots."

Hyde is a ferocious environmental libertarian. He is adamantly convinced that enlightened private-land owners are most likely better stewards of natural resources than are government bureaucrats, who are mainly responsive only to political, not ecological, realities. With that philosophy driving him, over the next two years Hyde accomplished some rather remarkable feats, including the following:

• He organized the Institute of Range and the American Mustang (IRAM).
• While making aerial surveys of several western states, he came across two land parcels suitable for mustang refuges in South Dakota. One was a 14,000-acre (5,670-hectare) site in the Black Hills west of Hot Springs. The other was a 35,000-acre (14,165-hectare) site about 150 miles (240 kilometers) to the east of the Hot Springs site.
• He raised money to get the 14,000-acre site, and then persuaded a group of private investors to acquire the other property for about $1.75 million and allow IRAM to use it.
• Hyde spent considerable time and energy in Washington, D.C., seeking congressional support for the wild-horse project.
• He reached an agreement with the BLM under which 2,000 wild horses would be placed on the two South Dakota sanctuaries. Hyde agreed to maintain the horses—providing supplemental feed and veterinarian care as necessary—and give BLM inspectors free access to the land and animals. In return, the feds promised to pay IRAM $1.00 a day for each of their horses on the sanctuary (now

Wild horses also thrive in areas of the United States outside the west. The wild horses that live on Cumberland Island, off the coast of Georgia, have become something of a tourist attraction.

raised to $1.34). Previously, feedlot operators received as much as $2.64 per day per head.

• He agreed that the nonprofit program would be self-sufficient by August 1991. That means he needs an endowment of about $6 million to cover operating expenses and interest and mortgage obligations. To that end, Hyde opened the 14,000-acre Hot Springs site to tourists in 1990, charging from $7.50 to $25.00 per person for specially guided tours.

In the fall of 1989, Hyde moved into a prefab cabin on the sanctuary, the equivalent of a cowboy line camp, to oversee refencing and to otherwise get the place in order. The Oregon ranch is now operated by his wife, Gerda, and one of his three sons.

"A neighbor of mine back in Oregon suggested that what I was doing—coming out here to take care of a bunch of decrepit horses—did not appear to be the action of a rational man," says Hyde. "Maybe so, but I believe that if your comforts and conveniences become so important that you can't or won't act on your passions, then it's probably time to pack it in. I hope I will if it comes to that."

The Adopt-A-Horse program has placed some 100,000 wild horses, many of them with families in the eastern part of the United States.

Physical and Psychic States

When the BLM horses began arriving from feedlots in 1989, many were in only fair condition or worse. Still, their psychic state was a more immediate concern to Hyde. He was particularly worried that if the horses were suddenly turned loose on the land, the confused, panicky animals, those that had spent months in cattle pens, would stampede and injure themselves on fences. "They only knew closed-in pens. They didn't understand our kind of fencing," says Hyde.

Therefore, horses were first released into small pastures, where Hyde and a group of volunteers could patrol the fences. Generally the mere presence of humans was enough to upset the horses, who, with good reason, probably have a low opinion of people. Gradually, more gates and pastures were opened to the horses. After about four weeks of acclimation, a horse is given more or less free access. Now there are about 1,800 mustangs on the two sites.

As they became more composed, the refugee mustangs, which had originally been taken on widely separated ranges, began to organize themselves into bands of 10 to 14 animals, coalescing, as they do in the wild, around dominant mares. Because the sanctuaries are meant to be a kind of limited-occupancy retirement community rather than a breeding farm, all stallions are gelded before arriving. The success of the IRAM model has encouraged the BLM to give its support to a second nonprofit, private wild-horse refuge in Oklahoma, which opened in September of 1989.

But in the past few months, things have soured for IRAM. The BLM says that Hyde has not raised sufficient funds to fulfill the requirements of his charter. When his original agreement expires, BLM officials say they will no longer support his operation. Hyde says that he has received assurances that a compromise will be negotiated, and that at least the smaller Black Hills sanctuary may continue to operate. In the meantime, there is no way to determine what the horses think.

"They Look Happy"

Slipping and sliding across the rough, roadless pastures in a battered pickup on a winter day, Dayton Hyde has no need to mention the obvious—that this is The World's Great Wild-Horse Sanctuary. But a visitor riding with him remarks that it is not only that, but also one of the Greater Places he has seen.

Basically the land is an undulating plateau above which rise sharp, tortured buttes. The flats are scored with narrow, serpentine ravines and canyons gouged out by spring-fed streams. Dark, dense stands of pine grow on the slopes of the canyons and buttes. There are signs of deer, turkeys, coyotes, and cougars. Above, against the slate-gray sky, hawks and eagles soar. Without horses, it is an uncommonly beautiful and interesting place. But when 50 mustangs—blacks, whites, grays, pintos, buckskins, and sorrels—appear, loping out of a ravine, long manes flying, condensed breath rising like smoke, the emotional impact of the scene is powerfully enhanced.

"They look happy," Hyde says, and that would seem to be accurate. "Sometimes out here, I imagine that I am running with them, feeling what they feel."

Unfortunately, the entrepreneurial activities on which the sanctuary and horses are dependent leave Hyde little time to spend alone on the land. His immediate concern is persuading the BLM to extend his operations, and he is always scrapping to raise funds to cover daily expenses (he says that the money the BLM pays is about $40,000 a year less than it costs to operate the sanctuary). He spends about three weeks each month traveling, talking to potential corporate, foundation, and private donors.

Because of the advanced age and infirmities of some of the horses, both the BLM and Hyde had estimated that by this spring, 150 to 200 of the horses would have died. However, they have thrived so well that fewer than 60 have died or were put down.

On this day, Hyde is looking for the oldest of the herd, a 35-year-old mare. By and by, her band appears and trots toward the truck, having learned that Hyde sometimes dispenses horse treats—handfuls of sweetened grain mash—on his visits. The mare comes up late, slow and a bit stiff, but apparently fit.

"In a way," Hyde remarks, "this is why I'm spending so much of the life left to me on planes and in airports, how I justify it. If I don't do this, there isn't any other place where she and the rest of them can live in a dignified way until they have lived long enough."

SAVING HAWAII'S BOTANICAL HERITAGE

by Jenny Tesar

Of the nearly 1,000 species of plants unique to the Hawaiian Islands, 63 species are endangered, and hundreds more stand threatened. In the race to save these plants, officials have set up botanical preserves (see map); some botanists have even taken to hand-pollinating rare species (left).

Suspended from a rope high above the Pacific Ocean, Ken Wood braces his feet against a sheer cliff. Nearby, also clinging to one of the world's tallest sea cliffs, is Steve Perlman. The men are not climbing rocks for fun and adventure. Rather, they are hoping to rescue a species from extinction. Using fine paintbrushes, the two botanists carefully place pollen in the flowers of a plant called *Brighamia*. This will result in reproduction and the production of seeds.

Once upon a time, an animal—perhaps a small moth—pollinated *Brighamia*. *Brighamia* has tubular flowers up to six inches (15 centimeters) deep. An animal that ventured deep into a flower in search of nectar would become coated with pollen. Later, when the animal crawled into another flower, some of

the pollen on its body would adhere to the flower's sticky reproductive organ.

But the unknown pollinator of *Brighamia* is apparently extinct. Without this natural pollinator, *Brighamia* also is threatened. Only about 120 plants remain in the wild. Wood and Perlman are trying to save these plants and the seedlings they produce. But aligned against the botanists—and against *Brighamia*—are some formidable opponents. Wild goats eat any young *Brighamia* plants they can reach. Aggressive weeds crowd out young plants in areas that are inaccessible to the goats.

Evolving in Isolation

Two species of *Brighamia*, one native to the island of Molokai, the other native to Kauai, are among the 63 Hawaiian plants listed as endangered by the U.S. Fish and Wildlife Service. Another 126 species of Hawaiian plants are scheduled to be listed. But getting on the federal list of endangered species is a slow process. Many botanists say the federal list does not reflect the true danger faced by Hawaiian plants. They believe that at least 30 percent of the approximately 1,000 known species of native Hawaiian plants are endangered. Because of the unique history of

The Loulu palm (left) is one of many palm species that grows nowhere else but in the Hawaii chain. Ohai plants differ from island to island; most are shrublike (above). The Hawaiian broadbean (below) was the first Hawaiian plant to gain endangered status.

the Hawaiian Islands, these plants are not native to any other place on Earth.

The Hawaiian Islands are located in the middle of the Pacific Ocean, some 2,000 miles (3,200 kilometers) or more from other land areas. Formed entirely by volcanic action, the islands became populated by plants thanks to water, wind, and birds, which brought seeds from distant places. As a result of the islands' isolation, the plants that survived and reproduced evolved along different lines than did their relatives elsewhere. Today similarities between related species are often very difficult to detect,

Only a few specimens of the once-common Hawaiian gardenia (left) have survived the invasion of alien plants and animals. A special botanical-garden cultivation program is all that stands between the Hawaiian yellow hibiscus (below) and extinction.

although modern genetic techniques applied to plant species are revealing some surprising evolutionary histories.

For instance, among the islands' most famous plants are the silverswords. The best-known species is characterized by a globular mass of silvery, bayonet-type leaves, which grows on the arid, alpine slopes of Maui's Haleakala Crater. There are some two dozen additional species, which have evolved into a wide variety of different forms and structures to enable them to survive in dramatically different ecosystems. Some silversword species form matlike clumps; others are shrubs or even trees; one is a vine. Some live in arid lava fields, others in dry forests, still others in rain forests. Amazingly, all this diversity is believed to have evolved from a single seed that reached the Hawaiian shore from California. Genetic studies indicate that silverswords probably evolved from an ancient species of tarweed. Anyone seeing a scrubby, insignificant cluster of tarweed in California's Sierra Nevada would probably never suspect that the plant is a "cousin" of Hawaii's elegant silversword.

The Arrival of Aliens

Ever since their arrival on the islands, people have changed natural habitats to suit their needs. They have cleared forests for farms and ranches, cut down trees to export the lumber, turned coastal areas and lowlands into towns and cities, and destroyed coastal sand dunes to build resorts. All these activities have depleted populations of native plants. Equally harmful, however, has been the impact of the organisms that people brought with them.

The first people arrived in Hawaii about 1,600 years ago. They were Polynesians, who brought with them the plants and animals they needed for survival. They introduced food crops such as taro, bananas, and sugarcane. They introduced the first mammals to the islands: pigs, dogs, and, inadvertently, rats. "With the introduction of mammals, many plants—as well as birds and other creatures—have had a hard time trying to survive," says David Orr, a research assistant at Waimea Arboretum and Botanical Garden.

The native plants had no defense mechanisms, such as thorns or poisons, to protect themselves against mammals. Rats soon were causing considerable damage, and they remain a destructive force. Rats eat large numbers of birds and insects, many of which are plant pollinators. They strip bark from young koa trees, inhibiting or stopping plant growth. They eat seeds, therefore preventing reproduction; loulu palms are endangered in part because of rats' love of seeds.

In 1778 Europeans made their discovery of Hawaii. They, too, brought mammals: cattle, goats, sheep, and big European boars. These animals were brought so that there would be a source of fresh meat for future visits. But as the animals multiplied, many of them became feral (wild). Their numbers grew dramatically, and their foraging and trampling resulted not only in a decline of native vegetation in many areas, but also in severe erosion of watersheds.

Feral animals continue to degrade habitats. Wild pigs and goats are particular problems. Pigs trample through rain forests, destroying seeds and seedlings of koa, tree ferns, and other plants. Foraging by goats threatens silverswords and the Hawaiian jack bean, among other botanical wonders.

People have accidentally introduced harmful insects to the islands, too. An introduced twig borer is partially responsible for the rarity of the Hawaiian gardenia. An introduced Argentine ant destroyed the wasps that pollinated silverswords.

Plants that are welcome in their native lands have become pests in Hawaii. An introduced *Passiflora* vine is spreading rapidly through the forests of Kauai, strangling and smothering native plants. Kikuyu grass, a native of Africa, is invading areas adjacent to pastureland, forming such dense growth that native seeds cannot germinate.

The end of the alien invasion is not in sight. "It's estimated that a dozen different types of plants and insects come into the islands every week," says Orr.

Protecting and Preserving

Hoping that it isn't too late, a highly dedicated group of people are using various tactics in an effort to save Hawaii's native plants. Some people are focusing their efforts on saving natural habitats. Preserving entire habitats, with their complex diversity of species, helps ensure that all those species have a chance to survive—and helps ensure that one extinction does not lead to a whole series of extinctions.

Some of Hawaii's most spectacular habitats are protected by the National Park Service. Haleakala National Park encompasses 28,665 acres (11,600 hectares). Much of the park has been designated as wilderness under the U.S. Wilderness Act, giving added protection to one of Hawaii's largest tracts of relatively undisturbed native vegetation.

The Nature Conservancy, a private conservation organization, protects more than 48,000 acres (19,425 hectares) of Hawaiian habitats, including Pelekunu Preserve on Molokai, where Wood and Perlman are helping some of the last *Brighamia* in their struggle for survival. The state's Natural Area Reserves System protects 19 areas covering over 109,000 acres (44,100 hectares). Among these habitats is Kaena Point Natural Area Preserve on Oahu, one of the few coastal areas in Hawaii where the rare ohai, *Sesbania tomentosa*, still grows. For many years the windswept dunes of Kaena Point were favorite playgrounds for people with off-road vehicles. Driving the vehicles down dunes and across flat expanses of sand gouged the fragile habitat, killing ohai and other vegetation in the area. Since the late 1980s, Kaena Point has been off-limits to vehicles. As a result, the population of ohai, a low-growing shrub with silvery leaves and red flowers, is slowly increasing.

Another tactic in the battle to save endangered plants is to grow them in botanical gardens. The National Tropical Botanical Garden on Kauai maintains a repository of hundreds of rare species of Hawaiian plants, either as seeds or as live specimens growing on the grounds or in greenhouses. Among the plants being propagated there are *Brighamias*; they are being raised from seeds harvested from some of the plants hand-pollinated by Wood and Perlman.

In a valley on Oahu's north shore is Waimea Arboretum and Botanical Garden, part of a privately owned 1,800-acre (730-hectare) facility founded in 1973. Waimea has 36 major botanical collections, including bromeliads, bamboos, cannas, gingers, and heliconias from tropical regions around the world. Among the endangered Hawaiian species growing at Waimea is *Hibiscus brackenridgei*. This lovely shrub with large yellow flowers is the state flower of Hawaii. Also growing at Waimea is the Hawaiian gardenia, *Gardenia brighamii*, a shrub that once grew on most of the Hawaiian islands, but which today is almost extinct in the wild.

"Our ultimate goal, like that of other botanical gardens trying to save endangered species, is to return the species to the wild, where they might flourish on their own," says Orr. He also points out that many native plants do well in people's homes and gardens: "Instead of raising imported plants, we try to encourage people to grow native plants."

Of course, botanists do not want people to dig up plants from natural habitats. But many garden shops sell native plants that have been raised in gardens and greenhouses. And several botanical gardens, including the National Tropical Botanical Garden, occasionally have native-plant give-aways.

The Aim Is Delisting

Once a plant is listed by the federal government as endangered, the Fish and Wildlife Service is required to develop a recovery plan for the species. "Such a plan lists all the steps believed to be needed to enable the species to recover," explains Derral R. Herbst, a botanist in the Fish and Wildlife Service's Honolulu office. One of the first such plans is for *Gouania hillebrandii*, a shrub with small white flowers, now found only on Maui. Threats to the remaining plants include browsing by domestic livestock, insect infestation, fire, and competition from alien plants. So far, attempts to grow *G. hillebrandii* from seed have not succeeded. However, transplanted seedlings do grow well in cultivation and produce seed.

One of the first steps being taken under the recovery plan is to protect *G. hillebrandii*'s two known habitats. Permits for cattle grazing will be revoked, and fences will be constructed and maintained to keep out feral animals. After the areas are secured, they will be weeded periodically to keep out alien plants. Infestations of hibiscus snow scale,

On Borrowed Roots

Cooke's kokia, *Kokia cookei*, is a small tree with large bright-red flowers that may well be the rarest plant in the Hawaiian Islands. A member of the hibiscus family, it once lived in the forests on Molokai. But cattle, goats, and other grazing animals found Cooke's kokia very tasty. By the 1970s only one tree was known to remain. Keith Wooliams, director of the Waimea Arboretum and Botanical Garden, collected seeds from the tree, but they didn't germinate. In 1974 he took cuttings from the tree and grafted them onto a close relative, *Kokia kauaiensis*. The cuttings grew, eventually providing material for new cuttings and new grafts.

Without Wooliams's grafting efforts, Cooke's kokia would now be extinct. That one tree that was left living in the wild was destroyed by fire in 1975. The only existing members of the species are those that have been grafted onto *K. kauaiensis*.

In 1991 Cooke's kokia returned to the wild. Some of Waimea's grafted trees were planted on former kokia land on Molokai, in a fenced area where they will be safe from grazing animals. It is hoped that the trees will produce viable seeds that will develop into healthy new trees.

Unfortunately, because they all are descended from that one tree, today's Cooke's kokias are genetically identical. This lack of genetic diversity within the species limits the ability of the species to combat disease and to adapt to a changing environment. Thus, it is likely that the species' survival will long depend on help from people like Wooliams.

an alien insect, will be monitored, and pesticides will be applied if necessary.

Studies will be conducted to learn more about *G. hillebrandii*'s life history, its reproductive patterns, and its environmental needs. Once scientists better understand its needs, they will try to locate new sites into which the species could be introduced. Then they will have to determine how the introduction site should be prepared, how the plants should be transplanted, and how they should be cared for once they are planted.

"Another important goal of the recovery plan is to locate any additional populations that may exist elsewhere in the species' historical range," says Herbst. "If such populations are found, the recovery plan will be amended to include measures for protecting them."

Pulling a species back from the brink of extinction does not come cheaply, however. The recovery plan for *G. hillebrandii* covers a 15-year period, from 1990 through 2004, at which time it is hoped that the species will have recovered sufficiently to remove it from the endangered list. The estimated cost of the plan: $306,000. But there is no doubt in Herbst's mind that saving this—or any other endangered species—is worth the cost. "It's absolutely critical that we maintain the biological diversity of our planet," he says. "Not doing so means we lose many possible but still-unknown economic and medical benefits. But more importantly, I believe, once we've lost diversity, we've lost what's interesting in the world."

Botanists theorize that the various species of the now-rare silversword (above) evolved from a single seed.

Pleasant Surprises

The hidden-petaled abutilon, *Abutilon eremitopetalum*, was believed to be extinct; no one had seen one of these shrubs for 40 years. Then, in 1987, some hidden-petaled abutilons were found in a remote gulch on Lanai. Another plant believed to be extinct was *Isodendrion pyrifolium*, a member of the violet family. It was last collected in 1870. But in 1991 Kenneth Nagata of the University of Hawaii's Harold L. Lyon Arboretum found some specimens on the island of Hawaii.

Finding a species believed to be extinct is exciting—it's as if we've been given a second chance to save our fellow species rather than heedlessly destroying them. But second chances are very, very rare; almost all the Hawaiian species that have become extinct as a result of people's careless destruction of natural habitats will never be seen again. And unless there is much greater support for the efforts of today's conservationists and botanists, many of the species that are now hanging on to existence by the barest of threads will soon join those extinct species.

PRIMATES OF THE SEA

by Alex Kerstitch

> *The tiger can only devour you. The devil fish—oh, horror!—sucks you in. It draws you to it and into it, and bound, glued and powerless, you feel yourself slowly absorbed into the frightful sack that is the monster itself. Beyond the terror of being eaten alive is the inexpressible terror of being imbibed alive.*
> **—Victor Hugo,** *Toilers of the Sea*

The "devil fish" Victor Hugo was writing about—presumably not from personal experience—was the octopus, and Hugo nicely summed up the fear and loathing humans have always felt for the creature. No animal more terrified ancient sailors, or, for that matter, 1950s moviegoers—in *20,000 Leagues Under the Sea,* the octopus played the part of the psychotic, serial-killing invertebrate.

Any marine biologist, though, will tell you that these are insults to the noble mollusk. The real octopus is far from the sea vampire of fiction; in fact, it is a reclusive creature that prefers to hide in a rocky crevice or an empty shell. It will certainly never attack a diver or swimmer, even if provoked. As diving expert Max Gene Nohl once wrote, "The chance of a diver being attacked by an octopus is as remote as the possibility of a hunter in the woods being attacked by a rabbit." There's no clearer and sadder proof of this than the behavior of the giant octopuses that live in Puget Sound off the coast of Seattle, Washington: when divers try to wrestle with the supposed beasts, the 20-foot-long monsters timidly retreat.

But while octopuses are shy, they're certainly not stupid. They have highly developed brains, and are by far the most intelligent of all invertebrates. Indeed, their capacity for learning and problem solving has earned them the sobriquet "primates of the sea."

In a classic experiment, researchers presented an octopus with a lobster sealed in-

Octopuses tend to inspire both fear and disgust. Scientific experiments have revealed, however, that these animals possess a high intelligence worthy of respect.

side a corked bottle. Without hesitation, the octopus wrapped its arms around the bottle, then clamped an arm on the cork. Within minutes, it had removed the stopper and slid inside the bottle, engulfing its prey. (Actually, the ability to squeeze through the bottle's small opening is a feat in itself. Since it lacks a skeleton, an octopus can slip its entire body through an opening the size of one of its smaller arms. A 3-foot [1-meter] octopus can crawl through a hole less than 1 inch [2.54 centimeters] in diameter.)

An octopus not only learns quickly, but also remembers what it learns. This has been demonstrated by a series of experiments in which researchers would offer an octopus a crab, and at the same time would either place or not place a white plastic disk in the tank. When the disk wasn't present, the octopus could attack the crab with impunity. But when the disk was in the water, the octopus would receive a mild electric shock every time it reached for the crab. The octopus soon recognized the white disk as a warning, and left the crab alone when the disk was around. Researchers would wait two to three weeks to retest the octopus, and it would still remember what the disk meant.

Experiments like these have sometimes revealed more than intelligence—they have also triggered reactions that are perhaps easiest to describe as emotions, for lack of a better term. Octopuses have shown signs of fear, irritability, aggression, and, surprisingly, rage. One octopus, after receiving a mild shock, retreated to its den, where it became flushed and went through a rapid kaleidoscope of colors. Next the animal shot out pulsating jets of water and flailed its tentacles about like enraged snakes. Dashing from one corner of the aquarium to the other, it bit everything in sight, itself included.

The toxin of a female blue-ring octopus (above and below) becomes even more venomous when she begins brooding her eggs. One bite can kill a human. The Mexican Octopus bimaculatus *(top) shows off its suction cups, which help it fasten to prey.*

The beak of an octopus is as hard as a parrot's and can pierce crustacean shells. This octopus tore at its own skin until it was in tatters. Soon it grew weak and turned pale. Eventually it died. Such self-devouring, known as *autophagy,* is rare among animals, but when it does occur, it is not necessarily fatal. The beak of an octopus, however, helps to ensure that an autophagous outburst is a deadly one.

Defense Mechanisms

The beak has an added feature: like a snake's fangs, it can deliver poison to prey. The toxin of most octopus species isn't very harmful to humans, but there is a notable exception known as the blue-ring octopus. Some populations of blue-rings that live in the Coral Sea off the coast of Australia can kill an adult human in minutes.

Normally, both male and female blue-rings are equally poisonous, but when a female is brooding her eggs, her venom becomes more toxic. Her power doesn't last long, though. Like most other female octopuses, she stops eating once she deposits her eggs, and soon after they hatch, she dies. This kind of starvation, which may function as a population-control mechanism, is clearly a hormonal process. When researchers remove endocrine glands from the female octopuses, they suddenly abandon their eggs, start eating again, gain weight, and continue on with their normal existence.

Like all recluses, many of the 150 to 200 species of octopus have come up with a variety of ways to be left alone. Usually the animals rely on their most famous protection strategy: squirting ink that ranges in color from deep brown to black and serves as a liquid screen to disorient attackers while the octopus gets away. The ink also contains a chemical that shuts down a predator's sense of smell.

Some octopuses have come up with more-inventive means of defense. The banded octopus, for example, manages to pretend it's a fish. The adult animal has a 2-inch (5-centimeter) head and 8- to 10-inch (20- to 25-centimeter) arms that are usually as thick as a pencil. A network of banded membranes covers its arms, and when the octopus is startled, it unfurls this extra skin to create an awesome display that makes it look much like a lionfish. Lionfish are notorious carnivores, and they can deliver venomous stings with fins that extend from their body like brown-and-white spokes. By mimicking these killers, the banded octopus may put off such potential predators as moray eels, sharks, seals—and humans.

Male (below, right) and female blue-ring octopuses mating. After the female lays her eggs (right), she stops eating and soon dies. Scientists theorize that this type of starvation suicide may function as a population-control mechanism.

A MULTITUDE OF MOOSE

by James McCommons

For 20 minutes we've stood motionless. Flies and mosquitoes—apparently invulnerable to bug repellent—hover about my face and neck, inflicting their bites at will. I suppress a frantic urge to flail at them.

Rob Aho, garbed in forest-green camouflage, now communicates to me in hand signals. Cow No. 36, he signs, must be just ahead. At his feet sits the radio receiver that led us to this cedar swamp, but we're too close to use it.

Suddenly there's a burst of noise. Cow No. 36 gets to her feet, rising with a grunt and sucking sounds as she pulls her long legs from the muck. Standing just 30 feet (9 meters) away from us, she's nearly 7 feet (2.1 meters) at the shoulder. A radio-transmitter collar hangs from her thick neck, and the yellow tags fastened to her floppy ears wave to us like semaphores.

I stop breathing. Although I've seen moose up close before, I'm unnerved to be so near such an enormous animal, standing there in its full glory.

Aho, a moose biologist in Michigan's Upper Peninsula, tells me afterward that fear—even flight—is a prudent response. Last fall, No. 36 terrorized a deer hunter, pursuing him for hours until he found protection behind a pile of logs. Aho relates his own experiences of sprinting through the forest: listening, hoping not to hear the bulldozer-like pursuit of a moose.

Aggressiveness usually occurs when bulls look for mates in the fall or when cows protect their calves in the spring. But that's why we're here—to see whether No. 36 has a calf.

The moose saunters to higher ground, stopping occasionally to strip the leaves from a hardwood sapling. Finally she flops down in the midday heat and stares at us indifferently. It's evident she's not a mother. Perhaps she hadn't bred last fall, or maybe a black bear ate her calf.

Cow No. 36 never reproduced after being brought from Canada to Michigan in 1987. But others in the relocated herd have reproduced, and, combined with a natural migration of animals from Ontario, the Upper Peninsula may be home to 200 moose.

Moose Resurgence

Michigan's moose population is tiny compared to Maine's 22,000 or Minnesota's 10,000 animals, but it's indicative of a trend: during the 1980s, moose repopulated much of their former range in the north woods of the Midwest and New England. Moose are no longer a rare sight there, and biologists believe the population will grow even larger. Today's moose herd expanded from existing populations in Maine and Minnesota, the natural moose migration from Canada, and—in the case of Michigan—transplants of additional animals.

Prior to European settlement, moose were abundant. But unrestricted hunting and clearing of the mature forest in the 19th century extirpated moose from many areas or drove them into isolated wildernesses.

In the 1800s, states like Vermont lost 75 percent of their forest cover. This change of habitat forced white-tailed deer to extend their range far to the north. They brought a plague upon the remaining moose. Many deer carry a parasite called brainworm that's benign to whitetails, but usually fatal to moose. Deer pass brainworm larvae in their feces, which are then fed upon by snails and slugs. Foraging moose inadvertently consume the parasite-laden snails with their food. Once the larvae migrate up the spinal cord and attack the nervous system, moose display odd behavior: lack of fear, blindness, cocked head carriage, and a strange circling. Although moose disease had been observed for some time, the connection to brainworm wasn't made until the 1960s.

Biologists still disagree over the significance that brainworm plays in moose/deer population dynamics, but where ranges of these animals overlap, increased numbers of moose directly correspond to declines in deer densities.

In 1966 Vermont contained 250,000 deer and just 25 moose. During the 1970s the oversized deer herd depleted the hardwood browse in its winter yarding areas, literally eating the pantry bare. In the succeeding winters, tens of thousands starved, reducing the herd by two-thirds. Moose exploited the niche left by deer, and now more than 1,000 roam the Green Mountains and commercial forestlands in northern Vermont. "It wasn't until the deer population dropped that we showed any growth in the moose population," says Charles Willey of the Vermont Fish and Game Department. "In the last decade, moose increased about 10 percent a year, and we're nowhere near the biological carrying capacity of the land."

Hardwood Browse

Generally, moose and deer prefer the same habitat. Both species benefit from logging activities that create young hardwood browse—early successional species like aspen, birch, and maple. By creating large areas of regeneration, logging mimics the role fire once played in the mature forest.

In the 1980s extensive clear-cutting on commercial forestlands allowed moose to greatly expand their range. "What happened in New Hampshire was a wonderful patchwork of mature forest and generating forest," says Kristine Klein, moose project leader for the New Hampshire Fish and Game Department. "If you cut back on the forest industry in this state, moose numbers would go down." And as the companies harvested the softwoods that whitetails require for winter shelter and food, the deer density fell. Deer do not tolerate winter as well as do moose, which move well in deep snow and thrive on balsam fir, a tree deer consume only in desperation.

Moose prefer four- to five-year-old saplings that stand chest-high. If the tree is taller, they straddle the trunk and walk forward, bringing branches as high as 15 feet (4.6 meters) down to mouth level. A moose eats 40 to 60 pounds (18 to 27 kilograms) of

The moose population in New York state is so sparse that this lonely but amorous bull was forced to commandeer a herd of heifers for his own personal harem.

Moose Facts

The name *moose* comes from the Algonquian Indians, who called the animal *mongswa,* or "eater of twigs."

Moose *(Alces alces)* range in the northern forests and tundra of the Northern Hemisphere. North America contains four subspecies: *A.a. americana* (Maine to central Ontario), *A.a. andersoni* (Michigan and Minnesota north to the Northwest Territories), *A.a. shirasi* (Rocky Mountains), and *A.a. gigas* (Alaska and the Yukon).

Adults stand 6 to 7½ feet (1.8 to 2.3 meters) high at the shoulder, weighing between 600 and 1,700 pounds (270 to 770 kilograms). Their color ranges from tan to almost black. Moose fur is glossier in the summer than in the winter, and its color is richer. A moose's front legs are longer than its hind legs, giving it an awkward gait, but the difference is useful when traveling over fallen trees and through deep snow.

They aren't social creatures, but several individuals sometimes feed together at a preferred site. An animal's home range extends in a 2- to 10-mile (3- to 16-kilometer) radius. They are most active at night.

Moose prefer hardwood browse such as aspen, ash, dogwood, willow, maple, and birch—eating 40 to 60 pounds (18 to 27 kilograms) per day. They'll also eat balsam fir and hemlock, particularly in winter. They like thickets and wetlands where they can find shelter and graze on aquatic plants. Moose can swim a few miles and for up to two hours if necessary. On land, they can run 35 miles (55 kilometers) per hour.

In winter, they travel easily through snow less than 2 feet (60 centimeters) deep. When the snow is deeper or crusty, moose seek out higher, windswept elevations.

Beginning in March and April, bulls grow distinctive, palm-shaped antlers that may reach 4 to 5 feet (1.2 to 1.5 meters) across. A bull breeds at five years of age, a cow at two. During the rut each autumn, bulls mate with several females.

Born in May and June, calves weigh about 25 to 35 pounds (11 to 16 kilograms) at birth. If the female's diet is nutritious, twins are not unusual. Reddish in color, calves gain weight quickly, weighing 300 pounds (135 kilograms) by fall. Typically, a calf will nurse for nine to 12 months. The mother will drive a yearling calf away shortly before she gives birth the following spring.

Moose calves, born in May or June, weigh 25 to 35 pounds at birth. Within three months they can gain up to 300 pounds.

Healthy adults are sometimes attacked by wolves or grizzly and brown bears, but a moose's great size deters most predators. However, bears, wolves, and coyotes take a significant number of newborn calves.

Moose can live to age 20, but the normal life span is between 13 and 20 years. Adult moose usually die of starvation, disease, or infestation by parasites.

browse per day—10 times as much as a deer.

In addition to young hardwood browse, a good moose habitat usually contains lots of lakes, marshes, and beaver ponds. Water lilies, sedges, pondweed, and other aquatic species furnish the minerals and salts essential to a moose's diet.

Before aquatics sprout in late spring, moose are drawn to roadsides, where they drink from ditches and eat mud seasoned by road salt. In moose country, these salt licks are superb sites to see moose.

Although some calves are preyed upon by bears and coyotes, there's little natural

Moose have become a common sight in many populated areas. Tourists who mistake a moose's demeanor for indifference or docility are often shocked when these enormous animals exhibit aggressive behavior.

predation—except in northeast Minnesota, where wolves range freely. So, to control the herds in specific areas, Maine, Vermont, New Hampshire, and Minnesota run limited fall hunts.

When it began in 1979, Maine's moose hunt stirred great debate. Animal-rights activists and hunters faced off at roadside check stations, and the furor led to a statewide voter referendum in 1983, when a 60 percent majority approved a five-day hunt.

Maine issues 1,000 permits in a lottery of 80,000 applicants. Hunting success is more than 90 percent, yet hunters remove just 5 percent of the population.

Growing Range

When moose first appeared in New York state in the early 1980s—wandering in from New England and Canada—they caused a sensation, especially when a rutting bull in the Adirondacks claimed a herd of heifers as his harem, driving off dairy farmers and spectators alike. Because New York's moose are so few—perhaps just 20 animals—and so dispersed, the big bull couldn't find a mate. The amorous male had to be tranquilized and removed from farms several times by Dale Garner, a graduate student working in the Adirondack Wildlife Program of the State University of New York's College of Environmental Science and Forestry. Garner carries the further distinction of being the only person licensed to tranquilize moose in all of New York state.

In his master's thesis on the reintroduction of moose to the Adirondacks, Garner concluded it would take decades for natural migration and reproduction to establish a self-sustaining herd in the Adirondacks.

"I don't think New York will ever have a really big population, because a lot of the Adirondacks are set aside as forest preserves where there isn't any cutting. But there would be enough—maybe several hundred to a couple of thousand—so people could come to the Adirondacks and see moose," he contends. Bringing more moose to New York would speed the process.

Relocating moose is tricky and expensive—about $5,000 per animal. For the Michigan transplant operation, biologists tranquilized moose in Algonquin Provincial Park in Ontario and airlifted them with helicopters, then trucked them 600 miles (965 kilometers) to the release site. Once in Michigan the moose dispersed. One animal—nicknamed Gulliver's Travels—took up temporary residence near the village of Gulliver, 165 miles (265 kilometers) from the release point. Others ended up in Wisconsin. Biologists decided additional animals were needed to ensure that breeding would take place, so another moose-relocation operation was put into effect.

New York hopes to avoid a similar predicament by transplanting family groups (if they really exist) rather than random individuals, theorizing that social units may stay together in unfamiliar surroundings. Garner is currently in Algonquin Provincial Park, doing doctoral work on whether cows share their home range with offspring. Algonquin has one of the highest moose densities in all of North America. In 1985 and 1987, the park supplied Michigan with 25 bulls and 36 cows in two transplant operations. If New York and Ontario reach an agreement on transplanting moose to New York, Garner's study animals may end up in the Adirondacks.

On one of the coldest winter days in 1987, Michigan's Department of Natural Resources ran shuttle buses to the release site to accommodate more than 2,000 people—including the governor and groups of schoolchildren—who applauded as the moose bolted from their wooden crates and ran for the woods.

"People felt like they were part of a historic event. And when they saw the size of these animals, it just knocked their socks off," says Aho, who has kept tabs on the growing herd for the past five years.

"There was a lot of concern about poaching, but it just hasn't happened. The people of the Upper Peninsula [of Michigan] are proud of 'their' moose—and they're watching out for them."

Moose are not typically known as social creatures, although a small group of them may feed together at a preferred site. Experts are trying to determine if moose share their home range with family members.

At the National Fish and Wildlife Forensics Laboratory, scientists use the latest technology in the battle to curb traffic in wild animals and wildlife parts and products. Above, three staff members pose with a variety of illegal wildlife goods.

WILDLIFE CRIMEBUSTERS

by Nancy Shute

The Bengal tiger staring balefully from the top of the stainless-steel counter is in no condition to harm anyone who ventures near, since all that remains of him is his head. The creature, dead for eight months, was killed under suspicious circumstances in Texas, and now three criminal investigators at the National Fish and Wildlife Forensics Laboratory in Ashland, Oregon, are about to find out how. Nancy Thomas, a veterinary pathologist with the U.S. Fish and Wildlife Service (USFWS), steps up to the counter. Thomas is on loan from the National Wildlife Health Research Center in Madison, Wisconsin. Wearing a lab coat, a plastic apron, and latex gloves, she calmly peels back the big cat's orange-and-black fur, picks up a plastic-handled scalpel, and starts carving into the massive musculature. Ed Espinoza, the lab's chief criminalist, holds a 6-inch (15-centimeter) plastic ruler along the head for scale, while Kent Oakes, a senior forensic scientist, snaps pictures. A tape recorder protected by a plastic bag hums on the counter. "You see the perforation?" Thomas asks, pointing at the newly exposed ear canal of the tiger.

She removes the ear. Espinoza picks it up and peers in tentatively. "These guys had an awfully good aim if none of this is scratched," he observes.

"Well," Oakes says, "if the tiger was sleeping. . . ."

"They wouldn't bump into anything until they were pretty far down, where the ear canal would take a turn," Thomas comments, pointing to a bony ridge in the skull. "So somehow they found this and introduced the ice pick there, if that's what they did." She gently scrapes more flesh from the skull, revealing a tiny perforation in the bone. "We didn't do that," she says.

Oakes cranes over her shoulder. "It is an interesting hole in that it's not consistent with a stabbing type," he notes. "There's no fracture."

"Yes, there is, a bit," Thomas replies. She scrapes again. "Bingo!" she cries, pointing to a line of perforations running from the first. She stands back as Oakes takes a few more photographs. "What we've got to do is find out whether the animal was alive when these holes were made," Thomas says. "I don't know if we'll be able to see much damage to the brain." She reaches for her saw.

Half an hour later, the tiger's brain lies exposed on a plastic cutting board. "I think we've got it," Thomas declares, beaming. "These are the lateral ventricles, and they're all full of blood. Any jury member should be able to see this." The hemorrhaging visible there and elsewhere in the brain indicates that the animal was indeed alive when the holes were made. The doctor bounces on her heels like a five-year-old. "It's satisfying; it's just so satisfying."

Ed Espinoza looks at the remains of the tiger in the sink. "When they first brought him in, he looked so majestic."

Curbing Wildlife Contraband

Tiger necropsies are all in a day's work at the Oregon lab. Although it has been in operation for less than three years, this highly specialized facility is already recognized as a key resource in the battle to curb contraband traffic in wild animals and wildlife parts and products—a worldwide problem that's getting worse. The legal trade in everything from Australian kangaroo hides and live lovebirds from Tanzania to tropical fish from Southeast Asia and lizard shoes from Argentina is a big business, with annual revenues conservatively estimated at about $6 billion. The illicit traffic involves the smuggling and killing of endangered or protected species, and the illegal taking of other animals. It is thought to be generating somewhere between $2 billion and $3 billion a year. A "record book" elk head may bring $20,000, a tiger skin $10,000, and a bear's paw and gallbladder (used in Oriental medicines) up to $5,000. Objects fashioned from unlawfully killed creatures range from the beautiful to the bizarre: elephant-trunk ice buckets, jewel-encrusted sea-turtle heads, toad coin purses with zippers up the tummy, and stools made of elephant feet.

The lab gets involved in a case when federal, state, or international law-enforcement officials uncover evidence—perhaps a finished product or maybe just the remains of a dead animal—that requires identification or forensic analysis. "These cases are complicated and difficult to prosecute," says James Kilbourne, chief of the agency in the Department of Justice that handles violations of U.S. wildlife laws and international wildlife treaties. There are seldom any witnesses to such crimes; they usually take place in wild and remote areas; the evidence often does not turn up until long after the crime itself has been committed, and by then it may well have been damaged or radically altered. "The forensics lab gives us the capacity to become a great deal more sophisticated in our investigations," Kilbourne says.

Forensic science is the application of the physical sciences to the evaluation of evi-

The bald eagle at left is one of many discovered dead and sent to the lab. The remains are held in a freezer until they can be given to American Indians, who use the remains in the preparation of artifacts for religious ceremonies.

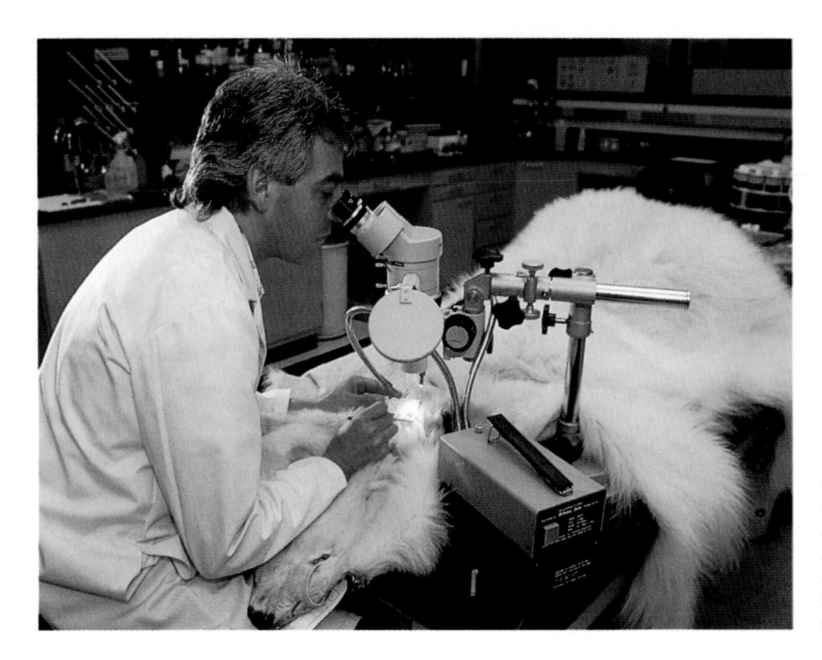

Careful examination of pelts can often reveal the cause of an animal's death. At left, a researcher uses a microscope to scrutinize a polar bear skin for incriminating evidence.

dence for judicial consideration. Its historical roots can be traced back to the origins of legal medicine in 6th-century China. However, it was Sir Arthur Conan Doyle's fictional detective Sherlock Holmes, according to Richard Saferstein's *Criminalistics: An Introduction to Forensic Science,* who "applied the newly developed principles of serology, fingerprinting, firearm identification, and questioned-document examination long before their value was first recognized and accepted by real-life criminal investigators."

Today there are about 350 federal, state, and local crime labs in the United States. They are able to analyze physical evidence ranging from a simple set of fingerprints on a glass to the complex genetic "fingerprint" contained within a human cell. "In traditional testimony, juries have to rely on the perceptions of others, but forensic evidence often proves irrefutable," says John Hicks, assistant director of the laboratory division of the Federal Bureau of Investigation (FBI), which pioneered the use of DNA fingerprinting in the United States.

Wildlife crimebusters have come to rely on this kind of evidence, too. Increasingly, they are finding themselves up against international networks armed with high-tech weaponry and electronic gear—poachers, for example, who use scientific tracking equipment to locate hibernating radio-collared bears in the Appalachians. Physical traces of such crimes are hard to come by and often fragmentary in nature. In the old days—five or 10 years ago—agents found themselves begging for forensic backup. "We relied on museums and universities," says John Neal, one of the 205 agents of the USFWS who investigate federal crimes against wildlife. "But they're not forensics experts." Some wildlife cases were lost because of delays in getting solid forensic evidence; others were never prosecuted.

A Biochemistry Major Turned Cop

In 1979 the USFWS hired Ken Goddard, then chief criminalist of the Huntington Beach Police Department in California, to set up a wildlife forensics lab. A criminalist is an

Some illegal wildlife products are truly grotesque. At right, the bejeweled hawksbill turtle head (left) and the beak of a young bald eagle have been made into paperweights.

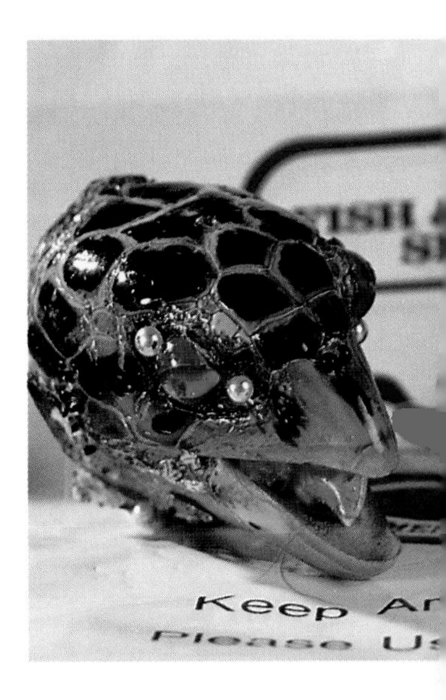

expert in the scientific study and evaluation of physical evidence in the commission of crimes. Before Goddard got into police work, the closest the genial biochemistry major had ever come to it was when he took a judo class from a Riverside County, California, deputy sheriff during college. After one particularly strenuous session, the deputy had to drive Goddard to the hospital. On the way, he leaned over and growled: "Ever think of becoming a cop?" He ended up chauffeuring his dazed recruit directly from the emergency room to a job interview.

Goddard spent the next 10 years scouring the sites of humanity's most heinous acts in search of the subtle clues that link victim, suspect, and crime scene, the holy trinity of detective work. He was intrigued by forensic science's heady mix of laboratory precision and true-crime thrills, but he never did master the macho imperatives of law enforcement. For example, helping other officers quell a disturbance in a saloon one day, he suddenly found himself facing an angry brawler wielding a beer bottle. "No, wait!" he yelled. "I'm the scientist!"

When he arrived in Washington, D.C., Goddard discovered that the USFWS had not yet asked Congress for funds to pay for the forensics lab. For eight long years, the joke at headquarters was, "Ken Goddard's briefcase is the forensics lab." He vented his frustrations by writing *Balefire* and *The Alchemist,* a pair of crime thrillers notable not only for their vivid scientific methodology, but also for an alarmingly high body count among law-enforcement officers.

It wasn't until the state of Oregon loaned land on the campus of Southern Oregon State College in Ashland, a bucolic little town of 16,000 tucked into the foothills of the Siskiyou Mountains, that Congress appropriated the $4.5 million needed for a state-of-the-art laboratory. The 23,000-square-foot (2,150-square-meter) building was completed in the fall of 1988; it took more than another year for Goddard to recruit the dozen scientists and support staff he would need to create a new discipline: wildlife forensic science.

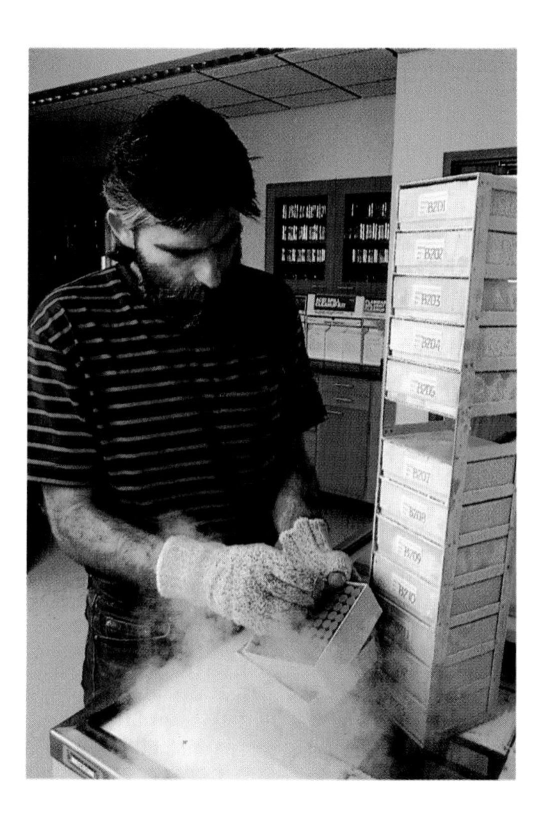

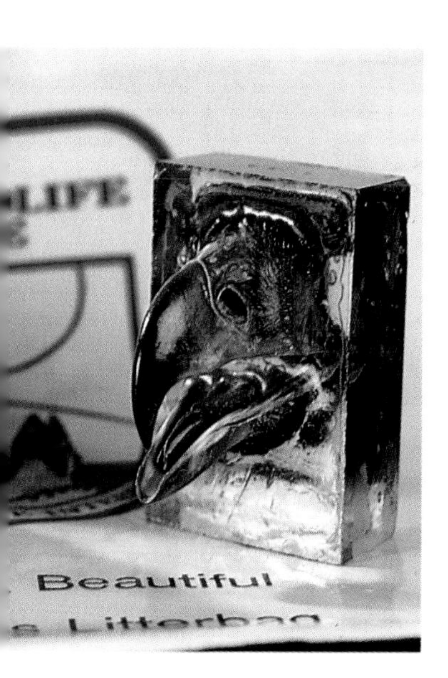

Using DNA fingerprinting (above, right), researchers can now identify an animal through a bit of blood or tissue or even just a few cells. Scientists hope to develop a "fingerprint bank" of endangered animals.

Half of the troops came from police crime labs; the other half, from museums and universities. "This had never been done before," says Goddard, a tall, white-haired 45-year-old who pads around the lab in jeans and sneakers, talking in overdrive. "It's fascinating research. The science is really cutting-edge. Then we immediately jump into the ice-cold real world of the judicial system. It's fun. It's a pure kick."

Wandering the laboratory's skylighted halls is a kick, too. In the conference room, Irene Brady, a nature writer and illustrator in her spare time, huddles over a table with two special agents sorting slides of oil-blackened otters and eagles. Brady will use them to design courtroom exhibits for a federal oil-spill lawsuit.

A Freezer Stacked with Eagles

The technical-support section is stocked with video cameras, tiny microphones, and homing devices for undercover operations. Tom Rayl, chief of the evidence-and-property section, presides over a cavernous warehouse at one end of the building that is cluttered with sea-turtle mandolins, eagle-head paperweights, and other gewgaws. Many of them have been confiscated from unwary tourists. A walk-in freezer is stacked to the ceiling with stiff bodies of bald and golden eagles. This is the National Eagle Repository. The birds, many of which were killed illegally or by accident, are kept here while they await distribution to native Americans, who convert them into religious artifacts.

The morphology lab is where wildlife and various parts and products are visually and microscopically identified. Chief morphologist Stephen Busack and geneticist Peter Dratch laugh and curse as they wrestle a 6-foot (2-meter)-long cougar out of a freezer. Bet Ann Sabo, a forensic ornithologist who apprenticed for eight years with Roxie Laybourne, the legendary feather expert at the Smithsonian's National Museum of Natural History, sits quietly in a corner, sorting a stack of red-tailed hawk feathers. "In a wildlife crime, the evidence tends to be part of your victim," Goddard explains. "The big job is to relate the parts and products to the individual species."

To that end, the morphology section is assembling a huge collection of animal feathers, skin, and bones that can be used to help identify crime victims. It is also searching zoos and universities around the world for known specimens of wild animals that have commercial or sporting value. "Crocodilian purses come in all the time, but I don't have a complete 'library' of crocodilian standards," explains Busack, a herpetologist. "We have to demonstrate beyond a reasonable doubt what species a purse is made from. If we can't do that, we're going to lose cases."

The serology section and its DNA unit identify wildlife, too, but by a bit of blood or tissue or even a few cells. When he's not wrestling with frozen cougars in the morphology lab, Peter Dratch, an authority on deer genetics, is assembling a bank of tissue samples of the large game animals that are often involved in hunting violations. From that bank, Steven Fain, a DNA expert, is creating a genetic database of DNA "fingerprints" that can be used to identify crime victims. He figures it may take five years of collection and research before he can fingerprint and interpret a dozen species.

The lab's criminalistics section employs forensic techniques that are familiar to any viewer of "Quincy, M.E." or other television crime dramas. These include detecting gunshot residues and mapping a bullet's trajectory through flesh.

The facility's first and greatest success involved the simplest of questions: What is ivory? In June 1989, a worldwide ban on trading ivory and ivory products from the endangered African elephant went into effect. A similar ban on ivory from Asian elephants had been established in 1976. This new ban

During an airport raid in Washington State, inspectors discovered a shocking selection of wildlife parts gathered during a safari in Zimbabwe.

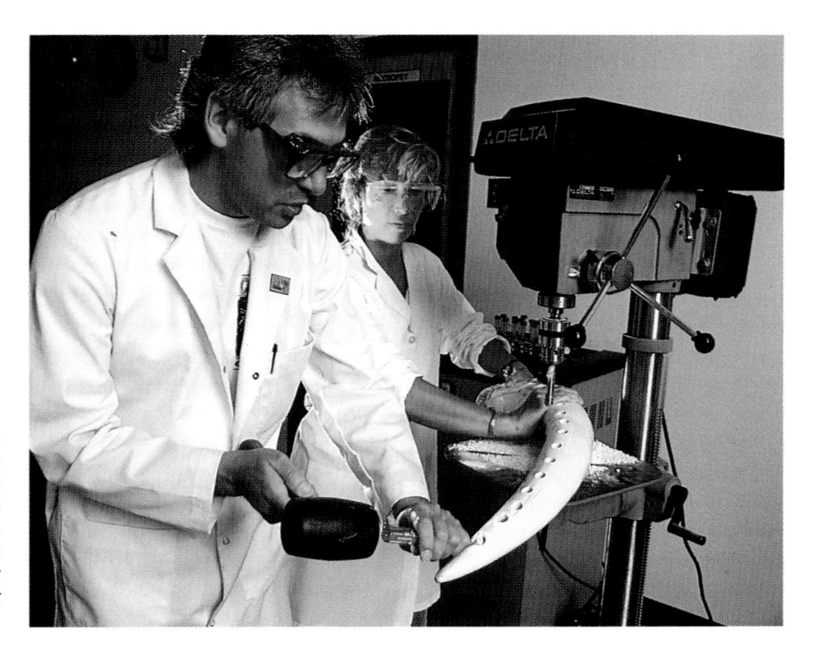

Now that the sale of any elephant product is illegal, the importation of ivory into the United States violates wildlife-protection laws. At right, researchers remove cores from a tusk to determine how old the elephant was when it was killed.

meant that law-enforcement officials had to be able to distinguish elephant ivory from other kinds of ivory.

"The law caught us with our pants down," says Ed Espinoza, a chemist and criminalist who taught forensic science and consulted in homicide investigations in northern California before joining the wildlife laboratory. "No one had ever defined ivory scientifically."

Ivory traders had it all figured out. Shortly after the ban went into effect, shipments of elephant ivory misleadingly labeled "mammoth" began showing up at ports around the world. The challenge for Espinoza and his colleagues: learn how to tell the difference between modern elephant ivory and Ice Age mammoth ivory, which is legal to trade. "We tried all kinds of stuff," Espinoza says, playing with the age-yellowed ivory carving of an old Chinese man that sits on his desk. "On very old samples, the outsides fluoresced under ultraviolet light. We were all excited about that—until we got a shipment in which the outsides had been removed."

Eventually Espinoza noticed differences in the angles of the grain lines, called Schreger lines, in cross sections of ivory from ancient and modern beasts. "The biggest problem was explaining why." For that, Espinoza turned to his colleague Mary-Jacque Mann, a senior forensic scientist who spent 10 years as a scanning-electron microscopist at the Smithsonian. Schreger lines are created by undulating rows of hollow tubes, called dentinal tubules. Mann's microscope showed that in ancient ivory, these tubules were packed much more densely than in modern elephant ivory. When mammoth ivory is carved, cross sections are exposed to reveal Schreger lines that form angles of 90 degrees or less. In modern elephant ivory, the Schreger lines form angles of 115 degrees or more. Measure the angles, and you've got your beast. Within a short five months, Espinoza and company had solved the mystery. "It's a very simple, nondestructive way of positively distinguishing between these two kinds of ivory," he says.

But the work didn't stop there. Espinoza's people went on to look at all of the different types of ivory and ivory substitutes, and came up with a set of physical descriptions that distinguish one type from another. The result: a handy little guide that can be used by just about anyone wishing to identify an ivory source.

When Rich McDonald got the call from Olympic National Park in Washington, he knew it was going to be bad. A ranger there had just found the carcass of a spotted owl. Nailed to a sign at the park entrance, the owl had a typewritten note attached to its breast: "If you think your parks and wilderness don't have enough of these suckers: plant this one." As a USFWS special agent covering the Pacific Northwest, McDonald had spent years patrolling the Olympic Peninsula's dense forests and remote logging towns. "It's one of the last strongholds of the old Wild West," he says. Killing the owl, a mottled brown-and-white bird that has become endangered as its old-growth forest habitat was logged, was a federal crime. Beyond that, it was an attempt to further inflame relations between the region's already-polarized loggers and preservationists.

The perpetrator didn't leave many clues: a Band-Aid, two beer cans, a soda can, a nail, a match. McDonald immediately shipped the evidence to Ashland. "It's just like an investigation of a human murder," the special agent says. "The only difference is, you can't go out and interview who the owl was drinking with the night before."

In the search for clues, the lab X-rayed and necropsied the owl carcass, tested the Band-Aid for blood, and scanned the note and cans for fingerprints. It reported back to McDonald that a high-powered rifle had been used to kill the bird, and told him that the creature, an immature male, had been kept in a freezer for some time. McDonald withheld the lab's findings from the press. He figured that a $16,000 reward might shake some information loose, and then maybe the lab's report could verify it. Time went by, but nothing happened.

McDonald knew what he had to do next: get in his battered Chevy Blazer and go work the small towns of the peninsula, just as he had so many times before. In his long career as a special agent, he had spent countless hours routinely checking hunters' licenses and bag limits, but he had also hidden in an attic crawl space above a room where an informant was talking to a suspected poacher who had threatened to blow that informant's head off. McDonald had also arranged undercover wildlife "buys" as elaborate as those involved in any big-time drug bust. In the process, he had been shot at, and his family had been threatened. It's a grim fact that wildlife agents are more likely to be assaulted in the line of duty than are other law-enforcement officers. "Everyone we contact is armed with some kind of weapon," McDonald says. "And we're usually interfering with somebody's 'God-given right' to get out and hunt and fish." Still, he wouldn't trade his job for the world. "It's 99 percent dull routine—and 1 percent sheer terror."

The plight of various American species of owl causes great concern among government wildlife specialists. Below, a forensic ornithologist determines that the feathers used in a craft item match those of a snowy owl. Away from the lab, special agents (right) keep an eye on the spotted owl, an endangered bird whose habitat in the Pacific Northwest is much in demand by loggers.

Elk Proteins on a Bloody Knife

Even when the lab is given next to nothing to work with, the experts there are often able to come up with amazing results. Take the case of the hunter who pleaded guilty to killing an elk out of season. A year later the case was overturned on a technicality and remanded for trial. But by then the evidence, the elk, had been disposed of. The authorities still had a bloodstained knife, which they sent to Ashland for analysis. Using immunodiffusion (a common crime-lab technique that compares antibodies) and electrophoresis (a standard genetic technique that identifies different forms of proteins), the scientists were able to establish that it was indeed elk blood on the knife, and the defendant was found guilty.

McDonald also turned to the lab when he needed help explaining to a grand jury an intricate undercover operation against Columbia River sturgeon poachers. For a year, he posed as an affable former logger who bought and sold illegally caught sturgeon out of the back of his truck, making deals 24 hours a day with a cellular phone and an answering machine. The gigantic fish, many 150 or 175 years old, are sold for their caviar at prices up to $3,100 each. Since the poachers concentrate on the older, breeding fish, the black market threatens the survival of the species.

When it was time to close down the operation, McDonald had to wear a concealed tape recorder and try to trick one of the bad guys into telling him where the records were kept. When 70 local, state, and federal officers finally closed in, he had to maintain his cover. "I'm getting out of here," he yelped. His fellow officers searched him and nabbed his driver's license before setting him free. Subsequently the lab prepared elaborate flow- charts tracking the black-market operation, based on the information provided by McDonald. Those exhibits helped make the complicated case clear to the jury, and were instrumental in obtaining convictions.

Notwithstanding the lab's input, the mystery of the spotted owl on the Olympic Peninsula has not been solved. In fact, two more of the birds have been killed, and two "look-alike" barred owls as well. On other fronts, however, progress is being made. The demand in the Orient for medicinals concocted from bear gallbladders has fueled a sharp rise in poaching in the U.S. In response to that, the folks in Ashland have developed a foolproof way to identify bear gallbladders, thus ensuring that enforcement activities will receive solid scientific support.

Wildlife scientists are working on compiling a "feather library" to help match feathers to specific bird species.

As for that Bengal tiger, the lab's investigation revealed that the drugged animal had been stabbed in the brain with an ice pick. The fastidiously documented findings backed up testimony that the defendant, once the proprietor of a business specializing in exotic wildlife decor, was killing tigers he had purchased from a petting zoo and selling the hides.

Faced with the evidence, the man pleaded guilty to a violation of the Endangered Species Act and was sentenced to a year in jail. "We're very happy," says agent John Neal. "The people at the lab made the case much, much stronger. Within two or three months, they had done their research, they had people ready to testify, and they had prepared effective courtroom displays. It was a very professional job."

REVIEW

Botany

A tissue-culture system that maximizes taxol production may soon make yew-tree nurseries (above) obsolete.

CELL CULTURE MAKES THE MOST OF YEW

A battle between human- and environmental-health interests may be settled by a new system for growing plant cells in a test tube. The conflict centers on the scrawny Pacific yew *(Taxus brevifolia)*, a relatively rare, slow-growing evergreen that grows on federal land in Oregon and Washington. The tree's bark contains small amounts of taxol, which researchers have found controls ovarian cancer in some patients.

The National Cancer Institute (NCI) wants to study taxol's anticancer properties, but it needs about 100 pounds (45 kilograms) of the compound annually—about 40,000 trees' worth—for testing. To supply the NCI, the U.S. Departments of Agriculture and the Interior have given Bristol-Myers Squibb the exclusive right to harvest the Pacific yew's bark and extract the 0.01 to 0.03 percent of taxol the dry bark holds.

Environmentalists are worried that the increased demand for taxol might threaten the 23 million Pacific yews in the northwest. They are especially displeased that Bristol-Myers Squibb is the only company harvesting and extracting taxol; they fear that the lack of competition will permit wasteful harvesting methods and make it impossible for others to develop more-efficient extraction methods. The traditional alternative to extraction—chemical synthesis—remains problematic because the compound is so complex.

A promising solution might be plant-cell tissue culture. The NCI is funding efforts to grow Pacific yew in a bioreactor that maximizes cell reproduction. In developing the bioreactor, which is the large-scale counterpart of a test tube, the researchers will explore what combination of nutrients and environmental conditions encourage the yew cells to produce the most taxol. They hope to have the system ready for commercial use by 1996.

PLEASANTER PESTICIDES

Concern about the dangers pesticides pose to humans, wildlife, and the environment have prompted researchers to look for safer alternatives. In many cases, they're looking to the plant kingdom.

Researchers at Cornell University are using an extract from one pest—quackgrass *(Agropyron repens)*—to control another pest: slugs. These slimy leaf eaters, which hide under decaying plant material, have found a happy home in the crop residues left on fields to suppress weeds and prevent soil erosion. The slug poisons currently on the market are expensive and toxic to mammals. The quackgrass extract has proved deadly for slugs, but easy on livestock, pets, and humans.

A battery of plant-based insecticides might also come from the neem tree *(Azadirachta indica)*, which grows in arid tropical regions. In a study by the U.S. Department of Agriculture (USDA), neem oil controlled, to varying degrees, 200 different insect species.

A different approach involves an effort to put insecticides into plants, rather than derive them from it. Researchers at Washington University in St. Louis are splicing a bacterium deadly to caterpillars into the genes of tobacco plants. The bacterium, *Bacillus thuringiensis*, is completely nontoxic to humans and has been available commercially for more than a decade. When a larva in the genus *Lepidoptera*—which includes moths and butterflies—digests the *B. thuringiensis*–laden leaf, its intestines are paralyzed.

Erin Hynes

Endangered Species

The decimated populations of the black-footed ferret (right) are making a comeback thanks to very successful captive-breeding programs.

You may not have heard of the razorback sucker, the jaguarundi, or Cooley's water willow, but they've all made the list—the list of endangered or threatened wildlife and plants, that is. While they don't have the fame of the northern spotted owl, the bald eagle, or other members of this growing group, they're all in serious trouble and in need of protection.

In this country, there are more than 500 endangered mammals, birds, reptiles, fishes, mollusks, and plants. Endangered species are in danger of extinction throughout all or a significant portion of their range. Approximately 161 species are threatened, meaning that they are likely to become endangered within the foreseeable future.

How does wildlife get accorded this protected status? Most often, it starts with the candidate list, which now includes close to 3,600 species. The U.S. Fish and Wildlife Service (or the National Marine Fisheries Service, for most marine species) prioritizes these candidates, proposes which species to list, and then publishes the proposed rule in the *Federal Register*. The public can comment, and a decision is then made based solely on the best available biological data. Due to time and financial constraints, the Fish and Wildlife Service lists only between 50 and 100 species a year. Once a species has been officially listed, the Endangered Species Act makes it "illegal to kill, harm, harass, possess, or remove protected animals from the wild." Experts then develop recovery plans to turn around a species decline and to ensure its long-term survival. These plans take time to create and put into motion, and so far, only 386 species have recovery plans.

BLACK-FOOTED FERRET

Recently some remarkable restoration stories show how wildlife can bounce back if given half a chance. The black-footed ferret *(Mustela nigripes)*, once common in 13 western states, teetered on the brink of extinction just a decade ago. Habitat loss and the decline of its main prey, the prairie dog, almost did in this creature from the weasel family. This species of ferret was thought to be extinct until several surviving black-footed ferrets were discovered in 1981; by 1985 they had been taken into captivity. Captive breeding programs upped their numbers to 325, and, in September 1991, the first of a group of 50 ferrets was released into the Wyoming wild.

The road ahead may be rocky for the black-footed ferrets, but Stephen Torbit, wildlife biologist for the Fish and Wildlife Service, thinks they are off to a good start. "Nobody's ever done this with ferrets before, so we're plowing new ground and learning a tremendous amount as we go." A survey in March 1992 found six to eight survivors from the original 50, better than the 90 percent mortality rate expected. The long-term goal is to establish 10 colonies of ferrets throughout the West.

RED WOLF

Another success story in the making involves the red wolf *(Canis niger)*. Exterminated from most of its range in the southeastern United States by the early part of this century, its numbers continued to drop with the increasing loss of prey and habitat. By 1975 it became clear that the red wolf's only hope was a captive-breeding program. Since 1987, 19 zoos and captive facilities have bred enough wolves to release 36 of them back into an eastern North Carolina site. At this site, which lies primarily in the Alligator River National Wildlife Refuge, at least 17 more wolves have been born. "The project has worked better than anybody expected," says Mike Phillips, wildlife biologist with the Red Wolf Recovery Project. "There is potential in the mountains for a significant number of red wolves."

Indeed, in the fall of 1991, a pair of wolves and their two pups were released in the Great Smoky Mountains National Park. So far, it is unclear whether this will make another good reintroduction site, but the goal is to establish three or more red-wolf populations in the wild.

FLORIDA PANTHER

The Florida panther sits in a more precarious position, with its very survival dependent upon a newly proposed captive-breeding program. One of the most critically endangered animals in the United States, the Florida panther *(Felis concolor coryi)* numbers between 30 and 50 individuals. Once found in portions of Alabama, Arkansas, Florida, Georgia, Louisiana, Mississippi, South Carolina, and Tennessee, the big cat's range is now limited to a small region of south Florida. Hunting, loss of habitat, getting hit by cars, and such environmental contaminants as mercury have reduced its numbers over the years.

One of the animal's biggest problems is its decline in genetic diversity. There are so few panthers that inbreeding continues to chip away at the genetic pool. On the present course, the Fish and Wildlife Service has determined that the Florida panther will become extinct in 25 to 40 years. They feel that by taking some kittens into captivity along with genetically important nonbreeding adults, they can selectively breed and maintain the greatest genetic diversity possible. Over time, they hope to reintroduce captive-bred panthers back into their historic range.

Linda J. Brown

While the Endangered Species Act has helped revive the lagging red-wolf population (above), it may be too late to help the Florida panther (right), whose extinction may come in 25 to 40 years.

Endangered Species Act Alert

The Endangered Species Act (ESA) is due for congressional reauthorization by the end of September 1992; it looks like there's a nasty fight brewing. Supporters want the act strengthened in order to make it work more quickly, to help more wildlife, and to save imperiled ecosystems. "We're facing the greatest extinction crisis since the disappearance of the dinosaurs, and the ESA is our front line defense," says Robert Irvin, attorney for the National Wildlife Federation (NWF). Opponents want the act weakened because they feel it stands in the way of progress and economic development.

One small bird, the northern spotted owl, has sharpened the focus of this debate in the past few years. Listed as threatened in 1990, this owl lives in the ancient forests of the Pacific Northwest. Great controversy rages because protecting the spotted owl will impact the timber industry.

When the ESA does come up for reauthorization, both sides will come out swinging. The fate of the spotted owl, the sockeye salmon, the delta smelt, and many other creatures may depend on the outcome.

Zoology

BUTTERFLY STING

The viceroy butterfly has long been heralded as the supreme biological con artist. For years, biology students have been taught that the viceroy has evolved to resemble the foul-tasting monarch butterfly in order to keep predators at bay—a textbook example of Batesian mimicry, named after British naturalist Henry Walter Bates. But a 1991 study published in the British journal *Nature* suggests that the viceroy may have pulled the ultimate sting operation—with the science community as the unlikely dupe.

Biologist David Ritland of Erskine College in South Carolina and Lincoln Brower of the University of Florida in Gainesville set up a taste test using a natural butterfly predator, the red-winged blackbird. In the study, 16 blackbirds were offered the abdomens from viceroys, monarchs, queen butterflies, and four other butterfly species known to be bird favorites. Just the abdomens were used so that the birds could not identify the individual species merely by sight.

Much to the researchers' surprise, birds that tasted a viceroy abdomen displayed obvious distaste by shaking their heads and becoming agitated. Any viceroys that were eaten were done so with great hesitancy. In the end, only 41 percent of the viceroy abdomens, 46 percent of the monarchs, and 68 percent of the queens were consumed. Conversely, the birds ate 98 percent of all the other butterflies with great gusto. Ritland and Brower concluded that the viceroys are as distasteful as the monarchs, and their apparent mimicry of the monarch butterfly is part of a more elaborate hustle than biologists had previously thought.

Zoologists now hypothesize that the viceroy exhibits Müllerian mimicry. In 1879 Brazilian zoologist Fritz Müller described a type of mimicry in which two equally foul-tasting species evolve to resemble each other so that each species will lose fewer individuals before predators learn to avoid them. And by doubling their numbers, each individual of both species runs a smaller chance of being attacked.

Contrary to a long-standing belief, the viceroy butterfly (top) may be just as foul-tasting to birds as is its look-alike cousin, the monarch butterfly.

Not everyone is convinced that this new study completely dispels the tasty-viceroy interpretation. Some critics propose that the viceroy both looks and smells like the monarch, and the birds in the taste test may have been put off by the butterfly's scent. In any case, scientists have learned that assumption can play no role in scientific theory, and they now are on the lookout for other zoological con artists like the viceroy.

THAR SHE BLOWS!

An elusive gray creature found in the Pacific Ocean off Peru has been identified as the first new species of whale to be discovered in 28 years. The newly discovered whale is the thirteenth species found of the genus *Mesoplodon*, which means "armed with a tooth in the middle of the jaw." Classified as *Mesoplodon peruvian-*

us, the newly discovered whale is the smallest known of its genus, resembling a dolphin more than its Moby Dick cousin. It has a dolphinlike snout, an elongated jaw, and few teeth. At birth the whale is about 5 feet (1.5 meters) long, and grows to about 12 feet (3.5 meters) by adulthood.

This particular species is seen so rarely that it took James G. Mead of the Smithsonian Institution, working with Julio C. Reyes and Koen Van Waerebeek of the Peruvian Center for Cetalogic Studies, 15 years to find enough specimens to substantiate the claim of a newly identified species. The first clue was found in 1976, when Mead found a partial cranium and vertebra in San Andes, Peru. Remains from nine additional whales were found between 1985 and 1989 on the central and southern Peruvian coasts.

Experts theorize that *M. peruvianus* has escaped notice until now because the whales typically lived far from shore, keeping to themselves and avoiding ships. In recent years the population has migrated nearer to shore, where they have garnered attention after getting caught in fishing nets or washing up on the beach when they die.

PROPOSED POLLINATOR

Most new species are identified from eyewitness accounts. But in an unusual turnaround, an entomologist has proposed the existence of an as-yet-undiscovered insect on the basis of a rare orchid found in the dense tropical jungle of Madagascar. Gene Kritsky of the College of Mount St. Joseph in Ohio hypothesizes that the existence of the orchid *Angrecum longicalcar*, which has a 16-inch (40-centimeter)-deep nectar tube, necessitates the existence of a giant moth with a 6-inch (15-centimeter) wingspan and a 15-inch (38-centimeter) proboscis. This is the only type of creature that could pollinate the orchid while dipping into its deep nectar tube.

According to Dr. Kritsky, "The orchid could only survive if it had a moth pollinator. The orchid and its pollinator must have co-evolved." Most likely over time, the nectar tube of the orchid grew a bit longer, and the moth's tongue followed suit. Eventually the relationship between the two became exclusive—the moth was assured of a nectar supply from an orchid that no other insect could penetrate, and the flower could depend on pollination by an insect that focuses its feeding efforts solely on *A. longicalcar*.

Dr. Kritsky's proposal follows closely on the heels of one by famed naturalist Charles Darwin. In 1862 Darwin proposed that a moth with an 11-inch (28-centimeter) tongue must exist in order to pollinate the 12-inch (30-centimeter)-long nectar tube of another Madagascan orchid, *Angrecum sesquipedale*. In 1903 a sphinx moth with a proboscis that matched Darwin's proposal was identified. Dr. Kritsky believes that his proposed pollinating moth may also be a member of the sphinx-moth family. But scientists will need to brave the wilds of the Madagascan jungle in order to find this hypothesized species. Unfortunately, the great expense of such an expedition and the current unstable political climate of the region tend to discourage even the most valiant scientist from combing the Madagascar jungle for the giant moth.

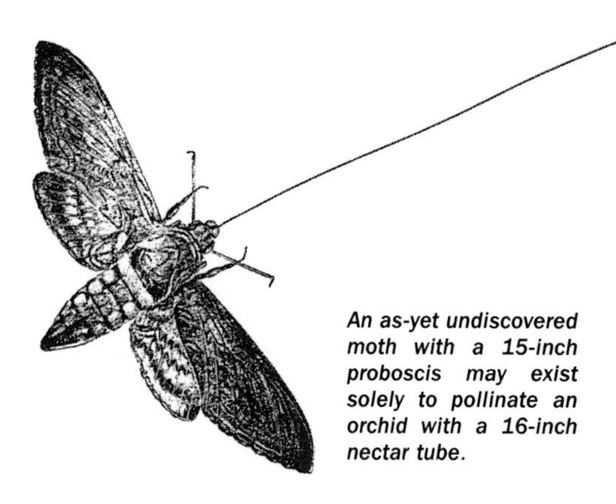

An as-yet undiscovered moth with a 15-inch proboscis may exist solely to pollinate an orchid with a 16-inch nectar tube.

A better understanding of how hibernating bears use calcium waste products to manufacture new bone tissue could someday lead to a significant breakthrough in the treatment of osteoporosis in humans.

THE BIG SLEEP

Hibernating bears have long puzzled zoologists for their amazing ability to seemingly defy some basic rules of mammalian physiology. For up to five months, bears are able to sleep comfortably without the need to eat, drink, defecate, or urinate. In all other animals, this prolonged inactivity without nutritional sustenance or waste disposal would result in such problems as bone-thinning and a fatal buildup of toxic urea, a waste product normally eliminated by the kidneys in urine.

Recently, Ralph A. Nelson, research director of the Carle Foundation in Urbana, Illinois, and colleagues have found that hibernating bears rely on internal recycling to survive. While bears sleep, their bones leak calcium. Instead of excreting this calcium, the bear's body is able to reuse it to build new bone. Furthermore, hibernating bears are able to reabsorb urea through the bladder wall and break it down into amino acids, which are then used to create new proteins.

This revelation of the internal chemistry of hibernating bears may have some wonderful applications for human health. Researchers hope to isolate chemicals from hibernating bear blood that may be effective in treating humans suffering from osteoporosis, a bone-thinning disease, as well as kidney failure, in which fatal levels of urinary toxins accumulate.

SOCIAL STATUS AND THE BRAIN

For the first time, scientists have uncovered persuasive evidence that social behavior can help sculpt the structure of the brain, which, in turn, can influence animal behavior. Studying the African cichlid fish, neurobiologist Russell Fernald of Stanford University found that domineering males who rule large territories while keeping encroaching male competitors away have brain cells in the hypothalamus that are six to eight times larger than the cells of their counterparts who have no social ranking. It is these brain cells that allow the fish to mate.

Should the domineering fish be challenged and defeated by a larger male, the nerve cells of the defeated fish rapidly shrink, Fernald found. Consequently the male's testes also become smaller, eventually robbing the fish of its desire and ability to breed.

Dr. Fernald suggests that these findings may be translated to the human brain as well. Since the brain cells studied are conserved across the evolutionary spectrum, the structure of the human brain may also be affected by a person's behavior. The findings of this study may have particular implications for Simon LeVay of the Salk Institute in La Jolla, California, who has discovered that the hypothalamuses of homosexual and heterosexual men apparently differ in size.

Fernald's study has forced scientists to conclude that the old nature-vs.-nurture controversy, which questions whether biological or environmental factors have the greater influence in determining an animal's behavior, is more complicated than was originally believed.

Lisa Holland

TECHNOLOGY

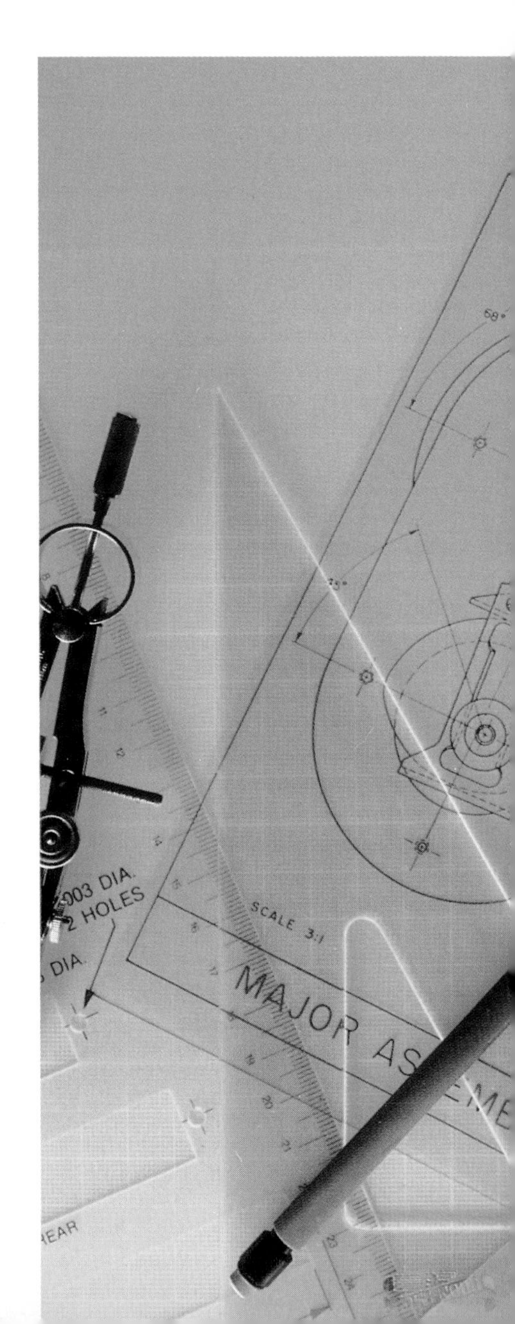

CONTENTS

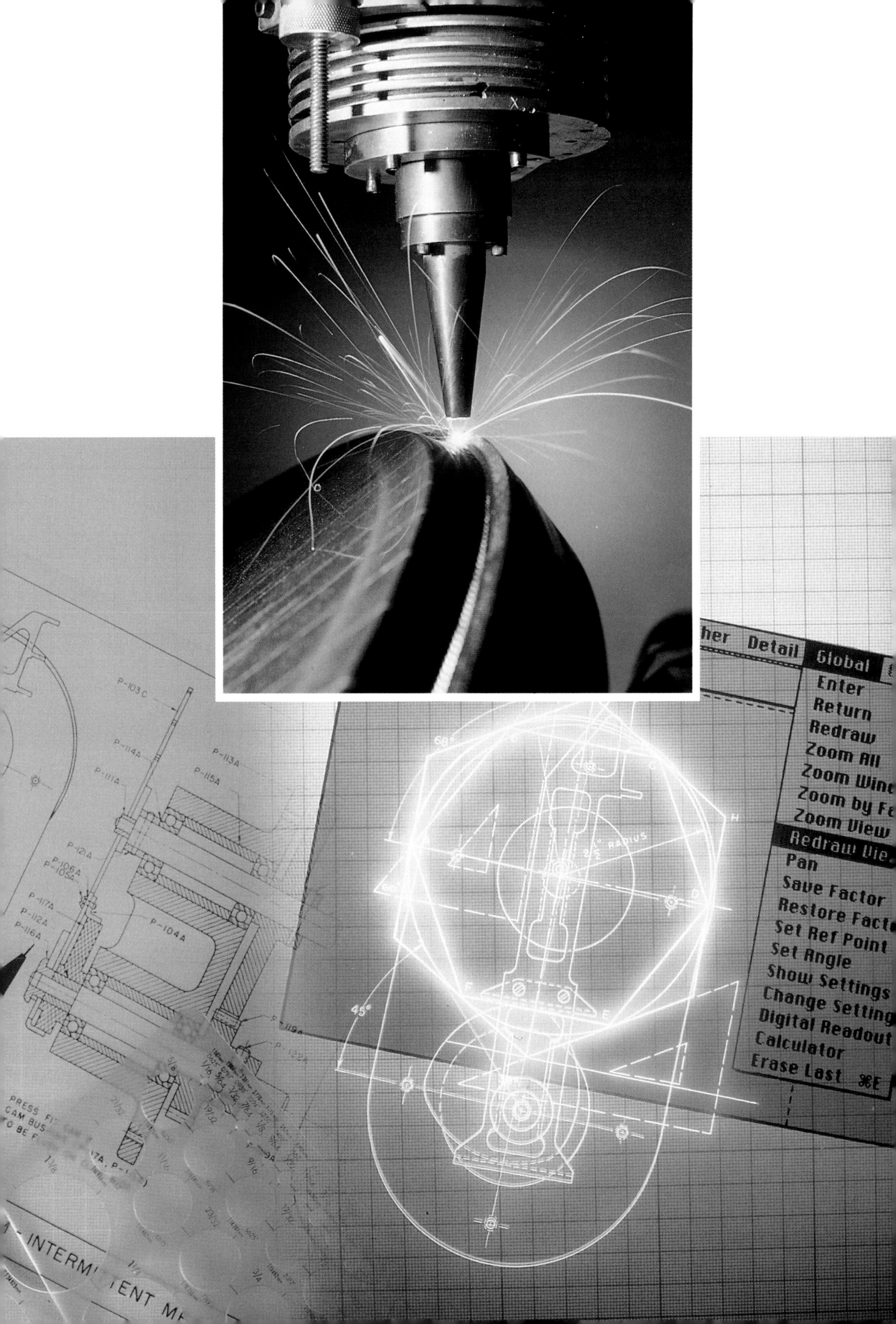
Detail
Global
Enter
Return
Redraw
Zoom All
Zoom View
Pan
Save Factor
Set Ref Point
Set Angle
Show Settings
Digital Readout
Calculator
Erase Last ⌘E
P-103C
P-114A
P-113A
P-115A
P-111A
P-121A
P-106A
P-105A
P-117A
P-112A
P-116A
P-104A
45°
2 1/2" RADIUS

THE NEW COMMUNICATIONS AGE

by James R. Chiles

The time is nearly 5:00 A.M. and two US West technicians are looking at me expectantly, their hands full of clotheslinelike ropes that lead into tall racks of brass-and-steel telephone equipment. Dennis Winter, the switching supervisor, is watching a big clock on the wall tick down the seconds. After months of planning, US West is about to cut this sturdy old "step-by-step" switching system off at the knees. "All right, it's time," Winter says. He nods to me to give the signal.

"Go ahead!" I say, and the technicians start yanking their ropes with all the energy of drowning men. Hundreds of tiny ceramic-coated circuit breakers, the size of necklace beads, fly from the panels and skitter across the brown linoleum floor. One of the workers flings his ropes aside and produces a small pair of scissors. He then begins snipping through rows of tiny blue wires as fast as he can. Every phone in the town of Fairfax, Minnesota, goes dead. For all we know, someone might be frantically trying to call the police right now, so each second counts.

Winter passes a few words over an intercom, and more technicians in a new building next door begin flipping plastic guards away from a brand-new bank of equipment. And with the touch of a few keys, a computer-operated switch takes over the duties of moving Fairfax's calls around. In the space of one minute, 1,280 phone owners in this farming community have taken a nearly 100-year leap in technology. Before, they had a motorized-gear switching system whose design originated with a Kansas City undertaker in the 1890s. Now they have a system to carry the town well into the 21st century, and it's linked to telephones throughout the world by fiber-optic cables.

This is just one scene in a quiet transformation that's under way across the entire length and breadth of a $250 billion-plus network that evolves not just daily but hourly under the aegis of America's 1,442 telephone companies. These changes are mostly out of sight, but your ears have probably noticed one of them: long-distance calls so static-free that you can't distinguish them from local calls. Nearly every time you phone long-distance these days, your voice travels in the cold, clear, digital language used by fiber optics, the fine glass strands that convey information as pulses of light. "All over the country, we're wiring up for the new communications age that's coming," says Frank Hawkinson, spokesman for the National Exchange Carriers Association, a telephone-industry cooperative.

The nation's phone system is doing the job that Alexander Graham Bell envisioned when his patent was issued in 1876—connecting nearly anyone's voice to anyone else's ear at an affordable price—and a lot more besides. At present the lines handle vast quantities of faxed documents and computer transmissions, but the network shows signs of evolving over the coming decades into an even more sophisticated conduit of information, one that can bring incredibly vivid pictures directly into the home.

The telephone has come a long way since Alexander Graham Bell invented the first crude transmitter in 1876 (facing page, top right). Today's phone systems can meet virtually every communications need.

When the Systems Break Down

Every transformation creates upheaval, however, and the advances have eroded some long-standing traditions of telephony. What was once a stable system of interlocking monopolies has grown nearly riotous with hundreds of competitors, equipment choices, and service options. And what had always been a remarkably reliable system has suffered several widespread failures in the past few years, attributable to the newest equipment. A problem in AT&T's long-distance network one day early in 1990 lost tens of millions of calls. A bad computer program in Baltimore once cut off nearly 7 million phones. A backhoe cut an MCI cable in 1990 and suspended the company's long-distance service to portions of five states.

Until something like this happens, most of us take the telephone for granted. You just pick the thing up, dial, and talk, right? No big deal. In fact, ensuring that caller and called are able to find each other through billions of miles of phone lines and billions of circuits is a big deal, and such arcane work that it has given rise to a profession that in some ways resembles a technological priesthood. Nearly all the people I interviewed for this story have spent their entire careers in phones, and appear happy to spend the remainder doing the same thing. With so many companies intent on downsizing and laying off employees these days in order to remain competitive, job security isn't what it once was. The work itself, however, continues to get more and more complicated.

That's because the phone business is no longer just a phone business. It's a computer-driven telecommunications network that in a very real sense constitutes the electronic lifeblood of countries all over the world. Everything depends on it—communications, transportation, education, businesses, hospitals, you name it.

To get an idea of how complex the modern-day communications business is, consider the problem of connecting just three or four phones: all you really need is a line from each phone to all the others. But connecting a hundred phones would require thousands of lines if you tried to hook each one to every other one. Instead, the network connects all the phones in a given neighborhood directly to a central point, called an exchange. The three-digit prefix on your local phone number is your exchange. Every exchange is in some kind of a building and has banks of connecting equipment known as a switch. When you call a neighbor down the street, your local exchange can handle the entire connection inside its own switch. But when you dial a friend with a different prefix, your exchange must pass your call to your friend's exchange, through high-capacity trunk lines stretching between the exchanges. The switch at the other exchange takes the number, checks to see if your friend's line is busy, and if not, rings the phone.

One by one, the antiquated circuit-breaker telephone systems in small towns are being dismantled (below) and replaced by computers.

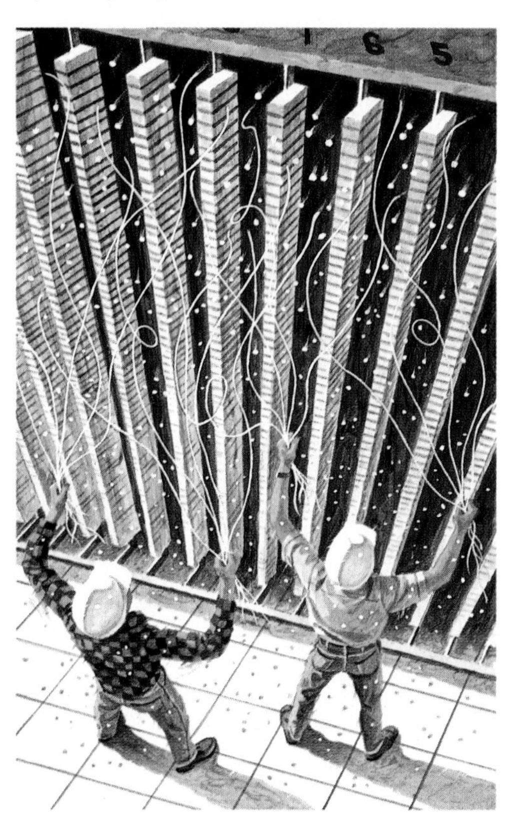

Interconnecting Exchanges Go the Distance

For calls crossing area codes, at least two more switches must enter into the transaction. These switches connect your call with cross-country trunk lines owned by well-advertised companies like MCI, AT&T, and US Sprint. Consider a long-distance call to New York City from an AT&T customer living in Fairfax, Minnesota. The call first travels via US West's underground fiber-optic cable to Marshall, Minnesota. There, US West hands it off to equipment owned by AT&T, which proceeds to send the call over its own microwave towers and fiber-optic cables to one of its long-distance switches located in Minneapolis.

US West does more to handle a call from someone who subscribes to a smaller long-distance company. If the call is from a subscriber in Fairfax who has signed up with MCI—which has fewer feeder transmission lines—US West will carry it all the way from Fairfax into Minneapolis and deliver it to MCI's switch downtown. If the subscriber belongs to US Sprint, US West delivers the call to Sprint's switch, which is located in nearby St. Paul.

All of the long-distance companies run the calls out of the Twin Cities to New York City on their own networks. AT&T owns the most transmission equipment, but the latecomers can charge competitive rates. Because AT&T has been around for such a long time, it has to serve a great many markets that do not generate much long-distance business. The newer, smaller companies are free to concentrate their resources on the best long-distance markets.

With the exception of AT&T, each of the many long-distance companies operating in this country is a relative youngster. In 1984 a court-ordered divestiture broke up AT&T (a.k.a. Ma Bell) into AT&T and Bell Laboratories on one side, and seven regional "Baby Bell" companies and the new research organization Bellcore on the other. AT&T now handles long-distance traffic; the Baby Bells carry local calls. But the judge's order also opened up long-distance service to competition. How has it worked out for the customer? People who make a lot of long-distance calls "are substantially better off," says Alfred Sikes, chairman of the Federal Communications Commission (FCC).

Most local telephone markets are still monopolized by one company, but here, too, competition is growing. In some cities, long-distance customers already can go around their local phone companies entirely by leasing lines from other carriers that connect them directly to a long-distance company. The monopolies might fade even more if the FCC agrees to open up local phone service to competition in delivering local calls. Last October a federal court gave the go-ahead for the regionally operated Baby Bell companies to begin offering computerized information services such as electronic Yellow Pages. Local phone companies also would like the government's permission to deliver high-quality television signals directly to the home over fiber-optic lines, enabling them to go head-to-head with local cable-TV companies.

The lively competition that resulted from the AT&T breakup has its origins in the early days of the phone system. At the turn of the century, some cities saw bitter rivalries develop between two and even three local phone companies. Each of the companies ran its own nonconnecting system.

In Bloomfield, Iowa, for example, Bell had the long-distance connections, but hardly any customers; the other phone company had the customers, but no long-distance link. Denied the use of his competitor's phones to notify Bell customers of incoming long-distance calls, the local Bell manager had to go out on the office balcony with a megaphone and bellow out the names of people who had calls waiting.

Nowadays the companies work together, and although we consumers seldom stop to think about it, the results are amazing. Typical calls take only a few seconds to connect across the country, even though they may pass through four companies and half a dozen types of switches. Each month, many local companies bill their customers and, in some cases, pass the long-distance receipts on to the long-distance companies, who then return some money to the same local phone companies to pay for the use of local lines. Specialized clearinghouses have been set up to handle this mass of accounting.

In the 1890s a Kansas City undertaker invented a motorized-gear switching system that many communities still rely on for telephone service.

The preferred route of a call across the nation is a straight line, but there are many exceptions. During busy times, such as the peak hour just before lunch each Monday on the East Coast, long-distance calls are sometimes required to take dramatic detours. When trunk lines between New York and Atlanta fill up, for example, calls between those cities may detour as far west as Los Angeles to find a spare line.

At AT&T's Worldwide Intelligent Network Operations Center in Bedminster, New Jersey, it's easy to get the feeling from the daily ebb and flow of these calls that the system has taken on a life of its own. Managers here, seated at mission-control-like consoles, keep an eye on a two-story-high wall of video screens. For a typical business day's volume of up to 140 million calls, AT&T's system is smart enough to run itself. When calls climb considerably above normal, though, managers stand ready to help the computers manage the traffic over the company's 118 switches.

AT&T's biggest day of the year for sheer number of long-distance calls is the Monday

Caller A phones neighbor B through her local exchange. But should A call across town to C, her exchange 1 switches to C's exchange 2. When A dials D or E, the call goes through exchanges 1 and 3, via trunk line 4, to exchange 5, then to 6 or 7. Caller A's call to F goes through overseas exchange 8 and then by satellite 9 or cable 10 to exchange 11 in France to produce a long-distance call that is so static-free it sounds like the caller is next door.

following Thanksgiving. The day on which people spend the most time on the phone is usually Mother's Day. And television can turn an ordinary day into an avalanche: when the MTV network offered to give away singer Jon Bon Jovi's house, in the space of just half an hour viewers tried to place 3 million calls to the advertised number. Employees at the operations center still remember the night when Eddie Murphy, hosting "Saturday Night Live," asked viewers nationwide to call in and vote whether he should drop a crustacean dubbed Larry the Lobster into a pot of boiling water or let Larry go. Nearly half a million people responded by phone. Most of them wanted Larry's life to be spared, but Murphy, ignoring his own poll, went ahead and dropped him in the pot anyway.

Within five minutes of the California earthquake in 1989, AT&T's wall of computer screens showed that the national calling volume was double the usual level for early evening. Network managers therefore used their computerized controls to suppress the calls flooding into California so that calls could come out.

When a mysterious force began to take AT&T's switches out of service on January 15, 1990, "we knew we had something out of the ordinary right away," says Jim Nelson, district manager at the Network Operations Center. Nelson speaks in a tense voice reminiscent of actor Al Pacino's. Around AT&T he's known as the Doomsday Man, because a message from him may mean trouble brewing on the network. Without knowing exactly what the gremlin was, AT&T's experts were able to get a grip on their unruly system within a few hours by diverting calls from the distressed cities. Engineers isolated the difficulty: a faulty computer program was "confusing" some switches and causing them to shut down temporarily.

Driving a "Pig" Through a "Pipe"

In the same way that they have devoted themselves to state-of-the-art switches, which are really just huge computers, AT&T and its competitors have cast their lot almost entirely with laser-driven fiber-optic cables for the transmission of calls. Most of US Sprint's 23,000-mile (37,000-kilometer) fiber network lies buried along railroad rights-of-way. WilTel, of Tulsa, Oklahoma, owns 11,000 miles (17,700 kilometers) of communications network, and nearly all of it is fiber cable. A subsidiary of a petroleum pipeline company, WilTel has uncovered an ideal location for almost a third of all its cable: obsolete pipelines, including one that stretches from Kansas City to Los Angeles. The technique: workmen cut a hole in the pipe every mile or so, and use compressed air or water to drive a ball, or "pig," inside the pipe, between the holes. The pig drags flexible hose behind it. Fiber cable goes inside the hose. Installation is faster than automatic trenching machines, which have to cut a deep slot in the ground. And because a pipeline is already in place, WilTel can avoid months of searching out and negotiating with new landowners for rights-of-way.

Fiber-optic cable's present call capacity is only a fraction of its potential. "If we had all the necessary electronic equipment at the ends," says Frank Gratzer, a manager at Bellcore, "you could carry the entire telephone traffic in the United States with one fiber, and that's at the busy hour, combining both voice and data traffic. The capacity is essentially unlimited." He adds that it might be decades before we can invent the electronics to harness this capacity, however.

All telephone companies do their utmost to prevent damage from occurring to the cables. They plant thousands of signs warning of cables nearby and advertising a toll-free number to call before digging. Every day, a fleet of airplanes informally called the AT&T Air Force patrols 40,000 miles (64,000 kilometers) of cable routes, as well as the stretches of continental shelf where transoceanic cables come within a few hundred feet of the surface. A shrimp boat spotted working near a cable will receive a flurry of message tubes from above, warning the skipper away.

Gil Broyles, a WilTel spokesman, recounts a 1990 incident in which a company inspector in Missouri, making his rounds, noticed a farmer operating a tractor with a power auger, digging postholes along a cable route. The inspector stopped his car and walked over to offer a warning. "The farmer was only on his second posthole when the inspector came up," says Broyles. "Unfortunately, he cut our cable on the first posthole."

Fiber optics' potential is promising enough to trigger hours of daydreaming about

A cable ship lays fiber-optic transoceanic lines. The yellow "repeater" on the cable regenerates the telephone signals. The fiber-optic lines can carry many more calls than can the older coaxial cables (inset).

what the future might hold if every house had a hookup in the wall. How about the entire contents of the Library of Congress flashing through every few minutes, making any book or record or television show ever produced available for your computer to grab so that you can browse it at your leisure? Perhaps someday, but don't burn your local library card just yet. "Most changes in the system are quite incremental, because the invested base is so huge," says Frank Gratzer of Bellcore. "You just can't make changes overnight." Running fiber optics to every home in the country will require a major effort that could take decades.

There are many examples of new equipment operating side by side with the old. The onshore equipment for the new transatlantic fiber-optic cable, TAT-9, is housed in an addition to the same red-brick building, dated 1940, on the New Jersey coast that houses radio operators for AT&T's High Seas Radiotelephone Service. High Seas operators have been placing phone calls to and from ships via shortwave radio since 1929.

Another good place to see the old and new together is on Hudson Street in Manhattan. Here, in the basement of an old brick building, is what looks like a disaster waiting to happen. In this one place, the telecommunications links of six long-distance companies share a common exchange. If an accident of some kind—a fire, say, or a power failure—were to disable some or all of this equipment, the impact on phone and data traffic could be incalculable.

A Telephone in the Outhouse?

Then there's the telephone system in Wawina, Minnesota. It's served by Northern Telephone Company, which is headquartered in the basement of Bob Riddell's farmhouse. Northern's 34-year-old switching gear is in a concrete-block building across the county road from Riddell's mailbox. Northern's customers have had one-plus dialing and direct international dialing for years. With 33 paying phone lines, Northern is exactly one line larger than the smallest known American telephone company. Diminutive though it is,

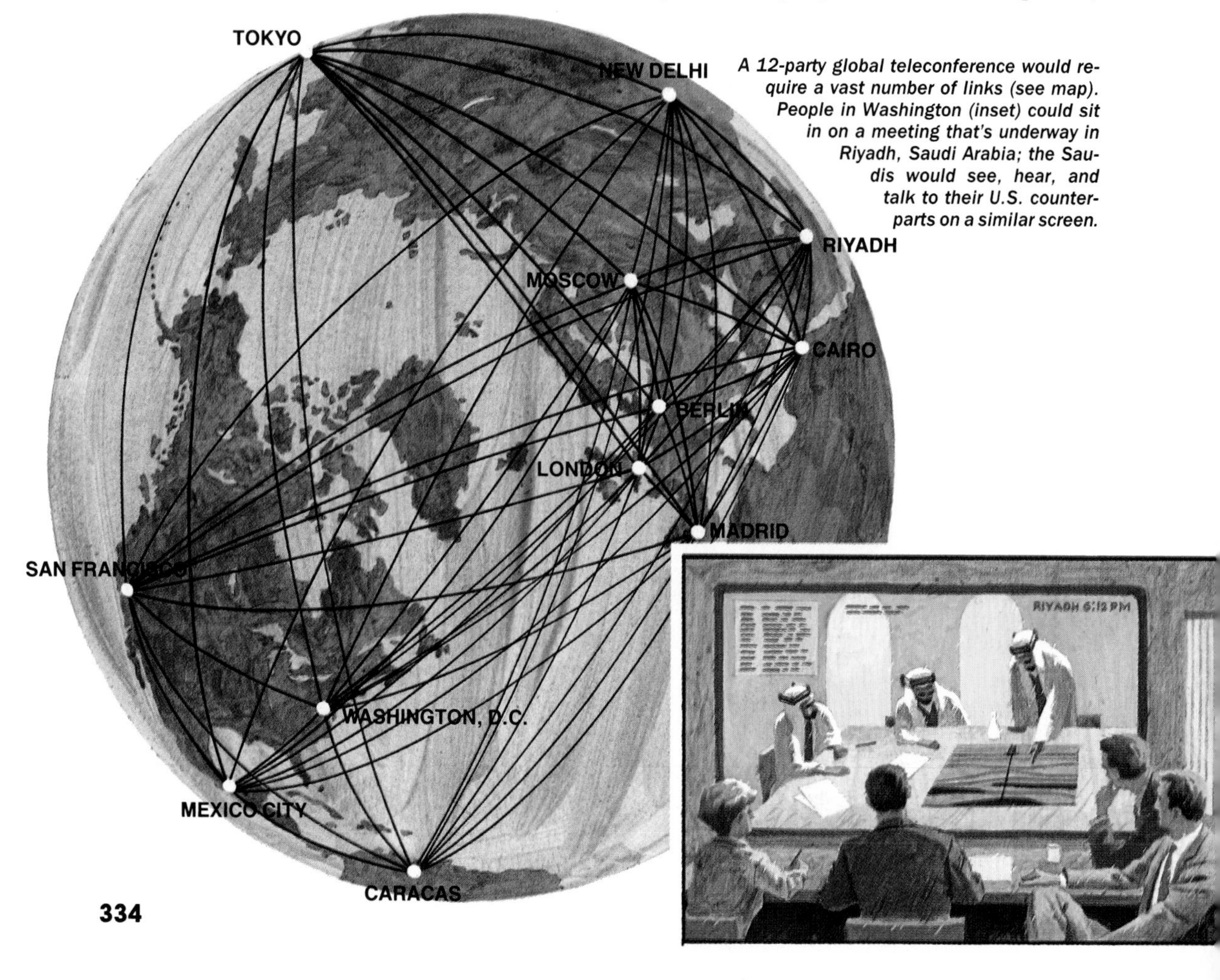

A 12-party global teleconference would require a vast number of links (see map). People in Washington (inset) could sit in on a meeting that's underway in Riyadh, Saudi Arabia; the Saudis would see, hear, and talk to their U.S. counterparts on a similar screen.

Riddell says Northern receives paperwork and product literature from telecommunications services and businesses by the boxload anyway. "They don't care if you have 30 subscribers or 30,000," he says. "But anything I don't understand goes in the black file in the other room." He explains that the "black file" is a wood stove. "If it's important, they'll call back, and I ask 'em to send another copy," he adds. Besides taking care of his subscribers' phones, Riddell also maintains 30 phones that are scattered around his farm. He can call home from virtually any location, including the clothesline, the outhouse, and the chicken coop. "I can even dial 117, 122, and 132 in the morning and get the kids out of bed," he says.

Riddell has no immediate plans to replace his old mechanical switch, because it still does the job. The new computerized switches are oversize for such a tiny phone system, he says, and they need more care and feeding than he can afford. "Once you buy one, you're always buying software upgrades."

In time, though, even Riddell's old equipment will pass on, making room for the latest gear. What might we expect in the coming decades? Phone companies are very excited about the prospect of something called the personal communication network, or PCN. This is a wireless pocket phone you could take to any city and be in constant touch. Just think: someday people might receive personal phone numbers at birth and carry them the rest of their lives. Much of the basic technology needed to create a global network of personal phones already exists.

Many in the industry think Americans might buy more than 100 million PCN telephones, says Ashok Ranade, a Bellcore manager in radio research. But expect some holdouts. "A lot of my friends would abhor the idea of having a phone in their pocket all the time," says John Davis, vice president of network development for Bell Laboratories.

In the years ahead, telephone companies would like to break free of their traditional reliance on handling voices, and leap over conventional television into the business of delivering high-definition television (HDTV). Already Japanese households can receive HDTV signals for an hour each day over tiny satellite dishes. Researchers at Bellcore are convinced that for this country, it makes more sense to pump HDTV into millions of homes over fiber-optic phone lines.

Scientists at Bellcore's different locations in New Jersey can chat with one another, face-to-face, through a "VideoWindow" designed by the company. I watch it work at a Bellcore building in Red Bank, New Jersey. It's an 8-foot (2.4-meter)-wide rear-projection TV screen, framed by wood, and set into a wall about midway between floor and ceiling. Four Bellcore employees walk up and sit at a table positioned in front of it, and look across the table at a life-size image of four of their coworkers taking their seats in a similar setting many miles away. The signals pass through a high-capacity phone line.

This is fine for old-fashioned meetings, but a lot of people do all of their work on computers. Sudhir Ahuja believes it will be possible to bring teams of these people and their computers together, through the telephone lines, in what he likes to think of as an imaginary conference room. Ahuja's testing ground is a fourth-floor workshop at AT&T Bell Laboratories, near Holmdel, New Jersey. Ahuja, smartly dressed and brimming with enthusiasm, is the head of Bell Labs' research department for integrated computer communications. His group is trying to work through all the details of a computer-telephone system that, in effect, will allow up to 12 participants anywhere in the world to work on the same project while looking over everybody's shoulders.

Ahuja demonstrates by sitting me down in front of an oversize computer screen while he calls up several other offices' phone numbers. A row of tiny video pictures gathers at the upper left to show the people in our conversation, each at their own computer. We can talk via speakerphone. We can bring new documents to the conversation and show them on the screen, including videotapes. And we can all make changes to columns of text on the screen.

In a full-blown conference, the participants would call in experts via the phone lines when needed. Those experts would in turn call still other consultants from around the country, who would throw in files of their own. All on that one screen. It leaves me feeling slightly dizzy. "Of course, the billing is going to be horrendous to figure out," Ahuja says, adding, "I hope somebody else can solve that problem."

Technological innovations have transformed the news media. When Hal Buell (above) joined the Associated Press 30 years ago, it took three hours to get news pictures to his editors. Today, it takes six minutes.

WHAT'S NEW IN THE NEWSROOM

by Dana Blankenhorn

The heart of the news is what's new. Technology now brings news to you faster and with more detail than ever before. Television networks are becoming libraries, selling their archives, while newspapers are using the telephone to deliver the news you want even faster than TV can. In today's newsroom, words and pictures move from the field to the press without touching paper. TV networks will soon put their video into the same computers that hold their scripts. And the past is far more accessible, not only to reporters, but to readers too.

Electronic Darkrooms

When Hal Buell joined the Associated Press (AP) 30 years ago, it took him a full three hours to move news pictures to his editors. He carried a complete darkroom to news sites, with two trunks of equipment. "You processed the film, made the prints, then wrapped prints around a drum linked to a telephone line," he recalls.

Now, as an assistant to the president of AP, Buell is introducing PhotoStream, which works with a new electronic darkroom that AP is installing at 1,000 member papers nationwide. "We transmit color pictures by scanning a negative. It moves to an editor's computer in just six minutes, where software reformats it onto a television screen." Then editors around the nation can work with the picture on their screens and output a finished page to the press.

"The larger future is to make pictures digitally," adds Buell, "with video still cameras." Add a portable satellite antenna, which fits in a suitcase, and AP will soon send pictures to affiliates almost as quickly as television networks do, all without chemicals. "Words have been handled electronically for 10 to 15 years. Now pictures are catching up."

USA Today was a pioneer in sending pages to printing presses by satellite almost nine years ago, and still sends them out that way, in a fax format. *The New York Times* will soon go beyond that, notes Lawrence Kingsley, publisher of the *Telepublishing Report,* based in Cambridge, Massachusetts. The *Times'* new system will use modems to send whole pages, in a computer format, to remote printing sites. "This lets the remote plant put in an ad or a local story without additional labor," notes Kingsley. The result will be national papers that look more like local papers—and, ultimately, more competition for smaller newspapers.

Computer-assisted Reporting

Computers can also help reporters make news. Elliot Jaspin, who won a Pulitzer Prize in 1979, teaches a technique called computer-assisted reporting at the University of Missouri School of Journalism.

Jaspin first used the technique in 1985 at the *Providence Journal-Bulletin.* For one story, he matched the names of 1,200 school-bus drivers with 600,000 driving and 500,000 criminal records, all in less than a single week's time.

Jaspin then confronted the drivers with what he'd learned. "That step is crucial. The worst story you can write is where you just analyze a tape and start writing," he says. Since joining the Missouri faculty, Jaspin has trained about 50 reporters. Others will soon analyze data from the Office of Thrift Supervision, which is overseeing the savings-and-loan bailout. The full course costs $500 to $1,000, depending on a paper's circulation or a TV station's market size.

TV stations aren't standing still, either. BASYS Inc. of Yonkers, New York, sells complete TV-news computer systems. "All of the systems today do text editing, schedules, and assignments, scanning the wire services and other sources, and linking to databases on command," says BASYS customer-support manager John Chapman.

But now the computers can control other machines, Chapman adds, "like the character generators used to put names and titles on the screen, the cameras which show the

Today's television meteorologists point to a blank screen to describe the movement of weather systems. At home, viewers see a composite of the meteorologist superimposed in front of a computer-generated map.

anchors, the still stores which create slides shown behind the anchor's head, the TelePrompTers holding the script, and closed-captioning systems for deaf viewers."

For Steve Schwaid, assistant news director at KYW-TV in Philadelphia, Pennsylvania, this means lower costs: "Instead of having people at three machines, I'll have someone at one terminal."

A typical BASYS user like Eric Gershon, weekend executive producer at the Cable News Network (CNN) in Atlanta, Georgia, may now spend half his time in front of a computer screen. First he'll read news wires, schedules of upcoming events, directives from the bosses, background from reporters in the field, and lists of stories available. Within two hours of airtime, he must print a list of stories to five or six writers, who will create TelePrompTer scripts. To meet this deadline, both a producer and his assistant must stay at their screens. While one decides on stories to emphasize, the other consults the computerized tape file.

At CNN, reporters and producers send two types of electronic mail: a one-line message for the top of their screens, or a longer note saved in an electronic mailbox. That holds true whether the colleague is in Atlanta, Georgia, Washington, D.C., or New York City—all CNN bureaus are linked to the same computer. The cacophony of the old newsroom has been replaced by the quieter hums of chitchat and terminals.

As airtime approaches, the producer can still see his plans bumped by breaking news from the assignment desk. On their terminals the writers can change scripts right up to airtime. Even while on the air, anchors have a computer screen beside them, which can de-

The busy CNN newsroom at left is not as noisy as it would seem. Much of the desk-to-desk and office-to-office communication takes place on the computer; each day producers and writers spend hours on end at their terminals. The information that they select for broadcast is edited down into brief scripts and transmitted to the anchors. The anchors then deliver the news; the audience rarely sees more than the broadcast studio (above).

liver fast-breaking stories or order them to introduce a live feed from a news site. Producers like to say they can "rock and roll" at a moment's notice.

Dream Systems

Despite all the advances, Don Cesare, vice president-operations for CBS News in New York City, says newsroom computing can go further. And he's putting CBS computer programmers to work on it. "Our goal is to develop a system that enables us to process all information through a computer—data, text, images, sound, and video," he says. "We have been working on a design that does not now exist." And they started the work several years ago.

Cesare says his dream system is still five years away, and will be built one piece at a time. Peter Doherty, operations producer for ABC News in Washington, D.C., agrees with the timetable, but issues this warning: "With a large system, if something breaks, you've lost everything, including the ability to run newsroom machines."

Once the day's news becomes history, news tapes are no longer erased. They're cataloged and filed for sale as clips. CNN gets hundreds of requests each week for pictures of the Capitol, the New York Stock Exchange, other news sites, and important events.

ABC News has gone further with its ABC News Interactive division, creating multimedia compact discs (CDs) from archive files. These CDs join words, pictures, and sound with computer interaction to teach such topics as the life of Martin Luther King, Jr.

The Associated Press is also offering old news through a service called GraphicsBank. Lee Perryman, deputy director of broadcast services in Washington, D.C., explains: "If you're sitting in a TV station and someone dies, you need a picture. You type the name into a PC [personal computer], and it dials up our database so you can buy images over the phone." GraphicsBank contains 65,000 images, including recent photos of government officials and corporate executives. Perryman will also sell AP wirephotos to TV stations for use in introducing stories.

CBS' Cesare says his newsroom can also make better use of its own archives. "Seventy percent of what goes on air has archival material," he says. "We want anyone, anywhere in the world, to access the archives. In the past, you'd call a broadcaster or have an archivist search the database. Now you'll search the database directly, and file the order."

Cesare also sees archives as a profit center. "We have deals with shows like 'Instant Recall' to use our archival material," he says. CBS wants to sell clips to businesses, but such sales must meet certain standards. "We've been reluctant to identify our product with commercials," Cesare says. Material such as cityscapes of Chicago is sold routinely, but CBS reports on events like the *Exxon Valdez* spill present a problem: "We decided against that sale, because we felt uncomfortable about controlling the future use of the material."

Telephone News

Besides using computers to create better products, newspapers are also using them to reinvent themselves. *USA Today* offers information by phone at 75 cents per minute, and offers an on-line computer service called Sports Center, which costs $29.95 for software and $4.50 per hour to use. The *Wall Street Journal* offers JournalPhone, which charges under $1 per minute for stock prices and business news. The *Washington Post* and *Atlanta Journal and Constitution* are among papers offering free phone services backed by ads.

The Atlanta paper's service, 222-2000, offers sports scores, soap-opera digests, and stock listings. It also lets buyers of classified and personal ads rent voice mailboxes and receive replies right away. Employers can use the mailboxes and 222-2000 to ask questions of job prospects, and give them replies within a day or so.

In its first year of operation, 222-2000 drew over 5 million calls. The *Atlanta Journal and Constitution* is now the fastest-growing urban daily newspaper in the United States, with morning circulation up 11 percent over the past year.

Chris Jennewein, director of information services for the newspaper, says phone services are putting papers back into the news business. "Back in the 1910s, papers erected billboards on corners. The idea was to get people the news fast, before radio and TV were around. With 222-2000, and your touch-tone phone, we can now give you the news you want more quickly than with radio and TV."

But the Atlanta paper goes further, helping reporters make news with "sound-off" voice mailboxes, which let them gauge reader reaction to news events. Jennewein also has operators using PCs and phones to conduct political polls and take orders for local merchants. "Because we had the operators, we were able to sell $100,000 worth of T-shirts displaying our front page on the day Atlanta won the Olympics," he adds.

Jennewein's newest innovation is Access Atlanta, a computer service accessed with a PC and modem. It offers news wires like AP and Reuters, back issues of the newspaper, and the text of classified ads before the ads are printed, as well as messaging services and other news wires the newspaper doesn't have room to print. Jennewein hopes to draw thousands of paid subscribers to Access Atlanta within the first year of operation, with prices starting at $6.95 per month.

Other local newspapers with on-line services include the *Fort Worth Star-Telegram,* which was the first to market with its StarText service, the *Omaha World-Herald,* and Long Island's *Newsday.*

Computers are changing the face of journalism. On the front end, they're bringing news in from the field faster than ever, making newspeople more productive and giving them new tools for news gathering. On the back end, they're creating new competition between urban and suburban papers, building new profit centers in old archives, and turning the daily paper into a veritable information pipeline.

Chris Jennewein might be guilty of understatement when he says, "This is an exciting time to be in journalism."

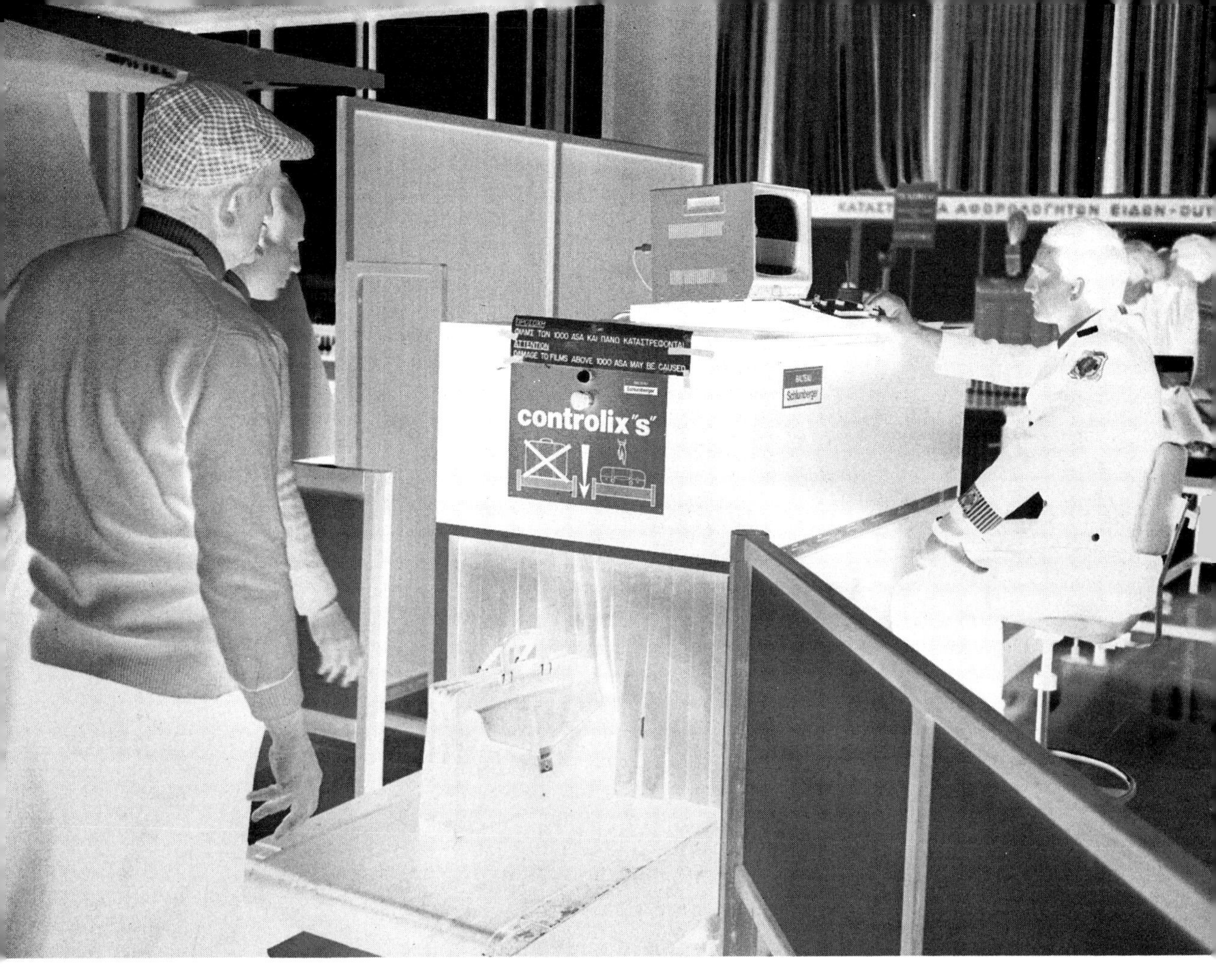

THE BOMB CATCHERS

The technology used by terrorists has long outpaced efforts by airlines to detect weapons and explosives in luggage. That should all change with the introduction of a variety of new bomb-detection innovations.

by Judith Yeaple

On May 18, 1990, a medical doctor boarded a British Airways flight from London to New York City. Although Dr. Jim Swire's baggage was examined at Heathrow Airport as is customary, the guard failed to discover the metal detonator and lump of marzipan (a dense, pastelike candy that can be used as an explosive imitator) stuffed in a radio cassette player.

Swire had constructed his "bomb" to closely resemble the one used by terrorists to down Pan Am flight 103 over Lockerbie, Scotland, on December 21, 1988. His motive: to show that security had not improved since the Pan Am bombing, the incident that killed his daughter and 269 others.

Airports Ill Equipped

Despite 1,379 deaths from 1975 to 1989 due to suspected terrorist bombings, the vast majority of U.S. and international airports are still equipped with the crudest of weapons-detection systems—magnetometers (metal detectors) and X-ray equipment—put in place to fend off a previous generation's form of terrorism. The guns and knives once brought on board for the purpose of hijacking an airliner have been joined by plastic explosives as terrorist weapons of choice.

HOW TO FIND AN "INVISIBLE" BOMB

BACKSCATTER X-RAY

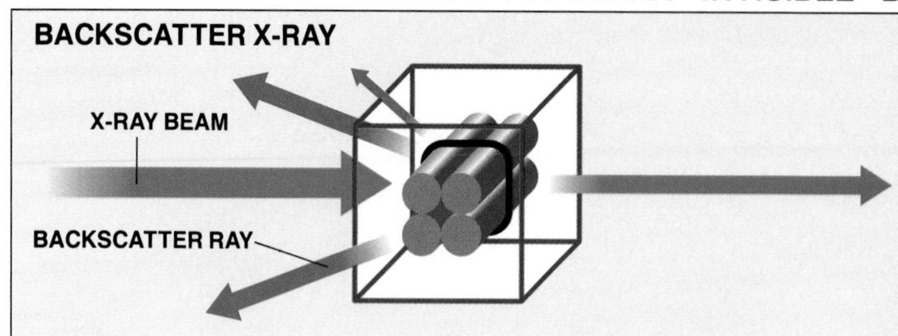

Plastic explosives are organic material. Advanced X-ray systems attempt to pry more information about organic items from bags than conventional systems can. One technique sends two X-ray beams—one high power, one low—then analyzes the characteristics. Another, called backscattering, detects both the rays that pass through the bag (green) and those that bounce back from organic compounds within the bag (red).

THERMAL NEUTRON ANALYSIS

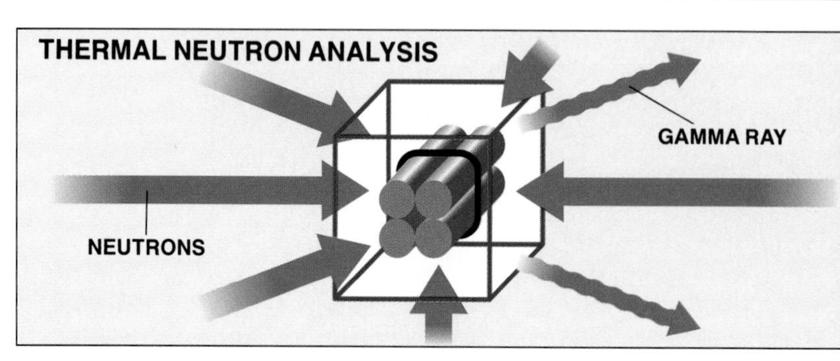

Some techniques bombard luggage with radioactive or electromagnetic probes to elicit a response from the specific elements that make up explosives. But many technologies are still in the labs; only thermal neutron analysis has been tested at airports. Baggage is enveloped in a cloud of low-energy neutrons (red), which push the atoms of nitrogen to a higher energy level. As they decay back, they release a telltale gamma ray (blue).

VAPOR DETECTION

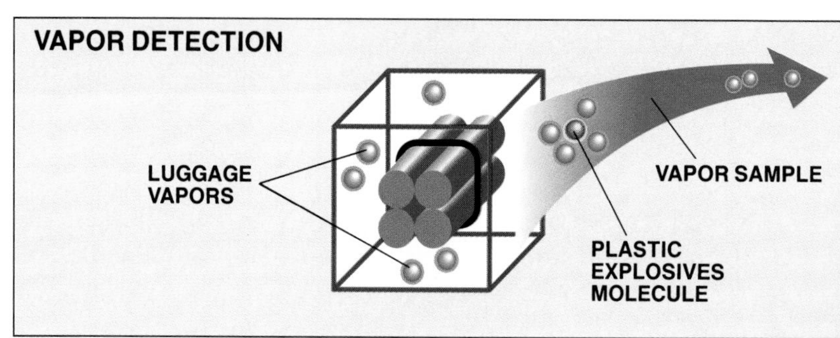

Hand-held vapor detectors, or bomb "sniffers," suck in air from the surface of the luggage. The air sample is then analyzed for traces of explosives. Because plastic explosives have a low vapor pressure, most of them will release only tiny amounts of vapor. These machines must recognize amounts at least as small as one part per trillion (blue); some experts believe the machines need to detect even smaller amounts.

According to the findings of numerous post-Lockerbie investigations, a new approach to weapons detection is required. The resulting U.S. Public Law 101-604—the Aviation Safety Improvement Act—dictates that security efforts must apply, not just to carry-on luggage, but also to checked bags, mail, and cargo loaded on commercial airliners. Additionally, standards for any given technology must be uniform, and those standards must be calibrated to the subtlety and effectiveness of today's terrorists and the technology they have at hand.

The bottom line is that plastic explosives are nearly impossible to detect. They are named "plastic" explosives, not because of their composition, but because they are malleable, like putty, and can be molded into disguising shapes, including thin sheets that slip behind the lining of luggage walls.

Plastic explosives lead a double life: when handled, they are remarkably stable. In fact, say experts at the Bureau of Alcohol, Tobacco, and Firearms, those used for military purposes must be able to withstand rifle fire without exploding. But in the presence of the proper explosive charge, they oxidize instantaneously and pack a powerful punch: they are roughly 25 percent more forceful than trinitrotoluene (TNT), and more than twice as powerful as dynamite. Many experts believe the bomb that destroyed Pan Am flight 103 could have been as small as 350 grams, or less than 1 pound.

Detecting Plastic Explosives

Furthermore, plastic explosives are organic materials made of carbon, oxygen, and an unusually generous helping of nitrogen. When screened by a conventional X-ray machine, they look like other dense nonmetallic materials, like marzipan.

Because of these characteristics, plastic explosives lend themselves particularly well to malevolent actions. "From a terrorist's point of view, a plastic-explosives weapon is ideal," remarks James Nemiah, general counsel for American Science and Engineering, detection-systems manufacturers in Cambridge, Massachusetts. "Plastic explosives are powerful, they're easy to work with, easy to hide, and, unfortunately, tremendously available."

There are only two ways, then, for a machine to detect plastic explosives. One method imitates the way bloodhounds find contraband, by sniffing for the illicit substance itself. The other capitalizes on a bomb's high concentration of nitrogen. Some machines try to locate items that are composed of a group of elements with approximately the same atomic number as nitrogen. Others may have the precision to pinpoint nitrogen atoms themselves.

According to John Wood, president of Thermedics in Woburn, Massachusetts, one of the companies that sells detection devices, "there are three keys to an effective explosives-detection system. One, the machine has to have the *sensitivity* to find the concealed explosive. Two, it must have the *selectivity* to ignore other things in the environment." A machine that too often mistakes a wool sweater for a bomb, for example, would bog down terminals with hundreds of bags inaccurately labeled as dangerous.

"And the third element," he continues, "is *speed*." During peak times, security systems face an avalanche of 600 or more bags per hour. If the system can't keep up, massive delays can result. Says one expert: "If we required the installation of a draconian system that made people show up three hours early to catch a flight to Chicago, that'd be inappropriate. The bad guys would have won."

To process baggage that quickly and accurately, machines must handle the lion's share of the job. "The concept we're working toward is security by exception," explains one high-ranking Federal Aviation Administration (FAA) official, who requested anonymity. "The routine stuff will be performed by a machine. Then, if there is a problem, a human will jump in to do what he or she does best: deal with the exception."

Joint Responsibility

Until 1985 the FAA had a research-and-development effort for explosives detection, but it was muddling along on a budget of about $1 million per year. But two terrorist incidents that year were pivotal: the three-week-long TWA flight 847 hijacking in which a U.S. sailor was killed, and the bombing of an Air India jet over the Atlantic Ocean. As Billie Vincent, former head of security for the FAA, explains, "Then Capitol Hill began throwing money at us."

With the boost in funds, the FAA pursued a few technologies, including thermal neutron analysis (TNA). By the middle of 1988, the FAA signed a contract for the deployment of five TNA devices. But these developments didn't come soon enough to protect the passengers of Pan Am flight 103, which was downed just four months later.

The tragedy turned the nation's attention again to aviation terrorism—and the blatant absence of a reliable means to detect these new bombs. President Bush named a commission that produced a fiery report in May 1990 charging that "the U.S. civil aviation security system is seriously flawed and has failed to provide the proper level of protection for the traveling public. This system needs major reform." The report added, "Only a massive effort now will bring our technology ahead of the destructive devices of terrorist adversaries."

Another report, written by the 10-member National Materials Advisory Board of the National Research Council, recommended technologies for priority funding. Among the board's favorites: improved X-ray explosive-detection systems and bomb "sniffers," as well as TNA and gamma resonance absorption, which both use state-of-the-art physics to spot nitrogen.

According to Public Law 101-604, the FAA must complete "an intensive review of threats to civil aviation." This includes types, amounts, and configurations of explosive materials that could damage an aircraft; those that could be detected by existing or "reasonably anticipated" detection technologies; and future detection technologies. Furthermore, the FAA must "develop and have in place [by November 1993] such new equipment and procedures as are needed to meet the technological challenges presented by terrorism."

The Federal Aviation Administration (FAA), in conjunction with scientists at the National Research Council, has adopted new, classified standards for explosives-detection systems. The FAA was required to establish such guidelines.

In 1989 the FAA had mandated that explosives-detection systems be placed in about 40 airport terminals; at that time, thermal neutron analysis (TNA) was the only technique approved—or available. Public Law 101-604 blocked this mandate, effectively halting the purchase of additional TNA machines until they are proven more effective. The FAA has not yet approved any explosives-detection systems under these guidelines.

DUAL-ENERGY X RAYS

How It Works:

There are several dual-energy systems. One of the most advanced, the Vivid Rapid Explosives System, uses two X-ray pulses: first a 140-kilovolt beam, then a weaker 70-kilovolt beam. A powerful computer uses the data from the pulses to determine the mass, density, and atomic number of items in the bag, then checks them against a database of explosives. The image highlights a match in red.

Comments:

Some experts question the low-energy beam's ability to penetrate thick pieces of luggage.

Status:

Operational tests of Vivid system prototypes are scheduled at American and foreign airports later this year. Vivid also plans to arrange testing with the FAA.

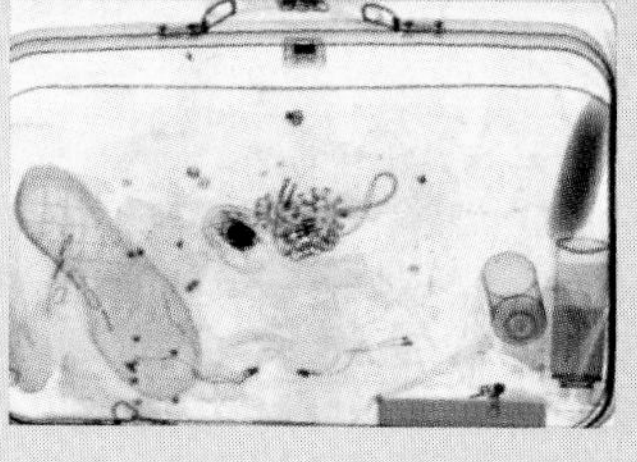

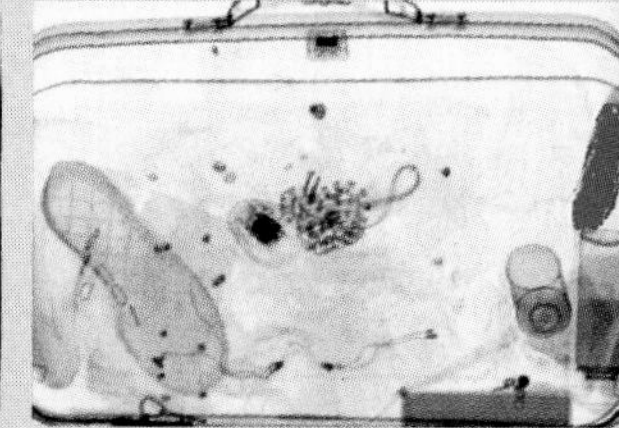

A nitrogen explosive that shows up red (below) on a dual-energy X-ray machine (left) looks like a shoe on a conventional detection system (below left).

Ultimately, aviation safety is a shared responsibility among the FAA, the air carriers, and the Federal Bureau of Investigation (FBI), which analyzes and monitors potential threats. The FAA dictates the standards for luggage and passenger screening, but the job itself falls on the shoulders of each airline.

FAA officials hope to create lists of technologies deemed appropriate for the levels of security established for each airport. Airlines could then choose among them. "We would like to leave that flexibility to the air carrier to select the machine most appropriate to the site," says an FAA official.

Today the FAA's research-and-development budget reflects the push for better technology. Funds jumped from $13 million in fiscal year 1990 to $30 million for fiscal year 1991. Sources say this funding should remain the same for 1992.

Not Quite Perfect

Most advanced X-ray systems get high marks for speed, but some might miss a bomb that is strategically placed in a suitcase or surrounded by certain materials, according to some experts. And because several X-ray-based systems require an operator to interpret what the scans reveal, human error is still a factor.

Bomb sniffers, which work like mechanical bloodhounds, have different advantages. "Our system looks for the explosive itself," explains Thermedics president John Wood. "We don't depend on some secondary characteristic," such as the explosives' interaction with X rays. But plastic explosives release only tiny amounts of vapor, making samples difficult to collect. And, adds former FAA head of security Vincent, even a machine that can detect one explosives mole-

VAPOR DETECTORS

How It Works:
A mailbox-size collector sucks in a 10.5-quart (10-liter) air sample from the bag's surface. The collector inserts the sample into a 4-foot (1.2-meter)-tall console that mixes the air with ozone in a dark chamber, a process called chemiluminescence. If it produces a weak light, explosives are present.

Comments:
The system detects molecules of explosives, not characteristics such as interactions with X rays. Plastic explosives have a low vapor pressure, so the sniffer must be very sensitive—detecting about one part per trillion. Screening requires two steps—collecting and analyzing. Inspectors may sample several bags at once. The device may be teamed with other systems.

German airports use a form of vapor detector to screen luggage. The U. S. State Department uses vapor detectors to screen incoming parcels at embassies around the world.

Status:
One method, Egis, is used by the U.S. State Department at embassies and by German airport security. Other sniffers under development at Sandia National and Oak Ridge National laboratories include supersensitive versions that may detect one part per quadrillion.

PFNA

How It Works:
The pulsed fast neutron analysis (PFNA) system sends high-energy neutrons that strike the contents of the luggage, producing signature gamma rays for elements, including carbon, oxygen, and nitrogen. The detector then assembles this information into a three-dimensional image of the suitcase contents. Security personnel would then view suspicious images and examine any luggage that might be carrying explosives.

Comments:
Because it analyzes gamma rays from carbon, oxygen, and nitrogen, PFNA should have fewer false alarms and greater sensitivity than thermal neutron analysis (TNA). The system may also be able to screen large numbers of bags simultaneously. PFNA machines would be heavier, slower, and more expensive than TNA.

Status:
Experts say prototypes won't appear before 1994; installation in airport terminals could be years beyond that.

cule in a trillion or hundred trillion may not be sensitive enough.

One of the most controversial methods, TNA, exemplifies the conflicting demands placed on detection systems. It is the only system in airports at this time that screens bags for explosives automatically—without relying on human judgment. But this nitrogen-detecting system can be fooled by innocent objects, like leather and wool; the smaller the amount of explosives it is set to detect, the more susceptible to false alarms it becomes. Moreover, airline carriers balk at the machine's price, size, and weight.

The Big Picture
"The first thing we should do is get advanced X-ray technology into the airport, because it makes it much easier to pick out potential threats," says MIT physicist and FAA consultant Lee Grodzins. Bomb sniffers, he adds, would be useful as well. "With reasonably trained people, these technologies can give us a very high measure of security. It may slow us down, but not by much."

While the more exotic technologies—TNA, pulsed fast neutron analysis (PFNA), and gamma resonance absorption—may ultimately perform better, they are most likely years away from becoming standard fixtures at airport terminals. Part of that reason, explains an FAA official, is that some of the necessary equipment isn't up to par. "A fast system implies that you have a high probe concentration—that is, you need a lot of gamma rays, or a lot of neutrons." Many accelerators aren't able to operate at the levels needed. And, he adds, practicality demands "a system that doesn't need six graduate students continual attention to keep it running."

A BOMB-DETECTION SURVEY

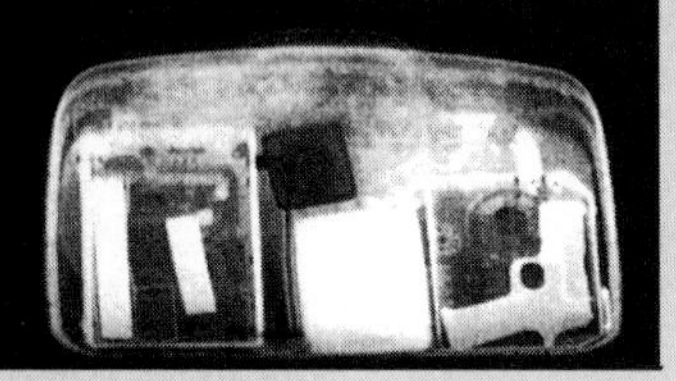

The maker of backscatter X-ray devices will soon introduce a machine that can screen 1,000 bags per hour.

BACKSCATTER X RAY

How It Works:
The machine sandwiches luggage between two X-ray detectors: one for rays traveling through the bag, and one for rays that bounce off, or backscatter. Low-atomic-number elements—like nitrogen—tend to backscatter X rays.

Comments:
The machine displays screens for each detector. Suspicious objects appear bright white on the backscatter screen. An upcoming fully automated prototype may process 1,000 bags per hour. To separate the bags with suspicious items from ones without, the machine will compare each X-ray image with a bag "profile"—the typical contents for a particular destination.

Status:
Operator-assisted versions have been available since 1986. The automated version is being examined by the FAA.

RESONANCE ABSORPTION

How It Works:
A beam of gamma rays, in an energy range that only nitrogen absorbs, strikes the luggage. Those rays that aren't absorbed by the luggage are measured by a detector on the other side. The machine then calculates the location within the luggage of nitrogen concentrations.

Comments:
Gamma-resonance systems could require less shielding than pulsed fast neutron analysis systems, making them less heavy. Experts believe the system would be adequately sensitive to detect small quantities of explosives.

Status:
Prototypes could be operating in airports by 1994.

Technology's Not Enough

"These developing technologies aren't going to bear fruit in the next six months or a year," notes Grodzins. "But terrorism isn't going to end then, either. It is going to continue as long as there are governments that feel terrorism furthers their own policies."

Technology alone, of course, won't make a perfect security system. The FAA is studying other aspects of airport security that may help suppress terrorist acts. "Work goes on in conventional weapons detection," says the unnamed FAA official, "because a terrorist can still use a gun to hijack an aircraft. And we're looking at ways to enhance the transaction between the screening personnel and the passenger. That includes the selection of personnel, the layout of the real estate, and how we can best move people through and yet act if something goes wrong."

That technology and human mixture, says Vincent, is a crucial aspect of any airport-security system. "I want to see a system with the right technology, but equally important, I want to see it balanced with the right people. It's fallacious to think you can keep the people out of the system. There has to be some human judgment."

Yet another tactic, adds the FAA source, is to minimize the destruction if a terrorist bomb does slip by the security net. FAA-backed research in aircraft design, such as hardened cargo containers, could help keep planes in the air after suffering the blow of an explosion. Discussions with airframe builders have begun, but conception, design, and manufacture will take years to show up on commercial jets.

If bomb detectors are getting ever better at finding explosives, security experts around

COMPUTER TOMOGRAPHY

A computer-tomagraphy security system creates a 3-D image with explosives highlighted in red.

How It Works:

Imatron's CTX 500 makes two-dimensional "scan projections" of bags, similar to X-ray images, which it feeds directly to the system's computer. The computer compares the density and composition of the bag's contents against a database of several hundred explosives. If there's a match, the machine orders computer tomography (CT) to assemble the data into a three-dimensional image and highlight the threat in red.

Comments:

Although it wasn't supported by FAA funds until a year ago, the system was favored by the National Research Council. Some bags require 40 or more CT slices, which may slow the machine, but Imatron claims it meets FAA standards for speed.

Status:

Imatron officials are working to arrange operational testing at airports; they would use the results to perfect the machine's software.

THERMAL NEUTRON ANALYSIS

How It Works:

A wash of low-energy neutrons interact with the nuclei of atoms inside the bag. These interactions produce gamma rays specific to nitrogen, which the machine detects. The system merges the thermal neutron analysis (TNA) data with an X-ray image, and then highlights suspicious items in the image.

Comments:

The manufacturer claims that TNA processes up to 600 bags per hour. The system uses radioactive californium 252 as the neutron source. While the manufacturer says this exposure won't harm medicine, food, or film, the machine is wrapped in thick shielding. At $1 million and 14 tons apiece, TNA is prohibitively expensive and heavy, say airlines. Moreover, it can be fooled by nitrogen-rich fabrics and plastics, creating a trade-off between sensitivity and the false-alarm rate.

Status:

One TNA system operates at Dulles Airport near Washington, D.C., and one at Kennedy Airport in New York City; a third was tested at Miami. Two are planned for San Francisco. In the United Kingdom, one unit operates at Gatwick Airport, and another is planned for Heathrow.

the world are working to make the explosives more detectable. Mike Catlett, chief of the explosives-technology branch at the Bureau of Alcohol, Tobacco, and Firearms, says that on March 1, 1991, at a conference sponsored by the International Civil Aviation Organization (ICAO) in Montreal, nearly 100 countries signed an agreement "in principle" to mark all plastic explosives with one or more chemical additives that have relatively high vapor pressures. These additives would make the explosives easier to detect by bomb sniffers.

"This is a pending treaty," Catlett explains. "The U.S. is still conducting tests to determine what combination of the four additives we want to use." Until then the Senate will be unable to ratify the treaty.

But the success of this treaty hinges on the destruction of existing plastic explosives. Once the treaty is in force, countries will have three years to destroy all commercial explosives, and 15 years to destroy military versions. However, stockpiles of explosives may be beyond the reach of the treaty. Czechoslovakia, for example, is said to have sold trainloads of the plastic explosive Semtex to nations backing terrorism.

It's a fine line, says MIT's Grodzins, between hurrying bomb-detection devices out of the labs and relying on immature machinery. "I wouldn't want any technology—however optimistic I may be—to be placed in an airport until it's received a skeptical going-over. If people believe they're on a safe airplane, then they better really be on a safe airplane. If the technology has a weak point, and terrorists know what it is, then it's worse than useless—it has created a sense of false confidence."

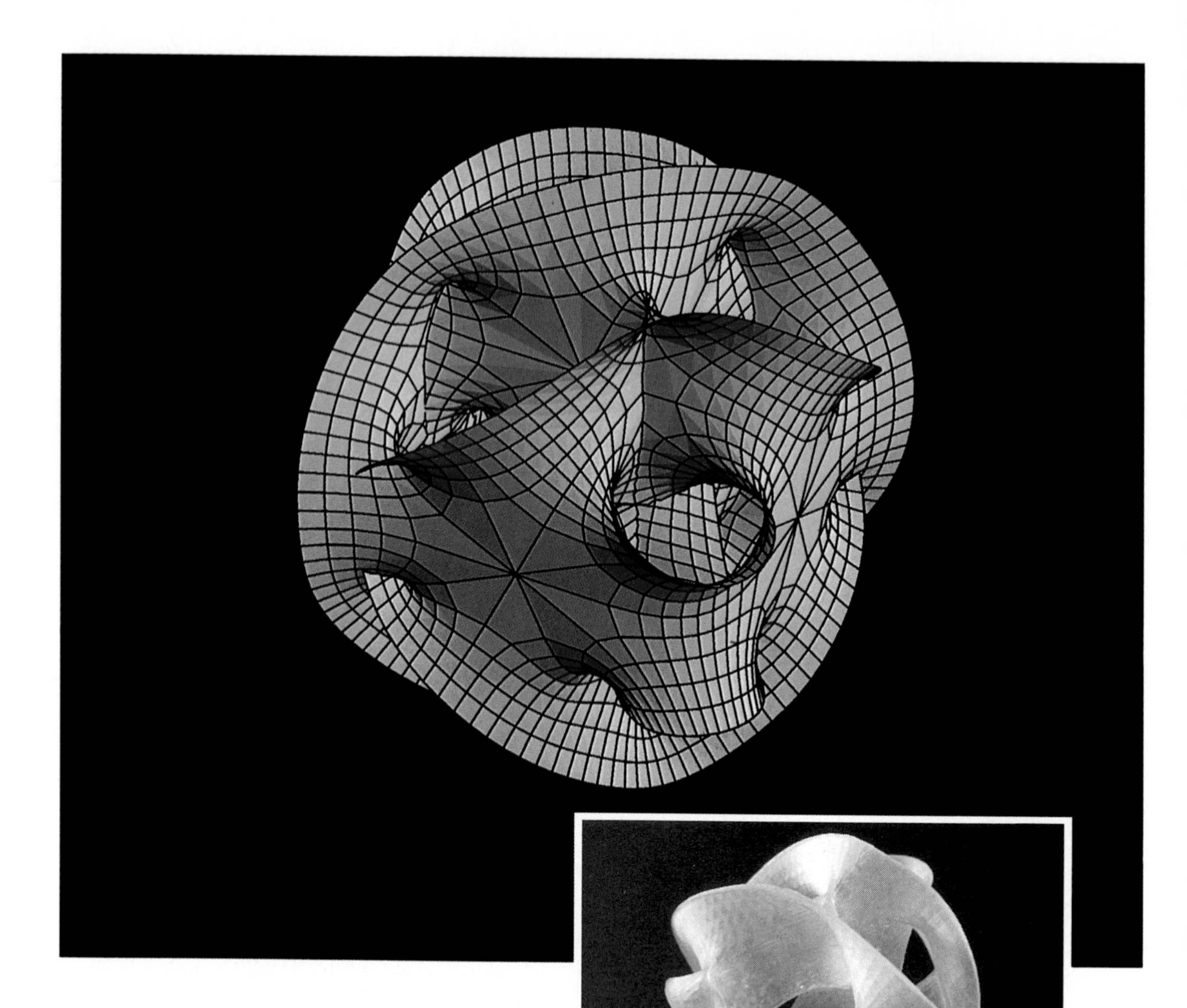

FROM GRAPHICS TO PLASTICS

by Ivars Peterson

The graphic images that Stewart Dickson creates on a computer screen are but pale shadows of the vivid, three-dimensional reality they represent. No amount of shading, color, animation, or other forms of graphic prestidigitation brings such images truly to life.

Dickson's frustration with the inherent flatness of two-dimensional displays spurred him to turn to sculpture and to new technologies for creating models of mathematical forms in all their three-dimensional splendor. "A picture of a three-dimensional object is an abstraction," he insists. "A sculpture is not."

A recently developed technology known as stereolithography has now provided Dickson with the tools he needs to realize his vision of sculpting mathematics. By linking

With the use of stereolithography, two-dimensional graphic images of mathematical forms can now be reproduced as plastic models. The surface shown here represents a visualization of the equation $x^5 + y^5 = z^5$.

the power of computer graphics to the direct formation of a solid, shaped object, the technique permits a user to design an object on a computer screen, then generate a three-dimensional plastic model. Already used for creating prototypes of commercial products ranging from perfume bottles to automobile wheels, stereolithography also shows promise as a method of visualizing mathematical shapes and scientific data.

A physical, three-dimensional model can help you understand a complicated mathematical surface better than a picture of it on a two-dimensional screen, says Clifford A. Pickover of the IBM Thomas J. Watson Research Center in Yorktown Heights, New York. "You can hold it at whatever angle you like; you can put your fingers into the nooks and crannies," he notes.

To demonstrate stereolithography's versatility, Dickson has produced an array of exotic shapes based on a variety of mathematical equations. He described his most recent efforts at the Physics Computing '91 conference, held in San Jose, California.

Turning to Solids

Currently a computer programmer specializing in graphics and animation at The Post Group, a postproduction video company in Los Angeles, California, Dickson started out in electrical engineering. Constantly sidetracked by his interest in sculpture and other aspects of the fine arts, he took seven years to complete his degree at the University of Delaware in Newark.

In the early 1980s, he worked for three years as an engineer at AT&T Bell Laboratories in Naperville, Illinois, assisting in the development of an electronic telephone-switching system. But he spent a lot of his spare time—often late at night—experimenting with sophisticated graphics devices and laser printers to produce a variety of images.

His outside activities weren't confined entirely to the visual arts. Dickson's interest in electronic music sparked a novel design for a stringed musical instrument, for which he received a patent in 1981.

By 1984 Dickson was deeply immersed in computer-generated imagery, working first for a design and film company in Illinois, then coming in 1988 to The Post Group, which provides special effects and other video-production services.

"It was also the year that stereolithography first hit the press, and I got right on it," Dickson says.

That year marked the introduction by 3D Systems, Inc., in Valencia, California, of commercial equipment suitable for building

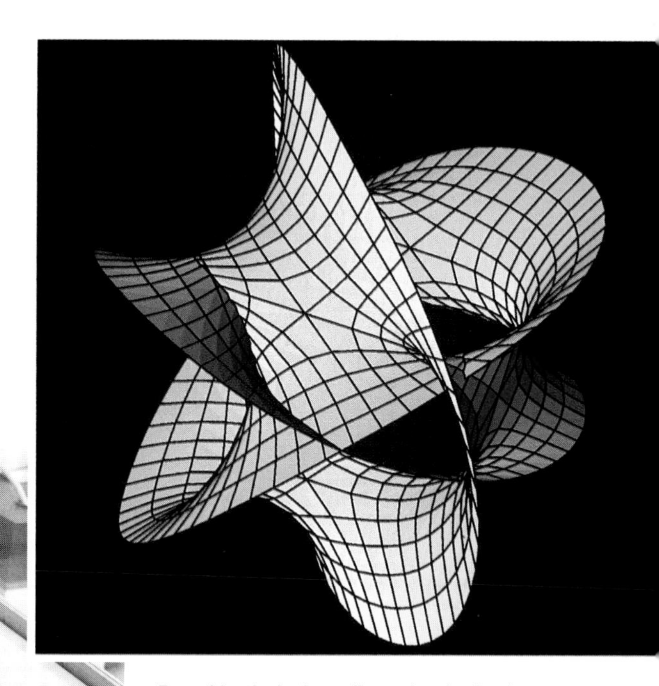

Four identical three-dimensional plastic models (left) of a mathematical surface (shown in a computer-generated graphic illustration above) emerge from a large vat of liquid polymer resin in an advanced stereolithographic machine used at the Hughes Aircraft Company.

Trinoid (Jorge-Meeks):

$$x = \mathrm{Re}\int_0^{re^{i\theta}} \frac{1-\zeta^4}{(\zeta^3-1)^2}\,d\zeta\ ;$$

$$y = \mathrm{Re}\int_0^{re^{i\theta}} \frac{i\left[1+\zeta^4\right]}{(\zeta^3-1)^2}\,d\zeta\ ;$$

$$z = \mathrm{Re}\int_0^{re^{i\theta}} \frac{2\zeta^2}{(\zeta^3-1)^2}\,d\zeta\ ;$$

$$r_0^4\ ;\ \theta_0^{2\pi}$$

A mathematical equation (above) generates a graphic image (far right). Stereolithography transforms that image into a plastic model (near right).

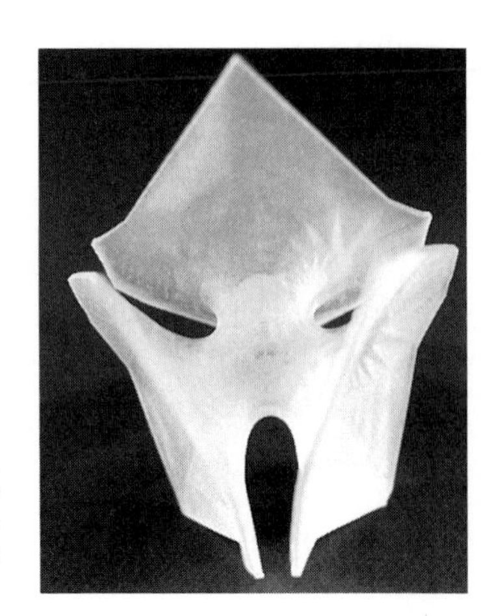

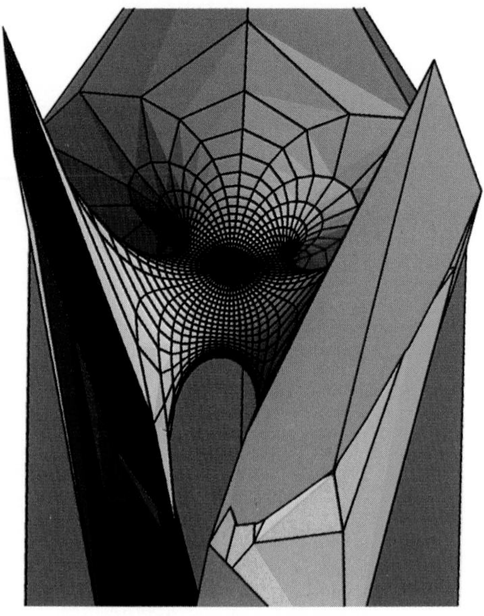

solid plastic objects from computer-aided designs. The process involves scanning a computer-guided beam of ultraviolet laser light across the surface of a photosensitive liquid, thereby turning to solid any areas exposed to the beam.

The creation of a three-dimensional plastic model begins in a vat of a liquid polymer resin. An overhead laser beam "draws" a layer, corresponding to a cross section through the object, on the liquid's surface, just a fraction of a millimeter above a submerged, movable platform. This initial layer hardens and sticks to the platform, which then descends a fraction of a millimeter. The laser then draws a second layer on the fresh liquid surface, the exposed liquid hardens on top of the original layer, and the platform sinks another fraction of a millimeter. The process repeats itself until the object is complete.

"Thickening" Zero Thickness

Through his persistence, Dickson persuaded initially reluctant customer-support engineers at 3D Systems to let him try generating several mathematical shapes using the new equipment.

The first object took nearly a week to build. "It was a solid object, and you had to scan its entire interior," Dickson says. "It was the first time this particular engineer had run into this [time] problem, but he succeeded in producing an object anyway."

Dickson's contribution lay in developing a method for converting a mathematical description of the surface of a three-dimensional object into terms that the stereolithographic equipment can understand. Because mathematical surfaces have zero thickness, one key step involves manipulating the data to "thicken" a surface so that it can exist as a physical object. Another involves testing if everything in the object is actually connected. Otherwise, the object would fall apart.

"It's the typical computer-graphics trade-off between accuracy and practicality," Dickson says.

Nowadays Dickson uses Mathematica, a software system for doing mathematics by computer, to design and specify his mathematical shapes and to create two-dimensional graphic images of the surfaces. A second computer program converts this output into the steps needed for building a three-dimensional plastic model.

Sculpting Mathematics

Stereolithography isn't for everyone—at least, not yet. Depending on its size, each of Dickson's convoluted mathematical objects takes from 24 to 36 hours to generate using the most advanced machine available. At this rate, he estimates that it costs a minimum of $1,000 to produce a model 8 inches (20 centimeters) high. Moreover, the choice of colors and types of plastics available to create three-dimensional models remains somewhat limited. And the equipment needed for producing the models is both expensive and bulky.

Researchers at 3D Systems and elsewhere hope to improve the process so that future stereolithographic machines can make larger parts faster and more accurately. The future may see the development of speedier, more-compact machines that create objects with finer detail.

At the same time, stereolithography isn't the only technology now available for building three-dimensional models out of a sequence of layers. In laminated-object manufacturing, a high-power laser cuts the outlines of an object's cross sections out of sheets of plastic, metal, or paper. These layers are then glued or welded together.

Ballistics-particle manufacturing constructs a layered object using a high-speed jet of tiny plastic beads that melt readily and bond together on impact. Laser sintering uses a high-power laser beam to fuse powdered plastic into a continuous solid, with fresh powder spread out atop the completed layers as the shape builds up.

Each scheme has its own commercial and creative advantages and disadvantages, Dickson says, but they all offer interesting possibilities for sculpting complex mathematical shapes.

Visual Speculation

Dickson's artworks—what he calls his "abstractions made concrete"—are starting to appear in public. Six of his finished pieces currently highlight an exhibition of electronic art in Finland. And about two dozen of Dickson's three-dimensional mathematical models are emerging one by one from a stereolithography machine on display at this summer's Heureka National Research Exhibition in Zurich, Switzerland.

At the same time, Dickson is exploring various ways of expanding his repertoire of mathematical shapes, looking at different types of equations and pondering complicated forms that few have ever attempted to visualize or model.

"A physical entity promotes a more profound understanding than a flat image on a display screen," Dickson asserts.

Using the new stereolithography technology, one can get very close approximations of objects that can be described mathematically, but can't actually exist in the ordinary physical world, he says. In essence, stereolithography allows one to put the results of an abstract human process into physical, three-dimensional, solid form.

"Dickson's work speaks to the future of [scientific] visualization and the future of art," Pickover says. "You can hypothesize that in the future, rather than looking at a computer screen to visualize a scientific model, you'll be able to hit a button and walk away with a physical model. It's interesting to speculate what that capability would do for science."

Stewart Dickson has devised a scheme to construct a complicated cubic lattice (computer graphic, right) by joining together single octahedral units like the plastic model shown below.

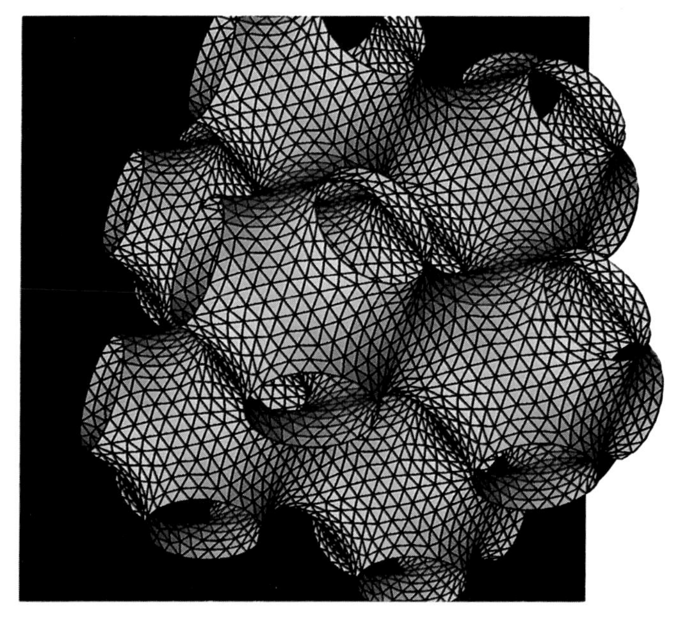

THE BLIMP BOWL

by John Grossmann

The Goodyear blimp does double-duty as a highly recognizable corporate symbol and as a high-altitude camera that imparts a sense of spectacle to sporting events. The blimp's pilot (above) uses both hands and feet to guide the aircraft and keep it steady in high winds.

Four hours to kickoff. Some 5 miles (8 kilometers) from Spartan Stadium in East Lansing, Michigan, where Michigan State will soon host number-one-ranked Notre Dame, another, less-publicized team goes through its regular pregame rituals. On a grassy island near a taxiway at Capital City Airport, a 16-man crew is transforming the Goodyear blimp *America* into an aerial camera platform.

Out of the gondola, or car, come passenger seats, unbolted from the floor. The rug is rolled up and carried away. Off comes a door-sized side panel behind the pilot's seat. On board are hoisted some 790 pounds (360 kilograms) of TV camera gear, monitors, and controls. The most striking piece of equipment is a huge white eyeball, which hangs below the car on a winch. Inside this eyeball is a gyro-stabilized, state-of-the art Ikegami

323P camera with a 44-to-1 zoom lens. Rack to lens, the system costs a cool $350,000. Goodyear owns three such systems, one for each of its airships.

Goodwill Ambassadors

Together, those three blimps make up perhaps the hardest-working corporate symbol in show biz. The *America,* which winters in Houston, and her sister ships—the *Columbia,* based in Los Angeles, and the newest blimp, the *Spirit of Akron,* based in Pompano Beach, Florida—each fly more than 100,000 miles (160,900 kilometers) a year. They fly passengers on half-hour joyrides and cruise with their night signs scrolling messages and graphics Times Square–style. Simply flying from one job to the next, the blimp (it's easier to think of them as a single entity) is at work as a head-turning, traffic-stopping, low-flying billboard. (Goodyear doesn't like to talk costs, but in the early 1980s, the company acknowledged an annual operating budget of $2.5 million per airship, and pegged the replacement cost for a model GZ-20—*America*'s 192-foot [58-meter]-long, 57.5-foot [17.5-meter]-high design—at $5.5 million. Both those numbers, clearly, have since risen.) The Goodyear Tire & Rubber Company, which has manufactured more than 300 airships since 1917, likes to call its lighter-than-air trio "goodwill ambassadors."

The blimp's most visible role is covering sports. Goodyear first assisted network sports coverage in December 1960, flying the airship *Mayflower* over the Orange Bowl Regatta (and apparently not, to the best of anyone's recollection, over the Orange Bowl game itself) with a CBS-provided camera. By 1970 the blimp's sporting assignments had grown to 20 events; by 1980, 50 events; last year the blimp racked up more than 75. Baseball play-offs, the World Series, key college football matchups, "Monday Night Football," the Indy 500, the U.S. Tennis Open, the Kentucky Derby, the Masters golf tournament—virtually all the big games have a blimp hovering overhead. Viewers have become accustomed to panoramic blimp shots of full stadiums—and a shot of the blimp itself. Following Goodyear's lead, Metropolitan Life, Sea World, and Fuji have all taken to the sky—and to sports coverage—to increase corporate visibility.

The competition is likely to increase, for Goodyear plans a major cutback in sports coverage for 1991. "In the current economic climate, we need to utilize our blimps more for direct product support, use the pulling power of the blimps to increase the traffic in our stores," says Goodyear spokesman Richard Sailer. "Consequently, we expect to cut back to 25 or 30 events." At each of those events, the role of the blimp will unfold much in the manner of airship *America*'s imminent role above the gridiron today.

Takeoff Time

As the blimp is readied for the Michigan–Notre Dame game, chief pilot John Moran stands nearby, crunching a few numbers in his head, mindful to stay well outside the arc traced in the grass by the single wheel beneath the cabin. *America* is moored to a 33-foot (10-meter) mast that's guy-wired like a

The blimp's cameraman can provide panoramic shots of full stadiums or zoom close to capture such fine detail as a ball leaving a bat. Sports fans have come to expect the unique views that the blimp's cameras supply.

big-top tent and erected by the ground crew with old-fashioned circus flair (and modern electric-stake drivers). Fastened by its nose only, the airship turns with the wind, and standing just outside the arc described by its wheel puts you in possible contact with the ship's pusher props. Seasoned pilots don't make this mistake. In his blue Goodyear uni-

Upon takeoff, ground-crew members give the airship a boost while the pilot guns the engines. A blimp heavily loaded with fuel usually requires a rolling takeoff down a runway to get airborne.

form, aviator-style sunglasses, and well-shined black shoes, Moran certainly looks seasoned, though the tiny gold-stud earring adorning his left earlobe is a surprising touch. He is 46 years old, and has been flying Goodyear blimps, including the discontinued overseas cousin, the *Europa*, for more than 20 years.

It's the pilot's job to specify the amount of fuel pumped aboard, taking into account both the amount needed by the twin 210-horsepower engines and the fuel's weight. Moran expects to encounter westerly winds of more than 15 miles (24 kilometers) per hour, and gusts likely to exceed 30 miles (48 kilometers) per hour. In other words, he's going to burn a lot of fuel, upwards of 10 gallons (38 liters) an hour.

Finally it's time for takeoff. Once ground-crew chief Jonah Carver gives the raised-thumb signal, the half-dozen or so crew members hanging on to the handrail below the gondola simply give the blimp a boost. Then Moran guns the engines. A heavily loaded blimp usually needs the momentum of a rolling takeoff down a runway to get airborne, but today, with takeoff seconds away, a powerful updraft grabs the ship, and *America* angles sharply toward the clouds.

Airborne Unicyclist

Moran levels off at the standard cruising altitude of about 1,000 feet (300 meters). To the right of his seat is the control for the ship's elevator, looking enough like the wheel on a wheelchair that more than one Goodyear-blimp passenger over the years has inquired about the pilot's disability. To Moran's left are the engine controls. Still, the busiest parts of a blimp pilot are his legs. Moran steers the *America* with his feet, which rest (in a manner of speaking only) on metal pedals angled a couple of inches off the floor. With the blimp constantly buffeted by wind and thermals, simply holding a course or staying put requires near-constant pedaling. Essentially, Moran works like an airborne unicyclist.

"No, I've never strapped on a pedometer," Moran says, suddenly yanking on the wheel with both hands and pointing the nose of the ship steeply down in response to a strong updraft. He smiles. "It's going to be a long day." The winds are now gusting to 35 miles (56 kilometers) per hour, and Moran's communications headset only slightly muffles the loud drone of the engines.

Behind Moran, cameraman Glenn Hampton warms up his skills on the brief flight to the Michigan State campus by following a red van on a road far below. Hampton faces three monitors—upper left, the view from the blimp camera, currently the red van motionless at a stoplight; upper right, a wave-form monitor that helps him maintain proper contrast in his picture; and below this, a small screen that provides, *sans* sound, the picture seen in millions of living rooms—right now, second-half action of the ABC telecast of the UCLA-University of Michigan game. When it ends, ABC will come live to East Lansing. With a glance at this monitor, Hampton will know when his shots are on the air.

He operates both camera and lens remotely, arcade-style, with hand controls mounted on a small panel just above lap height. His right hand, on a joystick, controls pan and vertical; left hand, zoom and focus.

A couple of turns of his left hand, and Hampton zooms in on the van close enough to all but read its license plate. He can, he says, actually follow a baseball leaving a bat, especially a grounder.

Out the gondola windows now, the view is festive. The band rehearses its halftime marches on a grassy practice field. Like a sunburst, motor homes and station wagons radiate in orderly rows from the stadium—tailgaters' heaven. The first specks of color appear in the stands as impatient ticket holders claim their seats.

Moran's headset, as well as Hampton's, carries the voice of Goodyear's manager of films and broadcasts, Mickey Wittman, who's already in the ABC Television truck, positioned at the elbow of director Larry Kamm. Wittman acts as liaison to the networks, helping them achieve the coverage they want from the blimp, and, conversely, sometimes trying to "sell" them, in TV jargon, on airing specific shots from the blimp. Look for the dish, Wittman reminds Moran. He means the microwave receiving dish responsible for capturing the camera signal beamed down from the blimp. This dish, about the size of a snow coaster, was positioned yesterday atop a building near the stadium and adjacent to the broadcast truck, to which it is wired. It's mounted on a tripod, and must be constantly swiveled by a Goodyear technician to follow the ship's movements in order to maintain peak reception.

When not airborne, the blimp is moored to a 33-foot mast by its nose only. Such a docking position allows the 192-foot long airship to sway about the mooring post with the changing directions of the wind.

Beauty Shots

The shot from the blimp, already appearing on one of the many monitors in the broadcast truck, is but one of 10 live views at Kamm's disposal. ABC has seven fixed cameras and two hand-held cameras at the ready; six of them, as well as the blimp camera, are potential replay sources.

Moran has a list of six "beauty shots" the blimp may be asked to cover during the telecast. The list includes the state capitol building in Lansing and various campus landmarks and new buildings—the latter apparently selected by ABC as much for alumni appeal as for architecture.

"O.K., blimp, give me Breslin Student Center," coaches Wittman.

Moran turns the blimp, and Hampton turns the camera, searching out one of the buildings he and Moran had pinpointed yesterday. Moran keeps tabs on the blimp camera by glancing at two monitors held together with duct tape. The left screen shows the blimp shot. The right screen shows the ABC telecast. Moran asks Wittman to relay a message to the director: "Tell him when he wants a shot, he's got to give me some time because of this wind."

Not only does a Goodyear pilot use both hands and both feet, but with his headset on during a telecast, he's got voices coming in both ears—lots of voices. In Moran's left ear: Wittman and director Kamm from the broadcast truck; also cameraman Hampton. Right ear: ground-crew technician at the dish, air-traffic control, and flight service. On occasion: a two-ear, six-voice circus.

"Hello, blimp; it's the truck," Wittman says closer to kickoff. "Just like that. That's a perfect location for 3:30."

From the truck comes the voice of Kamm, commanding his small army. The broadcast has begun. "Roll to 21. Twenty seconds to you, Ralph. Stay wide on the blimp, please. . . . Ready to dissolve to the blimp. . . ."

Moran's two screens bear a single image. The blimp camera has the nation's attention. Hampton's holding a wide shot of the jam-packed stadium framed by some of the campus.

"Push in," commands Kamm.

Hampton zooms in tighter.

"All the way in. . . . Dissolve to camera two."

Soon announcers Keith Jackson and former Miami Dolphins quarterback Bob Griese appear on the screen, setting the stage for the day's game. Blimp shot number two is a tight stadium view that widens to show the armada of parked cars. Kamm cuts to a tailgate party, then to commercial.

A Sense of Spectacle

Like all directors, Kamm is happy to have a blimp at his disposal. In fact, he'd like to have a blimp over every outdoor sports event he televises. "If the blimp can help us provide the viewer with a sense of place and a sense of spectacle, it's done its job," he says. He does not like to use the blimp for coverage of the action, with the exception of auto races where the blimp camera can follow the cars around the track, providing both a sense of speed and—in panning back, say, from a pair of leaders—a good perspective on how far ahead they are. "To cover a play from scrimmage from the blimp is disorienting and unfair to the viewer," Kamm says. "It's a device and a gimmick, and it can wind up hurting your telecast if not used properly. You try and weave it into the pattern. Make it part of the whole cloth."

Blimp shot three, coming in from another commercial, provides the first on-air mention of the blimp. Says announcer Jackson: "They're just certain they're going to have a new attendance record today at Spartan Stadium. There will be over 80,000 people. As we look down on it from the Goodyear blimp."

Moran, right hand off the wheel for a moment, flexes his fingers to release some of the tension starting to build in his arm. He admits the pain typically shoots well up the forearm long before a day's coverage ends.

Late in the first quarter, Kamm says, "I'll do a blimp pop after the exchange [of possession of the ball]." This rather interesting choice of words turns out to refer to a visual of the blimp itself. It appears with 26 seconds left on the clock, after Kamm has asked his cameramen, "O.K., who can see the blimp?", and has first established a smooth bridge by cutting to another shot from the blimp. "The Goodyear blimp *America* out of Houston, Texas, is bringing you that picture," says Jackson, as the blimp appears against the darkening sky. "The pilot is Captain John Moran of Spring, Texas. Hey John. Haven't seen John in a couple of years."

Although the Goodyear crews do enjoy friendly relations with the television folks, it's not simply friendship that begets such warm mentions and blimp pops. Contracts specify them. Similar arrangements are in effect with the competitors' airships. Broadcasting football, Kamm aims for one blimp pop in the first half, two in the second. And resounding pops they must be, for Goodyear accepts them as payment for services rendered. No money changes hands. Forget Wayne Gretzky to the L.A. Kings. This is the biggest trade in sports.

Aboard the blimp, there is very little fanlike spectating. There's little time. No big-screen views. No popcorn. No beer. No bathroom, for that matter. In the rear of the gondola, in a storage area behind a second camera positioned to take an inside shot of the blimp, one can, if one must, grab a specially outfitted small black hose.

By the end of the third quarter, Moran is eyeing the horizon behind the capitol dome. The ceiling is falling. A blanket of low clouds is sweeping in from the west. He checks his radar screen. No rain yet. In the truck, Wittman's thinking the same thing: "First thing you get on the radar, you tell me." Kamm has alerted everybody he wants to "go to inside the ship at the beginning of the fourth quarter," and Wittman's looking to sell a shot of the radar screen.

In preparation for the inside shot, Moran reaches for a satin jacket, which he asked for prior to boarding. He drapes it neatly over the back of the passenger seat beside him, the words "Goodyear" and "Houston, Texas" facing the camera. Moving to the back of the ship to operate his second camera, Hampton throws a switch that provides additional lighting.

The airship's propeller system provides airspeed and directional control. A blimp consumes as much as 10 gallons of fuel per hour simply to hover stationary over a sports arena.

A Notre Dame interception postpones the shot. Finally, midway through the period, Kamm works it in. "O.K., ready the blimp. Dissolve to the blimp." Hampton starts fairly tight on the blimp's radar screen, and then widens, cueing Moran: "Turn around, John." Moran's lips move in a classic "Hi, Mom" greeting. And it's back to the game, which Notre Dame manages to squeak out in the final minute, thanks to help from higher up than the blimp. A near-interception bounces off a Michigan State player and becomes a goal-line completion that sets up an improbable winning score. The last televised view from the blimp shows the still-full stadium, and across this image is superimposed the final score—Notre Dame 20, Michigan State 19—and then the sponsors' credits.

"Thank you, videotape. Thank you, Karen. . . ." One by one, Kamm thanks his team. "Thank you, blimp."

Mooring Time

Hampton starts the winch to raise the camera. "John, the ball is up," he announces. Moran points the ship back toward the airport and into a 30-mile (48-kilometer)-per-hour headwind. From a half-mile away, the ground crew can be seen on the runway. Moran angles the ship down, heading into the wind. A small crowd of onlookers stands by the airport security fence. Moran cuts the engines, then throws the props into reverse. Two trios of crewmen dressed in rain gear reach for the dangling nose lines and begin guiding the ship off the runway to the grass and the awaiting mooring mast. Moran sees four fingers held aloft (4 feet [1.2 meters] to coupling). Three fingers. Two. One. *America* is on the mast. Moran reaches for his logbook and enters the time of his return: 19:14. Under remarks, he writes: "MSU Notre Dame." He does not include the score.

Moran heads off in search of a bathroom. The crew begins removing the camera gear, which two days later will again be aloft, this time aboard the *Spirit of Akron,* serving "Monday Night Football" over Giants Stadium in New Jersey's Meadowlands. *America* will leave Lansing at 9:00 A.M. tomorrow. Destination: Cedar Rapids, Iowa, and the annual Farm Progress show.

By late Sunday morning, the crew has dismantled and stowed the mast, retrieved all the scattered ballast bags, stowed themselves inside the bus, van, and truck, and driven off. There will remain a reminder that the blimp was here. Visible to keen-eyed pilots is a circle nearly 170 feet (52 meters) in diameter worn into the grass—the tracing made by the ship's lone wheel as five days of changing winds pushed the airship, like a giant wind sock, through all points on the compass. The Goodyear blimp has left another footprint.

Scientists across the country are grappling with concepts of engineering and physiology in the quest to endow robotic "insects" with lifelike qualities.

ROBOTS GO BUGGY

by Elizabeth Pennisi

Last fall the U.S. Office of Naval Research (ONR) spent thousands of dollars trying to encourage university scientists to do what high school senior Christopher P. Stone did with less than $50: blend biology with robotics. Intractable problems in biological research and in technology—the complexity of the nervous system and the rather primitive state of autonomous robot locomotion compared with that of even simple animals—warrant such a blending, says Thomas McKenna, a neurobiologist with ONR in Arlington, Virginia.

At an ONR-sponsored workshop in September 1991 titled "Locomotion Control in Legged Invertebrates," about 15 biologists and 10 computer scientists and engineers gathered in Woods Hole, Massachusetts, to discuss their progress in studying the mechanics of how animals and machines move. The ONR wants these computer scientists to share ideas, learn from one another, and ultimately to make faster progress in understanding locomotion and building better robots.

Combining Biology and Technology

The workshop participants tend to view locomotion from quite different perspectives. Some care about making robots mobile. Some try to express movement as equations, which are then incorporated into computer simulations of locomotion. Others want to observe and understand a living creature. These differences mean that the investigators sometimes don't agree on what research they need to do, what questions need answering, or even how to go about answering those questions. Moreover, they tend not to exchange ideas very often.

But Stone, a 17-year-old tinkerer from Bloomington, Indiana, combines all these perspectives in his approach to research. As a high school freshman, he built an insect robot, in part by scavenging syringes for pistons and removing voltage converters from old computer printers for use in his robot. Last year, he got interested in how crayfish walk, and he began probing their nervous systems. For his junior-year science project, he pinpointed crayfish nerve cells involved in walking, and he connected them to his computer via very thin copper wires. He then modified the computer program that coordinated the robot's legs. With this new version, he stimulated a living crayfish by sending very tiny voltages down the copper wires. The voltages made the animal amble across a table.

At the International Science and Engineering Fair in May 1991, the crayfish project earned him an award from the scientific society Sigma Xi for the best use of scientific integration. It also won him an invitation to attend ONR's September 1991 robotics workshop.

"I want to help make [paralyzed] people walk, and for quadriplegics to sign their names," says Stone about his long-term goals.

Though his experiments may seem simplistic and his ideas a bit futuristic, Stone's desire to couple biological and technological knowledge may prove quite insightful. Several researchers—and ONR—think biology may provide strategies for making faster and more-versatile robots, McKenna says. At the same time, biologists find they need engineering and computing tools for their work, he adds.

Computational Neuroethology

Although many types of walking machines exist, engineers have not yet figured out the best way to make these machines handle unfamiliar situations. Even artificial intelligence, in which computers supposedly reason the way humans do, so far has proved inadequate for this task. "Even the simplest animals are much more versatile than the most sophisticated artificial-intelligence machine," says Randall D. Beer, a computer scientist at Case Western Reserve University in Cleveland, Ohio.

Yet the National Aeronautics and Space Administration (NASA) needs agile robots that can work in space or roam planetary surfaces to collect samples; the military wants robots that can maneuver along the ocean bottom or scour battlefields for mines. And to make robots do that, researchers need to develop more adaptable machines with better coordination.

Animals are adept at avoiding obstacles, escaping predators, and finding food. To understand how they do this, neurobiologists usually monitor nerve impulses and try to trace the pathways of nerve signals from the eye or other sensory organ, through the brain, and down to the muscle that actually moves the wing, fin, or foot. Other researchers investigate the tenets of biomechanics—the science that investigates theories of how joints and muscles move and how they keep an animal balanced.

But biological experimentation and observation alone have proved inadequate for figuring out all the intricate ingredients of animal movement. So in the past few years, some scientists have begun modeling these phenomena on computers. Their computer simulations help them make sense of experimental results.

A few have teamed up with other researchers to make artificial "insects" based on these computer models, or to create realistic environments that simulate on a computer what animals encounter in real life. "We're trying to understand how real critters organize their behavior," says David Zeltzer, a computer scientist at the Massachusetts

Eddie, a wall-climbing robot, relies on two vacuum-driven suction pads to ascend the glass partition separating it from its inventor, MIT undergraduate Chris Foley.

Institute of Technology (MIT) in Cambridge, Massachusetts.

In doing so, researchers have given birth to a field called computational neuroethology, in which computer and engineering experts create programs that try to do what animals do the way animals do it. "It allows you to get new insights into a system," Beer says.

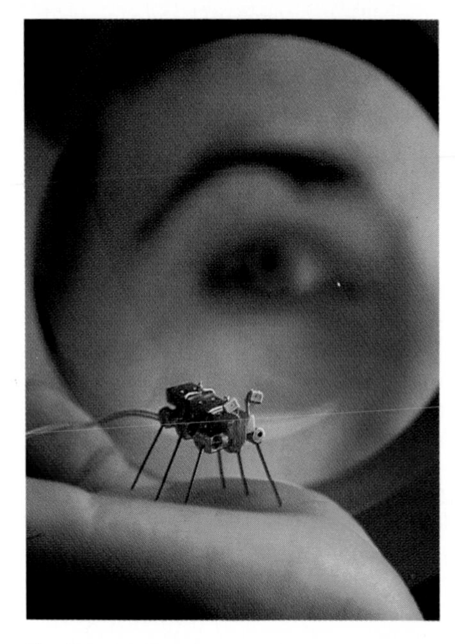

Tiny insect robots have come to be known as "gnats" by their inventors. Complex circuitry is necessary to enable the gnat above to make insectlike movements.

Locomotion

ONR wants to nudge the evolution of this discipline yet another step. It hopes that engineers interested primarily in building robots will link up with computational neuroethologists to learn how to propel robotics out of its primitive state locomotion-wise, says McKenna.

Unfortunately, the world of the engineer does not meld easily with the realm of biology. "The engineer wants to ignore as much of the biology as possible," says Beer, whereas "most biologists think every detail is important." The tension that inevitably rises between these two disparate perspectives can stimulate—or, perhaps more frequently, inhibit—collaboration, and it certainly makes life more complicated.

"Most engineers think neurons are on-off switches," says Allen Selverston, a neurobiologist at the University of California, San Diego. But neurons vary in their shape, their membrane composition, and the type of connections they form with other nerve cells—and all of these factors affect their responses. "All that detail is very important," Selverston notes, commenting that he spends a lot of time trying to change engineers' perspectives.

But would-be robot inventors prefer not to be constrained by all this detail, and, in many cases, they lack the technical know-how to replicate the complexities and subtleties of living systems.

Given these differences, it's no surprise that during most of the ONR workshop, talks divided along disciplinary lines. Neurobiologists described biological experiments; computer experts talked about their neat simulation programs; roboticists showed off their machines.

But exceptions did emerge, and McKenna thinks some collaborations may soon develop.

Four years ago, scientists at Case Western Reserve began to build a bridge between these very different worlds. Beer, then a graduate student in computer science, and neurobiologist Hillel J. Chiel decided to create *Periplaneta computatrix*—a make-believe cockroach with walking capabilities inspired by what neurobiologists knew about locomotion in *Periplaneta americana*, a real cockroach commonly found in the U.S.

On the computer, Beer and Chiel drew *P. computatrix* with a head, a mouth, two antennae, and a body with six legs. Each leg can move up or down, support the body, or push off to move or turn the insect. The limbs also can swing in midair. The mouth opens and closes and contains tactile and chemical sensors for detecting food. The tips of the antennae relay a message when they tap an object or bend. Beer programmed energy use into the simulated critter, as well as a way for it to tell when it needed to eat.

So that their computers could handle the computational load required for simulations, the researchers adapted and incorporated a commonly used model that simplifies neuron activity. In the model the simulated nerve cell fires depending on the cumulative effect of signals from all the neurons that connect with it. Some nerve cells send stimulating signals, and others send inhibitory signals. Collectively, the signals cause the model neuron to fire when they add up to a certain level of activity.

Beer then had to decide how many nerve cells the computer cockroach needed and how they would connect. These cells would form a locomotion controller, which would get the legs moving in a coordinated fashion. In a live cockroach, small groups of nerve cells—some of which fire rhythmically—make up a pattern generator, which controls leg movements. Other nerve cells monitor the position and load on each leg. Nerve cells in one leg can inhibit those in adjacent legs, keeping any two legs from swinging at once. For the simulated cockroach, the scientists made a neural network: for each leg, they put in three motor neurons, two sensory neurons, and a pacemaker neuron—one that generates signals regularly. A command neuron connected all six groups of nerve cells and moderated the overall activity level.

Once they completed the computer model, the researchers discovered they could make their insect walk in different ways at different speeds. To their surprise, it exhibited the range of gaits used by real insects, even though the researchers had not set up a centralized command center to coordinate the legs. Their insect also continued to walk even when the scientists made the computer disconnect parts of the circuitry. "If we hadn't taken the time [to look at the biology], we still would have made a locomotion controller," says Beer, "but I don't think it would have had these interesting properties."

From these studies, Beer and Chiel concluded that gaits develop as a result of interaction between the intrinsic rhythms and feedback from sensors that detect leg positions. Their results draw attention to the potential of biology for helping roboticists. "They've shown that these biologically realistic networks, at least in simulations, replicate the kinds of gaits that insects have," McKenna says.

The scientists went on to add more nerve circuitry to their computer creation so that it used its antennae to stay close to an edge as it walked around in its simulated world. They even made the computer insect seek out and eat food. Even with these bare-bones faculties—78 neurons with 156 connections—the insect did what its real counterpart tends to do, Beer says.

Response to Stimuli

Two years into the project, they decided to bring the simulated insect into the real world. Such an effort tests both the experimenters' models and the realism of the simulation. "It's one thing to say, 'I understand how something works,' but the proof is in the implementation," says McKenna. "The thing has to physically walk and not fall down."

Beer and Chiel asked Roger Quinn, a mechanical engineer at Case Western Reserve, to breathe life into their creation by building a robot that incorporated the simulation's neural network to control its locomotion. "I said, 'It will never work,' " Quinn recalls. Nonetheless, he decided to try.

Quinn and Kenneth S. Espenschied designed a six-legged, 2.2-pound (1-kilogram) machine in which each leg could move up or down and backward or forward. They then

A specialized "gnat" named Squirt seeks darkness, much like a cockroach. Scientists hope that someday such microrobots will be able to perform certain jobs that are impossible for humans to do. A tiny insect robot could, for example, crawl along an underground cable, searching for and repairing minute cracks.

Scoot, Scramble, and Roll

Invertebrate biologists have good reason to think their data could inspire even the best engineer. Real-life walking, sliding, inching, and somersaulting creatures showcase quite a few possibilities for robot locomotion.

''Nowhere is there greater diversity in locomotor designs than in arthropods,'' says Robert J. Full, a comparative physiologist at the University of California, Berkeley. These critters—crabs that scoot sideways just out of reach of waves, centipedes that scramble up tall walls, stomatopods that curl up and roll away backward—make their way through environments that would immobilize the most sophisticated machines.

Unfortunately, according to Full, ''[Conventionally designed] robots never move in a way that's similar to animals.''

To understand how arthropods scoot, scramble, and roll, he and others have begun breaking these movements down into their simplest mechanical components. They find, however, that deriving general principles for use in robot design requires ingenuity.

To get a better handle on crab scooting and cockroach running, for example, Full constructed a force plate—a platform with sensors underneath that connect to a computer. The sensors detect when a leg pushes down. As the crab or cockroach runs across the plate, Full films it with a high-speed video camera and then analyzes leg positions frame by frame, correlating the movements with force-plate measurements.

These and other studies show that no animal, not even a tumbling stomatopod, moves in the same way as a wheel does. Instead, animals' legs act like pendulums during slow walks, and like springs when the animal speeds up. Full has found that animals with two, four, six, and even eight legs all produce similar patterns on force plates, indicating that ''the body is being propelled alternately by two sets of legs,'' he says. ''The differences come when you look at individual legs.'' Thus, three legs of an insect, two legs of a poodle, and four legs of a crab act as units equivalent to one leg of a human being. But in each animal, the role of each leg varies. ''I believe the legs are positioned and develop forces to minimize the torque at all the joints,'' Full explains.

He found another parallel between animals with different numbers of legs when he examined the relationship between speed and stride. Trotting four-legged animals, for example, speed up by switching to a gallop: they take longer strides rather than move their legs faster to accelerate. A crab and a mouse of equal weight will switch gaits at the same speed—about 1 yard (1 meter) per second—and the same stride frequency, or number of times all the legs cycle per second.

''That suggests that there are general principles that can be applied to animals with tremendous differences in body form,'' he says. ''Those same concepts can be transferred to nonbiological systems.''

Such close examination has yielded other surprises about the versatility of moving arthropods. Scientists once thought that cockroaches always kept three legs on the ground, and so remained stable all the time as they moved. This static stability contrasts with the dynamic stability of running humans, who would fall over if stopped in midstride, but who stay upright because all the forces balance out as they move.

Full's videos, however, show that cockroaches can reach speeds of 1.5 meters per second—more than 3 miles per hour. As the roach accelerates, it leans farther back, first depending on the back four legs, and then, at top speeds, zooming along on just two. ''This rejects the notion that arthropods require static stability,'' says Full. It also shows that an animal or robot can have both dynamic and static stability, he says.

''The key is not to exactly mimic the biological system, but to take the concepts and see if they can be transferred to a design to make it better,'' says Full, who has provided MIT engineers with data about leg length and position for possible use in the design of new robots. He also sees the benefits: ''We can provide biological inspiration for them, and the questions they ask us will help us define how we need to quantify these systems in the animal world.''

E. Pennisi

When upside down, Genghis (the insect robot shown right side up in the photo on page 358) displays the complicated array of circuitry necessary for its form of locomotion.

constructed the insect robot using model-aircraft plywood. "The system didn't work perfectly at first," Quinn says. For example, the researchers discovered that the robot's electrical and mechanical devices slowed the signals, sometimes unevenly, and that the delays could make it walk inefficiently. "But we did not tune the network; we had to tune our implementation," Quinn adds. And in the end, the robotic insect proved itself capable of walking as well as the computer-simulated insect.

Now Beer and Chiel want to make an even more realistic simulation, one that will reenact a complex behavior called the startle, or escape, response.

When a cockroach detects a puff of wind, as might occur when a predator lunges toward it, the insect quickly turns and runs. In less than 60 milliseconds, it determines the wind direction and then acts. Beer would like the simulated cockroach to do the same when it encounters a simulated predator. "We're trying to make the simulated insect much more physiological," says Roy E. Ritzmann, a Case Western Reserve neurobiologist who studies startle response and works with Beer on this project.

But programming realism into the simulation is proving no easy task. "Just to build a model requires a tremendous amount of nagging; it requires day-to-day interaction," says Beer. "We need a lot of data."

Often neither he nor Ritzmann realizes what kind and how many biological details they lack until they get stumped designing the model. And sometimes the biologist has no answer. "For nine out of 10 questions, he just throws up his hands and says, 'It will take me a decade to find that out,' " says Beer. Then, together, they work out a solution. In this way, he and Ritzmann are learning about the startle response and what neurobiological experiments need to be done, they say.

In addition, they think their work can aid robot builders. It appears that neural networks, such as those that Quinn used, are more flexible and require less computing power than other types of control systems. And while neither Beer nor Ritzmann thinks robots need to be just like cockroaches to work, they do believe that "biology suggests different ways of designing control systems," says Beer. "The [systems] are more likely to be more versatile and robust."

GENETIC FINGERPRINTING

by Thomas H. Maugh II

When the 57-year-old Alzheimer's victim came home from a day-care center in Tacoma, Washington, her disheveled clothes made her daughters suspect that she had been raped, even though the woman remembered nothing. Their fears were confirmed when a medical examination showed the presence of semen. The police had an obvious suspect, 34-year-old Alan J. Haynes. Haynes drove the day-care center's van and was the only man who had been alone with the victim. But there were no witnesses and no fingerprints to link him to the crime.

Fortunately, police were able to take advantage of a newly developed technique called genetic fingerprinting. Authorities sent a sample of the semen from the woman and a blood sample from Haynes to a sophisticated laboratory, where technicians used the new technique to compare the DNA (deoxyribonucleic acid, the genetic blueprint of life) from the two samples. The test showed that the samples were identical. Confronted with this evidence, Haynes pleaded guilty and was sentenced to prison.

Developed only in the mid-1980s, genetic fingerprinting has rapidly become a

Comparing the unique DNA patterns of individuals has helped officials convict rapists, establish paternity or maternity, and even identify missing children.

widely used courtroom tool because it is "possibly the most powerful innovation in forensics since the development of fingerprinting in the last part of the 19th century," according to geneticist Daniel L. Hartl of the Washington University School of Medicine in St. Louis, Missouri. It has proved valuable, not only for convicting felons and exonerating the innocent, but also for establishing paternity or maternity and proving family relationships. More exotic uses include the identification of missing children in Argentina, soldiers killed in war, and even the body of Nazi physician Joseph Mengele, the so-called "Angel of Death."

Recently, however, some critics have begun to question the growing use of DNA fingerprinting. The problem is not the validity of the test itself: virtually all scientists agree that DNA comparisons can provide highly accurate identification of individuals. Rather, critics charge that some laboratories that conduct the tests, especially the Federal Bureau of Investigation (FBI) laboratory, do not have adequate quality-control procedures to ensure the test's reliability. To date, no one has been falsely convicted because of such poorly controlled testing, but, in a handful of cases, such questions have prompted courts to refuse to allow genetic fingerprinting evidence to be admitted. To a certain extent, the ferocity of the debate about the laboratory procedures has often overshadowed the usefulness of the test itself.

While the validity of DNA fingerprinting is well established, critics complain that sloppy lab procedures reduce the test's reliability.

Fingerprinting Procedure

The fundamental techniques involved in genetic fingerprinting were discovered serendipitously in 1984 by geneticist Alec J. Jeffreys of the University of Leicester in Great Britain while he was studying the gene for myoglobin, a protein that stores oxygen in muscle cells. He found that the myoglobin gene contains many segments that vary in size and composition from individual to individual and that have no apparent function. Jeffreys called these segments (which he found associated with most other genes as well) *minisatellites* because they are small and they surround the part of the gene that actually serves as a genetic blueprint. The minisatellites account for less than 1 percent of the total DNA of a human.

Jeffreys isolated several of these minisatellite genes and inserted each into bacteria, which produced large amounts of the DNA segments. These segments could then be purified and labeled with radioactive isotopes to produce genetic probes that are the key tool in producing genetic fingerprints.

The first step in producing a genetic fingerprint is to obtain a sample of DNA from such substances as blood, semen, hair roots, or saliva. Using newly developed biochemical techniques to multiply the amount of DNA present, researchers can work with as small a sample as one hair root. The individual cells from the sample are split open, and the DNA separated from the rest of the cellular debris. The DNA is then treated with specialized proteins called restriction enzymes, which cleave the DNA into smaller fragments by cutting it at specific sites. Because the minisatellites from any two individuals have different compositions, they are cleaved at different sites, producing fragments of different lengths.

The DNA fragments are then applied to one end of a thin, jellylike substance called

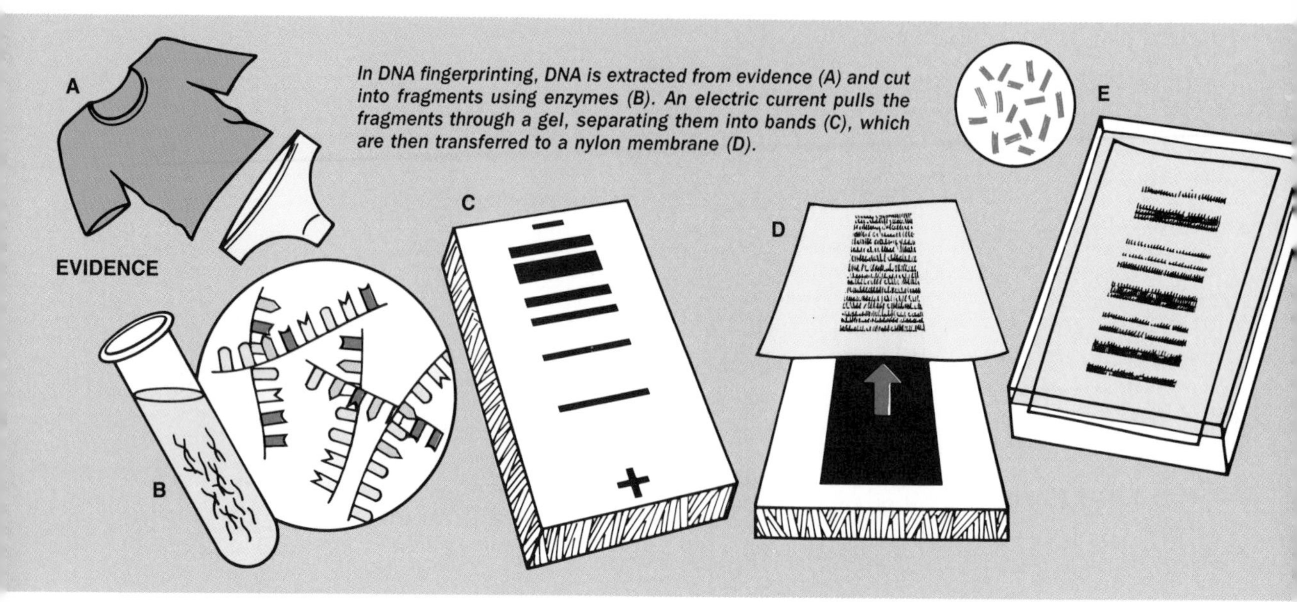

In DNA fingerprinting, DNA is extracted from evidence (A) and cut into fragments using enzymes (B). An electric current pulls the fragments through a gel, separating them into bands (C), which are then transferred to a nylon membrane (D).

an agarose gel, and an electric current is passed through the gel. Because the DNA fragments are negatively charged, they will migrate across the surface of the gel in response to the current, with the smaller, more mobile pieces traveling farther. The DNA is thus separated into individual bands, with the fragments in each one progressively smaller in size. Because the gel cannot be easily handled, a thin nylon membrane is laid over its surface and covered by a layer of paper towels. As the towels draw moisture from the gel, the DNA is transferred onto the surface of the nylon membrane, a process called blotting.

The bands of DNA are still invisible to the eye, and there are far too many of them to be useful. To solve both problems, a solution of the radioactive probes made from minisatellites is washed over the surface of the membrane. If any of the probes have the same composition as a part of a DNA fragment, they will bind to it. Typically, the probes will ignore the vast majority of the hundreds of bands present and will bind to anywhere from 6 to 20. To see the pattern of bands, the researchers place a sheet of photographic film on top of the membrane. The radioactive labels will expose the film, ultimately producing a pattern of thick-and-thin dark bands very similar to the bar codes used in supermarket checkout lanes. This pattern of bars is the genetic fingerprint. The entire process can require as long as four to six weeks in a commercial laboratory. But in November 1991, Jeffreys announced the development of a refined version of the test that is more amenable to automation and that allows results to be obtained much faster—in as little as two days.

Varied Applications

The likelihood that two individuals—other than identical twins—will have the same genetic fingerprint varies from about one in 800,000 to about one in 1 billion, depending on the number of probes that are used in the test. By comparison, the probability that two individuals will have the same conventional fingerprint is also about one in 1 billion. But the genetic fingerprint has many other advantages over conventional fingerprints. It is unusual for police to find a high-quality fingerprint at a crime scene, but much more likely that they will find blood or, in the case of a rape, semen. Furthermore, only a microscopically small sample is required for a positive identification.

DNA fingerprinting has many more applications than conventional fingerprinting. No relationship exists between the conventional fingerprints of parents and children, for example, but the genetic fingerprints are closely related because the child gets half of his or her genetic information from each parent. Thus, half the bands in the child's

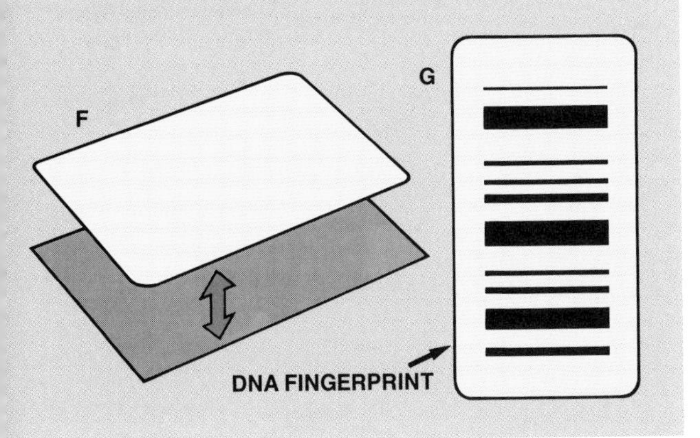

genetic fingerprint come from the mother, and half from the father. This similarity can be used to establish paternity (or maternity) with a much higher degree of certainty than is possible with other techniques, such as a blood test.

One of the first cases in which genetic fingerprinting was used by Jeffreys involved the case of a Ghanaian boy, born in England, who emigrated to Ghana to join his father. He subsequently attempted to return to England to rejoin his mother, brother, and two sisters. But British immigration officials refused a residence permit for the boy, fearing that he was either unrelated to the mother or was actually her sister's son. Conventional blood-typing techniques showed that the mother and the boy were probably related, but could not determine whether the woman was the boy's mother or aunt. Jeffreys used genetic fingerprinting to show that the woman was definitely his mother, and the boy was granted a residence permit. British authorities now routinely use genetic fingerprinting for this purpose.

A more recent application of the technique involves the effort to determine whether bones unearthed seven years ago in Argentina are those of Joseph Mengele, the chief physician at the Auschwitz concentration camp who supervised the systematic extermination of about 4 million Jewish men, women, and children during World War II. On behalf of the German government, Jeffreys extracted a small amount of DNA from one of the bones and compared it to DNA from Mengele's son, Rolf Jenckel, and Jenckel's mother. In April 1992, he reported that the bones

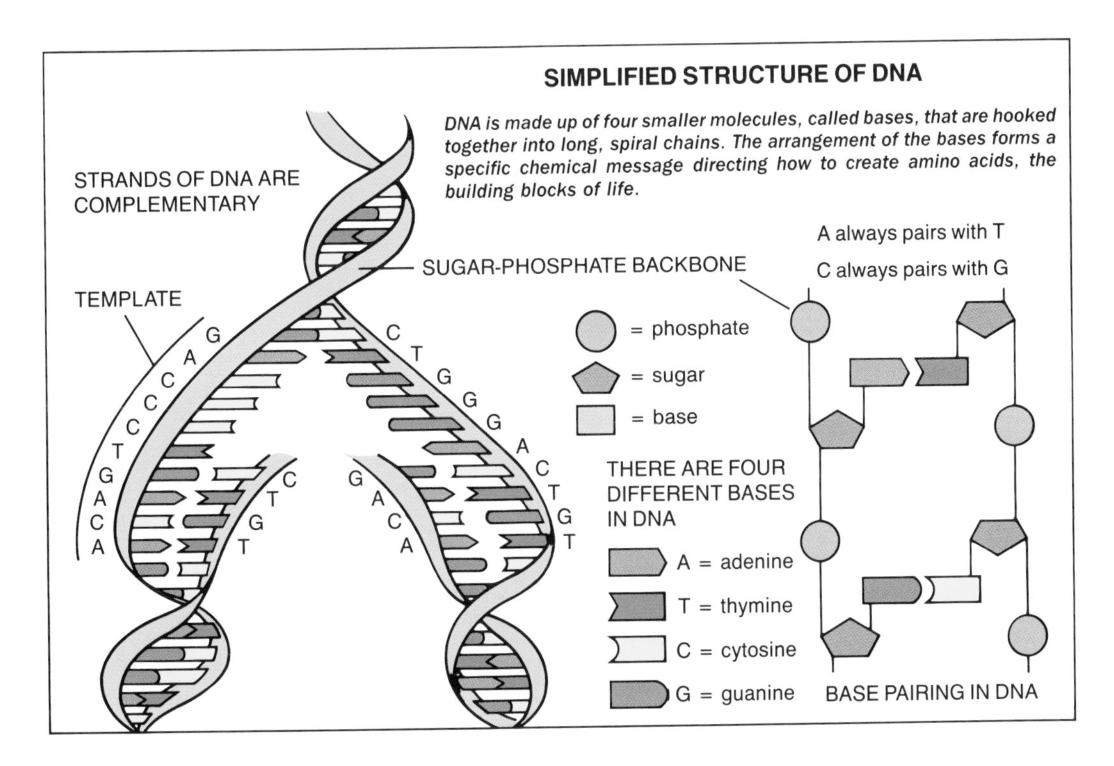

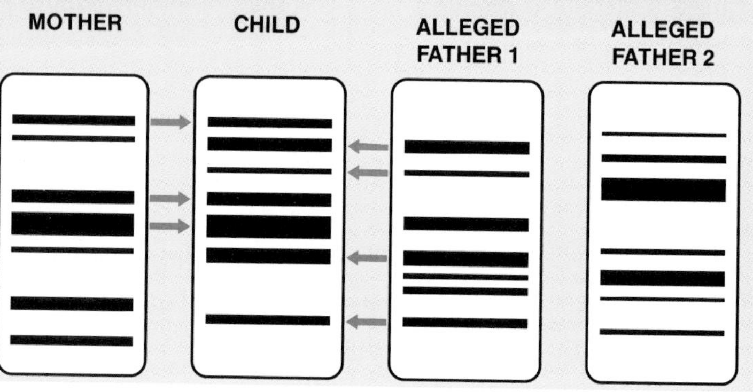

A child inherits half his or her DNA from the mother and half from the father, so DNA fingerprints can be used in paternity disputes (right). In a more exotic case, DNA was extracted from the bones of a corpse thought to be notorious Nazi physician Joseph Mengele (above). The DNA was positively matched with DNA samples from Mengele's son, thus closing the search for the war criminal.

were indeed Mengele's, finally closing the search for the sadistic physician.

Genetic relationships with grandparents can also be determined accurately. The ability to do so has been crucial in the work of geneticists, such as Mary-Claire King of the University of California, Berkeley, in reuniting the children of Argentina's *desaparicidos* (the disappeared ones) with their families. During the Argentine military's rule in the early 1980s, many government opponents were seized and killed, and their young children were given to military families or sold. After the military government was ousted, activist Argentine groups began trying to locate the children and return them to grandparents or other family members. Genetic fingerprinting has played a key role in many successful reunions.

In January 1992, the U.S. Department of Defense announced that it plans to establish a database of genetic fingerprints and blood samples for all 1.5 million people in the armed forces. The data would be used to identify the remains of soldiers killed in action, thereby preventing the recurrence of any more "unknown soldiers." Genetic fingerprinting was used during the Persian Gulf War to reassemble some bodies that had been blown apart, but it was not necessary for the identification of any soldiers, because the number of deaths was so low. Law-

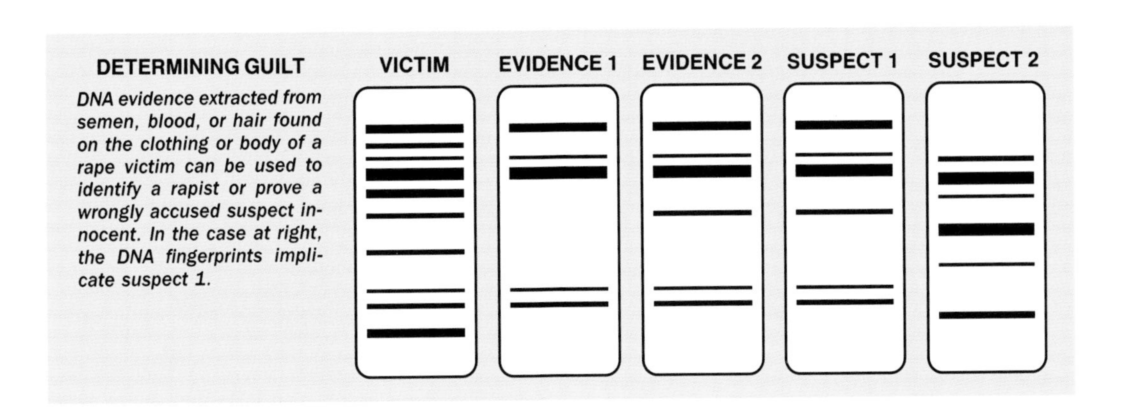

DETERMINING GUILT

DNA evidence extracted from semen, blood, or hair found on the clothing or body of a rape victim can be used to identify a rapist or prove a wrongly accused suspect innocent. In the case at right, the DNA fingerprints implicate suspect 1.

enforcement officials have also talked extensively about the prospect of establishing a similar database of known sex offenders that would serve, like the FBI's fingerprint files, to aid in the identification of unknown assailants. Critics have charged that the efforts, particularly the military's, raise questions about invasion of privacy and the possibility that the data could be misused, perhaps leading to new types of discrimination based on genetic predispositions to alcoholism, cancer, or other medical conditions. The projects seem likely to proceed, however.

Genetic fingerprinting is not restricted to humans. Biologists routinely use it, particularly in their efforts to protect endangered species. The Los Angeles and San Diego zoos, for example, have a captive-breeding program for the 36 California condors housed there—the last survivors of the species. Researchers have used genetic fingerprinting to establish the genealogy of each bird. This knowledge is then used to pair the most distantly related birds for mating, thereby preserving the maximum possible genetic diversity among the small number of birds, and reducing the risk of genetic weaknesses being perpetuated.

Criminal-Case Controversy

The greatest controversy about genetic fingerprinting involves its use in criminal cases. Since the technique was first introduced in a Florida case in 1988, it has been used in hundreds of trials. Defense attorneys have often tried to block its introduction, but have rarely succeeded. Their hopes were further derailed in a 1990 Ohio case, *United States vs. Yee*, in which three members of the Cleveland chapter of the Hell's Angels motorcycle club were accused of conspiring to kill a member of a rival club. A key piece of evidence was a blood sample at the crime scene that was matched to blood taken from one defendant, Johnny Ray Bonds.

Before the trial began, Magistrate James E. Carr of the federal district court convened an unusual six-week hearing in which 13 leading experts in molecular biology and genetics discussed the issues involved in genetic fingerprinting. Ultimately, Carr ruled that the genetic-fingerprinting evidence could be entered in the case, and that ruling has cleared the way for its use in a variety of other courts as well.

But the controversy did not end there. In late 1991 critics charged that some laboratories, especially the FBI's, were sloppy in their handling of DNA samples and in carrying out the tests. In at least one experimental study, one of the two large companies that carry out commercial genetic fingerprinting made a wrong match between two samples. Such cases have led some groups, such as the congressional Office of Technology Assessment (OTA) and the National Academy of Sciences (NAS), to call for increased proficiency testing of laboratories and the development of more-reliable techniques for handling samples. But virtually no one is arguing that the use of genetic fingerprinting should be restricted. "Any scientific procedure, no matter how reliable, can be misapplied," says law professor Paul C. Giannelli of Case Western Reserve University in Cleveland, Ohio. "Fingerprint testimony has sometimes failed, and yet no one would argue that such testimony should be excluded at trial." The same, he says, should be true for genetic fingerprinting.

Please Stand By: HDTV

by Christopher King

It's a rainy day, and you're spending the afternoon indoors, admiring your vivid reproduction of Van Gogh's famous painting *Starry Night*, which hangs in a large frame on the wall. Suddenly restless, you decide to see what's on TV. Grabbing your trusty remote control, you point at the painting and click a button. Instantly Van Gogh's swirling nightscape is replaced by the dynamic images of a football game in progress. As a running back gallops down the field, every detail of the picture—the player's uniform, individual faces of people in the stands, seemingly each blade of the manicured turf—shows with crystal clarity.

Football is not really your game, however, so you switch channels. A local station is showing an old movie from the early 1990s, *Dances with Wolves*, and your frame is now filled with stampeding buffalo, their stereophonic hoofbeats thundering through the room. The picture is so clean, the resolution so sharp, that it's difficult to believe that you're not sitting in a movie theater. Just then your Picturephone rings; it's a friend asking if you want to go out to see the latest sequel in a very successful film series: *Home Alone 12: Lost in Graduate School.* You switch off the movie and, on a whim, decide to replace your Van Gogh with something more modern. With a few more clicks of the remote, you leave one of Andy Warhol's famous soup cans showing proudly.

Is such a scenario possible? It is, according to proponents of a new form of TV known as high-definition television, or HDTV. This new format promises to deliver a dramatically sharper picture than is offered by current television technology—free of the static, the ghostlike double images, and other forms of interference that bedevil today's viewers. Sound will be in digital stereo.

High-definition television, or HDTV, technology produces screen images dramatically sharper (above left) than those of conventional TVs (right). Numerous legal issues must be ironed out before a nationwide HDTV system can be fully implemented.

But HDTV represents more than a new kind of television. It is one element in the ever-accelerating world of microelectronics—a world that is merging the technologies of entertainment, information, and education. Computers, telecommunications, and HDTV are all part of the mix. "We're not just talking about prettier television pictures here," says Richard E. Wiley, who chairs an advisory committee on advanced television systems for the Federal Communications Commission (FCC), the government body that will oversee HDTV's progression into the broadcasting marketplace. "We're talking about an imaging revolution that will touch many different industries: factory automation, medical devices, computer-aided engineering, defense products—many things."

However, before we can all watch razor-sharp images on wall-hung screens in our living rooms, HDTV faces many hurdles. Complex issues involving standards and licensing have yet to be worked out. It is by no means certain that consumers care sufficiently about a sharper picture to pay for a higher-priced TV set. And perhaps the most daunting question of all: Can over-the-air broadcast of HDTV really be made to work on a mass scale?

In a conventional TV, the picture consists of 525 lines. In HDTV, more than 1,000 lines make up the image. Special television cameras are needed to achieve this new high-density format.

What's Right with This Picture?

In a conventional television set, the picture consists of 525 lines, each line containing several hundred picture elements. As a signal is broadcast to the set's receiver, the lines are scanned instantaneously and continually. In effect the television receives 30 pictures a second; the human eye interprets this stream as one continuous image. In the cathode-ray tube (CRT) inside a TV set, the signal reaches the picture tube in the form of an electron beam that is fired at a phosphorus coating on the tube's surface. The screen then glows when struck by the electrons. Thus, the broadcast image is re-created with all its brightness and color intact. This 525-line signal is known as NTSC, after the National Television Systems Committee, which selected the format as the U.S. standard in television's infancy in the early 1950s.

The problem with all these formats is that the line structure in the picture remains all too evident on the television screen. "We don't want that structure to be visible," says Charles Rhodes, chief scientist at the Advanced Television Test Center (ATTC) in Virginia. "The world is not made of lines."

Instead of the conventional 30 pictures per second of today's standard, HDTV operates at nearly 60 pictures—nearly 1 billion bits of information—per second. More information means a much sharper image. Instead of 525 lines, there are more than 1,000. With HDTV, the picture matches the clarity and resolution of the 35-millimeter film images seen in movie theaters.

HDTV designers also intend to imitate the screen dimensions of a movie theater. Until the advent of television, most Hollywood movies were made in a square format, with an "aspect ratio" of 4 to 3, meaning four units of width to three of height. In the mid-1950s, when television threatened to keep moviegoers at home, Hollywood hit on a new attraction: the wide-screen format that more closely mimics a human's field of vision. HDTV developers are banking on this more visually pleasing format, which is closer to a movie screen's aspect ratio of 16 to 9. "The picture can now be much bigger without becoming coarse," says Rhodes. "There's five times the information in the picture, it has five times the resolution, there's no color error—no knobs to fiddle with—and it has digital sound."

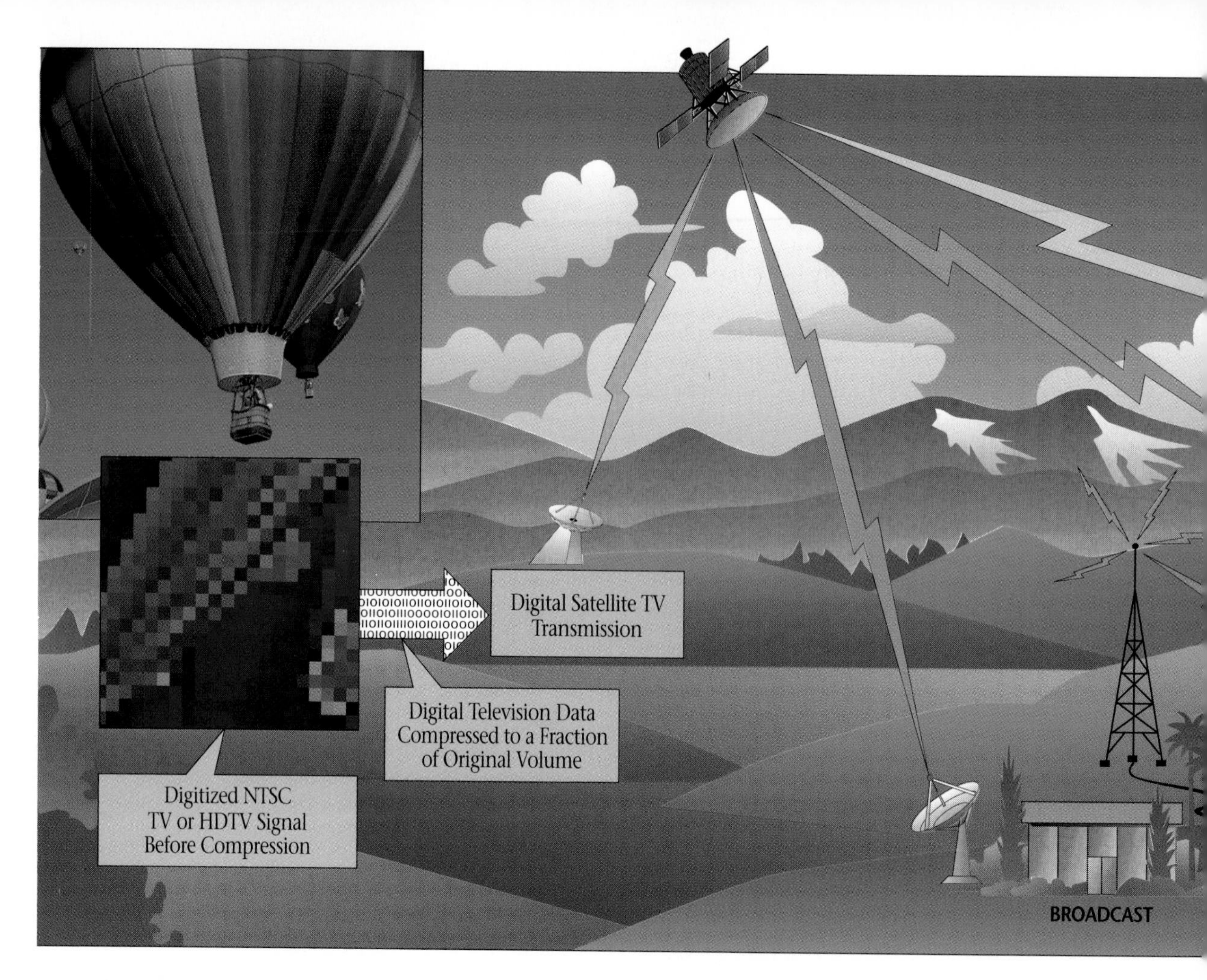

In a typical HDTV transmission scenario, the source material (video, film, etc.), is digitized prior to distribution to affiliates. The digital data stream is then compressed so that only a fraction of the bits—the essential data—is transmitted. The data stream is transmitted via satellite to local affiliates or to home satellite dishes.

Made in Japan—At First

HDTV was originally a Japanese innovation. In the late 1960s, engineers at Japan's national broadcasting company, NHK, began to explore a system that would surpass NTSC and become the new world standard. The first HDTV system was unveiled in the late 1970s, employing 1,125 scan lines and an aspect ratio of 5 to 3. Japanese designers soon confronted what remains a key problem for HDTV developers: squeezing the format's broader, information-laden signal into a workable bandwidth. The NHK system required 30 megahertz (MHz), as opposed to the 6 MHz on which conventional NTSC channels operate. Using a variety of techniques known collectively as "video compression," engineers were able to shrink the signal to more-manageable proportions—too wide for so-called "terrestrial broadcast" from ground-based transmitters, but narrow enough for satellite transmission. One-hour daily satellite broadcasts of HDTV were begun on a test basis in 1989. In the fall of 1991, the schedule was expanded to eight hours of HDTV programming a day.

Thus far, results in Japan have hardly triggered a stampede. The receivers required to process the HDTV signals are complex and, at this stage, very expensive. By early 1992, a single receiver cost $19,000, and fewer than 1,000 had been sold. The Japanese format was also based on what many observers view as a critical flaw: analog, as opposed to digital, signaling. Analog signals—the older format in which NTSC broadcasts currently reach TV receivers in the United States—are transmitted as wave forms.

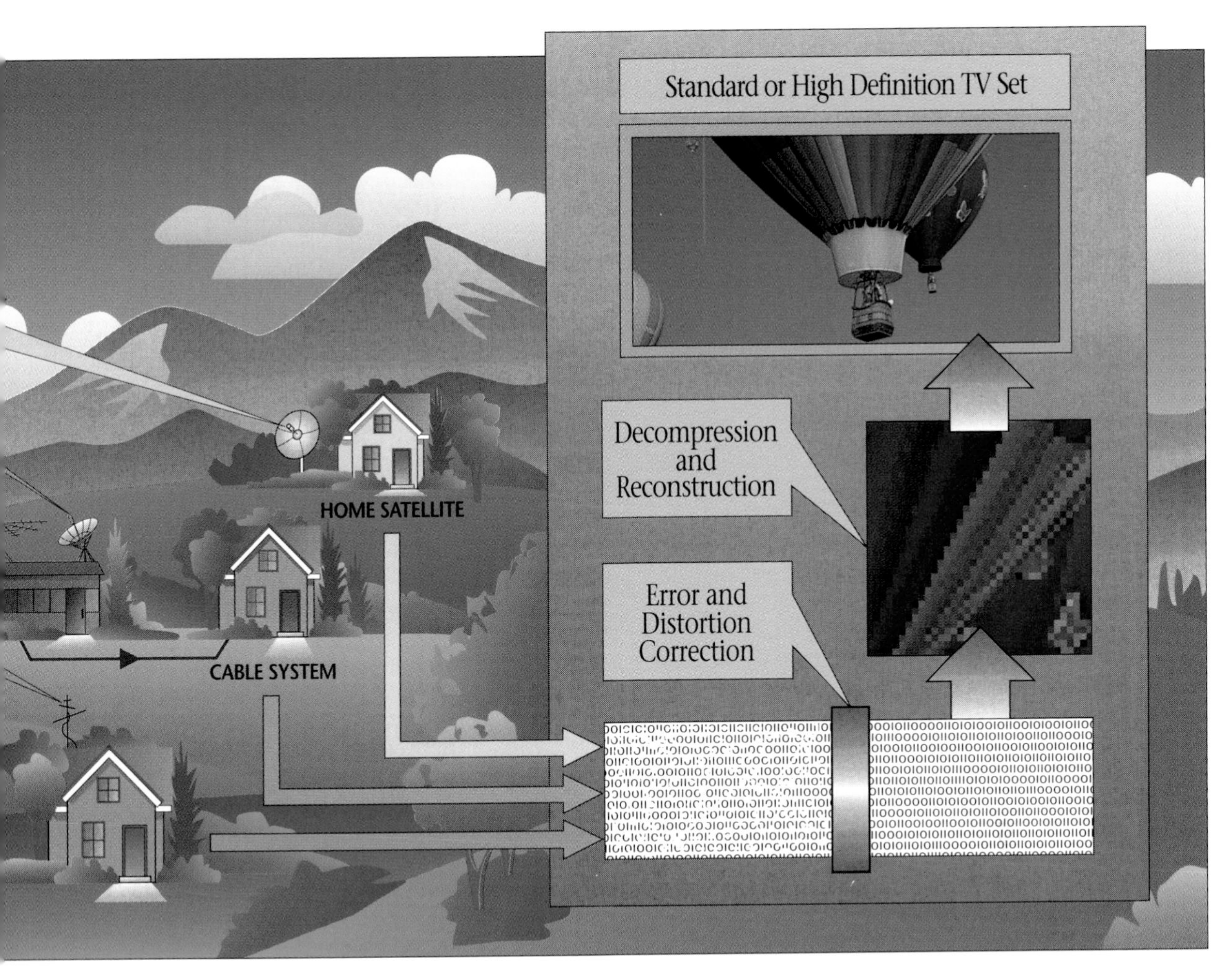

The digital television signal is broadcast over-the-air or via cables to homes. Error-correction and equalization systems help ensure accurate reception over a wide area. In the home, a data decompressor reconstructs the picture and converts it to analog form for display. The result: a studio-quality picture in the home.

Digital code, on the other hand, comprises the binary series of 1's and 0's that is the language of computers, compact discs, and other forms of telecommunications technology that have just started to emerge. U.S. developers of HDTV have pinned their hopes on digital transmission. The viability of over-the-air, digital broadcast has emerged as a critical issue in HDTV's odyssey from drawing board to living room.

Meanwhile, Back in the U.S.A. . . .

The United States and Europe were slower than the Japanese to enter the HDTV arena, not starting development efforts until the 1980s. By mid-decade, U.S. broadcasters began to perceive that HDTV might significantly affect their business if cable companies, satellite broadcasters, and the videocassette industry took the lead in delivering programming in the new format. Broadcasters had already seen enough of their audience defect to these alternative media. In 1987, at the behest of broadcasters, the FCC began its official inquiry into the possibility of offering HDTV programming.

The decisions made by the commission thus far have had enormous implications for designers, manufacturers, and broadcasters. One such decision was that the FCC would select a single standard for HDTV, in the same way that the NTSC standard had been set almost 40 years ago. The commission announced that it would accept competitive proposals for an HDTV system. After evaluating the proposed systems, according to the FCC plan, the commission would select its standard from among the entrants.

There were further FCC stipulations. As regulator of the nation's airwaves, the FCC has the job of allocating what is known as "spectrum," the limited range of frequencies that can carry broadcast signals. These frequencies, of course, are the lifeblood of a broadcaster's business, and competition for an FCC-licensed spot in the spectrum is fierce—particularly with the growing demands of the cellular-telephone industry, which also depends on the airwaves. The commission decreed that any new HDTV system must be completely different from the NTSC standard, yet must also fit within the conventional 6-MHz bandwidth. Furthermore, viewers who choose not to upgrade to HDTV would not be excluded. The FCC declared that all programming broadcast in HDTV must also be carried by NTSC, a process known as *simulcasting*. This would require broadcasters to have licenses for two channels instead of one, further complicating the scramble for available frequencies. But such a policy does allow for a period of transition between the old and new standards—a transition that might require 15 years, according to some experts.

Several major U.S. corporations, such as AT&T, Zenith, and General Instruments, as well as the Massachusetts Institute of Technology (MIT), have developed HDTV systems. Other countries—the Netherlands, France, and Japan—are also represented in the race to supply the new FCC standard. Originally, these fledgling HDTV systems represented a mix of approaches: combining analog and digital signals, for example, or simply augmenting an NTSC signal. Designers at General Instruments, however, pioneered a digital approach, simulating the first all-digital transmission of HDTV in the fall of 1990.

In July 1991, laboratory testing of HDTV formats began at the ATTC laboratory, a facility funded by the television networks, the National Association of Broadcasters (NAB), the Electronic Industries Association, and other groups. Identical TV footage was recorded in each of the competing formats, for evaluation in the lab as well as by test audiences (often college students) at facilities in Canada. By the spring of 1992, the field of proposed systems had narrowed from 23 to five—all but one digital. After the lab tests, according to the FCC schedule, a few finalists from among the entries will be selected for field tests, which will simulate broadcast over the air and via cable. With the recommendation of its advisory committee, the commission hopes to choose a standard by the end of 1993. But that target date might be considerably delayed—in part because some of the competitors switched to digital systems just before testing was to begin.

Terrestrial broadcast of digital HDTV presents many formidable obstacles. Maintaining the quality of the picture is one challenge, since digital signals degrade faster than analog. Finding a means to transmit the signal to a sufficiently wide area will be crucial. Sophisticated, computerlike microprocessors in HDTV sets will receive the digitally encoded signal, eliminating ghosts and presenting a flawlessly sharp picture. At least that's what broadcasters are hoping. But not everyone is convinced.

"In my opinion, digital transmission over the air is simply not going to work properly under actual broadcasting conditions," says William Schreiber, a consultant and former professor of electrical engineering at MIT who helped design a digital-analog system. "A lot of serious problems regarding the elimination of ghosts and other channel impairments are being swept under the rug." A more likely scenario, according to Schreiber, is that HDTV will be confined to satellite or cable broadcast.

Even apart from the technical uncertainties, the broadcast industry itself seems ambivalent about the new technology. "I think they're viewing it with some trepidation," says Wiley, whose advisory committee will play a crucial role in the FCC's decision on a standard. "They recognize that they've got to go into the new frontier, and yet they're going to be selling a new service for no new money, even while they're spending money to implement the new system. On the other hand, I think they're definitely committed to it." Two 1990 studies, by the CBS and PBS networks, each placed the cost of converting to HDTV transmission at $10 million to $12 million per station. A more recent NAB report suggested that the costs would likely be lower—particularly for local stations simply passing on an HDTV feed from a network.

Still another unknown quantity—and a considerable one at that—is the reaction of the American consumer. "There's no grassroots demand for a clearer picture," says

HDTV originated as a Japanese innovation. Even in Japan, however, where the 1992 Winter Olympics were broadcast in HDTV (right), the format has been slow to catch on, largely because of the prohibitive cost of the receivers (now approximately $19,000 apiece).

Schreiber. "The grass-roots demand is for better programs." As NAB vice president John D. Abel remarked in a 1992 speech to broadcasters: "Essentially, HDTV is technology-driven, not consumer-driven." In short, no one seems to know if people will want to spend what is projected to be upwards of $500–$1,000 extra for an HDTV-ready set, simply to receive a sharper picture.

Nevertheless, some experts predict that televisions incorporating HDTV receivers will become available as early as the mid-1990s. Gadget-hungry video enthusiasts and "more-affluent consumers," as Wiley refers to them, are expected to buy the first sets, and rising sales will prompt manufacturers to lower prices. Attempting to predict HDTV's commercial fortunes, some experts cite the early days of color TV in the 1950s, when a new set cost as much as a new car. As sales rose and manufacturers dropped prices, color sets spread to more and more homes. The hope is that history will repeat itself in the case of HDTV.

In the Foreseeable Future

The expectations for HDTV go far beyond entertainment. In many minds, this new TV technology has become a rallying point for American competitiveness—a means for regaining stature and leadership in the consumer-electronics industry. In Washington, politicians, lobbyists, and industry representatives have described HDTV and related systems—including high-definition computer imaging and displays, sensors, signal processing, and data storage—as critical technologies for American competitiveness.

Once the FCC selects a standard, the race to manufacture HDTV sets will be on. If the ultimate design is American, an economic windfall for U.S. companies could result from licensing fees, royalties, and the business of supplying the sets' computerized components—even if the sets themselves are actually manufactured overseas. Development of HDTV technology is expected to drive innovation in semiconductors and related fields. Display technologies, in particular, have become the focus of attention—especially from the defense community. The development of flat, high-resolution computer screens is seen as essential for many military applications, such as tactical display and systems for training that simulate combat. It is this technology that is expected to eliminate the CRT electron gun that now bulges from the back of conventional TV sets, replacing it, one day, with a flat panel that might be hung on the wall. With the insertion of a small, read-only memory (ROM) card, an entire museum wing's worth of art could be stored for display when the TV is "off."

But will HDTV fulfill such futuristic predictions? Will the technical challenges, the tangled process of government approval and supervision, and the perils of the marketplace combine to thwart the promise? Advisory committee chairman Wiley offers the best advice: "Stay tuned."

BOLDER BAR CODES *by Elizabeth Pennisi*

One day we may carry our medical histories on the back of our health-insurance cards. And labels on our cars, air conditioners, and computers will encode their maintenance records and list serial numbers for replacement parts. All these data will fit into a postage stamp-sized symbol, part of the next—and far more revealing—generation of bar codes and automatic identification technologies.

Bar codes are as familiar as a trip to buy groceries. Now part of almost every package that crosses the checkout counter, bar codes stand poised to move into many other facets of society—from hospital emergency rooms to hazardous-waste repositories, perhaps even battlefields. In their quest for ever better device identification, the U.S. Department of Defense and the National Aeronautics and Space Administration (NASA) plan to try out coding systems that pack in much more information than do current bar codes. Several companies have developed these new technologies, which promise more reliability as well as more-comprehensive data. As such, these codes could find their way onto far more products.

"The application in the commercial sector should be even more extensive [than in the military]," says Doug Mohr, mechanical engineer in program development at the Idaho National Engineering Laboratory in Idaho Falls. Developers of these technologies plan extensive promotion of the new scanning and coding devices, citing increased demand from manufacturers for ways to pinpoint where and when parts were produced.

However, if used to encode data about people, these new codes could threaten an individual's control over personal informa-

Bar codes are a familiar sight at checkout counters everywhere. Less familiar are the tiny bar codes that carry vastly more information than the old ones.

tion. "A technology like this, when it's made commercially available without protection, seems to lead to a further erosion of privacy," says Rob E. Kling, a computer scientist at the University of California, Irvine. An expert in issues of computers and privacy, he worries that people will not be able to read these codes or easily correct mistakes. "In effect, people might not know what they're carrying around," he adds. "The control [of personal records] shifts to the institution and away from the individual."

Two-dimensional Bar Codes

Today's bar codes work like license plates. They usually represent a number, and carry no information per se. Instead, a cashier scans across the code with a laser. The code's colored bars absorb light, while the white spaces bounce back light, creating a specific pattern of electrical pulses that a decoder translates into a number. That number calls up specific product information in the cash register's computer or in a centralized database, just as a license plate lets police look up information about a particular car and driver.

New "two-dimensional" bar codes recently developed can squeeze in enough information to fit the Gettysburg Address into a 2-inch (5-centimeter) square. But to pack that much data into a small space, they work much differently than those codes on supermarket products. To get more information using the old technology requires longer symbols—and these can require more room than the product's size allows. The new technology adds data along the vertical axis as well as along the horizontal one, and relies on computer programs and different scanning devices to read and interpret the codes.

"It's a technology that will open up a whole range of applications," says Richard Bravman, vice president of marketing for Symbol Technologies, Inc., in Bohemia, New York, one firm with a new bar-code system.

Miniaturized, some of these new codes can identify electronic components, jewelry, or even medical devices. "It represents a giant step in component traceability," says Robert S. Anselmo, president of Veritec, Inc., in Chatsworth, California. He boasts that his company's symbols could fit on a grain of rice. Others say they can make their codes invisible to the eye, but still readable to a scanner.

This next generation of identification codes needs no centralized database. Instead, the symbol itself can contain all the necessary information, says Bravman. Thus,

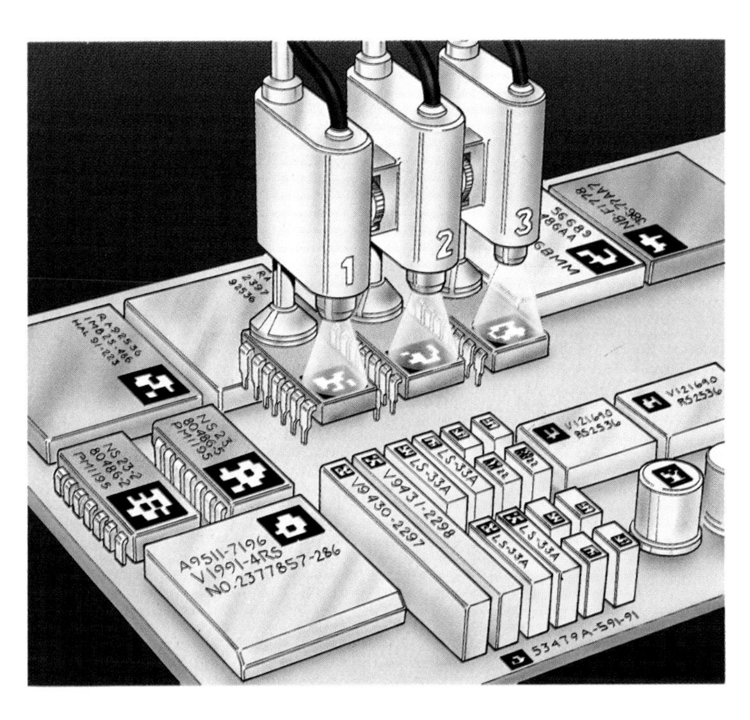

Unlike the conventional "consumer" bar codes (above right), the new, two-dimensional bar codes have found widespread application in business and industry. In some electronics factories, bar codes are attached to the components of computer circuit boards (right) to provide data about the latest revisions performed on them.

these codes can help companies and the military keep better track of products that "cross organizational boundaries," he adds. When the device, substance, or person travels to a new warehouse, store, hospital, or location, all its data go along, in compact form, accessible to anyone with a machine that can read the symbol.

"It's a portable data file," says Mohr, who is evaluating these technologies for use by the federal government. "It reduces considerably the amount of paperwork attached to that piece of equipment. For example, a soldier carrying a hand-held scanner could read the code of a piece of equipment to find out its maintenance schedule, specifications, even the serial numbers for replacement parts," he speculates.

Many Applications

No one uses these two-dimensional codes yet, but that will soon change. The U.S. Air Force has begun two projects to evaluate the technology's potential. Also, Anselmo says NASA is considering using Veritec's codes on the space shuttle's heat-resistant tiles. Mohr is doing a study for the Departments of Energy and Defense to determine the best ways to attach labels, and how well the new bar codes stand up to the harsh environments associated with hazardous materials.

At the Wilford Hall U.S. Air Force Medical Center in San Antonio, Texas, hospital administrators expect that within a year, patients there will carry ID cards with medical histories and personal data encoded on the back. The hospital had evaluated other types of codes, including current bar codes, but discovered with the two-dimensional format that "we didn't need to tie up our database memory," says Lieutenant Colonel Frank J. Criddle, an emergency-room physician at Wilford Hall. Also, other systems, such as smart cards with a microchip built into a wallet-size card, cost too much money per person to issue or update cards, he adds.

Instead, once the hospital has installed the new bar-code system, it will print out labels for any incoming patient, and then simply scan that label if and when the person returns. Criddle and others think these codes will help make Wilford Hall's emergency room—which treats about 56,000 cases a year—run more efficiently. "The quicker you can check in a patient, the quicker the patient can be seen," Criddle says.

In another application, the Air Force plans to put the codes on hazardous-materials containers during the next year. By federal law, safety data and handling instructions must accompany these materials, and by having that information encoded right on the package, manufacturers can ensure the data will not get lost during distribution.

As soon as the material reaches a central warehouse, a scanner can read the label and load all the information into a computer, where anyone tied into the computer's network can readily obtain it, says Colonel Phillip Brown, an industrial hygienist at Wright-

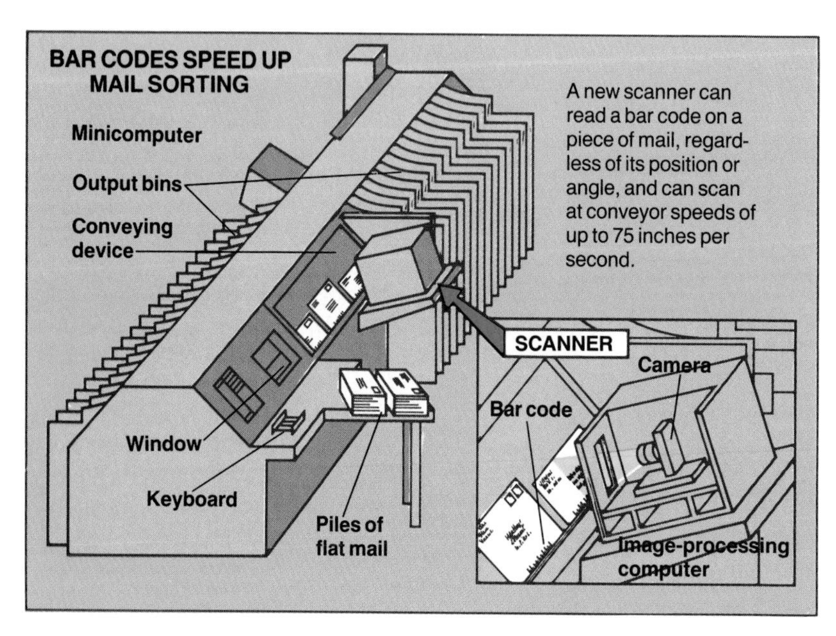

Bar-code-reading scanners will help the U.S. Postal Service process large-size mail, such as catalogs and magazines. These new machines can spot bar codes on mail whizzing by on a conveyor belt at the rapid rate of 75 inches a second. The machines interpret the codes, and direct the mail into the proper sorting bin.

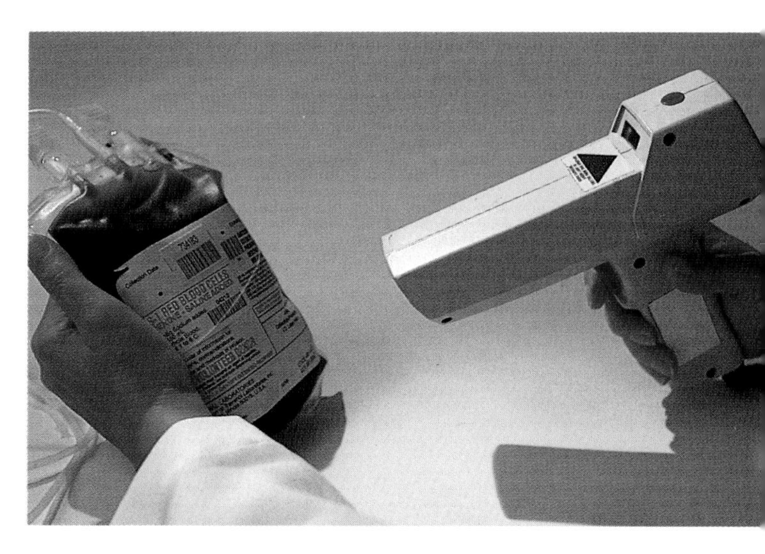

New bar codes are small enough to be laser-etched on a dime (above). Bar codes affixed to blood bags (right) provide a portable data file that keeps track of vital information as the product is transferred.

Patterson Air Force Base in Ohio. Otherwise, copies of the information must be passed along with the material to specific users. "I think it will save a lot of trouble and time," Brown adds.

Encoded Redundancy

Several different two-dimensional coding technologies vie for these applications. Anselmo's system, called Vericode, relies on a matrix of data cells. Each cell represents a binary bit, so the symbols wind up looking like a maze, with the numbers and words encoded in the pattern.

Other two-dimensional codes consist of tiers of squat bar codes. For example, one type called Code 49, developed by Intermec Corporation in Lynnwood, Washington, stacks rows of 17 spaces and 18 bars on top of each other, with stop and start codes at each end. Another stacked bar code, PDF417 by Symbol Technologies, uses thin dark-and-light spaces grouped into more than 900 "code words," each of which can represent about two characters. That adds up to 1,750 characters per coded symbol. Both companies can increase the information by adding additional symbols.

Both the matrix and stacked bar codes seem more reliable than current bar codes, Mohr says. Existing bar codes read in just one direction—horizontally—like the words in this sentence. The widths of the lines and spaces encode numbers, while the height adds redundancy and makes the pattern easier to read. But these codes become indecipherable if a smudge or scratch disrupts the scanner's ability to read these widths. In contrast, the new bar codes fit in enough data that important information is repeated, and therefore the scanner can retrieve it even if part of the symbol is destroyed. Also, the scanner software often can "reason" what the missing data should be, and fill in what's missing.

Two-dimensional codes have yet to stand the test of time in real applications, however. And companies with competing technologies can identify potential weak points in their competitors' approaches. Anselmo argues that stacked bar codes may work well on paper, but not when etched into glass, plastic, or metals, because these materials make it difficult for scanners to accurately read very thin lines. Also, as the density of information increases, the lines get thinner and the signal weakens, increasing the chances of error.

Those companies with stacked bar codes quickly point out that their technology blends in better with existing bar-code readers. The addition of new software allows conventional bar-code scanners to read the tiered bar codes, but not matrix ones. "The matrix approach potentially gives you .the highest reading density, but it needs an imaging system," says Bravman.

Whatever their strengths—or weaknesses—these new bar codes will likely start showing up on containers and in places where one least expects them. "You can increase your efficiency and capability with these two-dimensional codes," says Mohr. "I think they will catch on very fast."

REVIEW

Computers

PORTABLE COMPUTERS ARE HERE TO STAY!

Portable computers are no longer perceived as a luxury in the world of business computing. Corporate employees in highly competitive industries often travel extensively to maintain the corporate "bottom line." As a result, they need to stay productive while on the road—that means easy access to powerful software applications and up-to-date corporate data, as well as electronic mail and personal schedules. Today all that power is widely available in notebook-sized computers, and, in abbreviated form, palmtop computers. Pen-based computers, or notepads, hailed recently as the hottest new portable technology, are still neither fast enough nor powerful enough to replace keyboard-based portables as the tool of choice for busy executives.

▶ **Portable Memory.** While portability has increased dramatically since the first compact computer appeared (weighing over 10 pounds, or 4.5 kilograms, compared to today's lightest 1-pound-or-less marvel), their versatility away from home has lagged behind. Business users wanting to access their company's main computer network from the road have had to carry additional modem hardware and cables. And users wanting to run complex applications simply haven't had the power to do so. Until now, that is.

The latest version of a new memory-and-expansion card developed by the Personal Computer Memory Card International Association (PCMCIA) and Japan Electronic Industry Development Association (JEIDA) is expected to revolutionize portable computing. These solid-state silicon cards are the size and thickness of a credit card. Designed specifically for portable computers, these sturdy cards can store data and programs as well as act as devices such as modems and network adapters that link a laptop to the rest of the computing world.

The contents of earlier versions of the card had to be downloaded into a PC's "memory" to work; this decreased the portable's computing power and limited the card's versatility. The latest version, however, allows a user to run programs directly from the card, freeing up "memory" for other applications and speeding up the computer's ability to process data. Card makers are now designing cards to contain a variety of applications and memory types, including ROM, RAM, and "flash." ROM cards are programmed once by the manufacturer, and cannot be changed by the end user; RAM cards allow a user to read and write data on the card, and often work in conjunction with a small battery to protect against power loss; flash memory cards use a quick new technology that allows a user to read and write data on the card, but requires no battery for data protection.

▶ **Notebooks.** Portable systems have quickly been refined and reduced in size, and today's computer notebooks now weigh, on average, less than 8 pounds (3.6 kilograms). In addition, they fit easily in a briefcase and can do almost anything that a desktop personal computer can. For example, a notebook can contain a list of business contacts; an agenda; word processing, spreadsheet, and database software; and even communications/fax capabilities.

A typical notebook-sized computer is approximately 12 inches (30.5 centimeters) long, 10 inches (25.4 centimeters) wide, and 2 inches (5 centimeters) thick. It is composed of a standard-width keyboard, black-and-white LCD-display screen (currently up to 10 inches, or 25.4 centimeters, in size), and contains a limited amount of memory and data storage. It runs on a rechargeable battery that lasts on the average a couple of hours, although future models hope to increase that time to a full working day.

Most notebooks run the DOS operating system, although Apple recently introduced the portable PowerBook, which is a Macintosh-based system. Notebooks also contain "ports" that allow them to be linked to printers and to local area networks (LANs) via modems and cables. Notebooks are so powerful and versatile, in fact, that lots of "everyday" computer users who don't travel the world are purchasing them as their main computer systems: they've discovered that they can have the power of a desktop computer in a rather convenient and small package. One industry specialist predicts that sales of computer notebooks will grow from nearly $700 million in 1990 to $13 billion in 1994!

▶ **Palmtops.** The smallest and newest portable computer is the palmtop, which ranges in size from a wallet to a pocket-sized daily scheduler. Naturally these computers don't weigh nearly as much as notebooks or laptops. In fact, one model reportedly weighs just a half a pound! The best palmtops available today combine the features of a full computer system in a tiny package that fits in your hand. For example, the Psion Series 3 palmtop contains a simple database manager, a word processor, a calendar-scheduler, a clock, a world map, and a calculator. The actual hardware incorporates a 66-key keyboard laid out in the "QWERTY" key format found on traditional typewriters, and includes special function keys that allow the user to switch between applications, much like the graphical user interface (GUI) software available for desktop computers. The LCD screen displays eight lines of text, 40 columns wide.

Palmtops generally run on AA batteries; they may also use a lithium battery supplied by the manufacturer as a backup. The AA batteries can provide up to 120 hours of computing, depending on the applications run on the palmtop. Like their larger cousin the notebook, palmtops can be connected to printers or to modems for communicating with desktop systems from afar.

Computer notebooks, far right, are fast becoming the primary tool of choice by consumers who want the power of a desktop computer in a small, portable package.

Palmtops do have certain limitations. Most use their own proprietary operating systems, which limits their compatibility with other computers and most software produced today, although a few run DOS or are DOS-compatible. In addition, a palmtop contains limited internal memory, which prohibits users from running powerful applications. Several of the palmtops currently on the market, however, support the use of the credit-card-sized PCMCIA memory cards discussed above, which provide the increased memory required by most software packages.

In addition to limited memory, a tiny keyboard, and incompatibility with other computers and software, palmtops are also somewhat expensive ($500 to $1,500) for what they offer. However, palmtops have just been introduced into a market that only a few years ago considered laptops and notebooks to be mere peripherals for desktop computers. Now notebooks are *replacing* desktop computers in some organizations!

For now, palmtops will serve as very portable communication links between employers and employees who don't need powerful software capabilities away from the office. In addition, they will supply convenient computing power for the quickly burgeoning personal-information-management industry by providing busy executives and others with pocket-sized electronic address books, schedulers, and note takers.

Abigail W. Grissom

Technology

ELECTRIC CARS

General Motors Corporation continued to develop its electric car, which will become available to consumers in the mid-1990s. Named the "Impact," the two-door vehicle has a sporty look and can accelerate from 0 to 60 miles (96 kilometers) per hour in 8 seconds. The Impact has traditional lead-acid batteries, which give out more power than newer batteries being developed by other carmakers; however, the lead-acid batteries suffer from having a limited capacity to store charge. The batteries power two induction motors, which have high efficiency and can run at high revolutions, eliminating the need for a transmission. Operating together, the two 57-horsepower motors provide power equivalent to that of today's four-cylinder gasoline engines. The Impact features "regenerative braking," in which the wheels of the car, when coasting or braking, feed power (through the motors) back into the batteries.

General Motors' Impact may not be the first electric vehicle to hit the road in a big way. Ford, Chrysler, and many foreign manufacturers have pressed forward in the development of electric vehicles. An Anglo-Swedish company called Clean Air Transport, of Göteborg, Sweden, contracted with the city of Los Angeles to begin selling an electric vehicle there starting in 1993. The car, a two-door hatchback, can accelerate from 0 to 50 miles (80 kilometers) per hour in 17 seconds, and can reach a top speed of 75 miles (120 kilometers) per hour. This new vehicle, called the "LA301," is a hybrid: in addition to its electrically powered motor, it contains a four-cylinder, internal-combustion engine meant to run the car during long trips. Otherwise, the batteries can power the car for 60 miles (96 kilometers) between chargings. Like the Impact, the LA301's batteries are of the lead-acid type, but unlike the GM car, the LA301 has only one 57-horsepower motor.

General Motors' Impact electric car (below) can accelerate from 0 to 60 in 8 seconds. The vehicle should be available to consumers by the mid-1990s.

Ford Motor Company continued to develop its "Ecostar" van, an electric vehicle based on the Escort station wagon. The Ecostar's sodium-sulfur batteries can power it for about 100 miles (160 kilometers) of city driving before needing to be recharged. Ford plans to begin selling the vans to utility companies in 1992. The Electric Power Research Institute has already sold a small number of electrically powered vans—based on GM's "Vandura" model—to utility companies. Mercedes-Benz has been using a sodium-nickel-chloride battery in its electric car now under development. The Mercedes vehicle was an official pace car in the New York City marathon of 1991.

OTHER TRANSPORTATION TECHNOLOGIES

Radisson Hotels International, of Minneapolis, contracted to build a novel cruise ship with a hull design based on that of the familiar and smaller catamaran. The new ship will have twin hulls, which will provide stability on rough waters and allow the ship to be wider than an ordinary

The novel twin-hull design of the Radisson Diamond cruise ship provides the stability needed to assure maximum passenger comfort in rough seas.

vessel. The design is known as "Swath," or "small water-plane area twin hull"—referring to the fact that the surfaces of twin hulls have less net contact with the water than do the surfaces of single hulls of comparably sized boats.

This feature accounts for the diminished effect of the water's motion on the motion of the ship. The result will be less rolling and pitching of the ship, and a more comfortable ride for passengers. The new ship, called the Radisson Diamond, will have a breadth of 105 feet (32 meters)—as wide as single-hulled ships that are twice as long. However, the Diamond will be slower than comparable single-hulled cruise ships, and because of the deeper draft of the twin hulls, will not be able to dock at certain shallow ports.

AUDIO AND VIDEO

American Telephone and Telegraph Company (AT&T) developed a video telephone that can plug into an ordinary telephone jack; it will cost about $1,500. The videophone's screen has dimensions of approximately 2.5 inches by 3.5 inches (6.3 by 8.90 centimeters), and offers color images, which are somewhat fuzzy. The technology was developed by Compression Labs, of San Jose, California. It employs signal compression, in which a scanned picture is converted into millions of pixels, or digital picture elements, and only certain important pixels are chosen to be transmitted over the phone lines. The receiving videophone reassembles the pixels into a full image. The AT&T system transmits between two and 10 picture frames per second, which results in a jerky motion for moving images.

Two electronics companies introduced competing audio-and-video compact-disc systems for use in the home. Commodore International, Ltd., of West Chester, Pennsylvania, unveiled its "CDTV" system, which plays compact discs that store both sound and images. Discs that can be played on the system include the texts of the complete works of William Shakespeare, along with woodcut drawings; a cookbook of more than 450 recipes (with vocal comments by authors); and a world atlas that includes maps, photographs, text, and spoken comments. The Commodore players cost about $1,000, and individual discs cost between $30 and $100.

The Dutch firm Philips N.V. unveiled its "CD-I" system ("I" for interactive), which similarly plays compact discs containing audio and video. One disc program for the Philips system allows the user to tour the Smithsonian museums—at least on the home television screen—choosing from among the thousands of exhibits there. Like that of the Commodore system, the Philips technology is an outgrowth of CD-ROM technology, in which large amounts of data are stored on optical discs. The newer systems can store and read a wider variety of digitally encoded words and images, in addition to speech and sounds.

Creo Products Inc., of British Columbia, offered an optical-tape system for storing data, essentially a hybrid of laser-disc systems and older magnetic-tape systems. In these "digital-paper" systems, data is stored on traditional polyester tape as pits in a thin layer of embedded metal. As in optical-disc technology, the data is read by a small laser beam, which scans the tape (disc) as it moves past. A reel of optical tape can store more information than a comparative volume of optical discs. One drawback of the optical-tape system is its tendency to inscribe errors along with proper data, although the rate of error is small.

Donald Cunningham

In Memoriam

ABRAMSON, MAXWELL (55), U.S. physician who specialized in research on diseases of the middle ear. He served as chairman of the department of otolaryngology at the College of Physicians and Surgeons of Columbia University in New York City. d. Kona, Hawaii, May 3.

AGURSKY, MIKHAIL (58), Soviet scientist who specialized in physics and cybernetics. He served as an adviser to the military before becoming disillusioned with communism and opposing the Soviet system. He became an Israeli citizen in 1975 and was an active supporter of Soviet Jews. d. Moscow, Aug. 27.

ANDERSON, CARL D. (85), U.S. physicist who, in 1936 at the age of 31, won the Nobel Prize in Physics for discovering the positron, one of the tiniest particles in nature, sometimes called a positive electron. He taught at the California Institute of Technology from 1933 to 1976, and headed its division of physics, mathematics, and astronomy from 1962 to 1970. d. San Marino, Calif., Jan. 11.

ANDERSON, THOMAS F. (80), U.S. biophysical chemist and geneticist who developed a method of preparing specimens for the electron microscope. d. Philadelphia, Pa., Aug. 11.

BARDEEN, JOHN (82), U.S. physicist who co-invented the transistor and developed the theory of low-temperature superconductivity; he was the first person to twice win the Nobel Prize in the same field (physics). In 1947, working with scientists Walter Brattain (d. 1987) and William P. Shockley (d. 1989) at American Telephone and Telegraph Company's Bell Laboratories, Bardeen helped develop the first transistor, a tiny device that was to revolutionize electronics. Transistors were first used commercially in 1952, and now are used in virtually all electronic equipment from computers and radios to garage door openers and spaceships. Bardeen shared the Nobel Prize in Physics in 1956 for this discovery. Bardeen's work on superconductivity—the ability to carry electricity with virtually no resistance—began in 1951 and led to his second Nobel Prize in 1972, shared with two of his students, Leon Cooper and J. Robert Schrieffer. Bardeen was awarded the National Medal of Science in 1965, and the Presidential Medal of Freedom in 1976. d. Boston, Mass., Jan. 30.

BERGALIS, KIMBERLY (23), U.S. student who contracted AIDS from her dentist and crusaded in favor of legislation requiring mandatory testing of health workers for the AIDS virus. Bergalis, near death, made an emotional and dramatic personal appeal to Congress for support for such legislation. She died 15 months after her dentist, who apparently infected at least four other patients. d. Fort Pierce, Fla., Dec. 8.

BOGART, LARRY (77), a leading U.S. critic of nuclear power who published newsletters and organized opposition to commercial nuclear-power plants. Bogart complained that the plants were too complex and too dangerous, but his conclusions were rejected as scientifically invalid by academia, government, and industry. d. East Orange, N.J., Aug. 19.

BROWNING, IBEN (73), U.S. climatologist who used data about tides and gravity to correctly predict the San Francisco earthquake in 1989 and the eruption of Mount St. Helens in 1980. He caused a minor panic when he predicted that in 1990 a major earthquake would strike the New Madrid fault line that stretches 120 miles (200 kilometers) from Illinois to Arkansas. Although this fault had been the epicenter of one of North America's largest earthquakes in 1811, the 1990 one failed to materialize. d. Albuquerque, N.Mex., July 18.

BUTCHER, DEVEREUX (84), U.S. author and conservationist who served as executive director of the National Parks Association from 1942 to 1950, founded *National Parks* magazine and published the *National Wildlands News,* wrote four books about the park system, and served as a member of the Interior Department's committee on conservation. d. Gladwyne, Pa., May 22.

CARTER, MANLEY L. (43), U.S. physician who in 1989 spent 120 hours in space while on a secret military mission aboard the space shuttle *Discovery.* He was killed in a commuter-airline crash in Georgia. d. April 6.

CHANG, M. C. (82), Chinese-born U.S. scientist who was a co-developer of the birth-control pill and was a pioneer in in-vitro fertilization. After extensive experimentation with laboratory animals in the 1940s and 1950s that confirmed that certain chemicals were preventing release of an ovum, Chang and his associates began testing the compounds on women in 1956. In 1960 the Food and Drug Administration approved sale of "Envoid" as a contraceptive. In the 1950s Chang was able to produce a "test-tube bunny" by fertilizing a rabbit egg outside the body, and then replacing the fertilized egg in the mother. Several years later the technique was successfully applied to human beings. d. Worcester, Mass., June 5.

DAMMIN, GUSTAVE J. (80), U.S. pathologist who helped perform the first successful human kidney transplant and was an authority on the crucial factors in successful organ transplants: tissue matching and immune response or rejection. He also was a leading researcher on Lyme disease; the tick that causes the illness, *Ixodes dammini,* was named in his honor. d. Boston, Mass., Oct. 11.

DONELIAN, KHATCHIK O. (77), Turkish-born U.S. engineer who worked with the Manhattan Project developing the atomic bomb and later helped design nuclear-power plants. He also invented a smoke detector that was the prototype of those used today. d. New Smyrna Beach, Fla., Oct. 25.

DOWNS, WILBUR G. (77), U.S. clinical epidemiologist and tropical-medicine expert who identified, isolated, and characterized many viruses responsible for diseases in tropical climates. He set up research projects in the Caribbean, Africa, and Asia, and served as an adviser to several national and international scientific and health organizations. d. Branford, Conn., Feb. 17.

EGGAN, FRED RUSSELL (84), U.S. anthropologist who studied the Hopi Indians and the indigenous people of the Philippines; he became a campaigner for Indians of the Southwest and their claims against the government. d. Santa Fe, N.Mex., May 7.

ELSASSER, WALTER M. (87), German-born U.S. geophysicist whose research on planetary magnetism won him a National Medal of Science in 1987. He was one of the first to state that movements within the Earth's core might be responsible for the Earth's magnetic field; he also helped define theories of plate tectonics and continental drift. d. Baltimore, Md., Oct. 14.

FLETCHER, JAMES C. (72), U.S. physicist who twice headed the National Aeronautics and Space Administration (1971-77, 1986-89) and initiated or was involved in every major project undertaken by NASA during that time, including three Skylab missions, two Viking missions to Mars, the Voyager probe, and the *Apollo-Soyuz* joint project with the Soviets. He aggressively campaigned for the space shuttle, promising it would be cost-effective. He returned to NASA after the 1986 explosion of the *Challenger* to restructure and revitalize the agency. Before joining NASA, Fletcher was president of the University of Utah from 1964 to 1971. d. Washington, D.C., Dec. 22.

FRIEDMAN, MAURICE H. (87), U.S. physician who developed a once-popular test for pregnancy that is widely known as the "rabbit test," because it involves injecting a woman's urine sample into a female rabbit. If the woman is pregnant, distinctive formations appear in the rabbit's ovaries. It is a common misconception that the rabbit dies if the woman is pregnant; in fact, the rabbit is killed to examine its ovaries. New tests are able to detect pregnancy without using animals. d. Sarasota, Fla., March 8.

FULLER, FRED H. (64), U.S. inventor who developed a voice analyzer that has been used as a lie detector. d. Wood-Ridge, N.J., May 30.

GOLLNICK, PHILIP (56), U.S. physiologist who studied how muscles are affected by activity, and who researched the effect of weightlessness in space on muscle tissue of astronauts. d. Pullman, Wash., June 24.

GREENE, JOHN H. (45), chief operating officer of Outward Bound, an organization that builds self-confidence and teamwork through challenging physical tasks like mountain climbing and backpacking. Greene died of a heart attack while hiking. d. Loch Eil, Scotland, June 13.

GRONQUIST, CARL H. (87), U.S. civil engineer who helped design the Golden Gate Bridge, the Henry Hudson Bridge, the Sault Ste. Marie International Bridge, and one of the longest single-unit suspension bridges in the world, the Mackinac Straits Bridge in Michigan. d. Summit, N.J., June 16.

HARKER, DAVID (84), U.S. scientist who in 1967 used X rays to determine the atomic structure of a complex enzyme made up of more than 1,000 atoms. His methodology has since been used to determine the structure of many hormones and drugs, allowing scientists to make minor structural alterations to produce more-effective substances. d. Buffalo, N.Y., Feb. 27.

HASSLER, ALFRED (81), U.S. pacifist and a leader in the early environmental movement. d. Suffern, N.Y., June 5.

HEIDELBERGER, MICHAEL (103), U.S. pathologist whose study of antibodies made him known as the father of modern immunology. He helped change the field from a descriptive science to a precise, quantitative one. He discovered that antibodies are proteins, and developed methods of quantitatively analyzing them. d. New York, N.Y., June 25.

HENKE, WERNER (75), German-born U.S. petroleum engineer who developed a widely used and highly effective pollution-control device and invented a means of extracting additional oil from previously tapped wells. d. Lafayette, La., July 27.

JENSEN, HOMER (77), U.S. physicist who pioneered aerial mapping and co-invented the magnetometer, an instrument that detects slight changes in the Earth's magnetic field. He used this device to help in aerial searches for hidden underground deposits of oil and minerals. d. Wyncote, Pa., Oct. 5.

KAPLAN, JOSEPH (89), Austro-Hungarian-born U.S. physicist who researched such atmospheric phenomena as auroras and faint airglows. He also co-founded the Institute of Geophysics at the University of California at Los Angeles in 1944, was chairman of the National Committee for the International Geophysical Year from 1953 to 1962, served as a military adviser in World War II, and as a science adviser to Presidents Eisenhower and Nixon. d. Santa Monica, Calif., Oct. 3.

KIDWELL, C. HAROLD (95), U.S. chemical engineer who developed a method of preparing extremely finely ground powders for use in paints and cosmetics, and for fabricating lenses and other precision optical equipment such as bombsights. d. Summit, N.J., May 17.

KLOPSTEG, PAUL E. (101), U.S. scientist who co-founded the National Science Foundation and chaired its committee on prosthetic devices and its committee on atmospheric sciences. He was director of the American Association for the Advancement of Science from 1949 to 1959 and its president in 1959. He founded the American Association of Physics Teachers. d. Laguna Beach, Calif., April 29.

KOLCHIN, ELLIS R. (75), U.S. mathematician who pioneered the field of differential algebra. He twice was a Guggenheim Fellow and was visiting professor at a number of universities in the United States, Europe, and the Soviet Union. d. New York, N.Y., Oct. 20.

LAND, EDWIN H. (81), U.S. physicist who revolutionized photography by inventing the instant camera. He conceived the idea in 1943 in response to his three-year-old daughter's question of why cameras couldn't instantly produce photographs. In 1948 his company, Polaroid, introduced the Polaroid Land Camera. In 1959 Polaroid developed an instant-color-photograph system and began marketing it in 1963. By the mid-1960s, company officials estimated that nearly half the households in America owned a Polaroid camera. In the mid-1970s, however, the company's fortunes changed with the troubled SX-70 camera and the short-lived Polavision instant-movie system. In 1980, five years after stepping down as Polaroid's president, Land formed the Rowland Institute for Science, a research organization. In 1955, as an adviser to the government, Land recommended developing reconnaissance satellites, the first of which was launched four years later. Though he twice dropped out of Harvard University, he was later awarded honorary doctorates from Harvard and several other universities. Land won numerous honors and awards, including the Presidential Medal of Freedom and the National Medal of Science. d. Cambridge, Mass., March 1.

LAUTMAN, DON A. (61), U.S. astronomer who developed a method of tracking and predicting the path of satellites. He successfully put his theories to the test in tracking the Soviet satellite Sputnik in 1957. d. Cambridge, Mass., Aug. 9.

LAWRENCE, JOHN H. (87), U.S. physician whose radiation research led to its use in diagnosing and treating cancer and other diseases. In 1936 he created the Donner Laboratory at the University of California and began treating diseases with radioactive isotopes. He discovered that neutrons have a destructive effect on cells, and in 1939 he began using neutron beams on cancer. He was awarded the Enrico Fermi Award in 1983. d. Berkeley, Calif., Sept. 7.

LEVI, DORO (93), Italian archaeologist who worked with Allied officials during World War II to spare Italian monuments from bombing. d. Rome, July 3.

LOVELESS, MARY H. (92), U.S. physician who in the 1940s developed a method of isolating venom from bees or wasps and injecting it into people allergic to the poison. This helped build up an immunity to what otherwise could be a fatal reaction to a bee or wasp sting. Though controversial, the technique was approved in 1979 by the federal Food and Drug Administration. She also discovered antibodies that could prevent pollen allergies and hay fever. d. Westport, Conn., June 2.

LURIA, SALVADOR E. (78), Italian-born U.S. biologist and physician who shared the 1969 Nobel Prize in Medicine with Max Delbruck and Alfred D. Hershey for their work on the genetic structure and reproduction of viruses. He was active in the peace movement and was one of 48 scientists blacklisted in 1969 by the National Institutes of Health. In 1976 he and a group of other scientists called for an end to further nuclear-power-plant construction. He also established a microbiology program at the Massachusetts Institute of Technology and founded the MIT Center for Cancer Research. Luria encouraged his students to be well rounded, to study the humanities as well as the sciences. He won the 1974 National Book Award for *Life: The Unfinished Experiment*. d. Lexington, Mass., Feb. 6.

MAGUIRE, BASSETT (86), U.S. botanist who studied remote areas of South America and advocated responsible land-use policies that would protect these ecosystems of the tropics. In 1954 he discovered the "Mountain of the Clouds" deep in Brazil at the Venezuela border. In 1965 he was awarded the David Livingstone Centenary Medal by the American Geographical Society for his discovery and study of this isolated and pristine mountain. d. New York, N.Y., Feb. 6.

McMILLAN, EDWIN M. (83), U.S. chemist who won the 1951 Nobel Prize in Chemistry for discovering the elements plutonium and neptunium. McMillan co-discovered oxygen 15 in 1934, beryllium 10 in 1940, and, while studying uranium fission in 1940, co-discovered neptunium, the first element beyond the naturally occurring 92 elements. It is believed he also discovered carbon 14. His research to confirm the existence of plutonium was completed by Glenn T. Seaborg, leading to their sharing of the Nobel Prize. McMillan headed the Lawrence Berkeley Laboratory from 1968 to 1973, developed the concept of phase stability and applied it in his synchrotron, a particle accelerator that kept particles controlled and led to the discovery of new elements that do not occur naturally in nature. He and Vladimir I. Veksler, a Russian scientist who independently developed the same idea at the same time, were awarded the 1963 Atoms for Peace Award. d. El Cerrito, Calif., Sept. 7.

MILLMAN, JACOB (80), Russian-born U.S. electrical engineer who helped develop radar. d. Longboat Key, Fla., May 22.

NICHOLSON, THOMAS D. (68), U.S. astronomer who was director of New York City's American Museum of Natural History from 1969 to 1989. During his tenure, he doubled the research staff to 200, increased the annual endowment fund threefold to $143 million, and increased annual attendance by 50 percent to 3.1 million. d. Woodcliff Lake, N.J., July 9.

NORTHRUP, DOYLE (84), U.S. physicist who developed a method for detecting atomic-bomb testing and used it to confirm a Soviet test explosion in 1949. In the 1950s Northrup served as an adviser at talks in Geneva, Switzerland, on limiting tests and weapons. In 1959 he was awarded the President's Award for distinguished service. d. Melbourne, Fla., Dec. 15.

PAGE, IRVINE H. (90), U.S. physician whose research into high blood pressure led to the discovery of compounds affecting blood pressure and to the now widely held idea that high blood pressure can be caused by a complex interaction of factors. He recognized hypertension as a treatable disease and a contributor to strokes and other circulatory complications. Previously, high blood pressure was thought to be relatively unimportant, the heart's way of dealing with aging and less-efficient arteries. d. Hyannisport, Mass., June 10.

PAGE, RALPH E. (95), U.S. inventor who worked for 43 years with International Business Machines Corporation and developed such devices as a fingerprint-sorting machine, keypunchers, electric typewriters, and the first Braille typewriter. He held more than 50 patents. d. Hyde Park, N.Y., Feb. 3.

PARR, ALBERT E. (90), Norwegian-born U.S. marine biologist who was director of the American Museum of Natural History from 1942 to 1959. Several major exhibit halls opened during his tenure, which was marked by the desire to make museums interpretative, not just "dead circuses." After retiring in 1959, Parr became senior scientist at the museum. In 1980 he received the first Distinguished Service Medal of the American Association of Museums. d. Wilder, Vt., July 17.

PENNEY, LORD WILLIAM G. (81), British scientist known as father of the British atomic bomb. He was the chief British scientist working on the development of the U.S. atomic bomb in 1944 and 1945, a role for which he received the Medal of Freedom. After developing and testing a similar device in Britain in 1952, he was knighted. d. East Hendred, England, March 3.

PERLMAN, ISADOR (76), U.S. nuclear chemist who developed several artificial isotopes, researched the decay of alpha particles, and increased the understanding of the structure of radioactive elements. As a professor of archaeology and chemistry at Hebrew University in Jerusalem from 1975 to 1985, he developed a method of using neutrons to analyze ancient artifacts. d. Los Alamitos, Calif., Aug. 3.

RACKER, EFRAIM (78), Polish-born U.S. physician who found that cancer cells have more of a certain enzyme than do normal cells, causing an imbalance of lactic acid that is believed to be a factor in the spread of cancer cells. He was awarded the National Medal of Science in 1977 and, for the past 25 years, had been the Albert Einstein professor of biochemistry at Cornell University in New York. d. Syracuse, N.Y., Sept. 9.

ROSSI, IRVING (101), U.S. metallurgical engineer who in the 1960s introduced continuous steel casting to the United States, having perfected a German design from the 1920s. He also advanced electron-beam technology for welding certain metals and for precision drilling. d. Harding Township, N.J., June 14.

SEIBERT, FLORENCE B. (93), U.S. physician who developed a reliable skin test for tuberculosis that is still in worldwide use today. She also developed a means of distilling water used for intravenous injections so that it was germ-free. d. St. Petersburg, Fla., Aug. 23.

SMITH, HARLAN J. (67), U.S. astronomer who helped develop large telescopes and discovered the effect of solar winds on radio emissions from Jupiter, the variability of quasars, and the existence of variable stars. He served as director of the University of Texas's McDonald Observatory from 1963 to 1989. d. Austin, Tex., Oct. 17.

SPERTI, GEORGE S. (91), U.S. scientist with many inventions to his credit, including the sunlamp, various over-the-counter medications such as Preparation H and Aspercreme, a means of freeze-drying orange juice, and a method of irradiating milk to increase its vitamin-D content. d. Cincinnati, Ohio, April 30.

SUITS, CHAUNCEY G. (86), U.S. scientist who created synthetic diamonds by compressing carbon at extremely high temperatures. As director of research with General Electric, he studied and measured electric arcs. He foresaw microwave ovens and supported development of nuclear power. d. Pilot Knob, N.Y., Aug. 14.

TAYLOR, EDWARD S. (88), U.S. engineer who helped develop the gas turbine and reciprocating engines and whose work was instrumental in the development of the turbofan, turbine-driven propellers and jet engines of today's aircraft. He founded the Gas Turbine Laboratory at the Massachusetts Institute of Technology. d. Lincoln, Mass., Feb. 2.

TOMKINS, SILVAN S. (80), U.S. psychologist who published *Affect, Imagery, Consciousness* (1962), in which his "Affect Theory" was introduced. This theory suggests that motivation is influenced more by emotion than by biological drives. He developed the Tomkins-Horn Picture Arrangement Test and the Thematic Apperception Test, both designed to reveal aspects of personality. d. Somers Point, N.J., June 10.

WAGLEY, CHARLES (78), U.S. anthropologist who pioneered studies of Latin and South American cultures, particularly those of Brazil's Amazon Basin. Two of his books, *Welcome of Tears* and *Amazon Town,* describe how people in these areas adapted to the tropical environment. d. Gainesville, Fla., Nov. 25.

WIDLAR, ROBERT J. (53), U.S. designer of linear integrated circuits used widely today in electronic equipment from consumer products to scientific instruments. He worked for National Semiconductor Corporation of Santa Clara from 1966 to 1970, helping to make it the leader in linear circuits. He made a fortune from his stock ownership in the company and, at the age of 32, retired. He continued providing consulting services to the company, however, and in 1981 started Linear Technology Corporation. d. Puerto Vallarta, Mexico, Feb. 27.

WILLIAMS, CARROLL M. (74), U.S. physician and biologist who discovered and explained how insects develop from eggs through larval and pupal stages into adults. Considered a pioneer in insect physiology, he discovered the hormone that controls this development, and traced its source to a specific region in the insect's brain. He also researched communication among insects and the chemical means used by some plants to repel them. He made advances in insect surgery and invented an insect anesthesia. d. Watertown, Mass., Oct. 11.

WILSON, ALLAN C. (56), New Zealand-born U.S. biochemist whose genetic research identified a common DNA trait passed from women to their offspring, leading to the conclusion that all human beings descended from an African woman who lived 200,000 years ago. This 1987 theory, known as the African Eve hypothesis, also stated that descendants from this woman spread out through Europe and Asia 50,000 to 100,000 years ago, replacing all other archaic humans. This theory is disputed by paleontologists who claim that fossil research shows that modern human beings evolved in different parts of the world at different times. Wilson proposed that there is a "molecular clock" that can measure human evolution by analyzing the changes that occur in proteins and genes over time. Wilson and Vincent M. Sarich also proposed that human beings and apes had a common ancestry that had split 5 million years ago. d. Seattle, Wash., July 21.

WOODHOUSE, JOHN C. (92), U.S. scientist who helped develop a means of processing uranium fuel for use in nuclear-power reactors and who held more than 70 patents for inventions running the gamut from soft contact lenses to hydraulic brake fluid to sudsless detergent. d. Upland, Pa., Feb. 17.

Index

C

D

E

F

G

H

I

J

K

L

M

N

O

P

Q

R

S

T

U

V

W

X

Y

Z

Acknowledgments

Sources of articles appear below, including those reprinted with the kind permission of publications and organizations.

THE WORLD AT ZERO G, Page 35: Reprinted by permission of the author; article originally appeared in the September 1991 issue of *Air & Space/Smithsonian.*

ASTEROID THREAT, Page 41: Clark R. Chapman and David Morrison/© 1991 *Discover* Magazine.

ANIMALS IN SPACE, Page 46: Reprinted by permission of the author; article originally appeared in the May 1991 issue of *Ad Astra.*

COOLNESS UNDER FIRE, Page 58: Adapted from *Bechtel Briefs,* Bechtel Group, Inc.

THE SECRETS OF SNOW, Page 64: Copyright 1992 by the National Wildlife Federation. Reprinted from the December 1991/January 1992 issue of *National Wildlife.*

NATURE BLOOMS IN THE INNER CITY, Page 74: Reprinted by permission of the author; article originally appeared in the July/August 1991 issue of *Audubon,* the magazine of the National Audubon Society.

STORM CHASERS, Page 87: Copyright © 1992 by The New York Times Company. Reprinted by permission.

OFFSHORE OIL: HOW DEEP CAN WE GO?, Page 91: Reprinted from *Popular Science* with permission © 1992 Times Mirror Magazines, Inc.

HAIR: PERSONAL STATEMENT TO PERSONAL PROBLEM, Page 118: Reprinted from *FDA Consumer.*

THE AGE OF GENES, Page 124: Copyright, November 4, 1991, *U.S. News & World Report.*

COMPANIONABLE CANINES, Page 133: Reprinted by permission of the author; article originally appeared in the January 1992 issue of *Smithsonian.*

THE ATHLETE'S DILEMMA, Page 146: Jared Diamond/© 1991 *Discover* Magazine.

THE QUIET COMEBACK OF ELECTROSHOCK THERAPY, Page 154: Copyright © 1990 by The New York Times Company. Reprinted by permission.

TV OR NOT TV, Page 158: Reprinted from *In Health.* Copyright © 1992.

FEARS YOU CAN TURN OFF, Page 164: Reprinted from *In Health.* Copyright © 1992.

TAKING BACK THE PAST, Page 188: Reprinted by permission from the April 1991 issue of *Outside.*

WHAT HAPPENED TO THE PASSENGER PIGEON?, Page 202: Reprinted by permission of the author; article originally appeared in the March/April 1992 issue of *Wildlife Conservation.*

THE RISE OF THE INTERSTATES, Page 206: Reprinted by permission of AMERICAN HERITAGE Magazine, a division of Forbes, Inc., © 1991 Forbes Inc., 1992.

SALEM'S DARKEST HOUR, Page 215: Reprinted by permission of the author; article originally appeared in the April 1992 issue of *Smithsonian.*

THE GREAT BRIDGE CONTROVERSY, Page 226: David Berreby/© 1991 *Discover* Magazine.

BUCKYBALLS: THE MAGIC MOLECULES, Page 235: Reprinted from *Popular Science* with permission © 1991 Times Mirror Magazines, Inc.

VOLATILE VACUUMS, Page 242: Reprinted by permission of *OMNI,* © 1991, OMNI Publications International Ltd.

THE MYSTERY OF MAKING SCENTS, Page 248: Reprinted by permission of the author; article originally appeared in the June 1991 issue of *Smithsonian.*

THE PHYSICS OF CAR ACCIDENTS, Page 256: Tim Folger/© 1991 *Discover* Magazine.

THE MAGIC OF CARD SHUFFLING, Page 261: Tim Folger/© 1991 *Discover* Magazine.

NIKOLA TESLA: THE FORGOTTEN PHYSICIST, Page 264: Reprinted by permission from the Winter 1992 issue of *American Heritage Invention & Technology.*

THE ANCIENT SPELL OF THE SEA UNICORN, Page 278: Copyright 1989 by the National Wildlife Federation. Reprinted from the November/December issue of *International Wildlife.*

THE WORLD'S BIGGEST FUNGUS, Page 284: Copyright © 1992 by The New York Times Company. Reprinted by permission.

THE ENIGMATIC SLIME MOLD, Page 287: Reprinted by permission of the author; article originally appeared in the July 1991 issue of *Smithsonian.*

WILD HORSE REFUGE, Page 291: Adaptation of this article reprinted courtesy of *Sports Illustrated* from the June 17, 1991, issue. Copyright © 1991, The Time Inc. Magazine Company. All rights reserved.

PRIMATES OF THE SEA, Page 304: Alex Kerstitch/© 1991 *Discover* Magazine.

A MULTITUDE OF MOOSE, Page 307: Originally published in *COUNTRY JOURNAL,* May/June 1991, © Cowles Magazines, Inc.

WILDLIFE CRIMEBUSTERS, Page 312: Reprinted by permission of the author; article originally appeared in the March 1992 issue of *Smithsonian.*

THE NEW COMMUNICATIONS AGE, Page 328: Reprinted by permission of the author; article originally appeared in the February 1992 issue of *Smithsonian.*

WHAT'S NEW IN THE NEWSROOM?, Page 336: Reprinted by permission of the author through the courtesy of Halsey Publishing Company, publishers of Delta Air Lines' *SKY* magazine.

THE BOMB CATCHERS, Page 341: Reprinted from *Popular Science* with permission © 1991 Times Mirror Magazines, Inc.

FROM GRAPHICS TO PLASTICS, Page 348: Reprinted with permission from *Science News*, the weekly newsmagazine of science, copyright 1991 by Science Service, Inc.

THE BLIMP BOWL, Page 352: Reprinted by permission of the author; article originally appeared in the February 1991 issue of *Air & Space/Smithsonian.*

ROBOTS GO BUGGY, Page 358: Reprinted with permission from *Science News*, the weekly newsmagazine of science, copyright 1991 by Science Service, Inc.

BOLDER BAR CODES, Page 376: Reprinted with permission from *Science News*, the weekly newsmagazine of science, copyright 1991 by Science Service, Inc.

Manufacturing Acknowledgments

We wish to thank the following for their services:
Typesetting, TSI Graphics; Color Separations, Colotone, Inc.;
Text Stock, printed on S.D. Warren's 60# Somerset Matte;
Cover Materials provided by Holliston Mills, Inc. and Decorative Specialties International, Inc.;
Printing and Binding, R.R. Donnelley & Sons Co.

ILLUSTRATION CREDITS

The following list acknowledges, according to page, the sources of illustrations used in this volume. The credits are listed illustration by illustration — top to bottom, left to right. Where necessary, the name of the photographer or artist has been listed with the source, the two separated by a slash. If two or more illustrations appear on the same page, their credits are separated by semicolons.

3 © Charly Franklin/FPG International
12-13 © Finley-Holiday/FPG International; inset: AP/Wide World
14 NASA
16 © Paul DiMare
17 NASA
18 All photos: NASA
20 © Francois Gohier/Photo Researchers; © Dennis Oda/Sipa
21 © Dennis Mammana
22 © Frank Zullo/Sipa
23 © Larry Reider/Sipa
24 The Granger Collection
27 © Frank Zullo/Sipa
28-29 © David Parker/SPL Science Source/Photo Researchers
30 Evans and Sutherland, Salt Lake City, UT
31 Adler Planetarium
32 Reuben H. Fleet Space Theater, San Diego, CA
33 Adler Planetarium
34 Learning Technologies, Inc., Cambridge, MA
35-40 All photos: NASA/Johnson Space Center
41 © David A. Hardy/Science Photo Library/Photo Researchers
43 © Michael Collier/*Discover* magazine
44 © Alfred Kamajian/*Discover* magazine
45 JPL
46 Tass/Sovfoto
47 NASA/Johnson Space Center; Sovfoto
48 Sovfoto
49 NASA/Johnson Space Center
50 Both photos: NASA
51 Dana Berry/STSCI
52 Both photos: AP/Wide World Photos
53 The Boeing Company
54 All photos: NASA
56-57 © Dewitt Jones/The Stock Market; inset: © Phil Degginger/Tony Stone Worldwide
58-59 Photo: © Peter Menzel; map: © 1992 Bechtel Corporation
61 All photos: © Peter Menzel
63 Both photos: © Peter Menzel
64 © C.C. Lockwood/Animals Animals
65 © J.G. Paren/Science Photo Library/Photo Researchers
66 © Scott Camazine/Photo Researchers
67 © Vladimir Sichov/Sipa; © Press Studio/Lehtikuva/Saba
68 © Savran/Eastlight/Saba; © Gamma Moscow/Gamma-Liaison
70 © Wojtek Laski/Sipa
71 © Wojtek Laski/Sipa
72 © Wojtek Laski/Sipa
73 © Wojtek Laski/Sipa
74-75 Both photos: © Danny Gonzalez
76 © Diane Allford; © Courtney Brown, Jr.
77-79 All photos: © Danny Gonzalez
81 © J.M. Petit/Publiphoto; inset: © Francis Lepine/Valan Photos
82-83 © J.M. Petit/Publiphoto; © Wayne Lankinen/Valan Photos; © John Fowler/Valan Photos
84 © D. Auclair/Publiphoto
85 Both photos: © J.M. Petit/Publiphoto
87 © Kent Wood/Science Source/Photo Researchers
88 © Carl Purcell/Science Source/Photo Researchers; © Howard Bluestein/Science Source/Photo Researchers
91 © William Hosner
92 © Jared Schneidman
93 Both illustrations: © Jared Schneidman
94 © Jared Schneidman
96 © Jared Schneidman
99 © Alan Zale/NYT Pictures
100 © Tom Stewart/The Stock Market
101 Reuters/Bettmann
102 © Harvey Lloyd/The Stock Market
104 © Reuters/Bettmann
105 © Bob Stern/*The Hartford Courant*
106 © Neal Hamberg/Sipa
108-109 © R. Rathe/FPG International; inset: © Ted Horowitz/The Stock Market
110 © Richard Hutchings/Science Source/Photo Researchers
112 UPI/Bettmann
113 © Adam Hart-Davis/Science Photo Library/Photo Researchers
114 © Jay Freis/The Image Bank
115 © Anthony A. Boccaccio/The Image Bank
116 © Chris Gierlich/Saba
118 © Ulvis Alberts/Sipa; © J. Barabe/Custom Medical Stock Photos
119 © Yoav Levy/Phototake
120 All photos: Mike Mahoney/American Hair Loss Council
121 All photos: Mike Mahoney/American Hair Loss Council
122 Courtesy of Upjohn Pharmaceuticals
123 © Gerry Schneiders/Unicorn Stock Photos; all photos: Mike Mahoney/American Hair Loss Council
124-125 Crossover art: © Rob Kemp; timeline art: © James Noel Smith
126 Timeline art: © James Noel Smith; text art: © Matt Zang/USN&WR
127 Text art: © Matt Zang/USN&WR; timeline illustrations: Cold Spring Harbor Laboratory Archive; © James Noel Smith
128 © Jackie Lewin/Science Photo Library/Photo Researchers; © James Noel Smith
129 Photo: © Jon Gordon/Phototake; art: © James Noel Smith
131 © Linda R. Creighton/USN&WR
133-138 All photos: © Liane Enkelis
140 © Alon Reininger/Contact Press Images/Woodfin Camp & Assoc.
141 UPI/Bettmann
143 UPI/Bettmann
144 AP/Wide World Photos
145 UPI/Bettmann; © Dan Connolly/Gamma-Liaison
146 © Manfred Mothes/The Image Bank
149 © Gerard Vandystadt/Allsport; © Richard Martin/Sipa Sport
150 © Gary W. Griffen/Animals Animals
151 © Richard Kolar/Animals Animals
153 © Patti Murray/Animals Animals
154 © James Wilson/Woodfin Camp and Assoc.
157 Photofest
158-165 Artwork by Al Hering
166 © Geoff Tompkinson/Aspect Picture Library
167 © Geoff Tompkinson/Aspect Picture Library
168 Artwork: © Carol Donner/*Discover* magazine; photo sequence: © Geoff Tompkinson/Aspect Picture Library
169 © Ed Kashi/*Discover* magazine; © Geoff Tompkinson/Aspect Picture Library
171 © Peter Steiner/The Stock Market; © Ken Lax/Photo Researchers
172 © D & I MacDonald/envision
173 © Wendy Stone/Gamma-Liaison
174 © Peter Menzel/Stock Boston
175 © Kevin Haislip/Gamma-Liaison
177 © Brooks Kraft/Sygma; © Nina Berman/Sipa
178 Copyright © 1991, The New York Times Company. Reprinted by permission.
179 Both photos: © Patrick Piel/Gamma-Liaison
181 © Bob Daemmrich/The Image Works
182-183 © Ralph Mercer/Tony Stone Worldwide; inset: © Mark Gottlieb/FPG International
184 Both illustrations: © Sygma
185 © Georges Ollen/Sygma
186 UPI/Bettmann
188 © Terrence Moore
189 © David Hiser/Photographers Aspen
190 Both photos: © Scott S. Warren
192 © David Hiser/Photographers Aspen
193 © Terrence Moore
195 Purdue University; inset: © Frank Fournier/Contact Press Images/Woodfin Camp & Assoc.
197 The Granger Collection; inset: Globe Photos
198 AP/Wide World Photos
200 Map inset: AP/Wide World Photos
202 © Glenn Wolff
203 The Granger Collection
204 © Glenn Wolff
205 The Granger Collection
206 © Peter Miller/The Image Bank
207 Culver Pictures
208 Both illustrations: Archive Photos
210 Both photos: Culver Pictures
211 Culver Pictures
212 © Van Bucher/Photo Researchers
213 © Mitchell Funk/The Image Bank
215-219 All illustrations: The Granger Collection
220 The Granger Collection
221 © Sygma
222 © Tom McHugh/Photo Researchers
223 IVR
224-225 © 1992, IBM Corporation; inset: © Addison Geary/Stock Boston
226-227 Photo © David Hamilton/The Image Bank, computer-enhanced by Laurie Grace
228-229 Far left: Steve Hill/*Discover* magazine; bridge sequence: © Farquarson, negative numbers 14E, 14D, and 12, Special Collections Division, University of Washington Libraries
230 Culver Pictures
231 © National Center for Earthquake Engineering Research, State University of New York at Buffalo
232 © Andy Freeberg/*Discover* magazine
233 © Grenier Inc./*Discover* magazine
235 © John B. Carnett/*Popular Science* magazine/L.A. Times Syndicate
236 All illustrations: © Steve Stankiewicz
237 © J. Bernholc, *et al.*, North Carolina University/Science Source/Photo Researchers
239 Top left illustrations: © Steve Stankiewicz; right: IBM Almaden Research Center
241 Hawkins, Robinson & Loren/University of California at Berkeley
242 © Sally Bensusen/Science Photo Library/Photo Researchers
244-245 Photo: © Zigy Kaluzny; art: © Jana Brenning

246-247 Photo: © Zigy Kaluzny; art: © Jana Brenning
248-255 All photos: © Michael Freeman
256-257 Both photos: © Mark Reinstein/Gamma-Liaison
258 © Roger Ressmeyer/Starlight
259 © Andrew Christie/*Discover* magazine
261 © Rob Wilson/The Image Bank
262 © Martin Rogers/FPG
263 © William J. Kennedy/The Image Bank
264 The Bettmann Archive; inset: UPI/Bettmann Newsphotos
265-267 All illustrations: Tesla Book Company, Coronado, CA
268 Westinghouse Historical Collection
269 McKinley Museum, Canton, OH
270 © L. Seghers/Photo Researchers
271 © Martin Rogers/Stock Market
272 © William Caram/Stock Market
273 © Patrick Robert/Sygma
274 © Novovitch/Gamma-Liaison
275 © David Parker/Science Photo Library/Photo Researchers
276-277 © Carl Roessler/FPG International; inset: © Steve Hix/FPG International
278-279 All photos: © Flip Nicklin
280 The Granger Collection
281 The Granger Collection
283 © Flip Nicklin
284 © Johann Bruhn, Ph.D., Michigan Technological University
285 Photo: © Johann Bruhn, Ph.D., Michigan Technological University
286 © Johann Bruhn, Ph.D., Michigan Technological University
287-289 All photos: © Sylvia Duran Sharnoff
291 © Jeff Foott
292 © Barbara Laing/Black Star
293 © Jeff Vanuga; © Phil Huber/*Sports Illustrated* magazine
294 © Bill Eppridge/*Sports Illustrated* magazine
295 © Fred Whitehead/Animals Animals; © Phil Huber/*Sports Illustrated* magazine
297 Photo: National Tropical Botanical Garden
298-302 All photos: © Robert Gustafson
303 © Jeff Gnass Photography
304 © Alex Kerstitch/*Discover* magazine
305 All photos: © Alex Kerstitch/*Discover* magazine
306 © Alex Kerstitch/*Discover* magazine; © Norbert Wu
307 © Johnny Johnson/Animals Animals
308 © George Robinson/f/Stop Pictures
309 © Johnny Johnson/Animals Animals
310 © Budd Tilton/f/Stop Pictures; © Fox Hollow Photography/f/Stop Pictures; © Paul Rezendes
311 © Tom Walker/f/Stop Pictures
312-318 All photos: © Randall Hyman
319 © David Weintraub/Photo Researchers
320 © Matthew McVay/Saba
321 © Steve Kaufman/Peter Arnold
322 © Tom McHugh/Photo Researchers; © Erwin & Peggy Bauer/Bruce Coleman
323 © Phil Savoie/Bruce Coleman
324 Gene Kritsky, College of Mount St. Joseph
325 © Wayne Lankinen/Bruce Coleman
326-327 © Jook Leung/FPG International; inset: © Ed Wheeler/The Stock Market
328-334 All illustrations: © Pierre Mion
336 © Richard Drew/AP/Wide World Photos
337 © Al Levine/NBC Photos
338-339 Both photos: © 1991 CNN, Inc.
341 © Reuters/Bettmann
342 © Eliot Bergman
344 All illustrations: Vivid Technologies, Waltham, MA
345 ThermedeTec, Inc., Woburn, MA
346 Both illustrations: American Science and Engineering, Inc., Cambridge, MA
347 Imatron Industrial Products, Inc., Foster City, CA
348 © Stewart Dickson; inset: © Stewart Dickson/Andrew Hanson
349 Both photos: © Stewart Dickson/Andrew Hanson
350 All illustrations: © Stewart Dickson
351 Both photos: © Stewart Dickson
352-357 All photos: © Chad Slattery
358-363 All photos: © Peter Menzel
364 © Stephen Ferry/Gamma-Liaison
365 © Carlos Goldin/Science Photo Library/Photo Researchers
366-367 Art reprinted by permission of *NEW SCIENTIST*
368 Photo: © Sipa Press; art reprinted by permission of *NEW SCIENTIST*
369 Art reprinted by permission of *NEW SCIENTIST*
370 Zenith Electronics Corporation
371 Zenith Electronics Corporation
372-373 General Instrument Corporation
375 © F. Henry/REA/Saba
376 Top: © Junebug Clark/Photo Researchers; others: Veritec, Inc.
377 Veritec, Inc.
378 *The New York Times*
379 © George Haling/Photo Researchers
381 © Robbie McClaran
382 © Barr/Gamma-Liaison
383 Radisson Hotels International, Minneapolis, MN

Photo credits for The New Book of Popular Science edition

Front cover:
1. © J.C. Paren/SPL/Photo Researchers
2. © Peter Menzel
3. © John Fowler/Valan Photos
4. © Peter Menzel
5. © Michael Freeman
6. © Finley-Holiday/FPG International
7. © Stephen Ferry/Gamma-Liaison

1	2	
3	4	5
6	7	

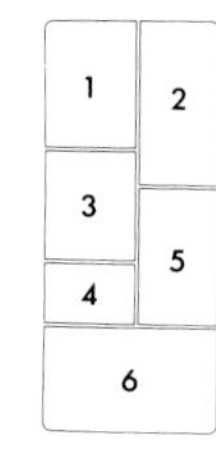

Back cover:
1. © Gary W. Griffen/Animals Animals
2. © Randall Hyman
3. © Sylvia Duran Sharnoff
4. © Fred Whitehead/Animals Animals
5. © David Hiser/Photographers Aspen
6. © Ed Kashi/*Discover* magazine

Cover photo credit for Encyclopedia Science Supplement: © 1992, IBM Corporation